SpringerWienNewYork

CISM COURSES AND LECTURES

The series presents lecture notes, monographs, edited works and proceedings in the field of Mechanics, Engineering, Computer Science and Applied Mathematics.
Purpose of the series is to make known in the international scientific and technical community results obtained in some of the activities organized by CISM, the International Centre for Mechanical Sciences.

INTERNATIONAL CENTRE FOR MECHANICAL SCIENCES

COURSES AND LECTURES - No. 485

NONSMOOTH MECHANICS OF SOLIDS

EDITED BY

JAROSLAV HASLINGER
CHARLES UNIVERSITY, PRAGUE, CZECH REPUBLIC

GEORGIOS E. STAVROULAKIS
TECHNICAL UNIVERSITY OF CRETE, CHANIA, GREECE
AND
TECHNICAL UNIVERSITY OF BRAUNSCHWEIG, GERMANY

SpringerWienNewYork

The publication of this volume was co-sponsored and co-financed by the UNESCO Venice Office - Regional Bureau for Science in Europe (ROSTE) and its content corresponds to a CISM Advanced Course supported by the same UNESCO Regional Bureau.

This volume contains 77 illustrations

Printed in Italy
SPIN 11908821

All contributions have been typeset by the authors.

ISBN-10 3-211-48241-5 SpringerWienNewYork
ISBN-13 978-3-211-48241-4 SpringerWienNewYork

PREFACE

No other science played such important role in the history of mathematics as mechanics. The development of many mathematical disciplines was influenced and motivated by needs to formulate and solve various problems of our daily life. Recall infinitesimal calculus, calculus of variations, theory of partial differential equations and last, but not least numerical methods (finite elements), among others. Originally, mechanics treated smooth objects. Physical quantities were coupled by smooth single-valued mappings. For example the force law expressed a force on the mechanical system in the form of the gradient of a smooth energy function. Technological progress in the past decades evoked the necessity to model and to solve numerically more complicated problems. To formulate unilateral and friction conditions, plasticity, delamination and adhesion, to describe properties of advanced materials, etc. one has to use more sophisticated mathematical tools compared with those ones we know from the classical mathematical analysis. This was a great challenge for mathematicians. In sixties a new branch of the mathematical analysis was developed, namely convex analysis and a differential calculus for convex functions. It turned out that multi-valued mappings defined by the subdifferential of convex functions represents an appropriate apparatus by means of which one could handle more complicated constraints. Basic constitutive laws for problems mentioned above can be expressed as an inclusion type problem involving subgradients of convex functions which do not need to be differentiable in a classical sense. The use of convex analysis has however certain limitations due to monotonicity of the subgradients of convex functions. In practice we often meet problems with nonmonotone constitutive laws. The mathematical model of such a type of problems has to utilize tools of a more general mathematical discipline called the set-valued analysis. It studies properties of abstract set-valued mappings including a certain generalized differential calculus which extends the notion of subdifferential from the convex case. Elements of the set-valued analysis can be used not only to model mechanical effects but also they can be used for developing new numerical methods.

From what it has been said until now it is clear that nonsmooth mechanics is a relatively complex field. For a deep understanding one has to have a good knowledge of mechanics and a good background in some parts of modern mathematics such as nonsmooth analysis, theory of partial differential equations, theory of approximations, numerical methods, etc. The present volume of lecture notes follows a very successful advanced school, which we had the honor to coordinate in Udine, October 4-8, 2004. It was the aim of this course to cover as much as possible all these aspects. Therefore one part of contributions concerns of mechanical aspects whereas the second part focuses mainly on the mathematical and numerical treatment of resulting mathematical models.

Chapter I written by M. Frémond is devoted to predictive theories of collisions of solids which cannot be restricted to points. A novel predictive theory is presented, which is simple and provides results which agree with the experiments. Moreover the thermal properties of collisions as dissipative phenomena are treated. Examples of collisions between balls with different temperatures and of collisions between solids and liquids are presented.

Chapter II is written by Ch. Glocker and deals with the modeling and solution of multi-contact impacts. The classical impact constitutive laws from non-smooth mechanics, including the Moreau's impact law, are reviewed. Several examples including the frictional reversible impact at a superball, Newton's cradle and a rocking rod demonstrate the theory.

Chapter III written by J. Haslinger deals with the approximation of variational and hemivariational inequalities of elliptic type. In the first section the primal, mixed and the dual variational formulation of elliptic inequalities of the first and the second kind and their approximations are revisited. An introduction to the theory of unconstrained and constrained hemivariational inequalities and their approximations is the subject of the next section. The rest of his contributions is devoted to applications of abstract results to the numerical realization of 2 and 3D contact problems with Coulomb's friction and to contact problems with a nonmonotone friction law.

Chapter IV written by Z. Naniewicz deals with semicoercive variational-hemivariational and hemivariational inequalities based on a recession technique. Problems defined on vector-valued function spaces and the resonant problem governed by the p-Laplacian are considered. Examples and applications from mechanics are given.

Chapter V by J.V. Outrata deals with the so-called mathematical programs with equilibrium constraints. A generalized equation similar to the ones arising in nonsmooth mechanics is introduced in the constraints of a mathematical programming problem. The necessary optimality conditions are studied and methods for the numerical solution are proposed. Typical applications arise in optimal design tasks for nonsmooth mechanical problems.

Chapter VI by G.E. Stavroulakis deals with nonsmooth modeling in computational mechanics, numerical methods based on nonsmooth optimization and complementarity techniques and related topics in optimal design and identification. An application of unilateral analysis on monumental masonry bridges demonstrates current trend in applications.

We thank the invited lecturers and the authors of this book for their contributions, the participants of the course for their active participation and the interesting discussions as well as the people of CISM for their hospitality and their well-known professional help.

Jaroslav Haslinger
Georgios E. Stavroulakis

CONTENTS

Collisions. Thermal effects. Collisions of deformable solids and collisions of solids and fluids

Michel Frémond

Laboratoire Lagrange,
LCPC, 58 boulevard Lefebvre, 75732 Paris, France
fremond@lcpc.fr

1 Introduction

These notes report lectures devoted to predictive theories of collisions, given at the Centre International des Sciences Mécaniques.

In most mechanics text books, collisions are either ignored or restricted to collisions of two points introducing the restitution coefficient. This concept appears to be misleading in case of collisions of solids which cannot be restricted to points. This very old topic is actually complex and, active and productive in terms of research, Glocker (2000), Pfeiffer and Glocker (1996), Pfeiffer (2001), Moreau (2003) and the survey book by Brogliatto (1999).

A novel predictive theory of collisions which may be assumed instantaneous, has the good and usual properties of mechanical theories: it is based on simple ideas; it is versatile; it relies on the basic laws of mechanics and the results of its more simple settings agree with the every day experiments, Frémond (2001), Frémond (2005).

The lectures give the thermal properties of collisions: they are dissipative phenomena and result in heat sources. The theory is described with the example of two colliding balls with different temperatures.

In order to show how the ideas of the theory apply, collisions of deformable solids and collisions of solids and liquids are described. Examples show that every day life results are accounted for by the theory, Cholet (1998 (a)), Dimnet (2000).

Let us also stress that, as any mechanical theory, collisions theory may be investigated by mathematical analysis and numerical analysis, Cholet (1998 (a)), Cholet (1998 (b)). This point is not studied in these lectures. The complete theory may be found in Frémond (2005), together with examples and applications. Solutions to the so-called paradoxes, Klein paradox, Painlevé paradox, are also given. Let us recall that these paradoxes claim that the equations of motion with Coulomb friction have in some situations no solution: it may be shown and proved that they have solutions within this predictive theory, Frémond (2005).

In the next section, an example, the collision of two balls or the collision of two points, gives the basic ideas for the mechanical phenomena and the thermal phenomena. A closed form solution shows that in a collision the temperatures of the balls tend to equalize and to increase. In order to show how the theory applies to more sophisticated phenomena: collisions of deformable solids (section 4) and collisions of solids and fluids (section 5) are described. The thermal consequences of collisions may be found in Frémond (2001). In the appendix, some useful properties of convex functions are briefly recalled.

2 Collisions of rigid bodies. An example: collision of two balls or two points

Let us consider two balls moving on a line. For the sake of simplicity, we assume they are points with mass m_i. They have position $x_i(t)$ and velocity $U_i(t) = dx_i/dt(t)$. Because the points can collide, we suppose that the velocities can be discontinuous, assuming the collisions to be instantaneous. The velocity before a collision at time t is $U_i^-(t)$ and the velocity after is $U_i^+(t)$. In terms of mathematics, it seems interesting to think that the velocities are functions of bounded variation with respect to time. In the sequel we will use virtual velocities: a set of virtual velocities $V = (V_1, V_2)$ is such that the two functions $t \rightarrow V_i(t)$ are bounded variation functions.

We denote $S(U, t_1, t_2)$ the set of discontinuity times of velocity $U = (U_1, U_2)$ between times t_1 and t_2 $(t_1 < t_2)$

$$S(U, t_1, t_2) = \left\{ t_k \in]t_1, t_2[\, \middle| U^+(t_k) \neq U^-(t_k) \right\}.$$

This set is numerable because the two functions U_i are functions of bounded variation. We denote

$$[U_i(t_k)] = U_i^+(t_k) - U_i^-(t_k),$$

the velocity discontinuity and in general $[A(t_i)] = A^+(t_i) - A^-(t_i)$ the discontinuity of function $t \rightarrow A(t)$.

More precisely, we assume that velocities U_i are special bounded variation functions, Ambosio et al. (2000), Attouch et al. (2004), Moreau (1988) : their differential

$$dU_i = \frac{dU_i}{dt} dt + \sum_{t_k \in S(U, t_1, t_2)} [U_i(t_k)] \, \delta(t - t_k),$$

is the sum of a Lebesgue measure, the smooth part of the differential, whose density is dU_i/dt and of a Dirac measure, the non-smooth part of the differential, whose density at point t_k is discontinuity $[U_i(t_k)]$.

2.1 The velocities and the velocities of deformation

It is easy to write the equations of motion for the two points. But in order to be a little bit more precise and have results which can be easily adapted to more general settings, we derive carefully these equations by using the principle of virtual power. Before we introduce it, let us remark that there is no reason to speak of deformation when dealing with an isolated point. But if we consider the system made of the two points, its form changes because the distance of the two points may change. Thus it is wise to consider that the system is deformable. A way to characterize the system velocity of deformation is to consider the relative velocity of the two points

$$D(U) = U_1 - U_2,$$

where $U = (U_1, U_2)$ is the set of the two actual velocities of the points. A rigid system set of velocities is such that the form of the system does not change: it is easy to see that they are characterized by

$$D(U) = U_1 - U_2 = 0,$$

because when the form of the system does not change, the velocities of the two points are equal.

2.2 The principle of virtual work

The principle is

The virtual work of the acceleration forces is equal to the sum of the virtual work of the interior forces and of the virtual work of the exterior forces. More precisely, for any time t_1, any time time $t_2 > t_1$ and any virtual velocity V

$$\mathcal{T}_{acc}(t_1, t_2, U, V) = \mathcal{T}_{int}(t_1, t_2, U, V) + \mathcal{T}_{ext}(t_1, t_2, U, V),$$

where $\mathcal{T}_{acc}(t_1, t_2, U, V)$ is the virtual work of the acceleration forces between times t_1 and t_2, $\mathcal{T}_{int}(t_1, t_2, U, V)$ and $\mathcal{T}_{ext}(t_1, t_2, U, V)$ are the virtual works of the interior and exterior forces between the same times. These works satisfy

1. the virtual work of the acceleration forces is a linear function of the virtual velocity V, when $S(V, t_1, t_2)$ is fixed. The actual work of the acceleration forces is equal to the variation of the kinetic energy between times t_1 and t_2

$$\mathcal{T}_{acc}(t_1, t_2, U, U) = \mathcal{K}(U^-(t_2)) - \mathcal{K}(U^+(t_1)),$$

   where $\mathcal{K}$ is the system kinetic energy. This property and the principle of virtual work with the actual velocities, i.e., $V = U$, are sometimes called the theorem of expanded energy (le théorème de l'énergie cinétique in French) ;
2. the virtual work of the interior forces is a linear function of the virtual velocity V, when $S(V, t_1, t_2)$ is fixed. It satisfies and

$$\mathcal{T}_{int}(t_1, t_2, U, V) = 0,$$

   for any rigid system velocity. This relationship is often called the axiom of the principle of virtual work, Germain (1973);
3. the virtual work of the exterior forces is a linear function of the virtual velocity V, when $S(V, t_1, t_2)$ is fixed.

Remark 1. The principle of virtual work we use here is not to be confused with the principle of virtual power where the velocities are understood as small displacements (this relationship is often also called the principle of virtual work).

2.3 The virtual works

Now we may introduce interior forces to the deformable system. A productive way is to define these generalized forces by their work (or by duality in terms of mathematics). The work of the interior forces of the system between times t_1 and t_2 ($t_1 < t_2$) is a linear function of the velocity of deformation. We choose as virtual work of the interior forces

$$\mathcal{T}_{int}(t_1, t_2, U, V)$$

$$= -\int_{t_1}^{t_2} R^{int}(\tau).D(V)(\tau)d\tau - \sum_{t_k \in S(U,t_1,t_2) \cup S(V,t_1,t_2)} \left\{ P^{int}(t_k).D\left(\frac{V^+(t_k) + V^-(t_k)}{2}\right)\right\},$$

which is a linear function of $D(V) = (V_1 - V_2)$ when $S(V, t_1, t_2)$ is fixed. Sign minus has no importance, it is chosen in accordance with habits of continuum mechanics. Its advantage will appear down below where a classical inequality has its classical sign. The virtual work defines an interior force $R^{int}(\tau)$ which intervene in the smooth evolution when the two points evolve with smooth interaction, for instance at a distance interaction (one may think that the two balls are connected by an elastic string). It defines also an interior percussion $P^{int}(t)$ which intervene when collisions occur.

Remark 2. The linear function of $D(V)$ we have chosen is not the more general. The choice involving

$$-P^{int+}(t_k).D(V^+)(t_k) - P^{int-}(t_k).D(V^-)(t_k),$$

does not give much more opportunities, Frémond (2001).

Let us note that $\mathcal{T}_{int}(t_1, t_2, U, V) = 0$ for any rigid system velocity, i.e., for $D(V) = (V_1 - V_2) = 0$.

The virtual work of the acceleration forces is

$$\mathcal{T}_{acc}(t_1, t_2, U, V)$$

$$= \sum_{i=1}^{2} \left\{ \int_{t_1}^{t_2} m_i \frac{dU_i}{dt}(\tau) V_i(\tau) d\tau + \sum_{t_k \in S(U,t_1,t_2) \cup S(V,t_1,t_2)} \left\{ m_i \left[U_i(t_k)\right] . \frac{V_i^+(t_k) + V_i^-(t_k)}{2} \right\} \right\},$$

where $[U(t_k)] = U^+(t_k) - U^-(t_k)$ is the velocity discontinuity. It is clear that this work is a linear function of V when $S(V, t_1, t_2)$ is fixed.

The theorem of expanded energy

Let us compute the actual work of the interior forces

$$
\begin{aligned}
&\mathcal{T}_{acc}(t_1,t_2,U,U) \\
&= \sum_{i=1}^{2}\left\{\int_{t_1}^{t_2} m_i\frac{dU_i}{dt}(\tau)U_i(\tau)d\tau + \sum_{t_k\in S(U,t_1,t_2)}\left\{m_i\left[U_i(t_k)\right].\frac{U_i^+(t_k)+U_i^-(t_k)}{2}\right\}\right\} \\
&= \sum_{i=1}^{2}\left\{\int_{t_1}^{t_2} m_i\frac{dU_i}{dt}(\tau)U_i(\tau)d\tau + \sum_{t_k\in S(U,t_1,t_2)}\frac{m_i}{2}\left[\left(U_i(t_k)\right)^2\right]\right\} \\
&= \sum_{i=1}^{2}\left\{\frac{m_i}{2}\left(U_i^-(t_2)\right)^2 - \frac{m_i}{2}\left(U_i^+(t_1)\right)^2\right\} = \mathcal{K}(U^-(t_2)) - \mathcal{K}(U^+(t_1)) \\
&= \mathcal{K}^-(t_2) - \mathcal{K}^+(t_1),
\end{aligned}
$$

which is equal to the variation of kinetic energy $\mathcal{K}$ between times t_1 and t_2. At any time we have kinetic energies $\mathcal{K}^-$ and $\mathcal{K}^+$ in case there is a collision.

The virtual work of the exterior forces is a linear function of the virtual velocities V when $S(V,t_1,t_2)$ is fixed, we choose

$$
\begin{aligned}
&\mathcal{T}_{ext}(t_1,t_2,U,V) \\
&= \sum_{i=1}^{2}\left\{\int_{t_1}^{t_2} f_i(\tau)V_i(\tau)d\tau + \sum_{t_k\in S(U,t_1,t_2)\cup S(V,t_1,t_2)}\left\{P_i^{ext}(t_k)\frac{V_i^+(t_k)+V_i^-(t_k)}{2}\right\}\right\}.
\end{aligned}
$$

Exterior force f_i may be the gravity force. Exterior percussion P_i^{ext} may represent hammer blows applied to the points.

Remark 3. It is not common that an exterior percussion is applied to a mechanical system at the same time a collision occurs. An example is in a pin-ball machine where a collision may produce an electrical blow.

2.4 The equations of motion

We assume that the interval $]t_1,t_2[$ is split in distinct intervals $]t_k,t_l[$ where there is no discontinuity of U, the two ends of the intervals being discontinuity times of U. By choosing smooth virtual velocities with compact support in interval $]t_k,t_l[$, we get

$$
\sum_{i=1}^{2}\int_{t_k}^{t_l} m_i\frac{dU_i}{dt}(\tau)V_i(\tau)d\tau = -\int_{t_k}^{t_l} R^{int}(\tau)D(V)(\tau)d\tau + \sum_{i=1}^{2}\int_{t_1}^{t_2} f_i(\tau)V_i(\tau)d\tau.
$$

The fundamental lemma of variation calculus gives

$$
\begin{aligned}
m_1\frac{dU_1}{dt} &= -R_1^{int} + f_1, \ \ a.e. \ in \ ]t_i,t_j[, \\
m_2\frac{dU_2}{dt} &= R_2^{int} + f_2, \ \ a.e. \ in \ ]t_i,t_j[.
\end{aligned}
\tag{1}
$$

Remark 4. *a.e.*means almost everywhere or almost always in this context.

Because points t_i are numerable, relationships (1) are satisfied almost everywhere in $]t_1, t_2[$. Thus the principle becomes

$$\sum_{i=1}^{2} \sum_{t_k \in S(U,t_1,t_2) \cup S(V,t_1,t_2)} \left\{ m_i \left[U_i(t_k)\right] \frac{V_i^+(t_k) + V_i^-(t_k)}{2} \right\}$$
$$= - \sum_{t_k \in S(U,t_1,t_2) \cup S(V,t_1,t_2)} \left\{ P^{int}(t_k) D\left(\frac{V^+(t_k) + V^-(t_k)}{2}\right) \right\}$$
$$+ \sum_{i=1}^{2} \sum_{t_k \in S(U,t_1,t_2) \cup S(V,t_1,t_2)} \left\{ P_i^{ext}(t_k) \frac{V_i^+(t_k) + V_i^-(t_k)}{2} \right\}.$$

At time t_j between intervals $]t_k, t_j[$ and $]t_j, t_l[$ by choosing virtual velocities with compact support in $]t_k, t_l[$, we have

$$\sum_{i=1}^{2} m_i \left[U_i(t_j)\right] \frac{V_i^+(t_j) + V_i^-(t_j)}{2}$$
$$= -P^{int}(t_j) D\left(\frac{V^+(t_k) + V^-(t_k)}{2}\right) + \sum_{i=1}^{2} P_i^{ext}(t_j) \frac{V_i^+(t_j) + V_i^-(t_j)}{2},$$

which gives immediately

$$m_1 \left[U_1(t_j)\right] = -P^{int}(t_j) + P_1^{ext}(t_j), \tag{2}$$
$$m_2 \left[U_2(t_j)\right] = P^{int}(t_j) + P_2^{ext}(t_j). \tag{3}$$

At time $t \in]t_k, t_l[$ by choosing virtual velocities with support in $]t_k, t_l[$ having a unique discontinuity point at time t, we get easily that relationships (2) and (3) are satisfied at time t. Thus, they are satisfied at any time

$$m_1 \left[U_1(t_j)\right] = -P^{int}(t_j) + P_1^{ext}(t_j),\ \forall t \in]t_1, t_2[,$$
$$m_2 \left[U_2(t_j)\right] = P^{int}(t_j) + P_2^{ext}(t_j),\ \forall t \in]t_1, t_2[.$$

A detailed derivation of the equations of motion is given in Frémond (2005).

2.5 The equations of the thermal phenomenon

Four temperatures appear in the equations : T_i^+ and T_i^- temperatures of the points after and before collision. We denote

$$[T_1] = T_1^+ - T_1^-,\ [T_2] = T_2^+ - T_2^-.$$

These differences are analog to the temperature time derivative dT/dt when the temperature is smooth. Let us recall that there is no dissipation with respect to this quantity because of Helmholtz relationship

$$s = -\frac{\partial \Psi}{\partial T}(T, \chi),$$

resulting from the definition of the free energy $\Psi(T, \chi)$

$$e(s, \chi) = Ts + \Psi(T, \chi),$$

where s is the entropy, $e(s, \chi)$ is the internal energy and χ are other quantities depending on the past. We denote some average temperatures

$$\bar{T}_1 = \frac{T_1^+ + T_1^-}{2},\ \bar{T}_2 = \frac{T_2^+ + T_2^-}{2}, \bar{\Theta} = \frac{T_1^+ + T_1^- + T_2^+ + T_2^-}{4} = \frac{\bar{T}_1 + \bar{T}_2}{2}.$$

The temperature difference

$$\delta\bar{T} = \bar{T}_2 - \bar{T}_1,$$

is the analog of the spatial thermal gradient in a smooth situation. Let us also recall that there is dissipation with respect to this quantity: the almost universal example is the Fourier law. In the following investigation of the thermal effects of collisions, analogous choices are made remembering that they sum up what occur during the duration of collisions when Fourier law and Helmholtz relationship are assumed.

In the sequel, we consider sums which are transformed in the following way

$$T^+B^+ + T^-B^- = \bar{T}\Sigma B + [T]\,\Delta B,$$

with

$$\bar{T} = \frac{T^+ + T^-}{2},\ [T] = T^+ - T^-,\ \Sigma B = B^+ + B^-,\ \Delta B = \frac{B^+ - B^-}{2}.$$

First law of thermodynamics for a point

The laws of thermodynamics are the same for the two points. For point 1, the first law is

$$[\mathcal{E}_1] + [\mathcal{K}_1] = \mathcal{T}_{ext1}(U) + \mathcal{C}_1,$$

where $\mathcal{E}_1$ is the free energy of point 1, $\mathcal{T}_{ext1}(U)$ is the actual work of the percussions which are exterior to point 1 and $\mathcal{C}_1$ is the amount of heat received by the point in the collision. This heat involves the heats T_1B_1 received from the exterior to the system and the heats T_1B_{12} received from the interior of the system, i.e., from the other point. We assume that these heats are received at temperatures T_1^+ or T_1^-

$$\begin{aligned}\mathcal{C}_1 &= T_1^+\left(B_1^+ + B_{12}^+\right) + T_1^-\left(B_1^- + B_{12}^-\right)\\ &= \bar{T}_1\left(\Sigma B_1 + \Sigma B_{12}\right) + [T_1]\left(\Delta B_1 + \Delta B_{12}\right).\end{aligned}$$

The theorem of expanded energy for point 1 is

$$[\mathcal{K}_1] = \mathcal{T}_{acc1}(U) = \mathcal{T}_{ext1}(U).$$

It gives with the energy balance

$$[\mathcal{E}_1] = \mathcal{C}_1 = \bar{T}_1 \left(\Sigma B_1 + \Sigma B_{12}\right) + [T_1] \left(\Delta B_1 + \Delta B_{12}\right). \tag{4}$$

In the right hand side, there is no mechanical quantity because a point has no interior force, the work of which could contribute to the internal energy evolution.

Second law of thermodynamics for a point

It is

$$[\mathcal{S}_1] \geq B_1^+ + B_1^- + B_{12}^+ + B_{12}^- = \Sigma B_1 + \Sigma B_{12},$$

where $\mathcal{S}_1$ is the entropy of point 1.

A useful inequality for a point

Second law and relationship (4) give

$$[\mathcal{E}_1] - \bar{T}_1 [\mathcal{S}_1] \leq [T_1] \left(\Delta B_1 + \Delta B_{12}\right),$$

or by introducing the free energy $\Psi = \mathcal{E} - T\mathcal{S}$

$$[\Psi_1] + \bar{\mathcal{S}}_1 [T_1] \leq [T_1] \Delta B_1 + [T_1] \Delta B_{12}.$$

We denote $[\Psi_1] + \bar{\mathcal{S}}_1 [T_1] = -S_1 [T_1]$ remembering that when $[T_1] \to 0$, we have $[\Psi_1] / [T_1] + \bar{\mathcal{S}}_1 \to 0$ because $\partial\Psi/\partial T + s = 0$. Then

$$0 \leq [T_1] \left(\Delta B_1 + \Delta B_{12} + S_1\right).$$

Due to Helmholtz relationship, it is reasonable to assume no dissipation with respect to $[T_1]$ which is, as already said, the analog to dT/dt

$$\Delta B_1 + \Delta B_{12} + S_1 = 0.$$

We deduce

$$[\mathcal{S}_1] = \Sigma B_1 + \Sigma B_{12}. \tag{5}$$

The first law for the system

The internal energy and entropy of the system are the sums of the internal energies and entropies of its components $\mathcal{E}_1$, $\mathcal{S}_1$ and $\mathcal{E}_2$, $\mathcal{S}_2$ to which interaction internal energy and entropy $\mathcal{E}^{int}$, $\mathcal{S}^{int}$ are added:

$$\mathcal{E} = \mathcal{E}_1 + \mathcal{E}_2 + \mathcal{E}^{int},$$
$$\mathcal{S} = \mathcal{S}_1 + \mathcal{S}_2 + \mathcal{S}^{int}.$$

The first law for the system is

$$[\mathcal{E}] + [\mathcal{K}] = \mathcal{T}_{ext}(U) + \mathcal{C},$$

where $\mathcal{C}$ is the heat received by the system in collision

$$\mathcal{C} = T_1^+ B_1^+ + T_1^- B_1^- + T_2^+ B_2^+ + T_2^- B_2^- .$$

The theorem of expanded energy is

$$[\mathcal{K}] = \mathcal{T}_{acc}(U) = \mathcal{T}_{int}(D(U)) + \mathcal{T}_{ext}(U).$$

It gives with the first law

$$[\mathcal{E}] = -\mathcal{T}_{int}(D(U)) + \mathcal{C}. \tag{6}$$

The first laws for each point (4) and for the system (6) give

$$\begin{aligned}
[\mathcal{E}] &= [\mathcal{E}_1] + [\mathcal{E}_2] + \left[\mathcal{E}^{int}\right] \\
&= \left[\mathcal{E}^{int}\right] + \bar{T}_1 \left(\Sigma B_1 + \Sigma B_{12}\right) + [T_1] \left(\Delta B_1 + \Delta B_{12}\right) \\
&\quad + \bar{T}_2 \left(\Sigma B_2 + \Sigma B_{21}\right) + [T_2] \left(\Delta B_2 + \Delta B_{21}\right) \\
&= P^{int} D \left(\frac{U^+ + U^-}{2}\right) + T_1^+ B_1^+ + T_1^- B_1^- + T_2^+ B_2^+ + T_2^- B_2^- \\
&= P^{int} D \left(\frac{U^+ + U^-}{2}\right) + \bar{T}_1 \Sigma B_1 + [T_1] \Delta B_1 + \bar{T}_2 \Sigma B_2 + [T_2] \Delta B_2.
\end{aligned}$$

Then

$$\left[\mathcal{E}^{int}\right] = P^{int} D \left(\frac{U^+ + U^-}{2}\right) - \bar{T}_1 \Sigma B_{12} - [T_1] \Delta B_{12} - \bar{T}_2 \Sigma B_{21} - [T_2] \Delta B_{21}. \tag{7}$$

This relationship is interesting because only interior quantities intervene.

The second law for the system

It is

$$[\mathcal{S}] \geq B_1^+ + B_1^- + B_2^+ + B_2^- = \Sigma B_1 + \Sigma B_2,$$

where

$$[\mathcal{S}] = [\mathcal{S}_1] + [\mathcal{S}_2] + \left[\mathcal{S}^{int}\right] \geq \Sigma B_1 + \Sigma B_2,$$

which gives with (5)

$$\left[\mathcal{S}^{int}\right] \geq -\Sigma B_{12} - \Sigma B_{21}. \tag{8}$$

In this relationship only interior heat exchanges intervene. Together with the first law (7), it gives an inequality coupling the mechanical and thermal dissipations.

A useful inequality for the system

We have with (7) and (8)

$$\begin{aligned}
\left[\mathcal{E}^{int}\right] - \bar{\Theta} \left[\mathcal{S}^{int}\right] &\leq P^{int} D \left(\frac{U^+ + U^-}{2}\right) - \bar{T}_1 \Sigma B_{12} - [T_1] \Delta B_{12} \\
&\quad - \bar{T}_2 \Sigma B_{21} - [T_2] \Delta B_{21} + \bar{\Theta} \left(\Sigma B_{12} + \Sigma B_{21}\right).
\end{aligned}$$

We define the interaction free energy: $\Psi^{int} = \mathcal{E}^{int} - \Theta\mathcal{S}^{int}$. It gives with the preceding relationship

$$\left[\Psi^{int}\right] + \bar{\mathcal{S}}^{int}\left[\Theta\right] \leq P^{int} D\left(\frac{U^+ + U^-}{2}\right)$$
$$-\bar{T}_1 \Sigma B_{12} - \left[T_1\right] \Delta B_{12} - \bar{T}_2 \Sigma B_{21} - \left[T_2\right] \Delta B_{21} + \bar{\Theta}\left(\Sigma B_{12} + \Sigma B_{21}\right),$$

with

$$\bar{\mathcal{S}}^{int} = \frac{\mathcal{S}^{int+} + \mathcal{S}^{int-}}{2}.$$

By denoting

$$\left[\Theta\right] = \frac{1}{2}\left(\left[T_1\right] + \left[T_2\right]\right),$$

we define

$$\left[\Psi^{int}\right] + \bar{\mathcal{S}}^{int}\left[\Theta\right] = -S_{int}\left[\Theta\right].$$

We obtain

$$0 \leq P^{int} D\left(\frac{U^+ + U^-}{2}\right) - \bar{T}_1 \Sigma B_{12} - \left[T_1\right]\left(\Delta B_{12} - \frac{S_{int}}{2}\right)$$
$$-\bar{T}_2 \Sigma B_{21} - \left[T_2\right]\left(\Delta B_{21} - \frac{S_{int}}{2}\right) + \bar{\Theta}\left(\Sigma B_{12} + \Sigma B_{21}\right).$$

It is reasonable not to have dissipation with respect to $[T_1]$ and $[T_2]$ which are temperature variations with respect to time and not with respect to space

$$\Delta B_{12} - \frac{S_{int}}{2} = 0, \ \Delta B_{21} - \frac{S_{int}}{2} = 0. \tag{9}$$

Then

$$0 \leq P^{int} D\left(\frac{U^+ + U^-}{2}\right) - \bar{T}_1 \Sigma B_{12} - \bar{T}_2 \Sigma B_{21} + \bar{\Theta}\left(\Sigma B_{12} + \Sigma B_{21}\right),$$

but

$$\bar{T}_1 \Sigma B_{12} + \bar{T}_2 \Sigma B_{21} = \bar{\Theta}\left(\Sigma B_{12} + \Sigma B_{21}\right) + \delta\bar{T}\frac{\Sigma B_{21} - \Sigma B_{12}}{2},$$

recalling that $\delta\bar{T} = \bar{T}_2 - \bar{T}_1$. Thus we get

$$0 \leq P^{int} D\left(\frac{U^+ + U^-}{2}\right) - \delta\bar{T}\frac{\Sigma B_{21} - \Sigma B_{12}}{2}. \tag{10}$$

This relationship is important due to the quantities which intervene : the one which characterizes the mechanical evolution $D\left((U^+ + U^-)/2\right)$, and the other one which characterizes the spatial thermal heterogeneity $\delta\bar{T}$. Let us stress its analogy with the spatial thermal gradient in smooth situation. Relationship (10) is a guide to choose the quantities which are to be related to the interior forces. Let us recall that on our way we have assumed no dissipation with respect to the temperature time variation.

2.6 The constitutive laws

Quantities which describe the mechanical evolution and thermal heterogeneity are $D\left((U^+ + U^-)/2\right)$ and $\delta\bar{T}$. We define the constitutive laws with a pseudo-potential of dissipation introduced by J. J. Moreau

$$\hat{\Phi}\left(D(\frac{U^+ + U^-}{2}), \delta\bar{T}, \chi\right).$$

The quantity $\chi = D\left(U^-/2\right)$ which depends on the past is used to take into account the non-interpenetration of the points. Then the constitutive laws are

$$\left(P^{int}, -\frac{\Sigma B_{21} - \Sigma B_{12}}{2}\right) \in \partial\hat{\Phi}\left(D(\frac{U^+ + U^-}{2}), \delta\bar{T}, D(\frac{U^-}{2})\right),$$

where the subdifferential set is computed with respect to the two first variables. It is easy to show that inequality (10) is satisfied.

2.7 An example of thermal effects due to collisions

We choose a pseudo-potential which does not couple the two first variables insuring that the mechanical problem is decoupled from the thermal one, for instance

$$\hat{\Phi}\left(D(\frac{U^+ + U^-}{2}), \delta\bar{T}, D(\frac{U^-}{2})\right)$$
$$= \hat{\Phi}_{meca}\left(D(\frac{U^+ + U^-}{2}), D(\frac{U^-}{2})\right) + \frac{\lambda}{4}\left(\delta\bar{T}\right)^2.$$

The pseudo-potential $\hat{\Phi}_{meca}$ gives the percussion P^{int}. The mechanical problem is not investigated in these notes. Complete investigation and results may be found in Frémond (2005). The thermal constitutive law is

$$\Sigma B_{21} - \Sigma B_{12} = -\lambda\delta\bar{T}.$$

We choose the simple free energy for the points, $\Psi = -CTLogT$, and a zero interaction free energy, $\Psi^{int} = 0$. This choice gives

$$S_{int} = 0,$$

and with (9)

$$\Delta B_{12} = 0,\ \Delta B_{21} = 0.$$

Let us assume that the mechanical problem is solved, i.e., the work $P^{int}D\left((U^+ + U^-)/2\right)$ is known. The thermal equations are

$$[S_1] = CLog\frac{T_1^+}{T_1^-} = \Sigma B_1 + \Sigma B_{12},$$

$$[S_2] = CLog\frac{T_2^+}{T_2^-} = \Sigma B_2 + \Sigma B_{21},$$

$$[\mathcal{E}_1] = C\,[T_1] = \bar{T}_1\,(\Sigma B_1 + \Sigma B_{12}) + [T_1]\,\Delta B_1,$$
$$[\mathcal{E}_2] = C\,[T_2] = \bar{T}_2\,(\Sigma B_2 + \Sigma B_{21}) + [T_2]\,\Delta B_2,$$

$$0 = \left[\mathcal{E}^{int}\right] = P^{int} D\left(\frac{U^+ + U^-}{2}\right) - \bar{T}_1 \Sigma B_{12} - \bar{T}_2 \Sigma B_{21}.$$

These equations are coupled with the description of the thermal exchanges with the exterior of the system. For instance, collision is isentropic

$$\Sigma B_1 = \Sigma B_2 = 0,$$

or adiabatic

$$T_1^+ B_1^+ + T_1^- B_1^- = \bar{T}_1 \Sigma B_1 + [T_1]\,\Delta B_1 = 0,$$
$$\bar{T}_2 \Sigma B_2 + [T_2]\,\Delta B_2 = 0.$$

The solution of these equations gives the temperatures T_i and the different entropy fluxes B_i and B_{ij}.

The isentropic case

The equations give

$$C Log\frac{T_1^+}{T_1^-} = \Sigma B_{12},\; C Log\frac{T_2^+}{T_2^-} = \Sigma B_{21},$$
$$C\,[T_1] = \bar{T}_1 \Sigma B_{12} + [T_1]\,\Delta B_1,\; C\,[T_2] = \bar{T}_2 \Sigma B_{21} + [T_2]\,\Delta B_2,$$
$$P^{int} D\left(\frac{U^+ + U^-}{2}\right) = \bar{T}_1 \Sigma B_{12} + \bar{T}_2 \Sigma B_{21},$$
$$\Sigma B_{21} - \Sigma B_{12} = -\lambda \delta \bar{T}.$$

Then we have

$$C\,[T_1] = \bar{T}_1 \Sigma B_{12} + [T_1]\,\Delta B_1,$$
$$C\,[T_2] = \bar{T}_2 \Sigma B_{21} + [T_2]\,\Delta B_2,$$
$$P^{int} D\left(\frac{U^+ + U^-}{2}\right) = \bar{T}_1 C Log\frac{T_1^+}{T_1^-} + \bar{T}_2 C Log\frac{T_2^+}{T_2^-},$$
$$C Log\frac{T_2^+}{T_2^-} + \lambda \bar{T}_2 = C Log\frac{T_1^+}{T_1^-} + \lambda \bar{T}_1.$$

The two last equations give the temperatures T_1^+ and T_2^+ after a collision. The first two give the thermal exchanges with the exterior : they are computed with ΔB_1 and ΔB_2.

The adiabatic case

The equations give

$$[S_1] = C Log\frac{T_1^+}{T_1^-} = \Sigma B_1 + \Sigma B_{12},$$
$$[S_2] = C Log\frac{T_2^+}{T_2^-} = \Sigma B_2 + \Sigma B_{21},$$

and

$$C\left[T_1\right] = \bar{T}_1 \Sigma B_{12},\ C\left[T_2\right] = \bar{T}_2 \Sigma B_{21},$$

and

$$P^{int} D\left(\frac{U^+ + U^-}{2}\right) = C\left[T_1\right] + C\left[T_2\right],$$

$$C\frac{[T_2]}{\bar{T}_2} - C\frac{[T_1]}{\bar{T}_1} = -\lambda\delta\bar{T}.$$

The two last equations give the temperatures T_1^+ and T_2^+ after a collision. The first two give the entropic exchanges ΣB_1 and ΣB_2 with the exterior.

A computation

Computations of temperatures T_1^+ and T_2^+ in the two preceding situations may be performed with numerics or with closed form solutions which are sophisticated. A case is easily solved: we assume that the collision occurs at the ambient temperature and that the temperature variations are negligible compared with the ambient temperature. We may say that we assume small perturbations. This is the case for an every day experiment. In the isentropic and adiabatic case, we have

$$Log\frac{T^+}{T^-} \simeq \frac{[T]}{T^-},\ \frac{[T]}{\bar{T}} \simeq \frac{[T]}{T^-}.$$

In both cases we have

$$P^{int} D\left(\frac{U^+ + U^-}{2}\right) = C\left[T_1\right] + C\left[T_2\right],$$

$$C\frac{[T_2]}{T_2^-} - C\frac{[T_1]}{T_1^-} = -\lambda\delta\bar{T}$$

$$= -\lambda(T_2^- - T_1^- + \frac{[T_2]}{2} - \frac{[T_1]}{2}).$$

These relationships give

$$P^{int} D\left(\frac{U^+ + U^-}{2}\right) = C\left[T_1\right] + C\left[T_2\right],$$

$$(\frac{C}{T_2^-} + \frac{\lambda}{2})\left[T_2\right] - (\frac{C}{T_1^-} + \frac{\lambda}{2})\left[T_1\right] = -\lambda(T_2^- - T_1^-).$$

To simplify again, let us assume that λ is small compared to C/T. It results

$$P^{int} D\left(\frac{U^+ + U^-}{2}\right) = C\left[T_1\right] + C\left[T_2\right],$$

$$\frac{C}{T_2^-}\left[T_2\right] - \frac{C}{T_1^-}\left[T_1\right] = -\lambda(T_2^- - T_1^-),$$

which gives

$$C[T_1] = \frac{T_1^-}{T_2^- + T_1^-}\left(P^{int} D\left(\frac{U^+ + U^-}{2}\right) + \lambda T_2^- (T_2^- - T_1^-)\right),$$
$$C[T_2] = \frac{T_2^-}{T_2^- + T_1^-}\left(P^{int} D\left(\frac{U^+ + U^-}{2}\right) - \lambda T_1^- (T_2^- - T_1^-)\right).$$

These formulas show that the mechanical work

$$\frac{T_i^-}{T_2^- + T_1^-} P^{int} D\left(\frac{U^+ + U^-}{2}\right) \geq 0,$$

tends to warm up the two points and that the conduction effect

$$\frac{T_i^-}{T_2^- + T_1^-}\left(\pm \lambda T_i^- (T_2^- - T_1^-)\right)$$

tends to equalize their temperatures.

2.8 Experimental results

As far as we know, there are almost no thermal measurements in collisions. Nevertheless some results are available. For instance, temperature of the back face of a metallic plate collided by a steel ball has been measured with an infrared camera, Pron et al. (2002), Pron (2000). A temperature increase up to $7^o C$ in less than a hundredth of a second is reported in Fig. 1. At the engineering scale, temperature is actually discontinuous, even if at a finer time scale it is continuous.

3 Collisions of deformable bodies

When a steel bar falls on a marble table, it bounces and vibrates. After the collision, there is a competition between the vibration velocity and the rising velocity. This competition results in about ten new collisions in a very small lapse of time of the order of a millionth of a second, before the bar moves away from the table. This is the micro-rebound phenomenon, Stoianovici and Hurmuzlu (1996). It is clearly a consequence from the deformability of the bar responsible for the vibrations. In order to establish a predictive theory, we keep the instantaneity of collisions and we sophisticate the description of the velocities of deformation. For the sake of simplicity, we assume that only one solid is deformable, the bar. The marble table being rigid and assumed to be firmly fixed in order that its velocity remains zero.

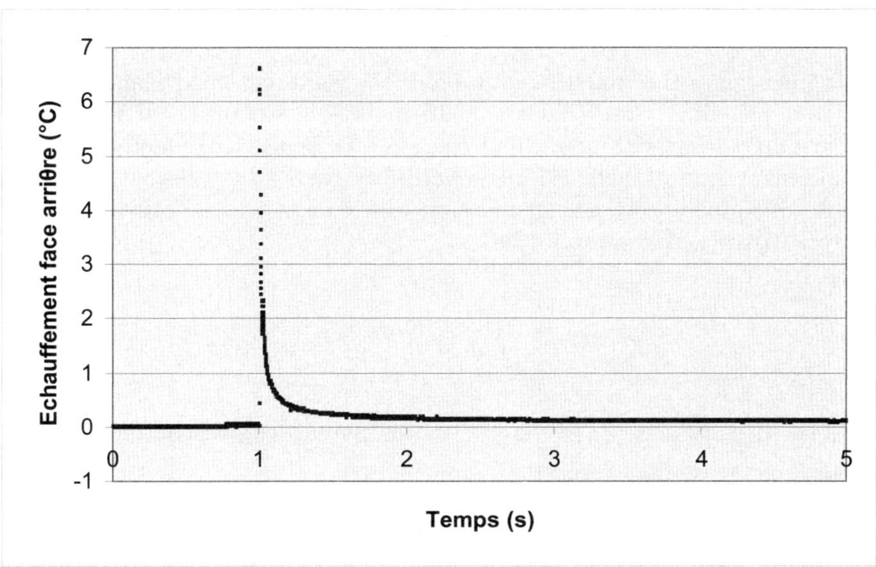

Figure 1. A metallic ball with 3 mm diameter collides a metallic plate with 0, 85 mm thickness. The kinetic energy of the ball is 0, 27 J. Temperature of the plate back face versus time is discontinuous at the time scale of the measurements, Pron et al. (2002).

3.1 The actual and virtual velocities

The deformable solid collides at time t the rigid and fixed plane on contact surface $\Gamma(t) = \Gamma$. Due to the kinematic incompatibilities, very large stresses and contact forces appear in solid $\Omega(t) = \Omega$ and on contact surface Γ in order that the velocities adapt to the obstacle. We assume that this phenomenon is very short and consider it is instantaneous. Thus the collision is characterized by time discontinuity of the velocities: $\overrightarrow{U}^-(\overrightarrow{x}, t)$ is the velocity before collision and $\overrightarrow{U}^+(\overrightarrow{x}, t)$ is the velocity after collision. Virtual velocities are denoted $\overrightarrow{V}^-(\overrightarrow{x})$ and $\overrightarrow{V}^+(\overrightarrow{x})$.

The deformation velocities

We use the velocity of the solid with respect to the plane to describe the deformation velocity of the solid-plane system

$$\frac{\overrightarrow{V}^+ + \overrightarrow{V}^-}{2}, \ on\ \Gamma.$$

In order to take into account the deformation velocities inside the solid, we introduce

$$D(\frac{\overrightarrow{V}^+ + \overrightarrow{V}^-}{2}), \ in\ \Omega,$$

where $D(\overrightarrow{V}) = (1/2)\,(\partial V_i/\partial x_j + \partial V_j/\partial x_i)$ is the regular strain rate.

3.2 The principle of virtual work

We do not investigate the smooth evolution of the solid and fixed plane, i.e., when they move without colliding. Note that the solid may slide smoothly on the plane. In this situation, contact either with or without friction is to be used. We focus on the collisions; i.e., on motions where a kinematic incompatibility results in a discontinuity of velocity. For the sake of simplicity we do not mention time t at which the collision occurs.

The work of the acceleration forces

It is

$$\mathcal{T}_{acc}(\overrightarrow{V}) = \int_{\Omega} \rho \left[\overrightarrow{U}\right] . \frac{\overrightarrow{V}^{+} + \overrightarrow{V}^{-}}{2} d\Omega,$$

where ρ is the density and $\left[\overrightarrow{U}\right] = \overrightarrow{U}^{+} - \overrightarrow{U}^{-}$ is the velocity discontinuity.

The work of the interior forces

The very large stresses in Ω resulting from the kinematic incompatibility result in percussion stresses $\Sigma(\overrightarrow{x})$ and the very large contact forces on Γ result in contact surface percussions $\overrightarrow{R}(\overrightarrow{x})$, their virtual work is

$$\mathcal{T}_{int}(\overrightarrow{V}) = -\int_{\Omega} \Sigma : D(\frac{\overrightarrow{V}^{+} + \overrightarrow{V}^{-}}{2}) d\Omega - \int_{\Gamma} \overrightarrow{R} . (\frac{\overrightarrow{V}^{+} + \overrightarrow{V}^{-}}{2}) d\Gamma.$$

The work of the exterior forces

For the sake of completeness, we assume that an external percussion $\overrightarrow{T}_{ext}$ concomitant to the collision, is applied on the other part $\partial\Omega\backslash\Gamma$ of the solid surface. Its virtual work is

$$\mathcal{T}_{ext}(\overrightarrow{V}) = \int_{\partial\Omega\backslash\Gamma} \overrightarrow{T}_{ext} . \frac{\overrightarrow{V}^{+} + \overrightarrow{V}^{-}}{2} d\Gamma.$$

For instance, it may be an hammer blow.

The principle

It is

$$\forall \overrightarrow{V}, \mathcal{T}_{acc}(\overrightarrow{V}) = \mathcal{T}_{int}(\overrightarrow{V}) + \mathcal{T}_{ext}(\overrightarrow{V}).$$

3.3 The equations of motion

They result easily from the principle

$$\rho \left[\overrightarrow{U}\right] = div\Sigma, \ in \ \Omega, \tag{11}$$

$$\Sigma\overrightarrow{N} = -\overrightarrow{R}, \ on \ \Gamma, \ \Sigma\overrightarrow{N} = \overrightarrow{T}_{ext}, \ on \ \partial\Omega\backslash\Gamma, \tag{12}$$

where $\overrightarrow{N}$ is the outward normal vector to Ω. Constitutive laws are needed for percussion stresses Σ and percussions $\overrightarrow{R}$.

3.4 The constitutive laws

They have to satisfy the inequalities resulting from the laws of thermodynamics. We assume the temperatures of the colliding solids are equal and are not affected by the collision. The case where the temperatures change is treated in Frémond (2001).

The first law

It is

$$\mathcal{K}^-(t) - \mathcal{K}^+(t) = \mathcal{T}_{ext}(\overrightarrow{U}) + TB^{ext}, \tag{13}$$

where $\mathcal{K}(t)$ is the kinetic energy, and TB^{ext} the amount of heat received from the exterior at time t. With the expanded energy theorem, (13), we get

$$0 = -\mathcal{T}_{int}(\overrightarrow{U}) + TB^{ext}. \tag{14}$$

The second law

It is

$$0 \geq B^{ext}. \tag{15}$$

It shows that a collision produces heat. The amount of heat is given by the first law.

A useful inequality

The preceding relationships (14) and (15) give

$$\mathcal{T}_{int}(\overrightarrow{U}) \leq 0.$$

It results

$$\begin{aligned} \Sigma : D(\frac{\overrightarrow{U}^+ + \overrightarrow{U}^-}{2}) &\geq 0, \\ \overrightarrow{R}.\frac{\overrightarrow{U}^+ + \overrightarrow{U}^-}{2} &\geq 0. \end{aligned} \tag{16}$$

Relationships (16) have to be satisfied by the constitutive laws. We assume they are defined with volume and surface pseudo-potential of dissipation Φ and Φ_s which insure that relationships (16) are satisfied. We choose

$$\Phi(D(\frac{\overrightarrow{U}^+ + \overrightarrow{U}^-}{2})) = k\left\{D(\frac{\overrightarrow{U}^+ + \overrightarrow{U}^-}{2})\right\}^2,$$

$$\Phi_s(\frac{\overrightarrow{U}^+ + \overrightarrow{U}^-}{2}, \frac{\overrightarrow{U}^-}{2}) = k_s(\frac{\overrightarrow{U}^+ + \overrightarrow{U}^-}{2})^2 + I_-(U_N^+),$$

where $U_N^+ = \overrightarrow{U}^+.\overrightarrow{N}$ is the normal velocity, I_- is the indicator function of the set of the negative numbers $\mathbb{R}^-$ and, k and k_s are positive physical parameters the values of which result from experiment, Dimnet (2000). The constitutive laws are

$$\Sigma = \frac{\partial \Phi}{\partial D((\overrightarrow{U}^+ + \overrightarrow{U}^-)/2)} = kD(\overrightarrow{U}^+ + \overrightarrow{U}^-), \tag{17}$$

$$\overrightarrow{R} = \frac{\partial \Phi_s}{\partial((\overrightarrow{U}^+ + \overrightarrow{U}^-)/2)} \in k_s(\overrightarrow{U}^+ + \overrightarrow{U}^-) + \partial I_-(U_N^+)\overrightarrow{N}. \tag{18}$$

The percussion which is an element of $\partial I_-(U_N^+)\overrightarrow{N}$ is the non-interpenetration reaction percussion. As a property of any non-interpenetration reaction, it is active, i.e., different from zero, only if all the other percussions are insufficient to prevent interpenetration. One may also say that it is active only if there is a risk of interpenetration, i.e., if normal velocity U_N^+ is zero.

3.5 The equations

Equations giving velocity $\overrightarrow{U}^+$ depending on velocity $\overrightarrow{U}^-$, result from equations of motion (11), and constitutive laws (17), (18). They are

$$\rho\left[\overrightarrow{U}\right] = kdivD(\overrightarrow{U}^+ + \overrightarrow{U}^-),\ in\ \Omega,$$
$$\Sigma\overrightarrow{N} + k_s(\overrightarrow{U}^+ + \overrightarrow{U}^-) + \partial I_-(U_N^+)\overrightarrow{N} \ni 0,\ on\ \Gamma,\ \Sigma\overrightarrow{N} = \overrightarrow{T}_{ext},\ on\ \partial\Omega\backslash\Gamma. \tag{19}$$

It is possible to show that they have good properties:

Theorem 5. If both the domain Ω and the contact surface Γ are smooth with $mes(\Gamma) > 0$, if $\overrightarrow{U}^- \in H^1(\Omega)$ and if $\overrightarrow{T}_{ext} \in L^2(\Gamma)$, then there exists a unique velocity after collision, $\overrightarrow{U}^+ \in H^1(\Omega)$, solution of equations (19).

Remark 6. Sobolev spaces $H^1(\Omega)$, $L^2(\Gamma)$,... give the smoothness of the functions (see Attouch et al. (2004) for instance).

3.6 The micro-rebounds

Let us consider a long solid as shown on Fig. 2 falling on a rigid plan, Dimnet (2000). We assume that this solid is elastic (Young modulus $2,5\ 10^{10}\ Pa$, Poisson coefficient 0.3 and density 7877 kg/m^3).

Equations of motion are

$$\rho_1\overrightarrow{U}^+ = k\overrightarrow{grad}(div\overrightarrow{U}^+) + k\Delta\overrightarrow{U}^+ + \rho_1\overrightarrow{U}^- + k\overrightarrow{grad}(div\overrightarrow{U}^-) + k\Delta\overrightarrow{U}^-,\ in\ \Omega,$$
$$\Sigma\overrightarrow{N} + k_s(\overrightarrow{U}^+ + \overrightarrow{U}^-) + \partial I_-(U_N^+)\overrightarrow{N} \ni 0,\ on\ \Gamma,\ \Sigma\overrightarrow{N} = \overrightarrow{T}_{ext},\ on\ \partial\Omega\backslash\Gamma. \tag{20}$$

Parameters of the dissipative constitutive laws are $k = 10^5\ Ns^2m^{-2}$ and $k_s = 8\ 10^6\ Ns^2m^{-3}$. Computations due to Eric Dimnet (Dimnet (2000)) show that the solid bounces and vibrates after collision. There is a competition between a slow ascending motion resulting from the rebound and the fast oscillations resulting from the elasticity of the solid. During the first oscillations, the solid is not very high above the plane, thus the oscillations produce new collisions, Dimnet (2000). The vertical position of a point at the bottom part of the solid versus time is shown on Fig. 3. Six micro-rebounds occur in less than a thousandth of a second, before the solid separates clearly from the plane. Experiments have shown such a behaviour, Stoianovici and Hurmuzlu (1996).

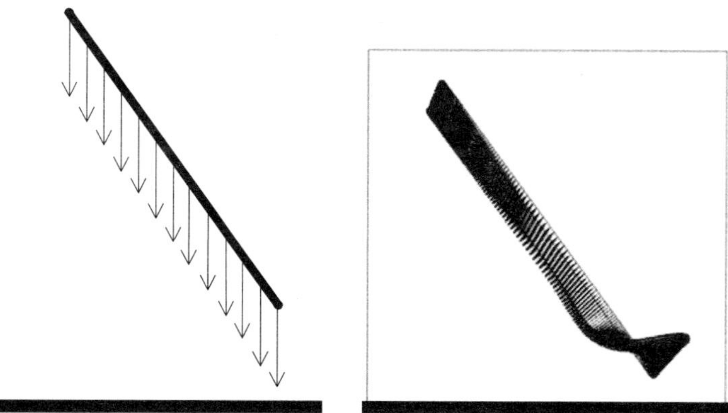

Figure 2. A rod falls with vertical velocity $-1\ ms^{-1}$ on a rigid plane. It has an angle $\theta = 54^o$ with the rigid plane. Its length is $0,6\ m$, its width is $0,012\ m$. Its two ends are round. Velocities of the rod are shown, Dimnet (2000).

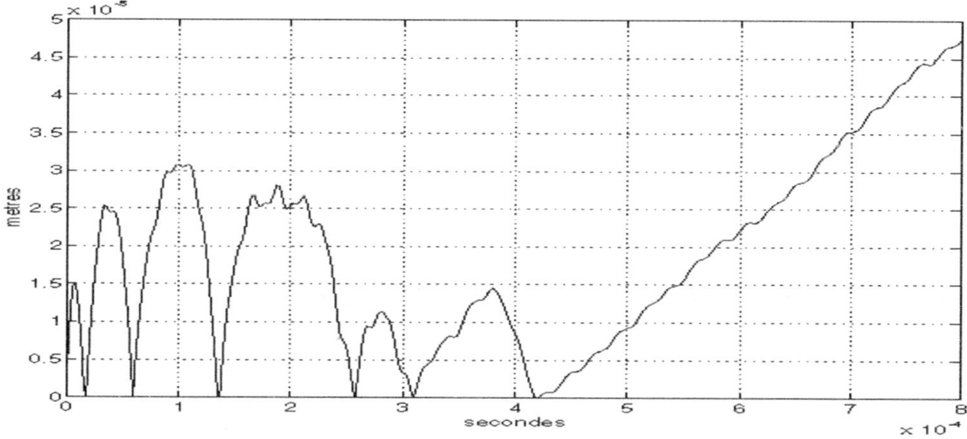

Figure 3. The vertical position versus time of a point of the lower part of a rod which collides with a rigid plane. After collision, the rod vibrates, micro-rebounds occur before the rod is sufficiently away from the plane, Dimnet (2000).

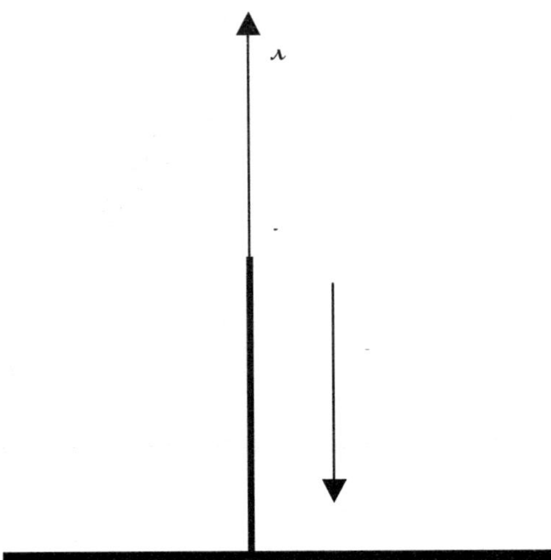

Figure 4. A vertical rod with length l falls with velocity U^- on an horizontal fixed plane.

3.7 A closed form solution: collision of a vertical deformable rod falling on a fixed plane

A vertical deformable rod with length l falls on a rigid fixed plane (Fig. 4). Before collision, the rod velocity is uniform

$$U^-(x) = U^-.$$

Let us compute its velocity after collision. It is a 1D problem.

The equations

They are the equations of motion

$$\rho [U] = \frac{\partial \Sigma}{\partial x}, in \ ]0, l[,$$
$$\Sigma(l) = 0, \ -\Sigma(0) = -R.$$

and the constitutive laws

$$\Sigma = k\frac{\partial(U^+ + U^-)}{\partial x} = k\frac{\partial U^+}{\partial x},$$
$$R = k_s(U^+(0) + U^-) + \partial I_+(U^+(0)).$$

There are those of previous paragraph but in $1D$. They are simple because they are linear besides the non-interpenetration condition. They give

$$\rho\left(U^+ - U^-\right) = k\frac{\partial^2 U^+}{\partial x^2}, \; in \;]0,l[, \tag{21}$$

$$\frac{\partial U^+}{\partial x}(l) = 0, \tag{22}$$

$$-k\frac{\partial U^+(0)}{\partial x} + k_s(U^+(0) + U^-) + \partial I_+(U^+(0)) \ni 0. \tag{23}$$

The closed form solution

Let us denote

$$\frac{\rho}{k} = \frac{1}{h^2}, \; \hat{k} = \frac{k_s}{k},$$

and

$$y(x) = U^+(x) - U^-.$$

Equations (21)–(23) become

$$\frac{y}{h^2} = \frac{\partial^2 y}{\partial x^2}, \; in \;]0,l[, \tag{24}$$

$$\frac{\partial y}{\partial x}(l) = 0, \tag{25}$$

$$-\frac{\partial y(0)}{\partial x} + \hat{k}(y(0) + 2U^-) + \partial I_+(y(0) + U^-) \ni 0. \tag{26}$$

Solution of equation (24) is

$$y(x) = ae^{x/h} + be^{-x/h},$$

where a and b are two constant to be fixed with the boundary conditions. Equation (25) give

$$\frac{a}{h}e^{l/h} - \frac{b}{h}e^{-l/h} = 0,$$

then

$$b = ae^{2l/h}.$$

We have

$$y(x) = a(e^{x/h} + e^{(2l-x)/h}).$$

Equation (26) becomes

$$-\frac{a}{h}(1 - e^{2l/h}) + \hat{k}(a(1 + e^{2l/h}) + 2U^-)) + \partial I_+(a(1 + e^{2l/h}) + U^-) \ni 0. \tag{27}$$

The different evolutions in collision

There are two possibilities for equation (27)

1. Case 1: Contact is maintained after collision: $U^+(0) = 0$ or $y(0) = -U^-$. We have

$$y(x) = \frac{-U^-}{1 + e^{2l/h}}(e^{x/h} + e^{(2l-x)/h}). \tag{28}$$

   It is a solution of equation (27) if

$$-\frac{\partial y(0)}{\partial x} + \hat{k}U^- \geq 0,$$

   or

$$\frac{U^-}{\left(1 + e^{2l/h}\right) h}(1 - e^{2l/h}) + \hat{k}U^- \geq 0.$$

   This condition is

$$\hat{k} \leq \frac{1 - e^{2l/h}}{\left(1 + e^{2l/h}\right) h}. \tag{29}$$

2. Case 2: The rod bounces or $\partial I_+(a(1 + e^{2l/h}) + U^-) = 0$.
   We have

$$-\frac{a}{h}(1 - e^{2l/h}) + \hat{k}(a(1 + e^{2l/h}) + 2U^-)) = 0,$$

   or

$$a = \frac{-2U^-\hat{k}}{\hat{k}\left(1 + e^{2l/h}\right) + \left(e^{2l/h} - 1\right)\frac{1}{h}}.$$

   We get

$$y(x) = \frac{-2U^-\hat{k}}{\hat{k}\left(1 + e^{2l/h}\right) + \left(e^{2l/h} - 1\right)\frac{1}{h}}(e^{x/h} + e^{(2l-x)/h}). \tag{30}$$

   It is solution of equation (27) if

$$y(0) \geq -U^-,$$

   or

$$\frac{-2U^-\hat{k}}{\hat{k}\left(1 + e^{2l/h}\right) + \left(e^{2l/h} - 1\right)\frac{1}{h}}\left(1 + e^{2l/h}\right) \geq -U^-.$$

   This condition is

$$\hat{k} \geq \frac{1 - e^{2l/h}}{\left(1 + e^{2l/h}\right) h}. \tag{31}$$

Because the function

$$h \to \frac{1-e^{2l/h}}{\left(1+e^{2l/h}\right)h},$$

is decreasing, we conclude that

- if h is small, this is the case either if the surface dissipation is small or if the rod is heavy, the rod does not bounce;
- if h is large, this is the case either if the surface dissipation is large or if the rod is not heavy, the rod bounces.

We retrieve a general property, a solid which is either light or very dissipative bounces and a solid which is either heavy or not very dissipative does not bounce.

The rod becomes rigid

When $k \to \infty$, the rod becomes rigid. Indeed, the pseudo-potential of dissipation becomes infinite except at the origin where it is zero. The solid becomes undeformable, i.e., rigid. The limit pseudo-potential of dissipation is

$$\lim_{k\to\infty} k\left(\frac{\partial(U^+ + U^-)}{\partial x}\right)^2 = I_0(\frac{\partial(U^+ + U^-)}{\partial x}),$$

where I_0 is the indicator function of the origin. This pseudo-potential is that of a rigid solid. Condition (31) becomes

$$\frac{k_s}{k} \geq \frac{1-e^{2l/h}}{\left(1+e^{2l/h}\right)h} \sim \frac{2l/h}{2h} = \frac{l}{h^2} = \frac{\rho l}{k},$$

or

$$k_s \geq \rho l = m,$$

where m is the rod mass. Formulas (28) and (30) show that

$$\lim_{k\to\infty} \frac{\partial y}{\partial x} \to 0.$$

The motion becomes that of a rigid solid. Moreover the equation of motion (21) gives

$$\int_0^l \rho\,[U]\,dx = -k\frac{\partial U^+}{\partial x}(0) \in -k_s(U^+(0) + U^-) - \partial I_+(U^+(0)),$$

and passing to the limit, we get

$$\rho l\,[U] \in -k_s(U^+ + U^-) - \partial I_+(U^+),$$

which is the equation for the collision of a rigid rod with the fixed plane, resulting from the following equation of motion and constitutive law

$$m\,[U] = -P,$$
$$P \in k_s(U^+ + U^-) + \partial I_+(U^+).$$

We retrieve that if $k_s \geq m$ the solid bounces and that if $k_s \leq m$ the solid does not bounce. Once again a solid either light or dissipative bounces and, a solid either heavy or not dissipative does not bounce.

This result is general. It applies to $3D$ solids: the previous predictive theory tends to the rigid body predictive theory for collisions, Frémond (2005).

4 Collisions of fluids and solids

Consider a swimmer diving in a swimming pool. When the diver impacts the water, he can be horizontal and do a belly flop. In this situation a violent collision occurs with the water. The second example consists of a fat boat in a rough sea.According to the creation of the waves, the boat rise up and in a small time interval its bottom hits the water. In eachof these problems a solid collides with a viscous incompressible fuid.

The hypothesis of instantaneousness of collisions is compatible with the fluid incompressibility. Further, in our case, in addition to the percussion on the contact surface between the fluid and solids there are also percussion stresses inside the fluid when the collision occurs. At this time, the velocity field is a discontinuous function with respect to time. There exist a velocity field before and after the collision. We denote them respectively as $\overrightarrow{U}^-$ and $\overrightarrow{U}^+$.

The first and second laws of thermodynamics lead to the assumption of the percussion stress depending on the strain rate $D(\overrightarrow{U}^+ + \overrightarrow{U}^-)$ where $D_{ij}(\overrightarrow{U}) = \frac{1}{2}(U_{i,j} + U_{j,i})$ is the usual strain rate. In the same way the percussion contact force depends on the relative velocity $\overrightarrow{U} - \overrightarrow{U}_s$ of the fluid with respect to the solid ($\overrightarrow{U}_s$ denotes the solid velocity).

Choosing the simplest constitutive law, i.e., linear constitutive laws, results in a set of partial differential equations which gives the velocities $\overrightarrow{U}^+$ and $\overrightarrow{U}_s^+$ after the collision depending on the velocities $\overrightarrow{U}^-$ and $\overrightarrow{U}_s^-$ before the collision. This set of partial differential equations is coherent in term of both mechanical and mathematical point of view. Moreover the two examples which are investigated below, show that the theory takes into accounts the basic physical properties. The results of this section are based on Frémond et al. (2003) and Dimnet and Gormaz ((to appear)).

4.1 The equation of motion

We focus on what occurs at collision time and do not investigate the smooth motions of fluid and solid. When a solid collides with water very large internal forces appear because of the incompatibility of the velocities of the solid and water: very large stresses inside and very large forces on the contact surface. In many circumstances the duration of the violent contact is very short compared to the duration of the free motion. For instance when skipping stones on the still water of a lake, the duration of the contact of the stone with the water is small compared to its time of flight. We decide as a schematization to consider that the collisions are instantaneous. Thus the velocities are discontinuous functions of time at time t of collision which is not mentioned in the sequel for the sake of simplicity: for the fluid, there is the velocity $\overrightarrow{U}^-(\overrightarrow{x})$ before the collision and the velocity $\overrightarrow{U}^+(\overrightarrow{x})$ after the collision; for the solid, there is the velocity $\overrightarrow{U}_s^-(\overrightarrow{x}) = \overrightarrow{X}^- + \overrightarrow{\varpi}^- \times (\overrightarrow{x} - \overrightarrow{x}_G)$

before collision and the velocity $\overrightarrow{U}_s^+(\overrightarrow{x}) = \overrightarrow{X}^+ + \overrightarrow{\varpi}^+ \times (\overrightarrow{x} - \overrightarrow{x}_G)$ after collision. The velocity of the center of mass $\overrightarrow{x}_G = \overrightarrow{x}_G(t)$ of the rigid body is $\overrightarrow{X}$ and $\overrightarrow{\varpi}$ is its angular velocity. Because the collision is instantaneous, the very large internal forces are also concentrated in time: thus they are percussions.

The principle of virtual work

Let us consider the system made of the fluid and solids. For instance, the solids are the container of the fluid and a moving obstacle (Fig. 5). For the sake of simplicity, we suppose there is only one solid (for instance the container). The fluid occupies the domain $\Omega = \Omega(t)$ and is in contact with the solid on the part $\Gamma_1 = \Gamma_1(t)$ of its boundary $\partial\Omega = \partial\Omega(t)$. At time t of collision, we consider the virtual velocities $(\overrightarrow{V}, \overrightarrow{Y}, \overrightarrow{\pi}) = \left(\overrightarrow{V}^+(\overrightarrow{x}), \overrightarrow{Y}^+, \overrightarrow{\pi}^+, \overrightarrow{V}^-(\overrightarrow{x}), \overrightarrow{Y}^-, \overrightarrow{\pi}^-\right)$. The velocity $\overrightarrow{Y}$ is a virtual velocity of the center of mass of the rigid body and $\overrightarrow{\pi}$ is a virtual angular velocity of the rigid body. It is denoted by $\overrightarrow{V}_s(\overrightarrow{x}) = \overrightarrow{Y} + \overrightarrow{\pi} \times (\overrightarrow{x} - \overrightarrow{x}_G)$, the virtual velocity at point $\overrightarrow{x}$ of the rigid body. The mass of the rigid body is M and its mass moment of inertia is I. The density of the fluid is ρ.

The virtual work of the acceleration forces

We choose the virtual work of the acceleration forces as

$$\mathcal{T}^{acc}(\overrightarrow{V}, \overrightarrow{Y}, \overrightarrow{\pi}) = \int_\Omega \rho \left[\overrightarrow{U}(\overrightarrow{x})\right] \cdot \frac{\overrightarrow{V}^+(\overrightarrow{x}) + \overrightarrow{V}^-(\overrightarrow{x})}{2} d\Omega$$
$$+ M \left[\overrightarrow{X}\right] \cdot \frac{\overrightarrow{Y}^+ + \overrightarrow{Y}^-}{2} + I\left[\overrightarrow{\varpi}\right] \cdot \frac{\overrightarrow{\pi}^+ + \overrightarrow{\pi}^-}{2},$$

where $[A] = A^+ - A^-$ is the discontinuity of the quantity A.

The theorem of expanded energy

Let us note that the actual work of acceleration forces is the variation of the kinetic energy $\mathcal{K}(t)$ in collision

$$\mathcal{K}^+ - \mathcal{K}^- = \mathcal{T}^{acc}(\overrightarrow{U}, \overrightarrow{X}, \overrightarrow{\varpi}).$$

The virtual work of the interior forces

The virtual work of the interior forces we choose, is a linear function of the virtual strain rate, $D_{ij}(\overrightarrow{V}) = (1/2)(V_{i,j} + V_{j,i})$, and of the difference of the velocities on the contact surface, $\overrightarrow{V} - \overrightarrow{V}_s$. They introduce percussion stress tensor Σ and interaction percussion $\overrightarrow{R}$ between the fluid and the solid. The percussion stress and the interaction percussion are generalized interior forces which appear when collisions occur. They are usual interior forces concentrated in a very short period of time. The virtual work of the interior forces we choose, is

$$\mathcal{T}^{int}(\overrightarrow{V}, \overrightarrow{Y}, \overrightarrow{\pi}) = -\int_\Omega \Sigma(\overrightarrow{x}) : \frac{D(\overrightarrow{V}^+)(\overrightarrow{x}) + D(\overrightarrow{V}^-)(\overrightarrow{x})}{2} d\Omega$$
$$- \int_{\Gamma_1} \overrightarrow{R}(\overrightarrow{x}). \frac{\overrightarrow{V}^+(\overrightarrow{x}) - \overrightarrow{V}_s^+(\overrightarrow{x}) + \overrightarrow{V}^-(\overrightarrow{x}) - \overrightarrow{V}_s^-(\overrightarrow{x})}{2} d\Gamma.$$

The work of the interior forces has to satisfy the axiom of the interior forces, Germain (1973):

the work of the interior forces is zero for any rigid system virtual velocities $(\overrightarrow{V}(\overrightarrow{x},t), \overrightarrow{Y}(t), \overrightarrow{\pi}(t))$.

The rigid system velocities are such that $D(\overrightarrow{V}) = 0$ and $\overrightarrow{V} - \overrightarrow{V_s} = 0$ on the boundary Γ_1. Thus it is obvious that $\mathcal{T}^{int}(\overrightarrow{V}, \overrightarrow{Y}, \overrightarrow{\pi}) = 0$ for such a set of velocities $(\overrightarrow{V}, \overrightarrow{Y}, \overrightarrow{\pi})$.

The virtual work of the exterior forces

The virtual work of the exterior forces involves the possibility to apply exterior percussions. For the sake of simplicity, we choose

$$\mathcal{T}^{ext}(\overrightarrow{V}, \overrightarrow{Y}, \overrightarrow{\pi}) = 0.$$

The equations of motion

They result from the principle of virtual work

$$\forall(\overrightarrow{V}, \overrightarrow{Y}, \overrightarrow{\pi}),$$
$$\mathcal{T}^{acc}(\overrightarrow{V}, \overrightarrow{Y}, \overrightarrow{\pi}) = \mathcal{T}^{int}(\overrightarrow{V}, \overrightarrow{Y}, \overrightarrow{\pi}).$$

They are

$$\begin{aligned}
\rho\left[\overrightarrow{U}(\overrightarrow{x})\right] &= div\, \Sigma(\overrightarrow{x}),\ in\ \Omega, \\
\Sigma(\overrightarrow{x})\overrightarrow{N} &= -\overrightarrow{R}(\overrightarrow{x}),\ in\ \Gamma_1, \\
\Sigma(\overrightarrow{x})\overrightarrow{N} &= 0,\ in\ \partial\Omega\backslash\Gamma_1, \\
M[\overrightarrow{X}] &= \int_{\Gamma_1} \overrightarrow{R}(\overrightarrow{x})d\Gamma, \\
I[\overrightarrow{\omega}] &= \int_{\Gamma_1} (\overrightarrow{x} - \overrightarrow{x}_G) \times \overrightarrow{R}(\overrightarrow{x})d\Gamma,
\end{aligned}$$

where $\overrightarrow{N}$ is the outward normal vector. The boundary conditions and the last two equations are valid for any solid if more than one are in contact with the fluid.

Constitutive laws are needed for $(\Sigma, \overrightarrow{R})$ in order to solve the equations. They result from experiments and are restricted by a basic relationship, namely the second principle of thermodynamics.

4.2 The constitutive laws

Because we intend to deal with incompressible fluids, we assume the temperature T to be constant and the internal energy and entropy to be also constant with respect to the time. A computation similar to the one performed in the previous chapter gives the useful inequality

$$\begin{aligned}
\Sigma : \frac{D(\overrightarrow{U}^+) + D(\overrightarrow{U}^-)}{2} &\geq 0 \\
\overrightarrow{R}.\frac{\overrightarrow{U}^+ - \overrightarrow{U}_s^+ + \overrightarrow{U}^- - \overrightarrow{U}_s^-}{2} &\geq 0.
\end{aligned} \tag{32}$$

Relationships (32) are a guide to choose the constitutive laws. It is used in the following paragraph to choose the constitutive laws for Σ and $\overrightarrow{R}$.

The constitutive laws for an incompressible fluid

Inequalities (32) and the expression of W^{int} suggest to assume Σ to depend on $D(\overrightarrow{U}^+ + \overrightarrow{U}^-)$ and $\overrightarrow{R}^d$ to depend on $(\overrightarrow{U}^+ - \overrightarrow{U}_s^+ + \overrightarrow{U}^- - \overrightarrow{U}_s^-)$. We define the constitutive laws with volume and pseudo-potentials of dissipation

$$\Phi(D(\frac{\overrightarrow{U}^+ + \overrightarrow{U}^-}{2})) = \gamma\left\{D(\frac{\overrightarrow{U}^+ + \overrightarrow{U}^-}{2})\right\}^2 + I_0(trD(\frac{\overrightarrow{U}^+ + \overrightarrow{U}^-}{2}),$$

$$\Phi_s(\frac{\overrightarrow{U}^+ - \overrightarrow{U}_s^+ + \overrightarrow{U}^- - \overrightarrow{U}_s^-}{2}, \frac{\overrightarrow{U}^- - \overrightarrow{U}_s^-}{2}) =$$
$$K(\frac{\overrightarrow{U}^+ - \overrightarrow{U}_s^+ + \overrightarrow{U}^- - \overrightarrow{U}_s^-}{2})^2 + I_-(U_N^+ - U_{sN}^+),$$

where $\gamma \geq 0$, is the dissipative volume viscosity and K the collision friction coefficient which depends on the nature of the colliding solid and fluid.. Function I_0 is the indicator function of the origin of $\mathbb{R}$ and the trace $trD(\overrightarrow{U}^+ + \overrightarrow{U}^-)$ of $D(\overrightarrow{U}^+ + \overrightarrow{U}^-)$ is $div(\overrightarrow{U}^+ + \overrightarrow{U}^-)$. Function I_- is the indicator function of the set of the negative numbers $\mathbb{R}^-$ and $U_N^+ = \overrightarrow{U}^+ . \overrightarrow{N}$ is the normal velocity. The constitutive laws are

$$\Sigma \in \partial\Phi(D(\frac{\overrightarrow{U}^+ + \overrightarrow{U}^-}{2})),$$
$$\overrightarrow{R} \in \partial\Phi_s(\frac{\overrightarrow{U}^+ - \overrightarrow{U}_s^+ + \overrightarrow{U}^- - \overrightarrow{U}_s^-}{2}, \frac{\overrightarrow{U}^- - \overrightarrow{U}_s^-}{2}),$$

where the volume subdifferential is computed with respcct to $D((\overrightarrow{U}^+ + \overrightarrow{U}^-)/2)$ and the surface subdifferential is computed with respect to $(\overrightarrow{U}^+ - \overrightarrow{U}_s^+ + \overrightarrow{U}^- - \overrightarrow{U}_s^-)/2$. We have

$$\Sigma = 2\gamma D(\overrightarrow{U}^+ + \overrightarrow{U}^-) - PI,$$
$$\overrightarrow{R} = K(\overrightarrow{U}^+ - \overrightarrow{U}_s^+ + \overrightarrow{U}^- - \overrightarrow{U}_s^-) + R^{reac}\overrightarrow{N},$$

with

$$-P \in \partial I_0(div(\overrightarrow{U}^+ + \overrightarrow{U}^-)), \quad (33)$$
$$R^{reac} \in \partial I_-(U_N^+ - U_{sN}^+), \quad (34)$$

Relationship (33) accounts for the incompressibility condition, $div(\overrightarrow{U}^+ + \overrightarrow{U}^-) = 0$ or $div\overrightarrow{U}^+ = 0$ because $div\overrightarrow{U}^- = 0$. It results in the percussion pressure P. Relationship (34) accounts for the impenetrability of the solid and fluid: $U_N^+ - U_{sN}^+ \leq 0$. It results in the impenetrability reaction percussion $R^{reac}\overrightarrow{N}$. The impenetrability reaction is active only when the risk of interpenetration is present, i.e., when the contact is persistent after the collision.

Let us emphasizes that the impenetrability condition (34) has two meanings. The first one is to imply that $U_n^+ - U_{sn}^+ \leq 0$ and the second one is to give the value of the reaction R^{reac} which is 0 when contact is not maintained after the collision and which is positive when contact is maintained after the collision.

At this point the predictive theory is completed. It is to be checked if it is consistent in terms of mathematics, numerics and if it has the ability to account for the more obvious experimental results. Let us stress that we have chosen the simplest constitutive laws which satisfy the basic requirements of mechanics. The collision constitutive laws are characterized only by two parameters γ which takes into account the volume phenomena occurring in a collision and K which describes the surface interactions between the surface of the solid and the fluid. We are convinced that this is the minimal number of information which are needed to predict what occurs after a collision.

4.3 The diver problem

For the sake of simplicity a 2D problem is investigated. It is assumed that the swimming pool is a fixed rectangle. The horizontal swimmer Γ_1 dives at the middle of the swimming pool (Fig. 5). We assume that the diver is symmetric, flat (Fig. 5) and its thickness is zero. Its velocities when hitting the water are

$$\overrightarrow{X}^- = (0, X^-),$$
$$\overrightarrow{\varpi}^- = \overrightarrow{0}.$$

It is assumed that no external load is applied. It result from the symmetries that the solution

$$\overrightarrow{U}^+(x,y) = \big(U_1^+(x,y), U_2^+(x,y)\big)$$

is such that $\overrightarrow{U}^+(-x,y) = \big(-U_1^+(x,y), U_2^+(x,y)\big)$ and that $\overrightarrow{X}^+ = (0, X^+)$ and $\overrightarrow{\varpi}^+ = \overrightarrow{0}$. All the points of the rigid body have the same vertical velocity, which is the velocity of the middle of Γ_1.

The equations

Let Ω be the swimming pool, an open subset in R^2 with outwards normal vector $\overrightarrow{N}$, (Fig. 5). This domain contains a viscous homogeneous incompressible fluid. We suppose that at collision time t the water is at rest: $\overrightarrow{U}^- = 0$. The diver schematized by a rigid body, collides the part Γ_1 of the boundary $\partial\Omega$. We suppose that the remaining boundary $\partial\Omega \setminus \Gamma_1$ is decomposed into two additional disjoint parts: a rigid boundary Γ_0 and a free boundary Γ_2. We are interested in finding the velocities $\overrightarrow{U}^+(\overrightarrow{x})$ and $\overrightarrow{X}^+$ after the collision. The equations are the equations of motion

$$\rho\overrightarrow{U}^+ = div\Sigma, \tag{35}$$

$$M[\overrightarrow{X}] = \int_{\Gamma_1} \overrightarrow{R}_1 d\Gamma, \tag{36}$$

$$\Sigma\overrightarrow{N} = -\overrightarrow{R}_0, \text{ on } \Gamma_0, \tag{37}$$

$$\Sigma\overrightarrow{N} = -\overrightarrow{R}_1, \text{ on } \Gamma_1, \tag{38}$$

$$\Sigma n\overrightarrow{N} = 0 \text{ on } \Gamma_2, \tag{39}$$

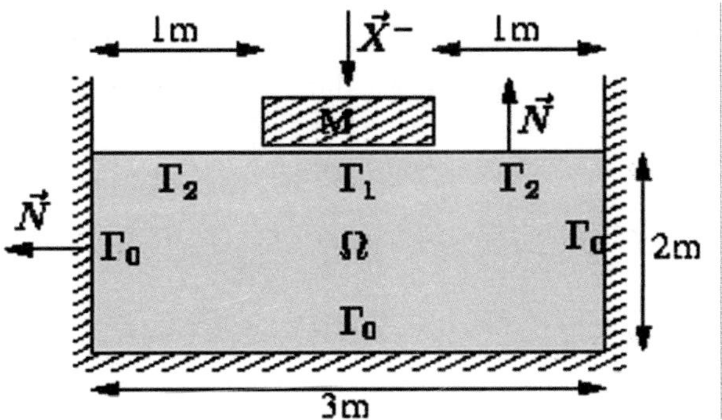

Figure 5. The diver and the swimming pool. The vertical velocity of the diver before he collides the water is negative, $U_N^- - U_{sN}^- = -X^- \leq 0$.

and the constitutive laws

$$\begin{aligned}\Sigma &= 2\gamma D(\overrightarrow{U}^+) - PI,\\ \overrightarrow{R}_0 &= K_0\overrightarrow{U}^+ + R_0^{reac}\overrightarrow{N},\ R_0^{reac} \in \partial I_-(U_N^+),\\ \overrightarrow{R}_1 &= K_1(\overrightarrow{U}^+ - \overrightarrow{X}^+ - \overrightarrow{X}^-) + R_1^{reac}\overrightarrow{N},\ R_1^{reac} \in \partial I_-(U_N^+ - X^+),\end{aligned} \tag{40}$$

where $U_N^+ - X^+$ is the normal velocity on Γ_1 and U_N^+ is the normal velocity on Γ_0. Both normal velocities are negative due to the impenetrability condition.

The variational formulation

The set of partial differential equations (35)–(40) gives the velocities $(\overrightarrow{U}^+, \overrightarrow{X}^+)$ after the collision depending on the velocities $(\overrightarrow{U}^-, \overrightarrow{X}^-)$ before the collision.

In this subsection we show that the equations fit in the variational inequalities theory. We begin by defining the functional spaces framework. For the fluid velocity field, we define the space

$$V = \left\{\overrightarrow{W} \in (H^1(\Omega))^2;\ div\overrightarrow{W} = 0\right\},$$

where $H^1(\Omega)$ is the usual Sobolev space. For the coupled system of fluid-rigid body we define the following convex cone for the kinematically admissible velocities

$$C = \left\{(\overrightarrow{W}, Y) \in V \times R : \overrightarrow{W}.\overrightarrow{N} \leq 0 \ on\ \Gamma_0, \overrightarrow{W}.\overrightarrow{N} - Y\overrightarrow{N} \leq 0 \ on\ \Gamma_1\right\}.$$

Let us consider $(\overrightarrow{W}, Y) \in C$, multiplying (35) by the function $\overrightarrow{W} - \overrightarrow{U}^+$ and (36) by the function $Y - X^+$, we prove that any classical solution must satisfy the equation

$$\begin{aligned}
&\int_\Omega \rho \overrightarrow{U}^+.(\overrightarrow{W} - \overrightarrow{U}^+) d\Omega + M(X^+ - X^-)(Y - X^+) \\
&+ \int_\Omega 2\gamma D(\overrightarrow{U}^+) : D(\overrightarrow{W} - \overrightarrow{U}^+) d\Omega + \int_{\Gamma_0} K_0 \overrightarrow{U}^+.(\overrightarrow{W} - \overrightarrow{U}^+) d\Gamma \\
&+ \int_{\Gamma_1} K_1(\overrightarrow{U}^+ - X^+\overrightarrow{N} - X^-\overrightarrow{N}).(\overrightarrow{W} - Y\overrightarrow{N} - (\overrightarrow{U}^+ - X^+\overrightarrow{N})) d\Gamma \\
&= \int_{\Gamma_0} -R_0^{reac}(W_n - U_n^+) d\Gamma - \int_{\Gamma_1} R_1^{reac}\{W_n - Y - (U_n^+ - X^+)\} d\Gamma.
\end{aligned}$$

Due to the properties of the subdifferential sets (see the Appendix), we have

$$\forall(\overrightarrow{W}, Y) \in C,\ R_0^{reac}(W_n - U_n^+) \leq 0,\ R_1^{reac}\{W_n - Y - (U_n^+ - X^+)\} \leq 0.$$

It follows

$$\begin{aligned}
&(\overrightarrow{U}^+, X^+) \in C,\ \forall(\overrightarrow{W}, Y) \in C, \\
&\int_\Omega \rho \overrightarrow{U}^+.(\overrightarrow{W} - \overrightarrow{U}^+) d\Omega + M(X^+ - X^-)(Y - X^+) \\
&+ \int_\Omega 2\gamma D(\overrightarrow{U}^+) : D(\overrightarrow{W} - \overrightarrow{U}^+) d\Omega + \int_{\Gamma_0} K_0 \overrightarrow{U}^+.(\overrightarrow{W} - \overrightarrow{U}^+) d\Gamma \\
&+ \int_{\Gamma_1} K_1(\overrightarrow{U}^+ - X^+\overrightarrow{N} - X^-\overrightarrow{N}).(\overrightarrow{W} - Y\overrightarrow{N} - (\overrightarrow{U}^+ - X^+\overrightarrow{N})) d\Gamma \geq 0. \qquad (41)
\end{aligned}$$

Conversely, it can be proved that a solution of the variational inequality (41), if it is smooth enough, satisfies the equation of motion (35)–(39) together with the constitutive laws (40).

An existence theorem

Let us define some notations

$$\begin{aligned}
a(\overrightarrow{V}, \overrightarrow{W}) &= \int_\Omega \rho \overrightarrow{V}.\overrightarrow{W} d\Omega + \int_\Omega 2\gamma D(\overrightarrow{V}) : D(\overrightarrow{W}) d\Omega + \int_{\Gamma_0} K_0 \overrightarrow{V}.\overrightarrow{W} d\Gamma, \\
b(Y, Z) &= MYZ, \\
c(\overrightarrow{V}, \overrightarrow{W}) &= \int_{\Gamma_1} K_1 \overrightarrow{V}.\overrightarrow{W} d\Gamma, \\
L(Y) &= M(X^-)Y, \\
L_1(\overrightarrow{W}) &= \int_{\Gamma_1} K_1(X^-\overrightarrow{N}).\overrightarrow{W} d\Gamma.
\end{aligned}$$

The equations (41) can be rewritten

$$\begin{aligned}
&(\overrightarrow{U}^+, X^+) \in C,\ \forall(\overrightarrow{W}, Y) \in C, \\
&a(\overrightarrow{U}^+, \overrightarrow{W} - \overrightarrow{U}^+) + b(X^+, Y - X^+) + c(\overrightarrow{U}^+ - X^+\overrightarrow{N}, \overrightarrow{W} - Y\overrightarrow{N}) \geq \\
&L(Y - X^+) + L_1(\overrightarrow{W} - \overrightarrow{U}^+) - L_1(Y\overrightarrow{N} - X^+\overrightarrow{N}). \qquad (42)
\end{aligned}$$

In this form, it is a variational inequality. It can be proved that this variational inequality (42) or (41) has a unique solution:

Theorem 7. The variational inequality (42) has a unique solution.

Proof. The bilinear form

$$A((\overrightarrow{W},Y),(\overrightarrow{V},Z))=a(\overrightarrow{W},\overrightarrow{V})+b(Y,Z)+c(\overrightarrow{W},\overrightarrow{V}),$$

is symmetric, continuous and coercive on $V\times R$. The linear form

$$B(((\overrightarrow{W},Y))=L(Y)+L_1(\overrightarrow{W})-L_1(Y\overrightarrow{N}),$$

is continuous on the same space. The set C is a non-empty closed convex subset of $V\times R$. It follows from the Lions-Stampacchia theorem, Lions and Stampacchia (1967), that the variational inequality (42) equivalent to

$$(\overrightarrow{U}^+,X^+)\in C,\ \forall(\overrightarrow{W},Y)\in C,$$
$$A((\overrightarrow{W},Y)-(\overrightarrow{U}^+,X^+),(\overrightarrow{U}^+,X^+))\geq B((\overrightarrow{W},Y)-(\overrightarrow{U}^+,X^+)),$$

has a unique solution

Numerical results

The mass of the diver is $M=100Kg$. Falling from $1m$, his vertical velocity is $U^-=-\sqrt{2g}=-4,47m/s$. The water density is $\rho=1000Kg/m^3$. The two parameters of the constitutive laws are

$$K_0=K_1=1Kg/m^2,\ \gamma=0,25Kg/m^2.$$

The equations may be solved by the classical numerical methods for variational inequalities, for instance the Uzawa method. In the present situation it is better to take advantage of the fact that the solution is unique: the equations are solved with the assumption that the contact is maintained on the parts Γ_0 and Γ_1 and it is checked that the reaction computed with the boundary conditions are negative (in case they are positive somewhere the solution is such that the contact is not maintained everywhere). The water splashes up on the two sides of the diver as shown on the Fig. 6. The velocity of the diver after the collision is the velocity of the water ($-0.75\ ms^{-1}$), it has been divided by 6, which seems to be in agreement with what occurs in a swimming pool.

Details are shown on the following Figs. 7 and 8: the velocity is drawn at different depth of the pool. The intensity of the vertical velocity is shown on Fig. 9. The maximum horizontal velocity is $1.96m/s$. The maximum vertical velocity is $1.95m/s$ and the minimum is $-0.75m/s$. The percussion pressure is shown on the Fig. 10. The average percussion stress $\left|\Sigma\overrightarrow{N}\right|$ under the diver is 373 Nm^2s^{-1} equal to the average percussion pressure 371 Nm^2s^{-1}.

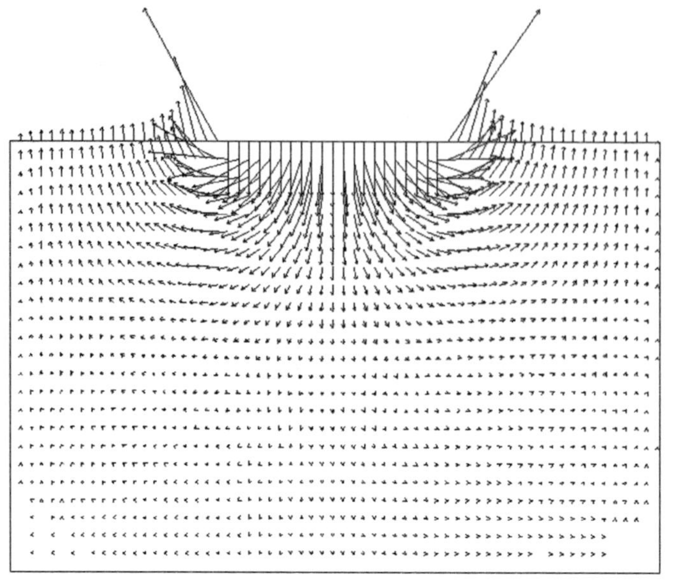

Figure 6. The velocity of the water which splashes up on the two sides of the diver. The arrows represent the velocities of water and diver.

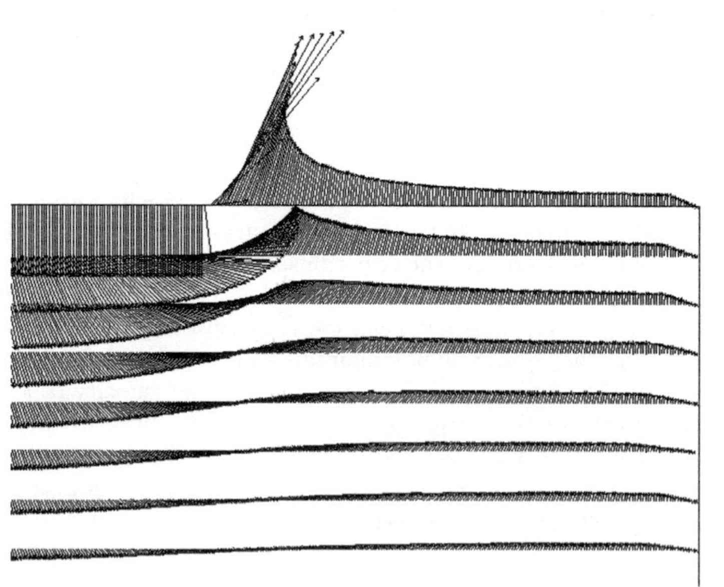

Figure 7. Water velocities at different depths.

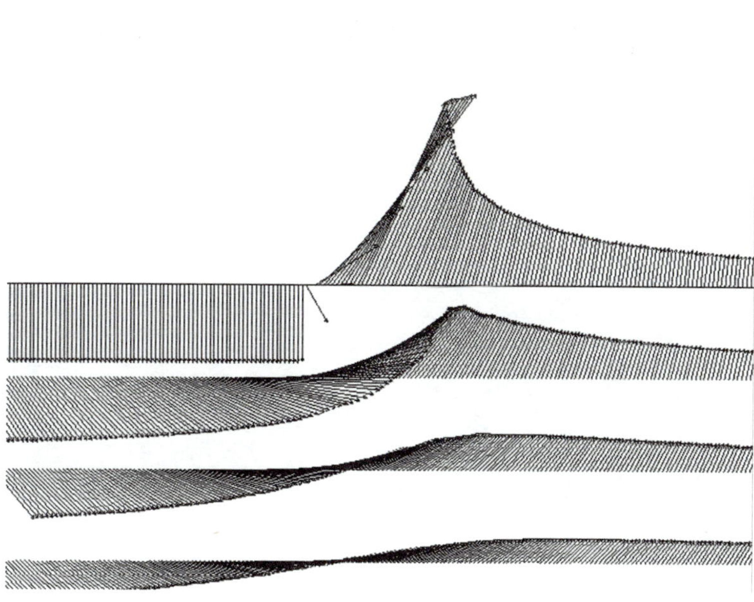

Figure 8. Details of the velocity of the water at different depths

Figure 9. The intensity of the vertical velocity. Red is upwards velocity. Blue is downwards velocity. Dark blue is large downwards velocities, light blue is low downwards velocity.

Figure 10. Pressure in the water. Red is a high pressure and blue a low pressure.

4.4 An other example: stone skipping

In the preceding problem, we assume the velocity of solid before collision has an horizontal component Z^-

$$\overrightarrow{X}^- = (Z^-, X^-),$$
$$\overrightarrow{\omega}^- = \overrightarrow{0}.$$

For instance, a stone colliding the still water of a lake with an horizontal velocity. At collision time the stone is horizontal. Fig. 11 shows what occurs when horizontal velocity Z^- is large: the stone bounces on the water. This not the case if the horizontal velocity is zero, Fig. 12. Thus the stone skipping phenomenon is accounted by this theory. Results are due to E. Dimnet and R. Gormaz, Dimnet and Gormaz ((to appear)). Constitutive laws giving surface percussion $\overrightarrow{R}_0$ and $\overrightarrow{R}_1$ involve different coefficients K for the normal friction, K_N and tangential friction, K_T

$$\overrightarrow{R}_0 = K_{0T}\overrightarrow{U}_T^+ + K_{0N}U_N^+ N + R_0^{reac}\overrightarrow{N},$$

where $\overrightarrow{U}_T$ and U_N are the tangential and normal velocities. The parameters are

$$K_{0N} = K_{1N} = 1Kg/m^2,\ K_{0T} = K_{1T} = 0,25Kg/m^2,$$
$$\gamma = 1,8\ Kg/m^2,\ M = 0,5Kg,\ \rho = 1000Kg/m^3.$$

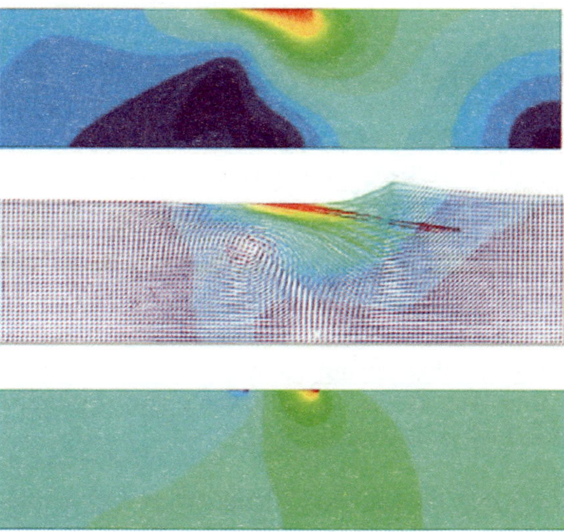

Figure 11. The stone collides water with a large horizontal velocity $\overrightarrow{X}^- = (5m/s, -1m/s)$, without angular velocity $\omega^- = 0$. It bounces and turns: $\overrightarrow{X}^+ = (2,46m/s,\ 0,09m/s)$ et $\omega^+ = -0,49rad/s$. From top to bottom, pictures show modulus of fluid velocity, fluid velocity and, percussion pressure.

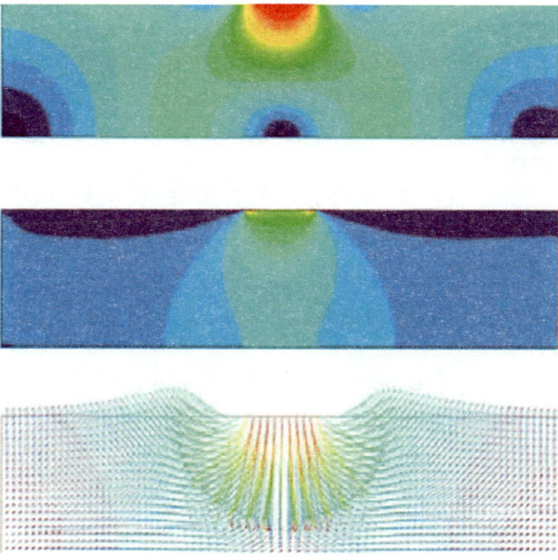

Figure 12. The stone falls vertically with velocity $\overrightarrow{X}^- = (0, -5m/s)$, without angular velocity $\omega^- = 0$. The stone does not bounce. It remains in contact with the fluid. Its velocity is $\overrightarrow{X}^+ = (0, -0,07m/s)$ and $\omega^+ = 0$. From top to bottom, the figures show modulus of the fluid velocity, percussion pressure and fluid velocity.

4.5 Conclusions

The theory of collisions of rigid bodies, of deformable solids and, of fluids and solids outlined is consistent from the thermal and mechanical points of view. It has good mathematical formulations to which usual numerical methods apply. The examples show that the basic physical properties are taken into account. The theory is general and versatile to apply to different practical settings.

Let us emphasize the basic idea which is to consider that the system made of the different solids and fluids is deformable. The system velocities of deformation result in interior forces which capture and sum up the very sophisticated physical phenomenons which intervene during the short duration of collisions. The interior forces are defined by duality i.e., by their work. In this setting the choice of the work of the interior forces of collisions is very crucial as well as the choice of the velocity of deformation, for instance $D(\overrightarrow{U}^+ + \overrightarrow{U}^-)$ to define percussion stress or $\overrightarrow{U}^+ - \overrightarrow{U}^+_s + \overrightarrow{U}^- - \overrightarrow{U}^-_s$ to define contact percussion.

Let us also emphasize that few parameters intervene and emphasize the simplicity of the constitutive laws which are linear, besides the reactions to the internal constraints which cannot be avoided.

5 Appendix

5.1 Convex Functions

Let $\bar{\mathbb{R}} = \mathbb{R} \cup \{+\infty\}$ where the addition has two new rules: $a + (+\infty) = +\infty$ and $+\infty + (+\infty) = +\infty$. The multiplication is restricted to multiplication by strictly positive numbers. The multiplication has the new rule: $\forall a > 0$, $a \times (+\infty) = +\infty$. In this context multiplication by negative numbers is forbidden, Moreau (1966), Nayroles (1973), Ekeland and Temam (1976) .

Let f be a function from the linear space X into $\bar{\mathbb{R}}$. It is a convex function if and only if

Definition 8. $\forall x \in V, \forall y \in V,\ \forall \theta \in]0,1[,$
$f(\theta x + (1-\theta)y) \leq \theta f(x) + (1-\theta) f(y).$

Remark 9. Because $\theta \in]0,1[$, f is multiplied by strictly positive numbers.

Example of convex functions
Let $V = \mathbb{R}$. It is easy to see that the functions

$$x \in \mathbb{R} \rightarrow f_p(x) = \frac{1}{p} |x|^p\,,\ with\ p \geq 1,$$

are convex.

Let the function I from $V = \mathbb{R}$ into $\bar{\mathbb{R}}$ be defined by

$$I(x) = 0,\ if\ x \in [0,1]\,,$$
$$I(x) = +\infty,\ if\ \ x \notin [0,1]\,.$$

This function is convex. It is called the indicator function of segment $[0,1]$ (Fig. 15). In the same way, we call indicator function of set $C \subset V$, the function I_C defined by

$$I_C(x) = 0,\ if\ x \in C,$$
$$I_C(x) = +\infty,\ if\ \ x \notin C.$$

Convex functions and convex sets are related by the following theorem

Theorem 10. A set $C \subset V$ is convex if and only if its indicator function I_C is convex.

Indicator functions may seem a little bit strange but they are used in mechanics to make precise relationships between quantities of constitutive laws. Let $V = \mathbb{R}$, the indicator functions I_+ (Fig.. 14) and I_- of the sets of positive and negative numbers are:

$$I_+(x) = 0,\ if\ x \geq 0,$$
$$I_+(x) = +\infty,\ if\ \ x < 0,$$

$$I_-(x) = 0,\ if\ x \leq 0,$$
$$I_-(x) = +\infty,\ if\ \ x > 0,$$

and the indicator function of the origin is

$$I_0(0) = 0,$$
$$I_0(x) = +\infty,\ if\ \ x \neq 0.$$

5.2 Linear spaces in duality

Definition 11. We say that linear spaces V and $V^*$are in duality if there exists a bilinear form $\langle \cdot, \cdot \rangle$ defined on $V \times V^*$ such that

$$for\ any\ x \in V,\ x \neq 0,\ there\ exists\ y \in V^*,\ such\ that\ \langle x, y \rangle \neq 0\ ;$$
$$for\ any\ y \in V^*,\ y \neq 0,\ there\ exists,\ x \in V,\ such\ that\ \langle x, y \rangle \neq 0.$$

Examples of linear spaces in duality

The spaces

$V = \mathbb{R}$, $V^* = \mathbb{R}$ are in duality with the bilinear form $\langle x, y \rangle = x.y$ which is the usual multiplication;

$V = \mathbb{R}^n$, $V^* = \mathbb{R}^n$ are in duality with the bilinear form which is the usual scalar product

$$\langle x, y \rangle = \overrightarrow{x}.\overrightarrow{y} = \sum_{i=1}^{i=n} x_i y_i, \tag{43}$$

where $\overrightarrow{x} = (x_i)$ is a vector of $\mathbb{R}^n$ with coordinates x_i ;

$V = S$, $V^* = S$ where S is the linear space of symmetric 3×3 matrices, are in duality with the bilinear from

$$e \in V,\ s \in V^*,\ \langle e, s \rangle = e : s = \sum_{i,j=1}^{i,j=3} e_{i,j} s_{i,j}$$

$$= e_{i,j} s_{i,j} = e_{11}s_{11} + e_{22}s_{22} + e_{33}s_{33} + 2e_{12}s_{12} + 2e_{13}s_{13} + 2e_{23}s_{23}. \tag{44}$$

Be careful, spaces $V = S$ and $V^* = S$ are in duality with the bilinear form

$$\langle\langle e, s \rangle\rangle = e_{i,j} s_{i,j} = e_{11}s_{11} + e_{22}s_{22} + e_{33}s_{33} + e_{12}s_{12} + e_{13}s_{13} + e_{23}s_{23},$$

which is different from the preceding one. Let us give two more examples. The linear space of velocities $\overrightarrow{U}$ of a material in the domain Ω of $\mathbb{R}^3$

$$V = \left\{ \overrightarrow{U}(x) \middle| \overrightarrow{U} \in L^2(\Omega) \right\},$$

is in duality with the linear space of the forces $\overrightarrow{f}$ applied to the material

$$V^* = \left\{ \overrightarrow{f}(x) \middle| \overrightarrow{f} \in L^2(\Omega) \right\},$$

with the bilinear form

$$\left\langle \overrightarrow{U}, \overrightarrow{f} \right\rangle = \int_\Omega \overrightarrow{U}(x).\overrightarrow{f}(x)dx,$$

which is the power. The linear space of the strain rates

$$V = \left\{ D = (D_{ij}(x)) \left| D_{ij} = D_{ji},\ D_{ij} \in L^2(\Omega) \right. \right\},$$

is in duality with the linear space of the stresses

$$V^* = \left\{ \sigma = (\sigma_{ij}(x)) \left| \sigma_{ij} = \sigma_{ji},\ \sigma_{ij} \in L^2(\Omega) \right. \right\},$$

by the bilinear form

$$\langle D, \sigma \rangle = \int_\Omega \sigma(x) : D(x)dx = \int_\Omega \sigma_{ij}(x) D_{ij}(x)dx,$$

which is minus the power of the stress and strain rate. Let us recall that the strain rate resulting from velocity $\overrightarrow{U}$ is

$$D_{ij}(\overrightarrow{U}) = \frac{1}{2}(\frac{\partial U_i}{\partial x_j} + \frac{\partial U_j}{\partial x_i}) = \frac{1}{2}(U_{i,j} + U_{j,i}),$$

by denoting $U_{i,,j} = \partial U_i / \partial x_j$.

5.3 Sub-gradients and sub-differential sets of convex functions

The convex function f_p given above is differentiable for $p > 1$ but it is not for $p = 1$: it is the absolute value function, $|x|$, which has no derivative for $x = 0$. In the same way the indicator function I has no derivative at points where it is equal to $+\infty$. It has no derivative at points $x = 0$ and $x = 1$ where it is equal to 0. Let us recall that the derivatives of smooth convex functions of one variable are increasing functions. Thus they satisfy

$$\left(\frac{df}{dx}(x) - \frac{df}{dx}(y)\right)(x - y) \geq 0. \tag{45}$$

It seems that we have to loose all properties related to derivatives of convex functions. Fortunately, this is not the case: it is possible to define a generalized derivative and to keep most of the related properties.

Let us consider the function f on Fig. 13. It is convex but is not differentiable because it has no derivative at point A. At a point where it has a derivative the curve representing function f is above the tangent line. At point A, there exist several lines which have this property.

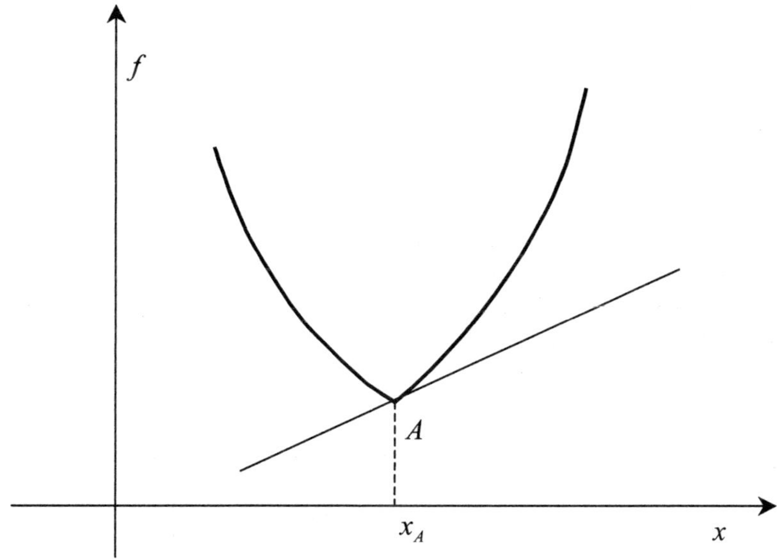

Figure 13. Convex function f has no derivative at point A. But it has generalized derivatives, slopes of lines originating at point A which are under the curve representing function f. These slopes are the sub-gradients which are the elements of the sub-differential set.

We define with the slopes of these lines, the sub-gradients which generalize the derivative:

Definition 12. Let convex function f defined on V in duality with V^*. A sub-gradient of f at point $x \in V$ is an element $y \in V^*$ which satisfies

$$\forall z \in V, \; \langle z - x, y \rangle + f(x) \leq f(z). \tag{46}$$

The set of the z which satisfy (46) is the sub-differential set of f at point x, denoted $\partial f(x)$.

Remark 13. The sub-gradient depends on f but also on the bilinear form $\langle \cdot, \cdot \rangle$.

Two properties of the sub-differential set

The sub-differential set keeps the usual properties of derivatives: a function has no derivative when its value is $+\infty$ and the derivative of a convex function satisfies relationship (45). Let us see the first property:

Theorem 14. Let a convex function $f \neq +\infty$. If this function is sub-differentiable at point x, i.e., $\partial f(x) \neq \emptyset$, then it is finite at this point : $f(x) < +\infty$.

Proof. Assume f is sub-differentiable at point x: let $y \in \partial f(x)$. Ab absurdo, let assume that f is not finite at this point, i.e., $f(x) = +\infty$. Let us have relationship (46) at point z where $f(z) < +\infty$. Such a point exists because $f \neq +\infty$. Then

$$+\infty = \langle z - x, y \rangle + f(x) \leq f(z).$$

It results that $+\infty = f(z)$. This is in contradiction with the assumption. Note that if the only point z where $f(z) < +\infty$ is x, then $f(x) < +\infty$ for function f not to be identical to $+\infty$.

This theorem applies in numerous mechanical constitutive laws where we have $R \in \partial I_-(U)$: because $\partial I_-(U)$ is not empty, we have $I_-(U) < +\infty$ which implies $I_-(U) = 0$ and $U \leq 0$. Relationship $R \in \partial I_-(U)$ implies that the velocity U is negative and satisfies some impenetrability condition.

Now let us show that relationship (45) is in some sense satisfied by the sub-differential set:

Theorem 15. Let us consider a convex function f. We have

$$\forall z \in \partial f(x), \; \forall w \in \partial f(y), \; \langle x - y, z - w \rangle \geq 0.$$

Proof. Write relationship (46) at points x and y.

Remark 16. We say that the sub-differential operator is a monotone operator.

Examples of indicator functions and sub-differential sets

Indicator Function of the set of the Positive Number R^+

Let I_+ be the function with value zero if $x \geq 0$, and value $+\infty$, if $x < 0$ (Fig.. 14). This function, the indicator function of the set R^+ of the positive numbers, is a convex function.

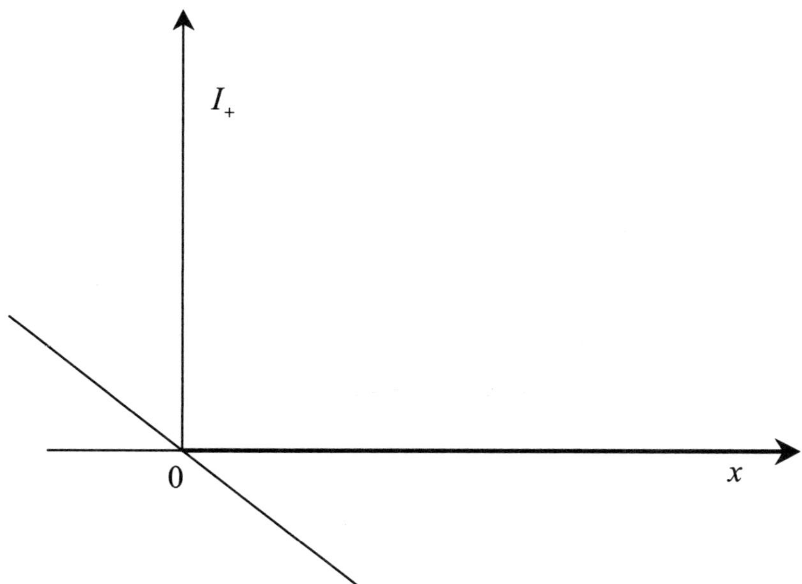

Figure 14. The indicator function of the set of the positive numbers. Its value $I_+(x)$ is zero if $x \geq 0$, and $+\infty$, if $x < 0$.

Its subdifferential set is the set $\{0\}$ if $x > 0$, and the set R^- of the negative numbers, if $x = 0$. The subdifferential set is empty, if $x < 0$, because $I_+(x) = +\infty$.

Indicator Function of the Set of the Negative Numbers R^-

Let I_- be the function with value zero if $x \leq 0$, and value $+\infty$, if $x > 0$. This function, the indicator function of the set R^- of the negative numbers, is a convex function. Its subdifferential set is the set $\{0\}$ if $x < 0$, and the set R^+ of the positive numbers, if $x = 0$. The subdifferential set is empty, if $x > 0$, because $I_-(x) = +\infty$.

Indicator Function of the Segment $[0, 1]$

Let I be the function with value zero, if $0 \leq x \leq 1$, and value $+\infty$, if either $x > 1$ or $x < 0$, Fig. 15. This function, the indicator function of the segment $[0, 1]$, is a convex function. Its subdifferential set is the set $\{0\}$, if $0 < x < 1$, the set R^+ of the positive numbers, if $x = 1$, and the set R^- of the negative numbers, if $x = 0$. The subdifferential set is empty, if $x > 1$ or $x < 0$, because $I(x) = +\infty$.

Indicator function of the origin of R

Let I_0 be the function with value zero, if $x = 0$, and value $+\infty$, if $x \neq 0$. This function, the indicator function of the origin of R, is a convex function. Its subdifferential set is the set R, if $x = 0$. The subdifferential set is empty if $x \neq 0$, because $I_0(x) = +\infty$.

The dual function of a convex function

Let linear spaces X and Y be in duality with bilinear form $\langle .,. \rangle$. Let f be a convex function from the linear space X into $\bar{R} = R \cup \{+\infty\}$, the dual function f^* from Y into $\bar{R}$ is defined by

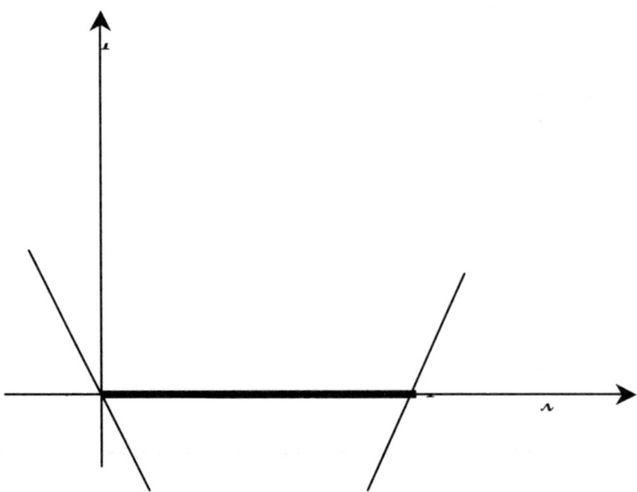

Figure 15. Indicator function of segment $[0,1]$. Its value $I(x)$ is zero if $0 \leq x \leq 1$, and $+\infty$, if either $x < 0$ or $x > 1$. Sub-gradients at points 0 and 1 are slopes of line which are under function I and are in contact with function I at points 0 and 1.

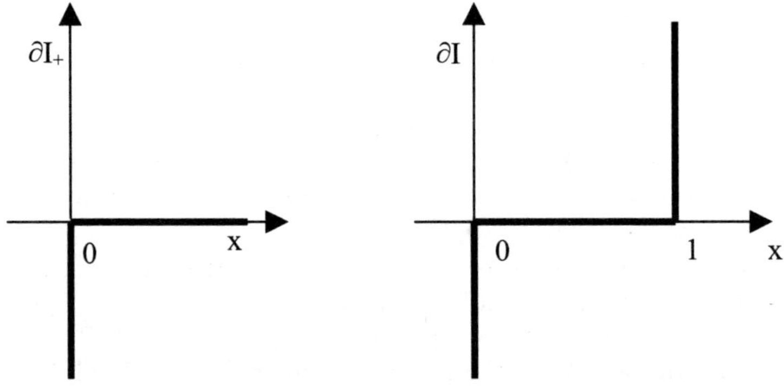

Figure 16. Sub-differential set ∂I_+ of the indicator function of the set of positive numbers $\mathbb{R}^+$: $\partial I_+(0) = \mathbb{R}^-$, $\partial I_+(x) = \{0\}$ for $x > 0$ and $\partial I_+(x) = \emptyset$ for $x < 0$. Sub-differential setl ∂I of indicator function I of segment $[0,1]$.

$$f^*(y) = sup\{\langle x, y\rangle - f(x)\,|x \in X\}.$$

It can be proved that the function f^* is convex. If the function f is subdifferentiable at point x, the properties

$$y \in \partial f(x),\ x \in \partial f^*(y), \text{and } f(x) + f^*(y) =< x, y >,$$

are equivalent. As an example, let us compute the dual function of the indicator function of the segment $[0, 1]$, the space Y is R and the duality mapping being the usual product, $\langle x, y\rangle = xy$. The dual function is defined by

$$I^*(y) = sup\{xy - I(x)\,|x \in X\} = sup\{xy\,|x \in [0, 1]\}.$$

It is called the support function of the convex set $[0, 1]$. This results in

$$I^*(y) = y,\ \text{if } y \geq 0, \text{and } I^*(y) = 0,\ \text{if } y \leq 0.$$

Let C be a convex set of the space S of the symmetric 3×3 matrices such that $0 \in C$. The space S is in duality with itself with the duality mapping $\langle \sigma, e\rangle = \sigma : \epsilon = \sigma_{ij}\epsilon_{ij}$. Let I_C be the indicator function of the convex set C. The dual function I^*, the support function of the set C, of the indicator function I_C is defined by

$$I^*(\epsilon) = sup\{\sigma : \epsilon\,|\sigma \in C\}.$$

Note that $I^*(\epsilon) \geq 0$. When C is the plasticity set, I^* is the plastic potential, Salençon (1988).

References

L. Ambrosio, N. Fusco, D. Pallara, 2000, Special functions of bounded variations and free discontinuity problems, Oxford University Press, Oxford.

H. Attouch, G. Buttazzo, G. Michaille, 2004, Variational analysis in Sobolev and BV spaces. Application to PDE and optimization, MPS/SIAM Series in Optimization.

B. Brogliato, 1999, Nonsmooth Mechanics, Springer Verlag, Berlin.

C. Cholet (a), 1998, Chocs de solides rigides, Thése de l'Université Pierre et Marie Curie, Paris.

C. Cholet (b), 1998, Collisions d'un point et d'un plan, C. R. Acad. Sci., Paris.

E. Dimnet, 2000, Collisions de solides déformables, thèse de l'Ecole nationale des ponts et chaussées, Paris.

E. Dimnet, R. Gormaz, Le ricochet d'une pierre sur l'eau d'un lac, à paraître.

I. Ekcland, R. Temam, 1976, Convex analysis and variational problems, North Holland, Amsterdam.

M. Frémond, 2001, Non smooth thermomechanics, Springer-Verlag, Heidelberg.

M. Frémond, 2005, Collisions, Dipartimento di Engigneria Civile dell'Università di Roma Tor Vergata, Consiglio Nazionale delle Richerche et l'Accademica, Istituto Poligrafico e Zecca dello Stato, Rome.

M. Frémond, R. Gormaz, J. San Martin, 2003, Collision of a solid with an uncompressible fluid, Theoretical and Computational fluid dynamics, 16, 405-420.

P. Germain, 1973, Mécanique des milieux continus, Masson, Paris.

F. G. Pfeiffer, C. Glocker, 1996, Multybody dynamics with unilateral contacts, Wiley Series in Nonlinear Sciences, Wiley.

C. Glocker, 2001, Set-valued force laws: Dynamics of nonsmooth systems. Lecture Notes in Applied and Computational Mechanics, Springer.

J.L. Lions, G. Stampacchia, 1967, Variational inequalities, Comm. Pure Appli. Math., 20, 493-519.

J. J. Moreau, 1966, Fonctionnelles convexes, Séminaire sur les équations aux dérivées partielles, Collége de France, Paris and, Università di Roma Tor Vergata, Dipartimento di Ingegneria Civile (2003).

J. J. Moreau, 2003, An introduction to Unilateral Dynamics, in Novel approaches in Civil Engineering, M. Frémond, F. Maceri eds, , Springer-Verlag, Heidelberg.

J. J. Moreau, 1988, Bounded variation in time dans Topics in non-smooth mechanics, J. J. Moreau, P. D. Panagiotopoulos, G. Strang, eds, Birkhauser, Basel, Chap. I, 1-71.

B. Nayroles, 1973, Point de vue algébrique. Convexité et intégrandes convexes en mécanique des solides, C.I.M.E., Bressanone, Ed. Cremonese.

J. Pérès, 1953, Mécanique générale, Masson, Paris.

F. G. Pfeiffer, 2001, Non-smooth mechanics, F.G. Pfeiffer ed., Philosophical Transactions of the Royal Society, Mathematical, Physical and Engineering Sciences, Londres, series A, 359, 1789.

H. Pron, J. F. Henry, R. Bouferra, C. Bissieux, J. L. Beaudouin, 2002, Etude par thermographie infrarouge du grenaillage de précontrainte ou shot-penning, Congrès français de thermique, SFT 2002, Vittel.

H. Pron, 2000, Application des effets photothermiques et thermomécaniques à l'analyse des contraintes appliquées et résiduelles, thèse, Université de Reims Champagne-Ardenne.

J. Salençon, 1988, Mécanique des milieux continus, Ellipses, Paris.

D. Stoianovici, Y. Hurmuzlu, (1996), A critical study of the applicability of rigid-body collision theory, ASME, J. Appl. Mech., 63:307-316.

An Introduction to Impacts

Christoph Glocker
IMES - Center of Mechanics, ETH Zurich, Zurich, Switzerland

Abstract Different methods to model and solve multi-contact collisions are presented in this report. The standard impact constitutive laws from non-smooth dynamics are reviewed for planar frictional collisions and formulated in terms of set-valued maps and linear complementarity. For the frictionless case, a geometric concept based on kinematic, kinetic and energetic compatibility is developed, which provides access to non-standard impact events as in Newton's cradle. Within this context, Moreau's impact law is reviewed and stated in various ways, providing even access to the collision problem at re-entrant corner points as an extension. Based on Moreau's law, a geometric classification of impacts is proposed. Several examples are presented, such as the frictional reversible impact at a superball, Newton's cradle and the rocking rod.

1 Introduction

For the lecture notes of this course, material has been taken from the sources (Glocker and Studer, 2005; Glocker, 2001b, 2004; Glocker and Aeberhard, 2005). Greater parts of this material have been revised, homogenized and equipped with detailed comments and additional examples to meet the requirements of course documents. Parts of the material are original and have never been published before. As an additional reading we recommend in particular Moreau (1988a) and Frémond (2002).

1.1 Some Remarks on Impacts

It has been known since Galileo Galilei (Discorsi, The Sixth Day) that impact forces can become unlimited. Huygens had been examining completely elastic collisions between two point masses since 1656. He recognized the fact, that besides conservation of momentum and kinetic energy the relative motion of two bodies has to be taken into account in order to be able to formulate a universally valid law of impact. His law $v - V = C - c$, describing the relative velocities inversion during the elastic impact, is extended by Newton in 1687 by the restitution coefficient ε in order to accommodate possible losses of energy during the collision. The form $\varepsilon\,(v - V) = (C - c)$ serves Newton as an experimental confirmation of his third law *actio* = *reactio*. By setting $\varepsilon = 1$ Huygens impact law for the elastic case is obtained, whereas $\varepsilon = 0$ describes the limiting case of maximum dissipation possible, such that the bodies do not penetrate after impact but keep moving with a common velocity. The restitution coefficient ε is regarded as a measure of dissipated energy during the impact in this context.

For systems composed of several bodies, conservation of linear and angular momentum does not lead to the target aimed at. Instead, the Newton-Euler equations have to be established for each body to obtain sufficient equations describing the dynamics. If one allows these bodies to have impacts at several contact points, one restitution coefficient is not enough any more to unambiguously determine all post-impact velocities, since it is not known how the kinetic energy is distributed among the individual degrees of freedom. One normally postulates ad-hoc impact laws that are more or less suited to describe reality in all cases. Known problems consist either in obtaining too many possible post-impact velocities while using the restitution coefficients provided by the impact laws, leading to energetic or kinematic inconsistency, or in not being able to reach certain velocities at all. In the latter case, behavior observed in experiments can sometimes not be reproduced in calculations by any choice of the impact parameters in use.

Another difficulty consists in a widely spread misunderstanding of the restitution coefficients. They are material-pairing-constants only in the fewest cases, but have to be generally understood as a measure of dissipation concerning the chosen spatial discretization level of the mechanical system. This is addressed by many people in saying that the restitution coefficients depend in some way on the geometry of the colliding bodies, meaning that they must somehow take into account the wave propagation process initiated by the collision. Since dissipation in mechanical systems has also to be understood as a transfer of energy to mechanical degrees of freedom which are not contained in the mathematical model, the restitution coefficients can be diminished being associated with a refinement of the discretization. In order to obtain the right dynamic behavior for the macroscopic degrees of freedom, the impact can finally be seen as completely inelastic when using a continuum model.

One reason for the mentioned difficulties in setting up the impact laws for multi-point-collisions is the fact, that it has not yet sufficiently been examined to what extend impact laws may be understood as independent equations, in order to avoid contradictions with the basic equations of kinetics and the kinematic and energetic restrictions. This question will be addressed in detail in this report. It will not yet attempt obtaining a complete parametrization free of contradictions in form of a general impact law, but propose an appropriate setting for a theoretical framework in which any impact law should reside. Only the impact itself will be examined. Pre- and post-impact motions will not be discussed. As an impact we will understand a velocity jump which occurs at a discrete point in time, and which is associated with infinite impulsive forces as a consequence of finite, non-disappearing masses in the system. Processes with rapidly changing velocities without discontinuities will not be understood as impacts, as e.g. models containing local stiffness in the contact points of the colliding bodies. Only collisions will be investigated. Impulsive forces applied from outside that can be regarded as external impact excitation, e.g. the impact from the queue on the ball when playing billiard, will not be examined in this context. All discussions will be limited to scleronomic systems. Even though explicit time-dependencies do not represent a major hurdle they will not be treated in favor of clearness. Influences of Coulomb friction during the collision will only be permitted in the first part of this report. They are later excluded in favor of perfect constraints. The concept of perfect constraints is essential for a general and structural procedure according to classical mechanics and will be extended to unilateral impulsive motion.

1.2 Discontinuity Events in Dynamics

In this section we briefly discuss some types of discontinuities that might occur in non-smooth dynamics. Even if only impacts are treated in the rest of this report, a slightly more general classification of possible discontinuities seems to be adequate in order to show the link to non-smooth impact-free motion as presented in Glocker (2001a). Discontinuity events in dynamics are in particular jumps in the velocities, usually associated with collisions and treated within impact theory, but also jumps in the accelerations as they appear, for example, at the transitions from sliding to stiction when dry friction is prevalent. Within the framework of integration we discuss the meaning of displacements, velocities and accelerations, and also their interconnections as stated in Moreau (1988a).

In order to motivate the use of absolutely continuous functions as the solution space of second order ordinary (measure-) differential inclusions in rigid body dynamics, some basic definitions and properties concerning absolute continuity, bounded variations, and differential measures are put together. This section has to be understood just as a scratch of these topics, but it should help the reader to accept the kinematical framework of non-smooth dynamics presented in Moreau (1988a). Most of the material has been taken from Moreau (1988b), but also Elstrodt (1996), Rudin (1981) and Monteiro Marques (1993) have been used. For an introduction to integration theory we refer to Rudin (1981). An advanced exposition may be found in Elstrodt (1996). Functions of bounded variations defined on $\mathbb{R}$ with values in $\mathbb{R}^f$ are treated in Moreau (1988b), and this is what one needs for finite-dimensional evolution problems, see also Glocker (2001a).

Time is denoted by t, and the Lebesgue measure on $\mathbb{R}$ by dt. Further, $I := [t_A, t_E]$ denotes a compact time interval on which the motion of the system is investigated. To allow for discontinuity events in the evolution process of the systems, the following function spaces adopted from the concepts in non-smooth dynamics (Moreau, 1988a) are used:

The velocities $\mathbf{u} : I \to \mathbb{R}^f$, $t \to \mathbf{u}(t)$ are assumed to be *functions of bounded variations.* As such, they have an at most countable number of finite discontinuities on I, allowing for velocity jumps to occur as in the collision problem of rigid bodies. The associated set of points at which $\mathbf{u}$ is discontinuous is denoted by $\{t_i\}$. Functions of bounded variations admit a right and a left limit everywhere, denoted by $\mathbf{u}^+$ and $\mathbf{u}^-$,

$$\mathbf{u}^+(t) = \lim_{\tau \downarrow 0} \mathbf{u}(t+\tau) \quad \text{and} \quad \mathbf{u}^-(t) = \lim_{\tau \uparrow 0} \mathbf{u}(t+\tau). \tag{1.1}$$

In the case of an impact, these magnitudes are called the *post- and pre-impact velocity*, respectively. Functions of bounded variation are differentiable almost everywhere, except for a countable set of points $\{t_j\}$ probably different from $\{t_i\}$. We denote their directional derivatives by $\dot{\mathbf{u}}^+$ and $\dot{\mathbf{u}}^-$,

$$\dot{\mathbf{u}}^+(t) = \lim_{\tau \downarrow 0} \frac{\mathbf{u}(t+\tau) - \mathbf{u}(t)}{\tau} \quad \text{and} \quad \dot{\mathbf{u}}^-(t) = \lim_{\tau \uparrow 0} \frac{\mathbf{u}(t+\tau) - \mathbf{u}(t)}{\tau} \tag{1.2}$$

if these limits exist with values in $\mathbb{R}^f$. They provide the *right and left accelerations* in mechanics. For $\mathbf{u}^+(t) = \mathbf{u}^-(t)$, $\mathbf{u}$ is continuous at t and, if in addition $\dot{\mathbf{u}}^+(t) = \dot{\mathbf{u}}^-(t)$, $\mathbf{u}$ is differentiable at t. The displacements $\mathbf{q}(t)$ of the mechanical system are obtained from $\mathbf{u}(t)$ by integration,

$$\forall t \in I: \quad \mathbf{q}(t) = \mathbf{q}(t_A) + \int\limits_{t_A}^{t} \mathbf{u}(\tau)\,\mathrm{d}\tau, \tag{1.3}$$

leading to *absolutely continuous* functions $\mathbf{q}(t)$ as a consequence on the bounded variation property of $\mathbf{u}(t)$. This means that jumps in the displacements are *not* allowed in this framework of modelling.

Due to the discontinuities of $\mathbf{u}(t)$, their time derivatives have *not* to be understood in the classical sense, but in the sense of measures when gaining back the full functions $\mathbf{u}(t)$ by integration of their derivatives is desired. Functions of bounded variations are known to admit a unique decomposition into an absolutely continuous part, a step function, and a singular part. The latter is assumed to vanish in the framework of this report. Thus, one obtains the velocity $\mathbf{u}(t)$ by integration of its associated *differential measure* $\mathrm{d}\mathbf{u}$,

$$\mathbf{u}^+(t) = \mathbf{u}^-(t_A) + \int\limits_{[t_A,t]} \mathrm{d}\mathbf{u}. \tag{1.4}$$

The integral in this expression can be split into a Lebesgue part and a purely atomic part

$$\int\limits_{[t_A,t]} \mathrm{d}\mathbf{u} = \int\limits_{[t_A,t]} \dot{\mathbf{u}}\,\mathrm{d}t + (\mathbf{u}^+ - \mathbf{u}^-)\,\mathrm{d}\eta, \tag{1.5}$$

providing after integration the absolutely continuous part and the step function of $\mathbf{u}(t)$, respectively. The measure $\mathrm{d}\eta$ in (1.5) turns out to be concentrated on the set of discontinuity points $\{t_i\}$ of $\mathbf{u}$. It is the sum of the Dirac point measures $\mathrm{d}\delta_i$,

$$\mathrm{d}\eta = \sum_i \mathrm{d}\delta_i \quad \text{with} \int\limits_{I_{kl}} \mathrm{d}\delta_i = \begin{cases} 1 & \text{if } t_i \in I_{kl} \\ 0 & \text{if } t_i \notin I_{kl}, \end{cases} \tag{1.6}$$

where I_{kl} is any one-dimensional cell with endpoints t_k and t_l in I, i.e. any open or closed or half-open interval (t_k, t_l), $[t_k, t_l]$, $[t_k, t_l)$, $(t_k, t_l]$. Note in particular that

$$\mathbf{u}^+(t) = \mathbf{u}^-(t) + \int\limits_{\{t\}} \mathrm{d}\mathbf{u} \tag{1.7}$$

when equation (1.4) is applied to a singleton $\{t\}$. Different values $\mathbf{u}^+ \neq \mathbf{u}^-$ are obtained as soon as t belongs to the discontinuity points $\{t_i\}$.

Once again, let us discuss how the discontinuity points $\{t_i\}$ of the velocities $\mathbf{u}$ and $\{t_j\}$ of the accelerations $\dot{\mathbf{u}}$ are connected together. For almost every t we have $\mathbf{u}^+ = \mathbf{u}^-$ and $\dot{\mathbf{u}}^+ = \dot{\mathbf{u}}^-$, that means continuity and differentiability. At the discontinuity points $\{t_i\}$ of the velocities we have $\mathbf{u}^+ \neq \mathbf{u}^-$. These velocity jumps may be accompanied by right and left accelerations that take identical values $\dot{\mathbf{u}}^+ = \dot{\mathbf{u}}^-$ or even different values $\dot{\mathbf{u}}^+ \neq \dot{\mathbf{u}}^-$, as it can already be seen by the most simple impact system: We consider the bouncing ball problem with completely elastic impact (Figure 1, left diagram). At the impact time the velocity $u^- < 0$ is reversed, $u^+ = -u^-$. The left and right accelerations at the impact time have the same values, because they are both obtained from the equation

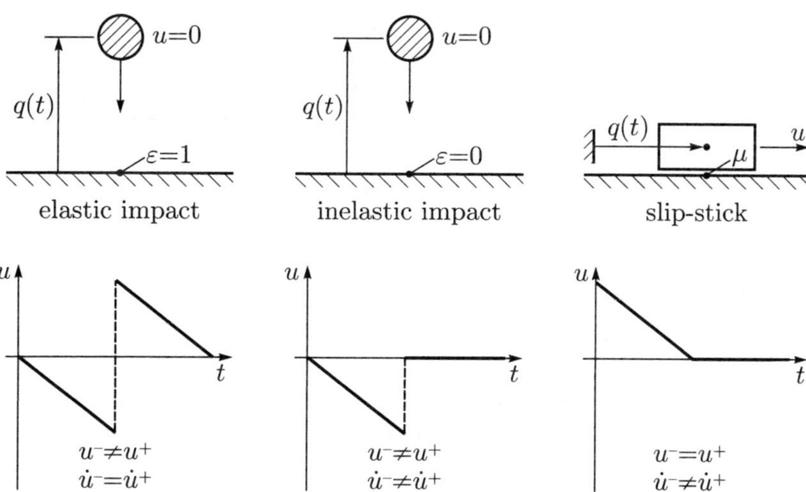

Figure 1. Discontinuities in the velocity u and acceleration $\dot{u}$.

of motion for the free-flight state $\dot{u} = \dot{u}^+ = \dot{u}^- = -\frac{g}{m}$, where g denotes gravity and m is the mass of the ball. Now we consider the same system but under the influence of a completely inelastic impact (Figure 1, middle diagram). The acceleration before the impact is $\dot{u}^- = -\frac{g}{m}$. At the impact the velocity $u^- < 0$ is changed to $u^+ = 0$. Afterwards, the ball remains on the impact surface and hence $\dot{u}^+ = 0$.

Suppose now that the velocities are continuous ($\mathbf{u} = \mathbf{u}^+ = \mathbf{u}^-$, i.e. $t \notin \{t_i\}$). In this case there might be still an acceleration jump $\dot{\mathbf{u}}^+ \neq \dot{\mathbf{u}}^-$ as it can be seen by the following example (Figure 1, right diagram): We consider a mass m sliding on a plane with a velocity $u > 0$ under the influence of Coulomb friction. The equations of motion during sliding are $\dot{u} = -\mu g$ where μ is the coefficient of friction and g denotes gravity as before. The mass is decelerated until it comes to a rest. After it has stopped moving we have $\dot{u} = 0$, and hence for the transition point $\dot{u}^- = -\mu g \neq 0 = \dot{u}^+$ without any velocity jump.

There are even situations for which the directional accelerations do not exist. Consider again the bouncing ball problem, now with an impact behavior that is neither elastic nor inelastic, but something in between (Figure 2). One recognizes that the time instances of impact constitute a geometric sequence with limit $t^\star$ if the impact law has been chosen to be $u^+ = -\varepsilon u^-$, $0 < \varepsilon < 1$. The right and left accelerations are still well defined by $\dot{u}^+ = -\frac{g}{m}$ for $t < t^\star$ and $\dot{u}^+ = 0$ for $t > t^\star$. At the accumulation point $t^\star$, however, one has $\dot{u}^+ = 0$, whereas the limit according to equation (1.2) does *not* exist for the left acceleration $\dot{u}^-$. One consequence of this particular situation is that the ball can decide at any instance to start bouncing off of the surface when we want to integrate *backwards* in time, even if the restitution coefficient $\varepsilon < 1$ has been chosen unequal to zero.

The main purpose of these lectures is to investigate impacts. In the following, we concentrate only on the discontinuities of the velocities and try to develop different concepts on how to obtain constitutive laws for impacts, in particular for collisions. If

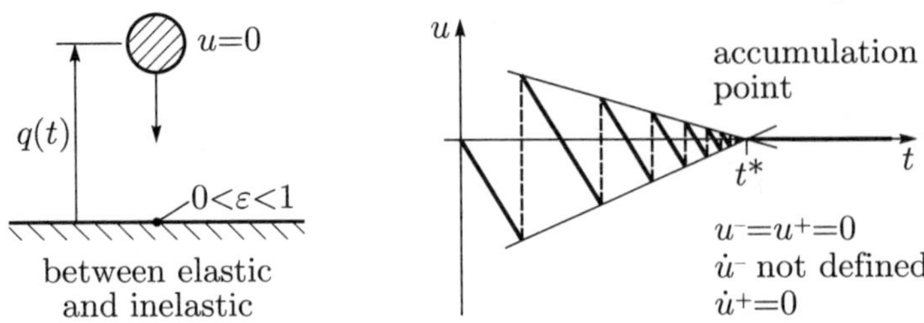

Figure 2. Non-existence of the left acceleration at an accumulation point.

we think about a motion of finite sort, i.e. an evolution problem which is composed of only a finite number of velocity jumps and impact-free motion in between, two things are needed: An impact law which tells us how to determine post-impact velocities $\mathbf{u}^+$ from given pre-impact velocities $\mathbf{u}^-$, and a method on how to compute from a pre- or post-impact state the associated directional accelerations of the system. The first matter is treated within these lectures, whereas the second can be found in great detail in Glocker (2001a).

2 Basic Set-Valued Elements

The use of set-valued maps as possible constitutive laws is one of the keys to non-smooth mechanics. Non-smooth mechanics is lead by the idea that different modes or states of a mechanical system are not treated in the sense of a Boolean distinction, but are incorporated as constitutive laws in the form of maximal set-valued maps, which describe in a physical manner the different states and the transitions between them. One helpful tool to understand how such systems can be solved is the idea of linear complementarity, which will briefly be reviewed in this section.

2.1 A Contact Problem in Statics

We consider a particle which is attached to a rigid wall by a linear spring with stiffness c, as shown in the left part of Figure 3. The displacement of the particle relative to the wall is denoted by x, and the displacement of the particle associated with the unloaded spring by x_0. Possible contact between the particle and the wall is assumed to be non-adhesive.

Depending on x_0, there are two possible states in which the system might be: Either, the particle is pressed against the wall by the spring, which results in a contact force $F_N > 0$ and an equilibrium position $x^\star = 0$. Or, the equilibrium is at $x^\star = x_0 \geq 0$, for which the contact is open and the contact force vanishes, $F_N = 0$. In order to obtain analytically those equilibrium positions, we do not treat the two contact states separately. We rather combine them into a single inclusion by using a set-valued map, which reflects the constitutive law of the contact. The decision of which of the contact states is valid has then been moved into the equilibrium inclusion, and pre-assumptions on the contact states are no longer necessary.

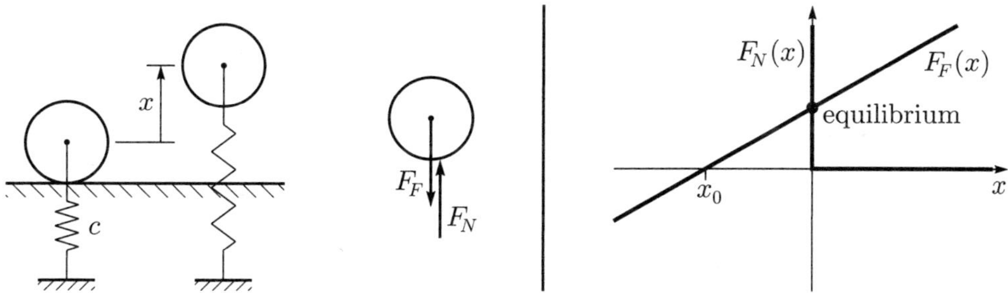

Figure 3. Static system and its equilibrium.

In the free body diagram both, the spring force F_F and the contact force F_N are taken into account. The question whether the contact force is really needed will be taken over by the constitutive law of the contact and decided later when the system will be solved. The equilibrium of forces thus reads

$$F_F = F_N. \tag{2.1}$$

The constitutive law of the linear spring is

$$F_F = c\,(x - x_0). \tag{2.2}$$

For the model of the unilateral contact, we have to demand the impenetrability condition $x \geq 0$. We further assume non-adhesive behavior $F_N \geq 0$ such that distant effects of the contact force are excluded. This means that a positive contact force can only appear if the contact is closed, which is expressed by the complementarity condition $F_N\,x = 0$. Our model of the contact may therefore be written in the form

$$F_N \geq 0,\ x \geq 0,\ F_N\,x = 0. \tag{2.3}$$

The graphs of both constitutive laws, (2.2) and (2.3), are depicted in the right diagram of Figure 3. By (2.1), equilibrium of the system is attained at the point at which the graphs intersect. One immediately recognizes that this point is unique as long as $c > 0$, and one gets $x^\star = 0$ for $x_0 < 0$ and $x^\star = x_0$ for $x_0 \geq 0$ as the equilibrium position. For $c = 0$, infinitely many solutions $x^\star \geq 0$ are obtained, whereas for a negative stiffness $c < 0$ one either gets two solutions or no solution at all, depending on the values of x_0.

When we eliminate F_F by (2.1) from (2.2), the two constitutive laws from (2.2) and (2.3) become

$$F_N = cx + b; \quad F_N \geq 0,\ x \geq 0,\ F_N\,x = 0 \tag{2.4}$$

with $b = -c\,x_0$. This is a one-dimensional linear complementarity problem which we write as $y = Ax + b$ together with $y \geq 0$, $x \geq 0$, $yx = 0$, and which will be generalized to the n-dimensional case in the next section. We finally introduce a set-valued map $x \to \mathrm{Upr}\,(x)$ to express the contact constitutive law (2.3) in the form $-F_N \in \mathrm{Upr}\,(x)$. The conditions for the equilibrium (2.4) may then be written as an inclusion,

$$-cx - b \in \mathrm{Upr}\,(x). \tag{2.5}$$

This inclusion still keeps all the information required to determine the equilibrium positions. The precise meaning of the map Upr, the unilateral primitive, will be explained in Section 2.3.

2.2 Linear Complementarity

A *linear complementarity problem* (LCP) is a problem of the following form, see e.g. Cottle et al. (1992) or Murty (1988) for a full account: For given $\mathbf{A} \in \mathbb{R}^{n,n}$ and $\mathbf{b} \in \mathbb{R}^n$, find $\mathbf{x} \in \mathbb{R}^n$ and $\mathbf{y} \in \mathbb{R}^n$ such that the linear equation $\mathbf{y} = \mathbf{A}\mathbf{x} + \mathbf{b}$ holds together with the complementarity conditions $y_i \geq 0$, $x_i \geq 0$, $y_i\, x_i = 0$ for $i = 1, \ldots, n$. The latter conditions are often written as $\mathbf{y} \succeq 0$, $\mathbf{x} \succeq 0$, $\mathbf{y}^\mathsf{T}\mathbf{x} = 0$ or, equivalently, as $0 \preceq \mathbf{y} \perp \mathbf{x} \succeq 0$. By moving $\mathbf{Ax}$ to the left-hand side, the LCP

$$\mathbf{y} = \mathbf{A}\mathbf{x} + \mathbf{b}; \quad 0 \preceq \mathbf{y} \perp \mathbf{x} \succeq 0 \tag{2.6}$$

takes the form

$$\left(\mathbf{e}_1, \ldots, \mathbf{e}_n, -\mathbf{a}_1, \ldots, -\mathbf{a}_n\right) \begin{pmatrix} \mathbf{y} \\ \mathbf{x} \end{pmatrix} = \mathbf{b}; \quad y_i \geq 0,\ x_i \geq 0,\ y_i\, x_i = 0. \tag{2.7}$$

From this representation one recognizes that the vector $\mathbf{b}$ is expressed by a non-negative linear combination of the canonical basis vectors $\mathbf{e}_i$ and the negative columns $-\mathbf{a}_i$ of $\mathbf{A}$. Moreover, for each i only one, either $\mathbf{e}_i$ or $-\mathbf{a}_i$, contributes due to the complementarity conditions $y_i\, x_i = 0$.

The set $\{\mathbf{e}_i, -\mathbf{a}_i\}$ is called the i-th complementary pair of vectors, and the set $\{y_i, x_i\}$ the i-th complementary pair of variables. Let now $\mathbf{c}_i \in \{\mathbf{e}_i, -\mathbf{a}_i\}$ and $z_i \in \{y_i, x_i\}$ such that $z_i = y_i$ if $\mathbf{c}_i = \mathbf{e}_i$ or $z_i = x_i$ if $\mathbf{c}_i = -\mathbf{a}_i$. The matrix $\mathbf{C}_k = (\mathbf{c}_1, \ldots, \mathbf{c}_n)$ is called the k-th complementary matrix, and the vector $\mathbf{z}_k = (z_1, \ldots, z_n)^\mathsf{T}$ the k-th complementary vector of variables. There are 2^n different matrices $\mathbf{C}_k$ and associated vectors $\mathbf{z}_k$. The set $\mathcal{K}_k = \{\mathbf{v}_k \mid \mathbf{v}_k = \mathbf{C}_k\, \mathbf{z}_k,\ \mathbf{z}_k \succeq 0\}$ is a closed convex cone, positively generated by the columns of $\mathbf{C}_k$, and is called the k-th complementary cone. There are 2^n different complementary cones $\mathcal{K}_k$. As a consequence, $\mathbf{b}$ can be expressed by a non-negative linear combination of the columns of those $\mathbf{C}_k$, for which the associated $\mathcal{K}_k$ contain $\mathbf{b}$. In order to solve (2.6), one therefore tries to solve each of the 2^n systems

$$\mathbf{C}_k\, \mathbf{z}_k = \mathbf{b}, \quad \mathbf{z}_k \succeq 0, \quad k = 1, \ldots, 2^n. \tag{2.8}$$

Every solution of (2.8) is then a solution of the LCP (2.6). Depending on the particular structure of the matrix $\mathbf{A}$ and the value of $\mathbf{b}$, the LCP may have one solution, a finite number of solutions, infinitely many solutions or no solution at all. Uniqueness of the solution, for example, is guaranteed independent of the value of $\mathbf{b}$, if $\mathbf{A}$ is a P-matrix. In this case, the 2^n complementary cones $\mathcal{K}_k$ partition $\mathbb{R}^n$.

As an example, we consider an LCP (2.6) with

$$\mathbf{A} = \begin{pmatrix} 2 & 1 \\ 1 & 2 \end{pmatrix}, \quad \mathbf{b} = \begin{pmatrix} 2 \\ -2 \end{pmatrix}, \tag{2.9}$$

which can be written by (2.7) in the form

$$y_1\, \mathbf{e}_1 + y_2\, \mathbf{e}_2 + x_1(-\mathbf{a}_1) + x_2(-\mathbf{a}_2) = \mathbf{b}; \quad y_i \geq 0,\ x_i \geq 0,\ y_i\, x_i = 0. \tag{2.10}$$

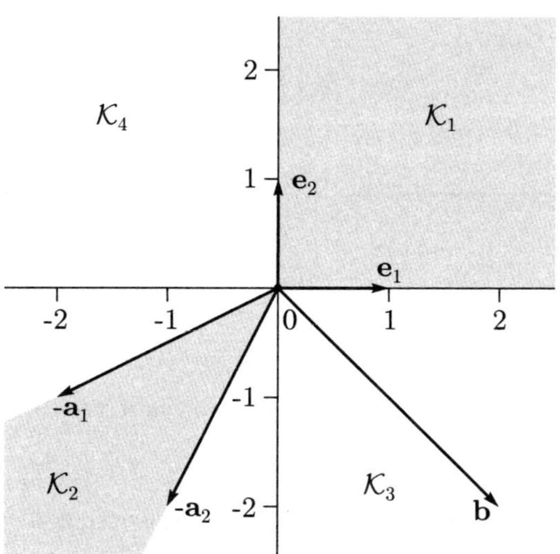

Figure 4. Complementary cones and solution of the LCP.

Figure 4 shows the four complementary cones $\mathcal{K}_k$, positively generated by the pairs $(\mathbf{e}_1, \mathbf{e}_2)$, $(-\mathbf{a}_1, -\mathbf{a}_2)$, $(\mathbf{e}_1, -\mathbf{a}_2)$ and $(\mathbf{e}_2, -\mathbf{a}_1)$. With $\mathbf{b} \in \mathcal{K}_3$, we obtain

$$\mathbf{b} = 3\mathbf{e}_1 - \mathbf{a}_2, \tag{2.11}$$

and therefore, as the solution of the LCP, $y_1 = 3$, $x_1 = 0$, $x_2 = 1$, $y_2 = 0$.

2.3 Unilateral Primitive and Relay Function

It is convenient to introduce a maximal monotone set-valued map defined on $\mathbb{R}^+$, which has been named in Glocker and Studer (2005) the *unilateral primitive* Upr, and which is the most important multifunction related to complementarity,

$$\operatorname{Upr}(x) := \begin{cases} \{0\} & \text{if } x > 0 \\ (-\infty, 0] & \text{if } x = 0 \end{cases}. \tag{2.12}$$

The graph of this map is depicted in the left part of Figure 5. Apparently, we are now able to express each complementarity condition of the LCP by one inclusion, since

$$-y \in \operatorname{Upr}(x) \;\Leftrightarrow\; y \geq 0,\; x \geq 0,\; xy = 0. \tag{2.13}$$

Unilateral primitives are used in mechanics on displacement and on velocity level to model unilateral geometric and kinematic constraints, such as free plays with stops, sprag clutches and the like. The associated set-valued force laws are conveniently stated as inclusions in the form (2.13).

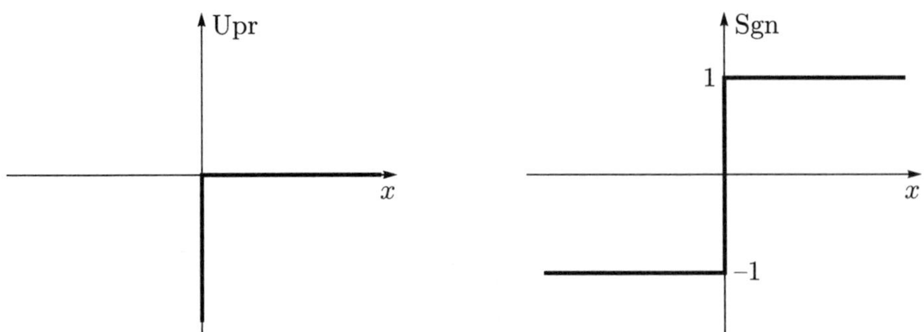

Figure 5. The maps $x \to \mathrm{Upr}\,(x)$ and $x \to \mathrm{Sgn}\,(x)$.

A second maximal monotone set-valued map frequently met in non-smooth systems is the filled-in *relay function* $\mathrm{Sgn}\,(x)$, defined by

$$\mathrm{Sgn}\,(x) := \begin{cases} \{+1\} & \text{if } x > 0 \\ [-1,+1] & \text{if } x = 0 \\ \{-1\} & \text{if } x < 0 \end{cases}, \tag{2.14}$$

see Figure 5 for the graph. Note the difference at $x = 0$ to the classical sgn-function, which is defined as $\mathrm{sgn}\,(x = 0) = 0$. In mechanics, relay functions on velocity level are used to model any kind of dry friction. On displacement level, they describe the behavior of pre-stressed springs. The relay function (2.14) can be represented by two unilateral primitives as indicated in Figure 6, which yields in terms of inclusions

$$-y \in \mathrm{Sgn}\,(x) \quad \Leftrightarrow \quad \exists\, x_R,\, x_L \text{ such that} \begin{cases} -y \in +\mathrm{Upr}\,(x_R) + 1 \\ -y \in -\mathrm{Upr}\,(x_L) - 1 \\ x = x_R - x_L \end{cases}. \tag{2.15}$$

More details on this decomposition can be found in Glocker (2001a) together with various applications of even more complex set-valued interaction laws and their representations via unilateral primitives and inclusions. By using (2.13), we may finally express (2.15) in terms of complementarities,

$$-y \in \mathrm{Sgn}\,(x) \quad \Leftrightarrow \quad \exists\, x_R,\, x_L \text{ such that} \begin{cases} 1 + y \ge 0,\ x_R \ge 0,\ (1+y)\,x_R = 0 \\ 1 - y \ge 0,\ x_L \ge 0,\ (1-y)\,x_L = 0 \\ x = x_R - x_L \end{cases}. \tag{2.16}$$

This representation has to be used when a problem involving Sgn-multifunctions is formulated as an LCP in standard form.

Both, the unilateral primitive as well as the relay function, may be expressed in terms of the subdifferential of certain convex functions. In particular, $\mathrm{Sgn}\,(x) = \partial|x|$ and $\mathrm{Upr}\,(x) = \partial I_{\mathbb{R}^+}(x)$, where $I_{\mathbb{R}^+}$ denotes the indicator function of the set of non-negative

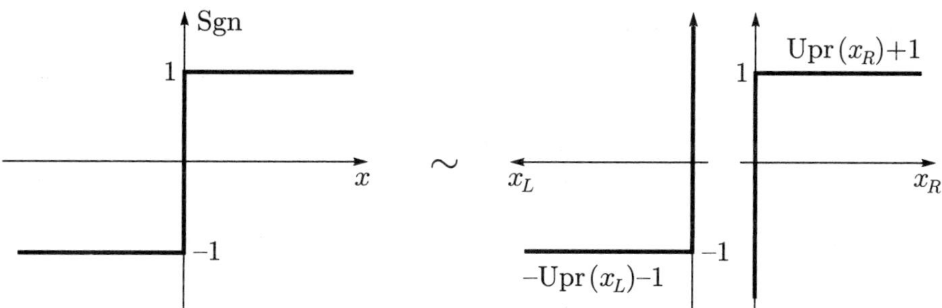

Figure 6. Decomposition of Sgn into Upr 's.

real numbers. The important fact that the subdifferential of lower semi-continuous proper convex functions provides maximal monotone set-valued maps links differential inclusions to variational analysis and optimization theory. We will finally state one important lemma that allows us to express complementarity conditions in one more different way.

Lemma 2.1. *Let $i = 1, \ldots, n$. The inequality and complementarity conditions*

$$y_i \geq 0, \quad x_i \geq 0, \quad y_i\, x_i = 0 \tag{2.17}$$

may be equivalently expressed by the variational inequality

$$\sum_{i=1}^{n} y_i(x_i^\star - x_i) \geq 0, \quad x_i \geq 0, \quad \forall x_i^\star \geq 0. \tag{2.18}$$

Proof. Suppose first that (2.17) holds. The terms $y_i\, x_i$ cancel out from (2.18) due to the complementarity condition in (2.17), and it remains $\sum_i y_i\, x_i^\star \geq 0$ for all $x_i^\star \geq 0$. This statement is true, because $y_i \geq 0$ by (2.17).

Suppose now that (2.18) holds. Fix one i and set $x_j^\star = x_j$ for all $j \neq i$. Equation (2.18) reduces then to $y_i(x_i^\star - x_i) \geq 0$, $\forall x_i^\star \geq 0$. Choose now $x_i^\star = 0$ and afterwards $x_i^\star = 2x_i$, which gives $-y_i\, x_i \leq 0$ and $y_i\, x_i \leq 0$, hence $y_i\, x_i = 0$ which proves the complementarity condition in (2.17). Suppose finally that $y_i \neq 0$ which yields $x_i = 0$, and set $x_i^\star = 1$. This gives $y_i \geq 0$ and proves the first inequality in (2.17). Proceed now in the same way for the remaining i in (2.18). □

3 Frictional Impacts of Newton Type

One of the key events in non-smooth dynamics are impacts caused by frictional collisions. In this section we describe in detail how the equations of motion of finite-dimensional planar frictional collision problems are obtained in a well-structured form. Special emphasis is placed on the formulation of and the physics behind the standard impact constitutive laws from non-smooth dynamics (Moreau, 1994), which admit a representation of the collision process as a linear complementarity problem. Two examples are presented which show both, the power and limitations of the chosen approach.

3.1 Equations of Motion for Non-Impulsive Dynamics

For all the rest of the report we restrict ourselves to a special class of mechanical systems as depicted in Figure 7. In particular, we treat multi-impacts in finite freedom dynamics. This comprises typical multibody systems which are composed of rigid bodies, but also deformable bodies after spatial discretization. Impacts are assumed to stem from collisions. Any other impact events, such as velocity jumps caused by sharp bends in the configuration manifold are excluded. We consider scleronomic systems only. Breathing configuration manifolds as well as additional (unilateral) constraints that depend explicitly on time are not considered. We do not allow for any kind of external excitation. In particular, kinematic excitation and excitation by impulsive forces is excluded. All joints in the system are assumed to be represented by perfect bilateral constraints. A finite number of unilateral frictional contact constraints is allowed, but only the planar contact case will be considered.

In order to proceed to the impact equations in an utmost straightforward way, we first consider classical Lagrangian systems within the setting of virtual power for episodes of impact-free motion,

$$\forall \delta\mathbf{u}: \quad 0 = \delta\mathbf{u}^{\mathsf{T}} \left[\frac{\mathrm{d}}{\mathrm{d}t} \left(\frac{\partial T}{\partial \mathbf{u}} \right)^{\mathsf{T}} - \left(\frac{\partial T}{\partial \mathbf{q}} \right)^{\mathsf{T}} + \left(\frac{\partial V}{\partial \mathbf{q}} \right)^{\mathsf{T}} - \mathbf{f} \right]. \tag{3.1}$$

Time is denoted in (3.1) by t. Further, $\mathbf{q} \in \mathbb{R}^f$ are local coordinates of the f-dimensional configuration manifold defined via the perfect bilateral constraints of the system, $\mathbf{u} = \dot{\mathbf{q}}$ are the generalized velocities, and $\delta\mathbf{u}$ the associated virtual velocities. The coordinates $\mathbf{q}$ constitute a set of minimal coordinates in the classical sense if all unilateral contacts are open, i.e. if the associated contact partners are strictly separated from each other in space. Kinetic and potential energy of the scleronomic system are denoted by $T(\mathbf{q}, \mathbf{u})$ and $V(\mathbf{q})$, respectively. All generalized forces that do not allow for a potential are summarized in $\mathbf{f}$. After having performed the differentiation process in (3.1), one arrives with a set of f ordinary second order differential equations,

$$\forall \delta\mathbf{u}: \quad 0 = \delta\mathbf{u}^{\mathsf{T}} \left[\mathbf{M}(\mathbf{q})\, \dot{\mathbf{u}} - \mathbf{h}(\mathbf{q}, \mathbf{u}) - \mathbf{f} \right] \tag{3.2}$$

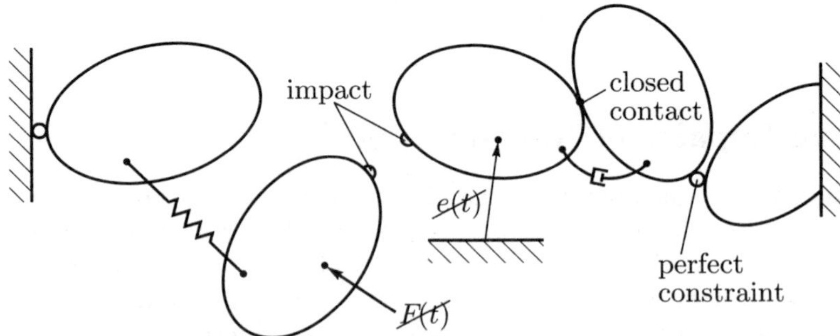

Figure 7. Class of mechanical systems considered in the collision theory.

which is still displayed within the framework of virtual power. $\mathbf{M}(\mathbf{q})$ denotes the symmetric and positive definite mass matrix of the system obtained from $\left(\frac{\partial T}{\partial \mathbf{u}}\right)^\mathsf{T} = \mathbf{M}(\mathbf{q})\,\mathbf{u}$, and $\mathbf{h}(\mathbf{q}, \mathbf{u})$ is the f-tuple of gyroscopic accelerations (Christoffel symbols) together with all finite-valued generalized forces that stem from the potential $V(\mathbf{q})$. Of particular interest are the contributions of the normal and tangential contact forces to $\mathbf{f}$, which will be elaborated based on a contact model in the next section.

3.2 Contact Model

In this section, we review the standard contact model for planar frictional contacts between two rigid bodies. Within this model, the gap function, the normal and tangential relative velocities as well as the contact forces are defined. Special attention is paid to the transformation, from which one obtains the contributions of the local contact forces to the generalized force vector $\mathbf{f}$ in (3.2). These contributions will be calculated by the invariance of the virtual power under coordinate transformations.

The left part of Figure 8 describes the kinematics of the contact model. The contact points P and Q are defined as the surface points that lie on a connecting line orthogonal to both of the bodies contours. In a first step, we derive all quantities that are related to the normal direction of the contact model. For $\mathbf{n}(\mathbf{q}) \in \mathbb{R}^3$ being the unit outward normal at point P of the left body, the distance $g_N(\mathbf{q})$ between the collision points becomes

$$g_N = \mathbf{n}^\mathsf{T}\,\mathbf{r}_{PQ}, \tag{3.3}$$

where $\mathbf{r}_{PQ}(\mathbf{q}) \in \mathbb{R}^3$ is the displacement of Q relative to P. For $g_N > 0$ both bodies are spatially separated from each other. The case $g_N = 0$ describes the situation, where both bodies touch each other at the points P and Q, as at the time instance of a collision. The case $g_N < 0$ corresponds to forbidden interpenetration or overlapping of the two bodies and has to be excluded. We therefore have $g_N(\mathbf{q}) \geq 0$, which constitutes an inequality constraint on the distance between the body contours. The function $g_N(\mathbf{q})$ is thus often called the *gap function.*

There are two equivalent ways on how to obtain the relative velocity γ_N in the normal direction of the two bodies. Either, one takes the velocities $\mathbf{v}_P$ and $\mathbf{v}_Q$ of points P and Q and projects their difference onto the unit normal $\mathbf{n}$

$$\gamma_N = \mathbf{n}^\mathsf{T}(\mathbf{v}_Q - \mathbf{v}_P), \tag{3.4}$$

Figure 8. Contact model: Unilateral contact with Coulomb friction.

or one differentiates (3.3) with respect to time. The latter yields $\gamma_N = \dot{g}_N = \mathbf{n}^\mathsf{T}\,\dot{\mathbf{r}}_{PQ} = \mathbf{n}^\mathsf{T}(\mathbf{v}_Q - \mathbf{v}_P)$, since $\dot{\mathbf{n}}^\mathsf{T}\,\mathbf{r}_{PQ} = 0$ because $\mathbf{n}$ is a unit vector and thus $\dot{\mathbf{n}} \perp \mathbf{r}_{PQ}$. For scleronomic systems, each velocity is a linear function of the generalized velocities $\mathbf{u}$. We may therefore write γ_N in the form

$$\gamma_N = \mathbf{w}_N^\mathsf{T}\,\mathbf{u}, \tag{3.5}$$

where $\mathbf{w}_N(\mathbf{q})$ can be identified from (3.4).

We also need the virtual normal contact velocity $\delta\gamma_N$, which we derive according to the principles from classical mechanics (Papastavridis, 2002). Within this local variational approach, one assumes a family of curves emanating from a given fixed position with different velocities. By applying this concept to (3.5), we obtain

$$\delta\gamma_N = \delta(\mathbf{n}^\mathsf{T}(\mathbf{v}_Q - \mathbf{v}_P)) = \mathbf{n}^\mathsf{T}\,\delta(\mathbf{v}_Q - \mathbf{v}_P). \tag{3.6}$$

Note that $\delta\mathbf{n} = 0$, because the normal $\mathbf{n}(\mathbf{q})$ depends only on displacements, which have to be kept fixed. In terms of generalized velocities, the expression in (3.6) becomes

$$\delta\gamma_N = \mathbf{w}_N^\mathsf{T}\,\delta\mathbf{u}. \tag{3.7}$$

At this point we just want to remark that *both*, the space of virtual velocities and actual velocities are *linear* for *scleronomic* systems, whereas the space of actual velocities is in general *linear affine* for *rheonomic* systems .

The right part of Figure 8 shows the contact forces of both bodies split into their normal and tangential parts. The contribution of the normal forces $\mathbf{F}_{NP}$ and $\mathbf{F}_{NQ}$ to the virtual power is

$$\delta P_N = \mathbf{F}_{NQ}^\mathsf{T}\,\delta\mathbf{v}_Q + \mathbf{F}_{NP}^\mathsf{T}\,\delta\mathbf{v}_P. \tag{3.8}$$

Since the normal forces $\mathbf{F}_{NP}$ and $\mathbf{F}_{NQ}$ are opposite to each other by the Principle of Interaction and directed towards the unit normal $\mathbf{n}$ by our contact model, we may express them in the form

$$\mathbf{F}_{NQ} = -\mathbf{F}_{NP} = \mathbf{n}\,\lambda_N. \tag{3.9}$$

The scalar value of the normal forces is thus determined by λ_N. With the help of (3.9) the virtual power expression (3.8) becomes

$$\delta P_N = \lambda_N\,\mathbf{n}^\mathsf{T}(\delta\mathbf{v}_Q - \delta\mathbf{v}_P), \tag{3.10}$$

and may be stated with (3.6) in terms of the scalar contact variables λ_N, γ_N as

$$\delta P_N = \lambda_N\,\delta\gamma_N. \tag{3.11}$$

We finally take into account (3.7) to proceed to the virtual power expression in terms of the generalized velocities

$$\delta P_N = \lambda_N\,\mathbf{w}_N^\mathsf{T}\,\delta\mathbf{u} =: \mathbf{f}_N^\mathsf{T}\,\delta\mathbf{u}, \tag{3.12}$$

which determines the generalized normal force

$$\mathbf{f}_N = \mathbf{w}_N\,\lambda_N. \tag{3.13}$$

The generalized normal force is thus determined by the product of the *generalized force direction* $\mathbf{w}_N(\mathbf{q})$ and the scalar normal force value λ_N.

For the tangential terms we proceed precisely in the same manner with the only difference that the tangential relative velocity γ_T has to be calculated from a projection as in (3.4), because a tangential "gap function" is not available in general. One obtains

$$\gamma_T = \mathbf{t}^\mathsf{T}(\mathbf{v}_Q - \mathbf{v}_P) = \mathbf{w}_T^\mathsf{T}\,\mathbf{u}, \quad \mathbf{f}_T = \mathbf{w}_T\,\lambda_T, \quad \mathbf{F}_{TQ} = -\mathbf{F}_{TP} = \mathbf{t}\,\lambda_T \tag{3.14}$$

with λ_T the scalar value of the tangential contact forces and $\mathbf{w}_T(\mathbf{q})$ the tangential generalized force direction.

3.3 Impact Equations and Kinematics

After having identified by (3.13), (3.14) the generalized contact forces $\mathbf{f}_N$ and $\mathbf{f}_T$ from our contact model, we may now write the virtual power equation (3.2) for a system with one planar frictional contact as

$$\forall \delta\mathbf{u}: \quad 0 = \delta\mathbf{u}^\mathsf{T}\left[\mathbf{M}(\mathbf{q})\,\dot{\mathbf{u}} - \mathbf{h}(\mathbf{q},\mathbf{u}) - \mathbf{w}_N(\mathbf{q})\,\lambda_N - \mathbf{w}_T(\mathbf{q})\,\lambda_T\right]. \tag{3.15}$$

Due to variation over all $\delta\mathbf{u}$, this expression only holds true if the term in brackets vanishes,

$$\mathbf{M}(\mathbf{q})\,\dot{\mathbf{u}} - \mathbf{h}(\mathbf{q},\mathbf{u}) - \mathbf{w}_N(\mathbf{q})\,\lambda_N - \mathbf{w}_T(\mathbf{q})\,\lambda_T = 0. \tag{3.16}$$

Note, however, that this equation still requires the existence of the velocities $\mathbf{u} = \dot{\mathbf{q}}$ and the accelerations $\dot{\mathbf{u}} = \ddot{\mathbf{q}}$, both being meaningless for the event of an impact: We may at best speak about the left and the right limit of the velocity at the time of impact, i.e. the pre- and post-impact velocity, but never about the velocity *at* the impact itself, a meaningless term already from the physical point of view. Similar reasons apply for the accelerations, as already pointed out in Section 1.2. Thus, in (3.16) precisely the points of interest, i.e. the discontinuity points of the velocities and accelerations are excluded. One has therefore to replace (3.16) by a more general formulation which gives access also to impacts. This step has been performed by Moreau (1988a) when he re-states (3.16) as an *equality of measures*, which provides a general framework for non-smooth rigid body dynamics,

$$\mathbf{M}(\mathbf{q})\,\mathrm{d}\mathbf{u} - \mathbf{h}(\mathbf{q},\mathbf{u})\,\mathrm{d}t - \mathbf{w}_N(\mathbf{q})\,\mathrm{d}\Lambda_N - \mathbf{w}_T(\mathbf{q})\,\mathrm{d}\Lambda_T = 0. \tag{3.17}$$

The velocities $\mathbf{u}(t)$ in (3.17) are assumed to be functions of bounded variations with associated absolutely continuous displacements $\mathbf{q}(t)$ obtained from integration, see (1.3). The probably "infinite values" of the "accelerations $\dot{\mathbf{u}}$ and the forces λ_{NT} at the impact" are taken into account by the differential measure $\mathrm{d}\mathbf{u}$ of $\mathbf{u}$ in (1.4) and by the force measures $\mathrm{d}\Lambda_{NT}$. Note that both, impact-free motion and impacts, are covered by (3.17) due to (1.5): If one takes the Lebesgue measure $\mathrm{d}t$, then $\mathrm{d}\Lambda_{NT} = \lambda_{NT}\,\mathrm{d}t$ with λ_{NT} being the contact forces from (3.16). If the Dirac point measure $\mathrm{d}\eta$ is chosen, then $\mathrm{d}\Lambda_{NT} = \Lambda_{NT}\,\mathrm{d}\eta$, where Λ_{NT} are then impulsive contact forces acting at some single point in time. In order to obtain now the impact equations we integrate (3.17) over a singleton $\{t_0\}$ and denote $\mathbf{q}_0 := \mathbf{q}(t_0)$. This yields

$$\mathbf{M}(\mathbf{q}_0)(\mathbf{u}^+(t_0) - \mathbf{u}^-(t_0)) - \mathbf{w}_N(\mathbf{q}_0)\,\Lambda_N(t_0) - \mathbf{w}_T(\mathbf{q}_0)\,\Lambda_T(t_0) = 0 \tag{3.18}$$

where $\mathbf{M}$, $\mathbf{h}$ and $\mathbf{w}$ as only functions of $\mathbf{q}$ and $\mathbf{u}$ are treated as constants at t_0 and do not affect integration, $\int_{t_0} \mathrm{d}\mathbf{u} = \mathbf{u}^+(t_0) - \mathbf{u}^-(t_0)$ by the property (1.7) of the differential measure, $\int_{t_0} \mathrm{d}t = 0$ because the Lebesgue measure is non-atomic, and $\int_{t_0} \mathrm{d}\Lambda_{NT} =: \Lambda_{NT}(t_0)$ defines the values of the impulsive contact forces at t_0. Note in particular that $\mathbf{u}$ at time t_0 is required in $\mathbf{h}$. Although undefined, this does not cause any problem. One may take any (finite) value $\mathbf{u}_0$, because this choice is immaterial and does not affect the Lebesgue measure $\mathrm{d}t$. The resulting expressions in (3.18) are called the equations of motion for the impact or, in short, the *impact equations* of the system.

Let us briefly return to the representation of the normal relative velocity in (3.4). For the same reasons as for $\dot{\mathbf{q}}$ we have choose another symbol for $\dot{g}_N$ in order to express that we are looking for a bounded variation function $\gamma_N(t)$ with possible discontinuities at the impact. For time t_0 both, the normal and tangential relative velocities $\gamma_{NT}^{\pm}$ are then induced by the associated values of $\mathbf{u}^+$ and $\mathbf{u}^-$,

$$\gamma_N^{\pm}(t_0) = \mathbf{w}_N^{\mathsf{T}}(\mathbf{q}_0)\,\mathbf{u}^{\pm}(t_0), \quad \gamma_T^{\pm}(t_0) = \mathbf{w}_T^{\mathsf{T}}(\mathbf{q}_0)\,\mathbf{u}^{\pm}(t_0). \tag{3.19}$$

These expressions define the *pre- and post impact contact relative velocities.*

In order to shorten notation we will write from now the impact equations (3.18) and the relative velocities (3.19) as

$$\begin{aligned} &\mathbf{M}\,(\mathbf{u}^+ - \mathbf{u}^-) = \mathbf{w}_N\,\Lambda_N + \mathbf{w}_T\,\Lambda_T \\ &\gamma_N^{\pm} = \mathbf{w}_N^{\mathsf{T}}\,\mathbf{u}^{\pm} \qquad \gamma_T^{\pm} = \mathbf{w}_T^{\mathsf{T}}\,\mathbf{u}^{\pm}. \end{aligned} \tag{3.20}$$

Note that the same vectors $\mathbf{w}_{NT}$ occur in the impact equations and the contact relative velocities. In general, this must not be taken for granted, because we have used different concepts to derive these formulas: A purely *geometrical* approach for the definition of the contact kinematics magnitudes which include the relative velocities, and some *physical imagination* to determine the point and the direction of interaction of the contact forces. Only within our idealized framework both terms coincide, which results in the simple virtual power expression $\delta P_{NT} = \Lambda_{NT}\,\delta\gamma_{NT}$, see (3.10). In this case, the virtual contact relative velocities are collinear to the associated contact forces.

There is more to keep in mind and to carefully check when modeling impact phenomena: Under the assumption of absolute continuous displacements $\mathbf{q}(t)$ and the existence of generalized force directions $\mathbf{w}_{NT}(\mathbf{q})$ as one-forms it is clear that neither $\mathbf{q}$ nor $\mathbf{w}_{NT}$ may vary during the time of impact when the latter is condensed to a singleton t_0 as in our approach. In reality, the duration of the impact extends to a small but non-vanishing time interval, accompanied by high contact stresses which may cause non-negligible deformations of the contact zone and, as a consequence, resultant mean contact force directions which may deviate considerably from the assumed ideal directions $\mathbf{w}_{NT}$. There are other kinds of constraints for which differentiability fails, for example sub-manifolds that are pasted together such that corners and edges appear. Velocity jumps may then be induced by instantaneous changes in the associated differentials $\mathbf{w}_{NT}$, primarily not related to any kind of collisions, see e.g. the example discussed in Glocker (1999). Further, the finite forces and gyroscopical accelerations contained in $\mathbf{h}\,dt$ may also contribute to some small amount to the impact. This effect may be numerically taken into account in a most sophisticated and consistent manner when some estimate of the impact duration is available, which is then used as the step width to pass the impact. This is the basic idea of all time-stepping algorithms, from which the first has been published in Moreau (1988a). Other numerical schemes may be found in the literature, such as the powerful Θ–method (Jean, 1999), an algorithm based on displacements with proven convergence (Paoli and Schatzman, 2002a,b), or several other well-developed codes described in Stewart and Trinkle (1996), Stewart (1998) and Anitescu et al. (1999).

3.4 Impact Constitutive Laws

Equations (3.20) do not yet completely describe the impact, because the impact laws are still missing. In this section we review the standard impact constitutive laws from non-smooth dynamics (Moreau, 1994) and state them as inclusions or likewise inequality conditions. As already a first part of the impact law we assume that non-vanishing impulsive forces Λ_{NT} occur only for a closed contact $g_N = 0$, which is a reasonable assumption for collisions. We further assume that the system arrived at $g_N = 0$ from an admissible state in the past, which necessarily requires a non-positive pre-impact normal relative velocity, $\gamma_N^- \leq 0$. Values $\gamma_N^- < 0$ correspond to an approaching process, at which the bodies touch under a non-vanishing velocity, whereas $\gamma_N^- = 0$ may correspond to a soft touchdown or to a contact that has been closed somewhere in the past

In a first step we state the impact law in the normal direction, which is actually a unilateral version of Newton's kinematic restitution law. The impulsive normal force, if there is any, should act as a compressive magnitude, $\Lambda_N \geq 0$. In the case of a non-vanishing impulsive force, we apply Newton's impact law as usual, i.e. we "invert" the normal relative velocity according to the rule

$$\Lambda_N > 0, \quad \gamma_N^+ = -\varepsilon_N \gamma_N^-. \tag{3.21}$$

The magnitude ε_N denotes Newton's coefficient of restitution with the usual values $0 \leq \varepsilon \leq 1$. The case $\varepsilon = 0$ corresponds to a completely inelastic impact with vanishing post-impact relative velocity ($\gamma^+ = 0$), whereas $\varepsilon = 1$ represents a completely elastic impact at which the relative velocity γ is inverted ($\gamma^+ = -\gamma^-$). Suppose now that, for any reason, the contact does *not* participate in the impact, i.e. that the value of the impulsive force is zero, although the contact is closed. This happens normally in multi-contact configurations, such as in the example in Figure 9. A rigid bar is resting on two unilateral obstacles and is hit by a rigid sphere. We expect an impulsive reaction

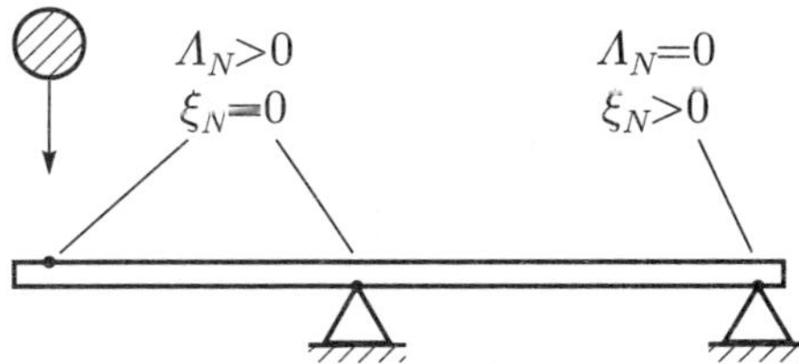

Figure 9. Two types of collision behavior in multi-contact configurations.

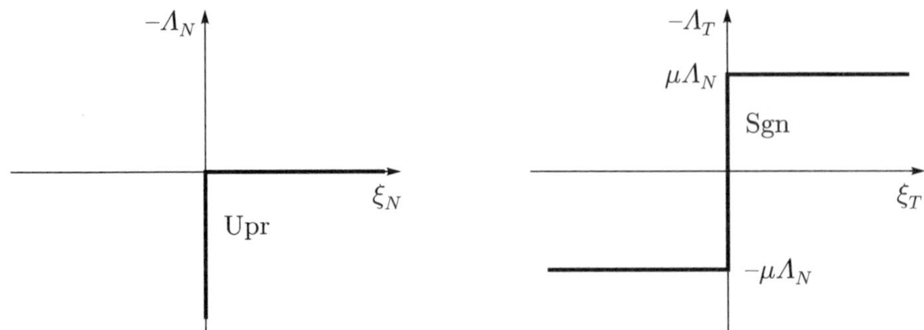

Figure 10. Constitutive collision laws in normal and tangential direction.

between the sphere and the bar, but also between the bar and the left obstacle. In contrast, the contact between the bar and the right obstacle is expected to open after the collision with a strictly positive relative velocity. This separation is solely induced by the overall impact configuration, and not supported by any local impulsive force. For this case we therefore allow post-impact relative velocities *higher* than what is prescribed by Newton's impact law, i.e.

$$\Lambda_N = 0, \quad \gamma_N^+ \geq -\varepsilon_N \, \gamma_N^-. \tag{3.22}$$

This condition expresses that the contact is superfluous and could have been removed without changing the contact-impact process. Both cases, (3.21) and (3.22), are obviously expressed by the Upr-inclusion

$$\xi_N := \gamma_N^+ + \varepsilon_N \, \gamma_N^-, \quad -\Lambda_N \in \mathrm{Upr}\,(\xi_N), \tag{3.23}$$

see (2.11) and (2.12). The graph of this constitutive law is depicted in the left part of Figure 10.

To model the collision behavior in the tangential direction, a combination of Coulomb type friction and the Newtonian impact law is applied. One first assumes that the value of the tangential impulsive force is restricted to $|\Lambda_T| \leq \mu \, \Lambda_N$, where μ denotes the coefficient of friction. For the stick case $|\Lambda_T| < \mu \, \Lambda_N$, pure Newtonian collision behavior is proposed,

$$|\Lambda_T| < \mu \, \Lambda_N, \quad \gamma_T^+ = -\varepsilon_T \, \gamma_T^-, \tag{3.24}$$

which yields a reversion of the tangential relative velocity in the same fashion as for the normal direction (3.21). Tangential restitution is taken into account by the coefficient ε_T with the usual values $0 \leq \varepsilon \leq 1$. Similar to (3.22), one completes now the the tangential impact law for the case $|\Lambda_T| = \mu \, \Lambda_N$ by setting

$$\begin{aligned} \Lambda_T &= -\mu \, \Lambda_N, \quad \gamma_T^+ \geq -\varepsilon_T \, \gamma_T^-, \\ \Lambda_T &= +\mu \, \Lambda_N, \quad \gamma_T^+ \leq -\varepsilon_T \, \gamma_T^-, \end{aligned} \tag{3.25}$$

which is called the slip state. One easily recognizes that Coulomb type friction is present, but also that the slip state still carries something from the Newtonian collision behavior.

Both cases, (3.24) and (3.25), may be expressed via the Sgn-multifunction (2.14) as

$$\xi_T := \gamma_T^+ + \varepsilon_T \gamma_T^-, \qquad -\Lambda_T \in \mu \Lambda_N \operatorname{Sgn}(\xi_T). \tag{3.26}$$

The graph of this inclusion is depicted in the right diagram of Figure 10. Note that pure Coulomb type friction is obtained when setting $\varepsilon_T = 0$. To get a better sense on how the tangential impact law behaves, we refer in particular to the paper of Payr and Glocker (2005), in which the frictional impact of a rigid bar against an inelastic half-space has been analyzed in full detail. Among other things, it has been demonstrated that the post-impact velocity ratio γ_T^+/γ_N^+ depends *continuously* on the pre-impact velocity ratio γ_T^-/γ_N^-, a fact that people are still not fully aware of.

We have stated the contact laws (3.23) and (3.26) only for the case of a collision. There is, of course, a consistent extension which covers also impact-free motion (Moreau, 1994), and which can be found in our notation in Glocker and Studer (2005). Within this more general setting, it can be easily shown that the tangential contact behavior reduces precisely to classical Coulomb friction for impact-free motion.

3.5 The Multi-Contact Impact Problem

Multi-contact configurations are, indeed, one of the standard problems in dynamics. It is thus necessary to extend the presented approach to the case of several unilateral constraints that appear as a consequence of contacting bodies. We allow single contacts between different bodies which includes that one body may have several contact partners, but we also allow several contacts at the same contacting couple. The impact equations in (3.20) are already stated in a form suitable for superposition. We assume a finite number of contacts, each of them represented on displacement level by its associated gap function $g_N^i(\mathbf{q}) \geq 0$. We denote by $\mathcal{H}$ the set of contacts that are closed at a probable impact event $(\mathbf{q}_0, t_0)$,

$$\mathcal{H}(\mathbf{q}_0) = \{i \mid g_N^i(\mathbf{q}_0) = 0\}, \tag{3.27}$$

and call it the *active set* of unilateral constraints. The contact force contributions of each member in the active set has then to be taken into account in (3.20) together with their associated impact laws (3.23) and (3.26),

$$\begin{array}{cc} \multicolumn{2}{c}{\mathbf{M}(\mathbf{u}^+ - \mathbf{u}^-) = \sum\limits_{i\in\mathcal{H}} (\mathbf{w}_N^i \Lambda_{Ni} + \mathbf{w}_T^i \Lambda_{Ti})} \\ \gamma_N^i = \mathbf{w}_N^{i\,\mathsf{T}} \mathbf{u} & \gamma_T^i = \mathbf{w}_T^{i\,\mathsf{T}} \mathbf{u} \\ \xi_N^i = \gamma_N^{i\,+} + \varepsilon_N^i \gamma_N^{i\,-} & \xi_T^i = \gamma_T^{i\,+} + \varepsilon_T^i \gamma_T^{i\,-} \\ -\Lambda_{Ni} \in \operatorname{Upr}(\xi_N^i) & -\Lambda_{Ti} \in \mu_i \Lambda_{Ni} \operatorname{Sgn}(\xi_T^i) \end{array} \tag{3.28}$$

The multi-contact collision problem is then fully determined by this set of equations, and one has obtained a complete description of the impact dynamics of the system.

This extension to multi-contact problems by just superposing the impact laws of the different contacts must not be taken without a critical study of the physical impact process. Impacts are complicated, highly dynamic processes which affect the entire structure under investigation. Our impact laws (3.23) and (3.26) have been designed as local contact interactions that do not allow the different contacts within the system to talk to each other. If, in addition, these laws are applied within a rigid body system and the physical collision behavior is such that wave effects are needed to properly describe it, then too much of the system's physics has been neglected in the model. The most prominent example of such a situation is Newton's cradle, for which the standard impact laws are not general enough to provide the right result when applied together with rigid models of the balls. This example will be investigate in Section 3.8 and, again, within a more general framework in Section 5.6. There are, however, dozens of application problems for which the presented approach yields excellent results; even more, it may help to gain fruitful structural knowledge, leading to a much better understanding of the impact process itself.

There are many different ways to solve the inequality problem (3.28), such as Gauß-Seidel iterations, the Augmented Lagrangian approach combined with non-smooth Newton methods or fixed point iterations, or subsequent iterative evaluations of the normal and tangential subproblem. Another possibility for planar frictional contacts, as in our case, is to formulate (3.28) as a linear complementarity problem and to invoke an appropriate LCP solver. For completeness, we will at least state the resulting LCP, but without going into detail on how all the required steps of reformulation of (3.28) have to be performed, see e.g. Glocker and Studer (2005). The main difficulty in formulating the LCP stems from the Sgn -functions that have to be decomposed into Upr 's according to (2.15), (2.16) to achieve the desired complementarity formulation. To set up the LCP, rigorous matrix notation is required. One defines

$$\begin{array}{lll} \mathbf{W}_N := (\mathbf{w}_N^{i_1}, \dots, \mathbf{w}_N^{i_k}), & \mathbf{W}_T := (\mathbf{w}_T^{i_1}, \dots, \mathbf{w}_T^{i_k}) & \in \mathbb{R}^{f,k} \\ \boldsymbol{\Lambda}_N := (\Lambda_{Ni_1}, \dots, \Lambda_{Ni_k})^\mathsf{T}, & \boldsymbol{\Lambda}_T := (\Lambda_{Ti_1}, \dots, \Lambda_{Ti_k})^\mathsf{T} & \in \mathbb{R}^{k} \\ \boldsymbol{\gamma}_N := (\gamma_N^{i_1}, \dots, \gamma_N^{i_k})^\mathsf{T}, & \boldsymbol{\gamma}_T := (\gamma_T^{i_1}, \dots, \gamma_T^{i_k})^\mathsf{T} & \in \mathbb{R}^{k} \\ \boldsymbol{\xi}_N := (\xi_N^{i_1}, \dots, \xi_N^{i_k})^\mathsf{T}, & \boldsymbol{\xi}_T := (\xi_T^{i_1}, \dots, \xi_T^{i_k})^\mathsf{T} & \in \mathbb{R}^{k} \\ \boldsymbol{\epsilon}_N := \operatorname{diag}(\varepsilon_N^{i_1}, \dots, \varepsilon_N^{i_k}), & \boldsymbol{\epsilon}_T := \operatorname{diag}(\varepsilon_T^{i_1}, \dots, \varepsilon_T^{i_k}) & \in \mathbb{R}^{k,k} \\ & \boldsymbol{\mu} := \operatorname{diag}(\mu_{i_1}, \dots, \mu_{i_k}) & \in \mathbb{R}^{k,k} \end{array} \tag{3.29}$$

and rewrites the first three lines in (3.28) according to these definitions. One then eliminates $\boldsymbol{\gamma}_N$ and $\boldsymbol{\gamma}_T$ from the third line with the help of the second line and introduces the quantities

$$\boldsymbol{\Lambda}_R := \boldsymbol{\mu}\boldsymbol{\Lambda}_N + \boldsymbol{\Lambda}_T, \quad \boldsymbol{\Lambda}_L := \boldsymbol{\mu}\boldsymbol{\Lambda}_N - \boldsymbol{\Lambda}_T, \quad \boldsymbol{\xi}_T = \boldsymbol{\xi}_R - \boldsymbol{\xi}_L \tag{3.30}$$

to split up the Sgn multifunctions. Subsequently, $\boldsymbol{\xi}_T$, $\boldsymbol{\Lambda}_T$, and $\mathbf{u}^+$ are then removed from the remaining equations. The result is

$$\begin{gathered} \begin{pmatrix} \boldsymbol{\xi}_N \\ \boldsymbol{\xi}_R \\ \boldsymbol{\Lambda}_L \end{pmatrix} = \begin{pmatrix} \mathbf{W}_N^\mathsf{T}\mathbf{M}^{-1}(\mathbf{W}_N - \mathbf{W}_T\boldsymbol{\mu}) & \mathbf{W}_N^\mathsf{T}\mathbf{M}^{-1}\mathbf{W}_T & 0 \\ \mathbf{W}_T^\mathsf{T}\mathbf{M}^{-1}(\mathbf{W}_N - \mathbf{W}_T\boldsymbol{\mu}) & \mathbf{W}_T^\mathsf{T}\mathbf{M}^{-1}\mathbf{W}_T & \mathbf{E} \\ 2\,\boldsymbol{\mu} & -\mathbf{E} & 0 \end{pmatrix} \begin{pmatrix} \boldsymbol{\Lambda}_N \\ \boldsymbol{\Lambda}_R \\ \boldsymbol{\xi}_L \end{pmatrix} + \begin{pmatrix} (\mathbf{E} + \boldsymbol{\epsilon}_N)\boldsymbol{\gamma}_N^- \\ (\mathbf{E} + \boldsymbol{\epsilon}_T)\boldsymbol{\gamma}_T^- \\ 0 \end{pmatrix} \\ 0 \preceq \begin{pmatrix} \boldsymbol{\xi}_N \\ \boldsymbol{\xi}_R \\ \boldsymbol{\Lambda}_L \end{pmatrix} \perp \begin{pmatrix} \boldsymbol{\Lambda}_N \\ \boldsymbol{\Lambda}_R \\ \boldsymbol{\xi}_L \end{pmatrix} \succeq 0 \end{gathered} \tag{3.31}$$

which constitutes a linear complementarity problem (2.6) of dimension $3k$ when k is the number of active contacts. Note the following property of the LCP: For friction that is independent of the impulsive normal forces $\mathbf{\Lambda}_N$, the terms $\boldsymbol{\mu}\mathbf{\Lambda}_N$ have to be replaced by constants $\mathbf{a}$ and move into the vector $\mathbf{b}$ of the LCP (2.6). The resulting matrix $\mathbf{A}$ is then bisymmetric, and the LCP states the Kuhn-Tucker conditions of an associated quadratic program with inequality constraints, see Glocker (2001a). Note also that the LCP does *not* bring any new physics into the problem! It is just another way to express what has already been said in (3.28).

3.6 Energetic Consistency

Energetic consistency of the scleronomic multi-contact impact problem is guaranteed if the restitution coefficients of the active contacts are all equal to each other, $\varepsilon := \varepsilon_N^i = \varepsilon_T^j \; \forall i, j \in \mathcal{H}(\mathbf{q})$. The associated proof can be found in Moreau (1994) and is based on the following strategy that has also been used in Pfeiffer and Glocker (1996) for Poisson's impact law: The difference in kinetic energy,

$$T^+ - T^- = \tfrac{1}{2}\,\mathbf{u}^{+\mathsf{T}}\mathbf{M}\,\mathbf{u}^+ - \tfrac{1}{2}\,\mathbf{u}^{-\mathsf{T}}\mathbf{M}\,\mathbf{u}^-, \tag{3.32}$$

is reformulated in terms of the local contact variables, i.e. in terms of those variables $(\Lambda_{NTi}, \xi_{NT}^i)$ that have been used to state the impact constitutive equations in the last line of (3.28). By using the whole set of equations (3.28), this results in an expression

$$T^+ - T^- = \tfrac{1}{1+\varepsilon}\,\mathbf{\Lambda}^\mathsf{T}\boldsymbol{\xi} - \tfrac{1}{2}\tfrac{1-\varepsilon}{1+\varepsilon}\,\mathbf{\Lambda}^\mathsf{T}(\mathbf{W}^\mathsf{T}\mathbf{M}^{-1}\mathbf{W})\mathbf{\Lambda} \tag{3.33}$$

with the matrix and vector quantities being composed of both, the normal and tangential contributions (3.29) of all $i \in \mathcal{H}$, i.e. $\mathbf{W} = (\mathbf{W}_N, \mathbf{W}_T)$, $\mathbf{\Lambda}^\mathsf{T} = (\mathbf{\Lambda}_N^\mathsf{T}, \mathbf{\Lambda}_T^\mathsf{T})$, $\boldsymbol{\xi}^\mathsf{T} = (\boldsymbol{\xi}_N^\mathsf{T}, \boldsymbol{\xi}_T^\mathsf{T})$. By the common restriction of the restitution coefficient ε to values $\varepsilon \in [0,1]$ one immediately recognizes that $\frac{1}{2}\frac{1-\varepsilon}{1+\varepsilon}\,\mathbf{\Lambda}^\mathsf{T}(\mathbf{W}^\mathsf{T}\mathbf{M}^{-1}\mathbf{W})\mathbf{\Lambda} \geq 0$, because $\mathbf{W}^\mathsf{T}\mathbf{M}^{-1}\mathbf{W}$ is (at least) positive semidefinite. For the first term on the right hand side in (3.33) it holds that

$$\mathbf{\Lambda}^\mathsf{T}\boldsymbol{\xi} = \sum_{i\in\mathcal{H}} (\Lambda_{Ni}\,\xi_N^i + \Lambda_{Ti}\,\xi_T^i) \leq 0. \tag{3.34}$$

This is a direct consequence on the constitutive equations in the last line of (3.28) and Figure 10, for which $\Lambda_{Ni}\,\xi_N^i = 0$ by complementarity, and $\Lambda_{Ti}\,\xi_T^i \leq 0$ due to the relay function and the restriction $\Lambda_{Ni} \geq 0$. Kinetic energy is preserved by the impact if, for example, $\varepsilon = 1$ and $\mu_i \gg 1$ has been chosen. For this case, the last term in (3.33) drops out and one has, in addition, $\Lambda_{Ti}\,\xi_T^i = 0$. For those contacts participating in the impact by non-vanishing impulsive normal forces $\Lambda_{Ni} > 0$, the relative velocities in both, the normal and tangential directions, are inverted according to $\xi_N^i = \xi_T^i = 0$, which gives $\gamma_N^{i+} = -\gamma_N^{i-}$ and $\gamma_T^{i+} = -\gamma_T^{i-}$ by the last two lines in (3.28).

3.7 Example: Frictional Impact of a Superball

In this example we investigate the frictional impact behavior of a highly elastic superball that is thrown against an inelastic half-space. We assume Coulomb friction with a

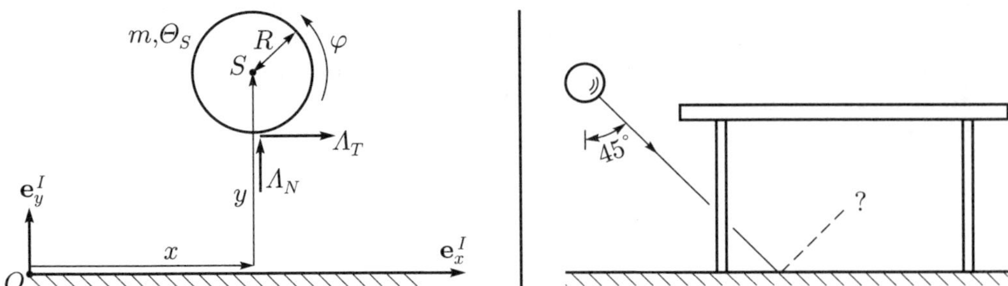

Figure 11. Mechanical model of a ball impacting against a rigid half-space.

friction coefficient μ that is big enough to keep the ball in the stick phase during the impact. Normal and tangential restitution is assumed to be perfectly elastic, $\varepsilon_N = \varepsilon_T = 1$. This example demonstrates the need of a tangential restitution coefficient to enforce the ball to bounce back after the second collision when it is thrown under a table.

The free body diagram of the ball (radius R, mass m, moment of inertia $\Theta_S = \frac{2}{5}mR^2$) is depicted in the left part of Figure 11. The absolute linear and angular displacements of the ball are denoted by x, y, φ, and the normal and tangential impulsive forces by Λ_N and Λ_T, respectively. The impact equations

$$m(\dot{x}^+ - \dot{x}^-) = \Lambda_T, \quad m(\dot{y}^+ - \dot{y}^-) = \Lambda_N, \quad \tfrac{2}{5}mR^2(\dot{\varphi}^+ - \dot{\varphi}^-) = R\Lambda_T \tag{3.35}$$

and the relative velocities in the normal and tangential directions

$$\gamma_N = \dot{y}, \quad \gamma_T = \dot{x} + R\dot{\varphi} \tag{3.36}$$

may be stated in matrix notation as

$$\mathbf{M}(\mathbf{u}^+ - \mathbf{u}^-) = \underbrace{\begin{pmatrix} \mathbf{w}_N & \mathbf{w}_T \end{pmatrix}}_{=:\mathbf{W}} \underbrace{\begin{pmatrix} \Lambda_N \\ \Lambda_T \end{pmatrix}}_{=:\mathbf{\Lambda}}, \qquad \underbrace{\begin{pmatrix} \gamma_N \\ \gamma_T \end{pmatrix}}_{=:\boldsymbol{\gamma}} = \underbrace{\begin{pmatrix} \mathbf{w}_N^\mathsf{T} \\ \mathbf{w}_T^\mathsf{T} \end{pmatrix}}_{=:\mathbf{W}^\mathsf{T}} \mathbf{u}, \tag{3.37}$$

from which one identifies the mass matrix $\mathbf{M}$ and the two generalized force directions $\mathbf{W}$ after the generalized velocities $\mathbf{u}$ have been specified,

$$\mathbf{u} := \begin{pmatrix} \dot{x} \\ \dot{y} \\ \dot{\varphi} \end{pmatrix} \quad \Rightarrow \quad \mathbf{M} = \begin{pmatrix} m & 0 & 0 \\ 0 & m & 0 \\ 0 & 0 & \frac{2}{5}mR^2 \end{pmatrix}; \quad \mathbf{W} = \begin{pmatrix} 0 & 1 \\ 1 & 0 \\ 0 & R \end{pmatrix}. \tag{3.38}$$

To proceed with the evaluation of the impact laws, it is convenient to state the whole impact problem in terms of the local contact variables $\boldsymbol{\gamma}$ and $\mathbf{\Lambda}$. This representation is obtained from (3.37)

$$\mathbf{M}(\mathbf{u}^+ - \mathbf{u}^-) = \mathbf{W}\mathbf{\Lambda}, \quad \boldsymbol{\gamma}^\pm = \mathbf{W}^\mathsf{T}\mathbf{u}^\pm \tag{3.39}$$

after elimination of $\mathbf{u}^+$ and $\mathbf{u}^-$, which results in

$$\boldsymbol{\gamma}^+ - \boldsymbol{\gamma}^- = \mathbf{W}^\mathsf{T}(\mathbf{u}^+ - \mathbf{u}^-) = \underbrace{\mathbf{W}^\mathsf{T}\mathbf{M}^{-1}\mathbf{W}}_{=:\mathbf{G}}\,\mathbf{\Lambda}. \tag{3.40}$$

One finally gets

$$\boldsymbol{\gamma}^+ - \boldsymbol{\gamma}^- = \mathbf{G}\boldsymbol{\Lambda} \quad \text{with } \mathbf{G} = \tfrac{1}{m}\begin{pmatrix} 1 & 0 \\ 0 & \frac{7}{2} \end{pmatrix}, \tag{3.41}$$

where $\mathbf{G}$ has been computed from the values of $\mathbf{M}$ and $\mathbf{W}$ in (3.38). For this particular example $\mathbf{G}$ is diagonal. This means that the problem is decoupled with respect to the inertias, but *not* with respect to the overall impact behavior, since the normal impulsive force still influences the tangential direction due to Coulombs law.

To evaluate now equations (3.41) together with the impact constitutive laws from the last two lines in (3.28), we first look at the normal behavior. For $\varepsilon_N = 1$, this is completely determined by

$$\gamma_N^+ - \gamma_N^- = \tfrac{1}{m}\Lambda_N, \quad \xi_N = \gamma_N^+ + \gamma_N^-, \quad -\Lambda_N \in \mathrm{Upr}\,(\xi_N). \tag{3.42}$$

We assume that the pre-impact normal relative velocity is strictly less than zero, $\gamma_N^- < 0$, which corresponds to the situation that the ball moves towards the floor before the impact. It is quite clear from physical intuition that there must be a strictly positive impulsive normal force at the impact. However, we will formally derive this result from equations (3.42) to demonstrate that the normal impact constitutive law automatically takes this decision via the unilateral primitive. Therefore, suppose first that $\Lambda_N = 0$. This gives by the first equation in (3.42) that $\gamma_N^+ = \gamma_N^-$. Hence, $\xi_N = 2\gamma_N^- < 0$ by the second equation, which contradicts the Upr-inclusion, because ξ_N is requested to be non-negative. We consider now the second case, i.e. $\Lambda_N > 0$. We immediately obtain by complementarity that $\xi_N = 0$. The first two equations in (3.42) yield

$$\gamma_N^+ = -\gamma_N^-, \quad \Lambda_N = -2m\gamma_N^-, \tag{3.43}$$

which is in accordance to our assumption $\Lambda_N > 0$, because $\gamma_N^- < 0$ has been presupposed. As a result, the normal relative velocity γ_N is inverted by the completely elastic impact law.

We look now at the tangential behavior, which is described by the second equation in (3.41) together with the constitutive laws from the lower right equations in (3.28) and $\varepsilon_T = 1$,

$$\gamma_T^+ - \gamma_T^- = \tfrac{7}{2m}\Lambda_T, \quad \xi_T = \gamma_T^+ + \gamma_T^-, \quad -\Lambda_T \in \mu\Lambda_N\,\mathrm{Sgn}\,(\xi_T). \tag{3.44}$$

Assumed is a friction coefficient μ sufficiently large to ensure stick, i.e. to ensure $|\Lambda_T| < \mu\Lambda_N$. The Sgn-inclusion in (3.44) then yields $\xi_T = 0$, and

$$\gamma_T^+ = -\gamma_T^-, \quad \Lambda_T = -\tfrac{4}{7}m\gamma_T^- \tag{3.45}$$

as a consequence on the first two equations in (3.44). One observes that the tangential relative velocity γ_T is also inverted without any losses.

By equations (3.43) and (3.45) we have found the post-impact relative velocities of the ball, $\boldsymbol{\gamma}^+ = -\boldsymbol{\gamma}^-$. In a final step we will now calculate the impact transition matrix, which relates the post- and pre-impact generalized velocities $\mathbf{u}^+$ and $\mathbf{u}^-$. This requires the impulsive forces $\boldsymbol{\Lambda}$ as functions of the pre-impact generalized velocities $\mathbf{u}^-$, which can be computed from (3.41) and $\boldsymbol{\gamma}^- = \mathbf{W}^\intercal\mathbf{u}^-$ from (3.39),

$$\boldsymbol{\gamma}^+ = -\boldsymbol{\gamma}^- \quad \Rightarrow \quad \boldsymbol{\Lambda} = -2\mathbf{G}^{-1}\boldsymbol{\gamma}^- = -2\mathbf{G}^{-1}\mathbf{W}^\intercal\mathbf{u}^-. \tag{3.46}$$

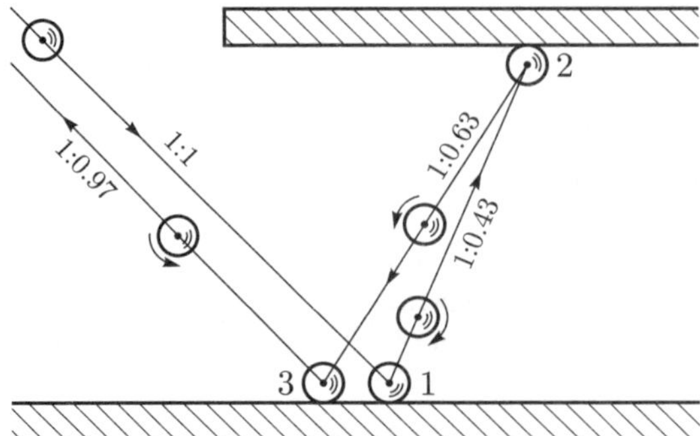

Figure 12. Trajectory of a highly elastic rubber ball hitting two parallel surfaces.

By putting this expression into the first equation of (3.37) and solving the latter for $\mathbf{u}^+$, one obtains the desired result

$$\mathbf{u}^+ = (\mathbf{I} - 2\mathbf{M}^{-1}\mathbf{W}\mathbf{G}^{-1}\mathbf{W}^\intercal)\mathbf{u}^-, \tag{3.47}$$

where $\mathbf{I}$ denotes the identity matrix. With $\mathbf{M}$, $\mathbf{W}$ from (3.38) and $\mathbf{G}$ from (3.41) one explicitly gets

$$\begin{pmatrix} \dot{x}^+ \\ \dot{y}^+ \\ \dot{\varphi}^+ \end{pmatrix} = \begin{pmatrix} \frac{3}{7} & 0 & \frac{\pm 4R}{7} \\ 0 & -1 & 0 \\ \frac{-10}{7R} & 0 & \frac{\pm 3}{7} \end{pmatrix} \begin{pmatrix} \dot{x}^- \\ \dot{y}^- \\ \dot{\varphi}^- \end{pmatrix}. \tag{3.48}$$

The double sign takes into account the two different cases that the ball bounces either against the floor $(-)$ or against the plate of the table $(+)$. For the collision against the table, the tangential relative velocity in (3.36) becomes $\gamma_T = \dot{x} - R\dot{\varphi}$ and the impulsive moment $R\Lambda_T$ in (3.35) has to be replaced by $-R\Lambda_T$.

Figure 12 shows the trajectory of the ball when it is thrown under 45° without any initial angular speed against the floor. By neglecting gravity, the motion is composed of phases of constant velocities, which are initiated and terminated by their associated impact events. One just has to evaluate sequentially the impact transition relation (3.48) to proceed from one collision to the next. The post-impact velocities of the actual collision enter then the problem as the pre-impact velocities of the succeeding one. This results in the amazing effect that a superball thrown under a table bounces back after the second collision. In contrast, the ball would resist to bounce back if the tangential restitution coefficient would have been chosen as $\varepsilon_T = 0$, as it can be observed for ping-pong balls. The associated trajectory may be found in Pfeiffer and Glocker (1996).

3.8 Example: Standard Impact Laws and Newton's Cradle

In this section we try to verify the standard impact laws (3.28) for frictionless collisions ($\mu_i = 0 \Rightarrow \Lambda_{Ti} = 0$) by means of Newton's cradle. A simplified version of Newton's cradle consists of three steel balls of equal masses m that are arranged along a line as depicted in the left part of Figure 13. We denote the linear absolute displacements of the bodies

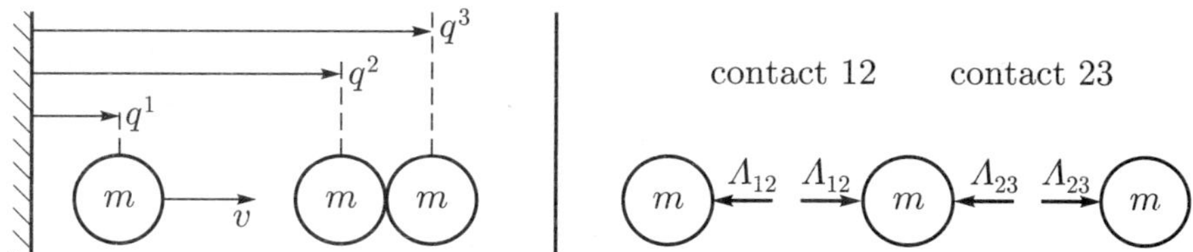

Figure 13. Mechanical model of Newton's cradle with three balls.

by q^i and introduce the associated generalized velocities as $\mathbf{u} := (\dot{q}^1, \dot{q}^2, \dot{q}^3)^\mathsf{T}$. Assume now a pre-impact configuration such that balls 2 and 3 are in contact but don't move, and ball 1 is shot against ball 2 with a velocity $v > 0$. We know by experiments that (nearly) all the velocity is transferred to ball 3 which leaves with a post-impact velocity v, whereas balls 1 and 2 are at rest after the collision.

Both contacts, 12 and 23, are closed at the impact and have thus to be considered as elements of the active set $\mathcal{H}$ in the impact equations. In the following we try to identify the restitution coefficients ε^{12} and ε^{23} of the contacts such that the observed pre- and post-impact velocities $\mathbf{u}^- = (v, 0, 0)^\mathsf{T}$ and $\mathbf{u}^+ = (0, 0, v)^\mathsf{T}$ are in accordance with the standard impact laws (3.28) for frictionless collisions. With the help of the free body diagram in the right part of Figure 13 we first evaluate the impact equations $\mathbf{M}(\mathbf{u}^+ - \mathbf{u}^-) = \sum_{i\in\{12,23\}} \mathbf{w}^i \Lambda_i$ for each ball, which yields

$$-mv = -\Lambda_{12}, \qquad 0 = \Lambda_{12} - \Lambda_{23}, \qquad mv = \Lambda_{23}. \tag{3.49}$$

The three equations in (3.49) may be uniquely solved for the two unknown impulsive forces

$$\Lambda_{12} = mv > 0 \qquad \text{and} \qquad \Lambda_{23} = mv > 0, \tag{3.50}$$

which tells us that the experimental observation is in accordance with the principles of mechanics. Furthermore, strictly positive impulsive forces of the same magnitude are obtained.

The values of the pre- and post-impact relative velocities $\gamma^{i\pm} = \mathbf{w}^{i\,\mathsf{T}}\,\mathbf{u}^\pm$ observed from the experiment are

$$\gamma^{12-} = -v, \quad \gamma^{12+} = 0 \qquad \text{and} \qquad \gamma^{23-} = 0, \quad \gamma^{23+} = v, \tag{3.51}$$

which yields by $\xi^i = \gamma^{i+} + \varepsilon^i\,\gamma^{i-}$

$$\xi^{12} = -v\,\varepsilon^{12} \qquad \text{and} \qquad \xi^{23} = v. \tag{3.52}$$

Note already that the expression obtained for ξ^{23} does *not* depend on the restitution coefficient ε^{23} of contact 23, since the associated pre-impact relative velocity γ^{23-} is equal to zero!

We finally have to evaluate the impact constitutive equations $-\Lambda_i \in \mathrm{Upr}\,(\xi^i)$ or, equivalently, the inequality-complementarity conditions $\Lambda_i \geq 0$, $\xi^i \geq 0$, $\Lambda_i\,\xi^i = 0$. For contact 12 it holds by (3.50) that $\Lambda_{12} > 0$, which causes $\xi^{12} = 0$ by complementarity.

With ξ^{12} from (3.52), the associated restitution coefficient becomes $\varepsilon^{12} = 0$, which corresponds to completely inelastic behavior of this contact. For contact 23 we have from (3.50) and (3.52) that $\Lambda_{23} > 0$ and $\xi^{23} = v > 0$, which is in contradiction to the complementarity condition $\Lambda_{23}\,\xi^{23} = 0$. As a result, there is no possible choice for the restitution coefficient ε^{23} to make (3.26) capable of treating Newton's cradle. Apparently, there are impact events that go beyond the standard impact constitutive laws (3.28). We will meet Newton's cradle again in Section 5.6 to discuss it in a much broader framework.

4 Cones in Non-Smooth Analysis

In this section some basic definitions, notations and properties of cones used in non-smooth analysis are put together. These cones characterize sets in the neighborhood of a chosen point and indicate whether convexification as the next step available after linearization is possible or not. As a standard tool for inequality systems, these cones will be used throughout the rest of this report; the related representations, variational formulations and decomposition rules as far as needed are presented in this section. The material is mainly taken from Aubin and Ekeland (1984) and Rockafellar (1972) but applied only to finite-dimensional real spaces.

4.1 Linear Spaces

We will shortly review some basic definitions and concepts from vector spaces. Only the finite-dimensional case is considered. Let $\mathcal{L}$ be a n-dimensional linear space over the field of real numbers $\mathbb{R}$. The *dual* of $\mathcal{L}$, denoted by $\mathcal{L}^\star$, is defined as the set of all linear maps that assign to each element of $\mathcal{L}$ a real number,

$$\mathcal{L}^\star := \{f : \mathcal{L} \to \mathbb{R}, \quad f \text{ linear}\}. \tag{4.1}$$

One can show that $\mathcal{L}^\star$ is also a linear space, and that it has the same dimension n as $\mathcal{L}$. The linear maps (4.1) are often written in the form $f(u) \equiv \langle f, u\rangle$, which is called the duality pairing as indicated in Figure 14.

A subspace $\mathcal{T}$ of a vector space $\mathcal{L}$ is a subset of elements of $\mathcal{L}$, which is itself a vector space. Let $\mathcal{T}$ be a m-dimensional subspace of $\mathcal{L}$. The set $\mathcal{N}$ in $\mathcal{L}^\star$ defined by

$$\mathcal{N} := \{f \in \mathcal{L}^\star \mid f(u) = 0 \quad \forall u \in \mathcal{T}\} \tag{4.2}$$

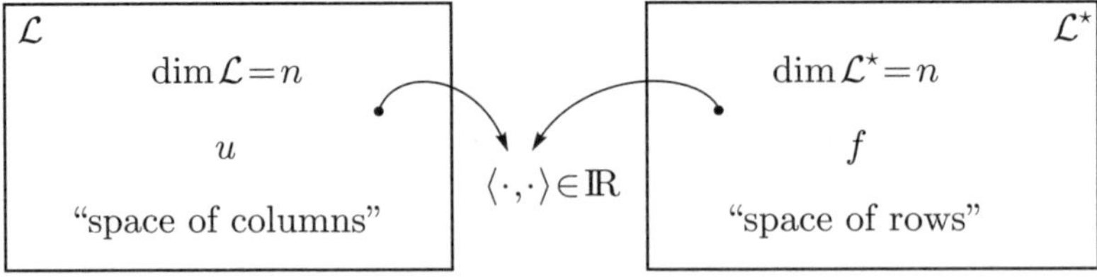

Figure 14. Pair of dual spaces $\mathcal{L}$ and $\mathcal{L}^\star$.

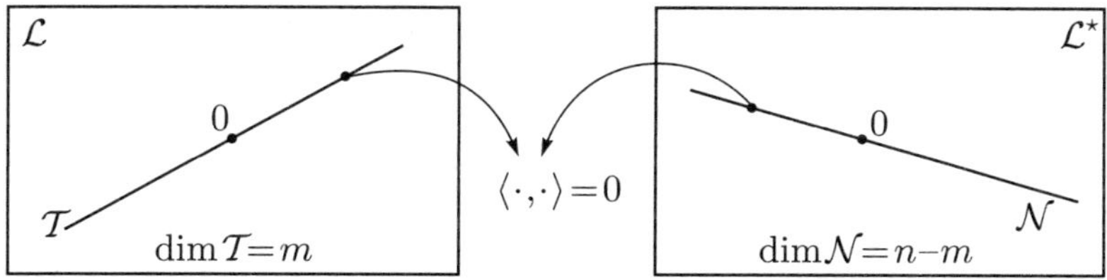

Figure 15. The annihilator $\mathcal{N} \in \mathcal{L}^\star$ of the subspace $\mathcal{T} \in \mathcal{L}$.

is called the *annihilator* of $\mathcal{T}$. It is a subspace of $\mathcal{L}^\star$, and its dimension is $(n-m)$, see Figure 15.

By just the definition of a linear space, geometrical concepts such as lengths and angles are not yet available. We therefore do not yet know what orthogonality means. In order to introduce these concepts, one has to chose a symmetric and positive definite bilinearform on $\mathcal{L}$

$$\Phi : \mathcal{L} \times \mathcal{L} \to \mathbb{R}, \quad (v,u) \to \Phi(v,u), \tag{4.3}$$

which defines an *inner product* on $\mathcal{L}$. By this inner product, one may further define the norm $\|u\|$ of a vector $u \in \mathcal{L}$ and the angle α between two vectors v and u of $\mathcal{L}$ as

$$\|u\| := \sqrt{\Phi(u,u)}, \qquad \cos\alpha = \tfrac{\Phi(v,u)}{\|v\|\,\|u\|}. \tag{4.4}$$

A vector $u \in \mathcal{L}$ is called normalized or a unit vector, if $\Phi(u,u) = 1$. Two vectors v and u of $\mathcal{L}$ are called orthogonal, if $\Phi(v,u) = 0$.

There is still another interesting question: Is there any way to map elements from $\mathcal{L}$ to its dual $\mathcal{L}^\star$ in a uniquely determined, natural way? This question may positively be answered, as soon as an inner product has been specified. For that, consider v as a parameter in (4.3). The resulting function f_v

$$f_v(u) := \Phi(v,u), \quad f_v : \mathcal{L} \to \mathbb{R} \tag{4.5}$$

maps u to a real number, and is itself therefore an element of $\mathcal{L}^\star$, $f_v \in \mathcal{L}^\star$. Now remember how we specified f_v: We picked up an arbitrary $v \in \mathcal{L}$ to get via (4.5) the associated $f_v(\cdot) \in \mathcal{L}^\star$. Precisely this map,

$$h : v \to f_v, \quad \mathcal{L} \to \mathcal{L}^\star, \tag{4.6}$$

is called the *natural isomorphism* from $\mathcal{L}$ to $\mathcal{L}^\star$. It is clearly a linear map, and it can be shown that it is one-to-one and has an inverse h^{-1}.

As for $\mathcal{L}$ in (4.3), we may also define an a inner product Ψ on $\mathcal{L}^\star$,

$$\Psi : \mathcal{L}^\star \times \mathcal{L}^\star \to \mathbb{R}, \quad (f,r) \to \Psi(f,r). \tag{4.7}$$

One possible choice — which we take — is to make it compatible with the one on $\mathcal{L}$ from (4.3) in the sense that norm and angles are preserved under the map (4.6),

$$\Psi(f,r) := \Phi(h^{-1}(f), h^{-1}(r)). \tag{4.8}$$

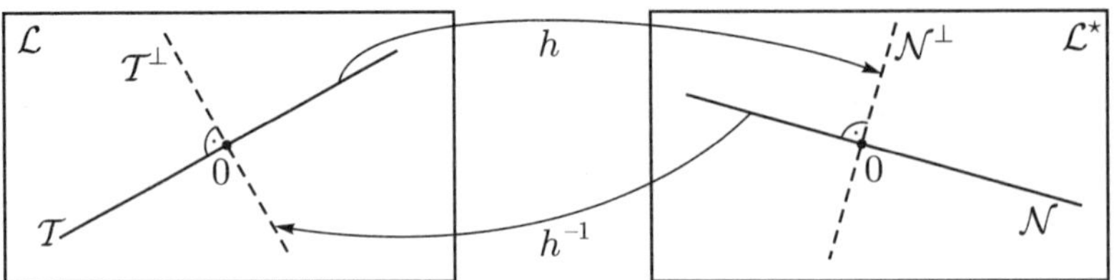

Figure 16. Orthogonal subspaces.

By the natural isomorphism (4.6) we have now different possibilities to express the duality pairing $f(u) = \langle f, u\rangle$. Let $v, u \in \mathcal{L}$ and $f = h(v)$, $r = h(u)$ the associated elements in $\mathcal{L}^\star$. Then

$$\langle f, u\rangle = \Phi(v, u) = \Psi(f, r), \tag{4.9}$$

which is obtained from (4.5), (4.6) and (4.8).

Let us return to the annihilator (4.2), from which we may now easily derive the concept of orthogonal subspaces. The pair of annihilating subspaces $\mathcal{T}$ and $\mathcal{N}$ fulfil likewise the conditions $\langle f, u\rangle = 0$ for all $u \in \mathcal{T} \subset \mathcal{L}$ and all $f \in \mathcal{N} \subset \mathcal{L}^\star$. Map now $\mathcal{T}$ to $\mathcal{L}^\star$ and $\mathcal{N}$ to $\mathcal{L}$ under the natural isomorphism (4.6) as shown in Figure 16 to obtain as images

$$\mathcal{N}^\perp := h(\mathcal{T}) \subset \mathcal{L}^\star, \qquad \mathcal{T}^\perp := h^{-1}(\mathcal{N}) \subset \mathcal{L}. \tag{4.10}$$

Since $\langle f, u\rangle = 0$ for all annihilator elements, it is immediately obvious from (4.9) that the elements of $\mathcal{T}$ and $\mathcal{T}^\perp$ are orthogonal to each other. We further see by the dimensions of $\mathcal{T}$ and $\mathcal{T}^\perp = h^{-1}(\mathcal{N})$ from Figure 15 that $\mathcal{L} = \mathcal{T} \oplus \mathcal{T}^\perp$. The same, of course, holds true for $\mathcal{N}$ and $\mathcal{N}^\perp$. The space $\mathcal{T}^\perp$ is called the *orthogonal complement* of $\mathcal{T}$ in $\mathcal{L}$, and $\mathcal{N}^\perp$ the orthogonal complement of $\mathcal{N}$ in $\mathcal{L}^\star$. We have thus obtained pairs of orthogonal spaces in $\mathcal{L}$ and $\mathcal{L}^\star$, which we denote by

$$\mathcal{T} \perp \mathcal{T}^\perp, \qquad \mathcal{N} \perp \mathcal{N}^\perp. \tag{4.11}$$

From now, the symbol $\perp$ is only used when orthogonality has explicitly been declared with respect to a certain inner product. Further, we only speak about orthogonality from now, if the elements to be considered are *from the same linear space*.

4.2 Coordinates

As soon as a basis e_i in $\mathcal{L}$ has been chosen, all the expressions from the preceding section may be evaluated in coordinates. In particular, we write the coordinate representation of $u \in \mathcal{L}$ as $u = \sum_i u^i\, e_i$ and denote the coordinates of u by $(u^i) =: \mathbf{u} \in \mathbb{R}^n$. The coordinate representation is a linear map $u \to \mathbf{u}$, $\mathcal{L} \to \mathbb{R}^n$. Instead of $\mathcal{L}$ and $\mathcal{L}^\star$, we have now to work with $\mathbb{R}^n$ and $\mathbb{R}^{n\star} = \mathbb{R}^{n\mathsf{T}}$. However, in order not to overload notation, we will stick to the same symbols as in Section 4.1 and will further write $\mathbf{u} \in \mathcal{L}$.

We start by evaluating the inner product (4.3) in terms of the coordinates of v and u. One obtains

$$\Phi(v, u) = \mathbf{v}^\mathsf{T}\mathbf{M}\mathbf{u} \tag{4.12}$$

where the entries m_{ij} of the symmetric and positive definite matrix $\mathbf{M}$ are calculated from the inner product of the associated basis vectors, $m_{ij} = \Phi(e_i, e_j)$. As a consequence, the norm and angle in (4.4) may also directly be computed from the coordinates of v and u,

$$\|\mathbf{u}\| := \|u\| = \sqrt{\mathbf{u}^\mathsf{T}\mathbf{M}\mathbf{u}}, \qquad \cos\alpha = \frac{\mathbf{v}^\mathsf{T}\mathbf{M}\mathbf{u}}{\|\mathbf{v}\|\,\|\mathbf{u}\|}. \tag{4.13}$$

In order to shorten the notation in (4.12), we will often just write $\Phi(v,u) = \mathbf{v}\cdot\mathbf{u}$. However, it must then be clear from the context which matrix has to be used.

Next, let us look how the natural isomorphism (4.6) works for coordinates. We take the bilinearform Φ in (4.12) now as a function of the *coordinates* $\mathbf{v}$ and $\mathbf{u}$, chose according to (4.5) the *coordinates* $\mathbf{v}$ as a parameter, and identify the associated $\mathbf{f}_\mathbf{v}^\mathsf{T}$,

$$\Phi(\mathbf{v},\mathbf{u}) = \underbrace{\mathbf{v}^\mathsf{T}\mathbf{M}}_{\mathbf{f}_\mathbf{v}^\mathsf{T}}\mathbf{u} \quad \Rightarrow \quad \mathbf{f}_\mathbf{v}^\mathsf{T} = \mathbf{v}^\mathsf{T}\mathbf{M}. \tag{4.14}$$

In this case, the map $h : \mathbf{v} \to \mathbf{f}_\mathbf{v}^\mathsf{T}$, $\mathcal{L} \to \mathcal{L}^\star$ is simply given by $\mathbf{f}_\mathbf{v}^\mathsf{T} = \mathbf{v}^\mathsf{T}\mathbf{M}$. We recognize that the result $\mathbf{f}_\mathbf{v}^\mathsf{T} \in \mathcal{L}^\star$ is a row. In coordinates, we may therefore think about $\mathcal{L}$ as the space of columns and $\mathcal{L}^\star$ as the associated space of rows as already indicated in Figure 14. As a consequence, the duality pairing takes the form

$$\langle \mathbf{f}, \mathbf{u}\rangle = \mathbf{f}^\mathsf{T}\,\mathbf{u}, \tag{4.15}$$

where $\mathbf{f}$ are the coordinates of f with respect to the *dual basis* ε^j of e_i, i.e. the basis ε^j in $\mathcal{L}^\star$ for which $\langle \varepsilon^j, e_i\rangle = \delta_i^j$.

It is impossible to write consistently elements from $\mathcal{L}$ as columns and elements from $\mathcal{L}^\star$ as transposed columns within matrix vector notation, which can already be seen at the inner product (4.12). The important thing is to know to which of the spaces a certain n-tuple belongs in order to apply the correct transformation rules. Let now be $\mathbf{f} = \mathbf{M}\mathbf{v}$ and $\mathbf{r} = \mathbf{M}\mathbf{u}$ with $\mathbf{v},\mathbf{u} \in \mathcal{L}$ and $\mathbf{f},\mathbf{r} \in \mathcal{L}^\star$. According to (4.8), the induced inner product on $\mathcal{L}^\star$ becomes

$$\Psi(\mathbf{f},\mathbf{r}) = \Phi(\mathbf{M}^{-1}\mathbf{f}, \mathbf{M}^{-1}\mathbf{r}) = \mathbf{f}^\mathsf{T}\mathbf{M}^{-1}\mathbf{M}\mathbf{M}^{-1}\mathbf{r} = \mathbf{f}^\mathsf{T}\mathbf{M}^{-1}\mathbf{r}, \tag{4.16}$$

and the duality pairing (4.15) may be equivalently written by (4.9) as

$$\mathbf{f}^\mathsf{T}\mathbf{u} = \mathbf{v}^\mathsf{T}\mathbf{M}\mathbf{u} = \mathbf{f}^\mathsf{T}\mathbf{M}^{-1}\mathbf{r} \quad \text{or, in short, as} \quad \mathbf{f}^\mathsf{T}\mathbf{u} = \mathbf{v}\cdot\mathbf{u} = \mathbf{f}\cdot\mathbf{r}. \tag{4.17}$$

These transformation rules are best known in physics, but also in optimization theory when switching from primal to dual formulations.

Let us finally state the orthogonal complements from (4.11) in coordinates, for which the annihilator (4.2) takes the form

$$\mathcal{N} = \{\mathbf{f}^\mathsf{T} \in \mathcal{L}^\star \mid \mathbf{f}^\mathsf{T}\,\mathbf{u} = 0 \quad \forall \mathbf{u} \in \mathcal{T}\}. \tag{4.18}$$

The subspaces $\mathcal{N}^\perp$ and $\mathcal{T}^\perp$ are then simply obtained from (4.10) as

$$\mathcal{N}^\perp = \mathbf{M}\mathcal{T}, \qquad \mathcal{T}^\perp = \mathbf{M}^{-1}\mathcal{N}. \tag{4.19}$$

As a result on the mutual orthogonality of $\mathcal{T}$ and $\mathcal{T}^\perp$, it holds that $\mathbf{u}\cdot\mathbf{v} = 0$ for all $\mathbf{u} \in \mathcal{T}$ and all $\mathbf{v} \in \mathcal{T}^\perp$ and, in symmetrical form on $\mathcal{L}^\star$, $\mathbf{r}\cdot\mathbf{f} = 0$ for all $\mathbf{r} \in \mathcal{N}^\perp$ and all $\mathbf{f} \in \mathcal{N}$.

4.3 Mutual Polarity and Pairs of Orthogonal Cones

In this section, we extend the concept of orthogonal spaces to pairs of orthogonal cones and state the associated orthogonal decomposition theorem. From now, we express everything in coordinates.

A subset $\mathcal{C}$ of a finite-dimensional vector space $\mathcal{L}$ is said to be *convex* if $(1-\lambda)\,\mathbf{x}+\lambda\,\mathbf{y} \in \mathcal{C}$ whenever $\mathbf{x} \in \mathcal{C}$, $\mathbf{y} \in \mathcal{C}$ and $0 < \lambda < 1$. Examples of convex sets are $\mathcal{L}$, every subspace of $\mathcal{L}$, the set $\{0\}$, half-spaces of $\mathcal{L}$, half-lines, and every translate and every intersection of convex sets. In particular by the translation and intersection property, we can easily generate new convex sets, such as polyhedral sets or the orthants of $\mathbb{R}^n$.

A subset $\mathcal{K}$ of $\mathcal{L}$ is called a *cone*, if $\lambda\mathbf{x} \in \mathcal{K}$ for $\mathbf{x} \in \mathcal{K}$ and $\lambda \geq 0$. Cones are therefore collections of half-lines emanating from the origin. Examples of cones are $\mathcal{L}$, every subspace of $\mathcal{L}$, the set $\{0\}$, half-spaces of $\mathcal{L}$, half-lines, and every *union* of cones. Note that the translate of a cone is in general no longer a cone, because the tip of the cone will be shifted away from the origin. Note also that cones are not necessarily pointed.

A *convex cone* is a convex set which is, at the same time, a cone. Examples of convex cones are $\mathcal{L}$, every subspace of $\mathcal{L}$, the set $\{0\}$, half-spaces of $\mathcal{L}$, half-lines, and every intersection of convex cones.

We may now extend the concept of annihilating subspaces (4.2), (4.18) to cones which are mutually polar to each other. For that, let $\mathcal{T}$ be a closed convex cone in $\mathcal{L}$. The set $\mathcal{N}$ in $\mathcal{L}^\star$ defined by

$$\mathcal{N} = \{\mathbf{f}^\mathsf{T} \in \mathcal{L}^\star \mid \mathbf{f}^\mathsf{T}\,\mathbf{u} \leq 0 \quad \forall \mathbf{u} \in \mathcal{T}\} \tag{4.20}$$

is called the *polar cone* of $\mathcal{T}$. It is a closed convex cone in $\mathcal{L}^\star$, see Figure 17. Note that (4.20) is indeed an extension of (4.18), because we may still chose for $\mathcal{T}$ a subspace of $\mathcal{L}$. For this particular case, $\mathbf{u} \in \mathcal{T}$ implies $-\mathbf{u} \in \mathcal{T}$, and (4.20) holds as an equality.

If an inner product is available on $\mathcal{L}$, then $\mathcal{T}$ and $\mathcal{N}$ may be mapped by the natural isomorphism to their associated duals

$$\mathcal{N}^\perp := \mathbf{M}\mathcal{T}, \qquad \mathcal{T}^\perp := \mathbf{M}^{-1}\mathcal{N}, \tag{4.21}$$

just in the same fashion as we did it for the annihilating subspaces in (4.10) and (4.19). We then obtain what we call a *pair of orthogonal cones* $\mathcal{T}, \mathcal{T}^\perp$ in $\mathcal{L}$, and another one $\mathcal{N}, \mathcal{N}^\perp$ in $\mathcal{L}^\star$, see Figure 18. As a consequence on (4.20), (4.21) it holds that

$$\begin{array}{ll} \forall \mathbf{u} \in \mathcal{T}, \quad \forall \mathbf{v} \in \mathcal{T}^\perp : & \forall \mathbf{r} \in \mathcal{N}^\perp, \quad \forall \mathbf{f} \in \mathcal{N} : \\ \mathbf{u}^\mathsf{T}\mathbf{M}\mathbf{v} \leq 0 \quad \text{or} \quad \mathbf{u}\cdot\mathbf{v} \leq 0 & \mathbf{r}^\mathsf{T}\mathbf{M}^{-1}\mathbf{f} \leq 0 \quad \text{or} \quad \mathbf{r}\cdot\mathbf{f} \leq 0. \end{array} \tag{4.22}$$

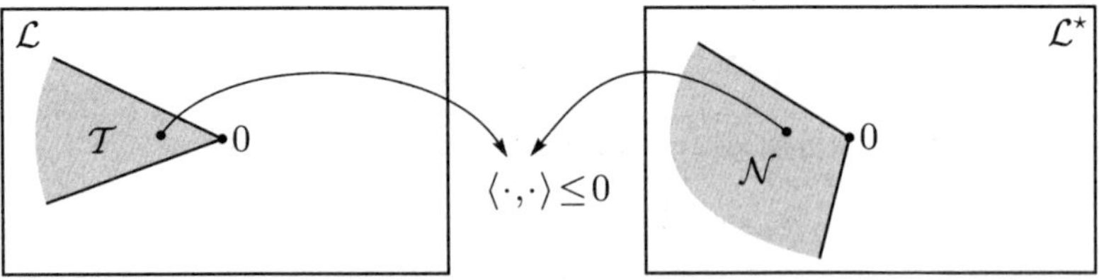

Figure 17. The polar cone $\mathcal{N} \in \mathcal{L}^\star$ of the convex cone $\mathcal{T} \in \mathcal{L}$.

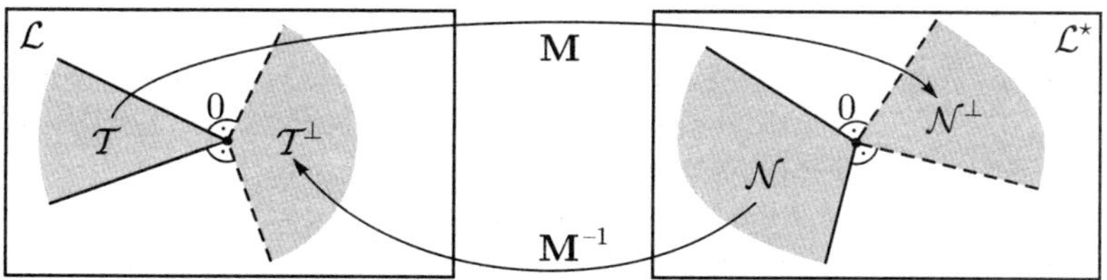

Figure 18. Pairs of orthogonal cones.

Based on (4.22), we review now the decomposition of a vector with respect to a pair of orthogonal cones. The following statement is contained as a special case in Moreau's Theorem, see e.g. Rockafellar (1972) for the full version and the proof, and generalizes the classical orthogonal vector decomposition.

Theorem 4.1. *Let $\mathcal{L}$ be a finite-dimensional real inner product space, and $\mathcal{T}, \mathcal{T}^\perp$ a pair of orthogonal closed convex cones in $\mathcal{L}$. Any $\mathbf{u} \in \mathcal{L}$ can then be decomposed uniquely into a sum $\mathbf{u} = \mathbf{v} + \mathbf{v}^\perp$ such that*

$$\mathbf{v} \in \mathcal{T}, \quad \mathbf{v}^\perp \in \mathcal{T}^\perp, \quad \mathbf{v} \cdot \mathbf{v}^\perp = 0. \tag{4.23}$$

This decomposition is depicted in the left part of Figure 19. For $\mathbf{u} \notin \mathcal{T} \cup \mathcal{T}^\perp$ one obtains for the decomposition $\mathbf{v} \neq 0$, $\mathbf{v}^\perp \neq 0$, whereas $\mathbf{v}^\perp = 0$ or $\mathbf{v} = 0$ as soon as $\mathbf{u} \in \mathcal{T}$ or $\mathbf{u} \in \mathcal{T}^\perp$. Apparently is $\mathbf{v}$ the nearest point to $\mathbf{u}$ in the set $\mathcal{T}$, which is usually denoted by $\mathbf{v} = \text{prox}_\mathcal{T}(\mathbf{u})$. The corresponding map $\mathbf{u} \to \text{prox}_\mathcal{T}(\mathbf{u})$ is called a proximation which is, in fact, a projection because $\text{prox}^2_\mathcal{T}(\mathbf{u}) = \text{prox}_\mathcal{T}(\mathbf{u})$. In addition, this projection is orthogonal since $\text{prox}_\mathcal{T}(\mathbf{u}) \cdot (\mathbf{u} - \text{prox}_\mathcal{T}(\mathbf{u})) = 0$. Of course, the same properties apply for the second term $\mathbf{v}^\perp$ in the decomposition due to symmetry, i.e. $\mathbf{v}^\perp = \text{prox}_{\mathcal{T}^\perp}(\mathbf{u})$. The proof of the orthogonal decomposition theorem will be provided in Section 4.5.

Contained in Theorem 4.1 as a special case is the decomposition of a vector with respect to a pair of orthogonal subspaces (4.11). For this particular situation, the orthogonality condition $\mathbf{v} \cdot \mathbf{v}^\perp = 0$ in (4.23) is automatically fulfilled as indicated in the right diagram of Figure 19.

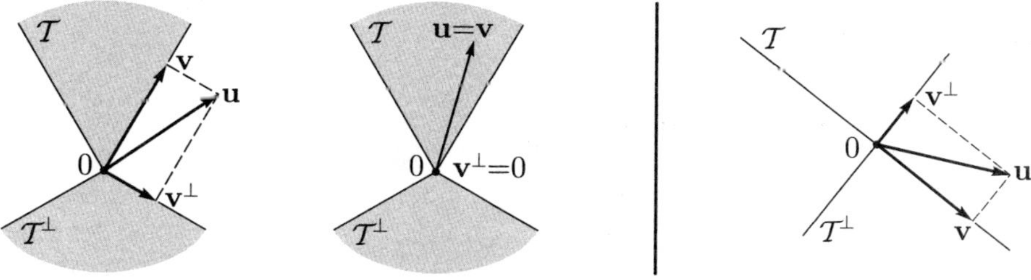

Figure 19. Orthogonal decomposition.

4.4 Cones to Sets

The concept of linearization plays a crucial role in analysis. However linearization fails, if the objects to deal with become non-smooth. The next step is to look for convexification. Instead of working with linear spaces one has to deal with cones which are convex at best. The cones introduced in the following characterize sets in the neighborhood of a chosen point and indicate whether convexification is possible.

We consider an n-dimensional real manifold $\mathcal{M}$ and denote by $\mathbf{q}$ the n-tuple of local variables in use. The tangent and cotangent space to $\mathcal{M}$ at a point $\mathbf{q}$ are denoted by $\mathcal{L}_{\mathcal{M}}(\mathbf{q})$ and $\mathcal{L}^{\star}_{\mathcal{M}}(\mathbf{q})$, respectively. Let $\mathcal{C}$ be a non-empty subset of $\mathcal{M}$. Following Aubin and Ekeland (1984) the *contingent cone* $\mathcal{K}_{\mathcal{C}}$ and the *tangent cone* $\mathcal{T}_{\mathcal{C}}$ to $\mathcal{C}$ at a point $\mathbf{q} \in \mathcal{C}$, defined in terms of sequences, are

$$\mathcal{K}_{\mathcal{C}}(\mathbf{q}) := \{\mathbf{u} \mid \exists \mu_n \downarrow 0 \quad \exists \mathbf{u}_n \to \mathbf{u} \quad \text{satisfying } \mathbf{q} + \mu_n \mathbf{u}_n \in \mathcal{C} \quad \forall n\}, \tag{4.24}$$

$$\mathcal{T}_{\mathcal{C}}(\mathbf{q}) := \{\mathbf{u} \mid \forall \mathbf{q}_n \overset{c}{\to} \mathbf{q} \quad \forall \mu_n \downarrow 0 \quad \exists \mathbf{u}_n \to \mathbf{u} \quad \text{satisfying } \mathbf{q}_n + \mu_n \mathbf{u}_n \in \mathcal{C} \quad \forall n\}, \tag{4.25}$$

where $\mathbf{q}_n \overset{c}{\to} \mathbf{q}$ denotes the convergence of $\mathbf{q}_n$ to $\mathbf{q}$ in $\mathcal{C}$. We set $\mathcal{K}_{\emptyset} = \emptyset$, $\mathcal{T}_{\emptyset} = \emptyset$ and $\mathcal{K}_{\mathcal{C}}(\mathbf{q}) = \emptyset$, $\mathcal{T}_{\mathcal{C}}(\mathbf{q}) = \emptyset$ if $\mathbf{q} \notin \mathcal{C}$. Some details on how to construct the cones $\mathcal{K}_{\mathcal{C}}(\mathbf{q})$ and $\mathcal{T}_{\mathcal{C}}(\mathbf{q})$ at a given point $\mathbf{q} \in \mathcal{C}$ are presented in Figure 20: We first translate the set $\mathcal{C}$ such that the point $\mathbf{q}$ is identified with the origin. According to (4.24), $\mathbf{u}$ belongs to contingent cone $\mathcal{K}_{\mathcal{C}}(\mathbf{q})$ if we are able to find any sequence $\mu_n \downarrow 0$, $\mathbf{u}_n \to \mathbf{u}$ such that $\mu_n \mathbf{u}_n \in \mathcal{C} - \mathbf{q}$. Such a sequence is depicted in the left diagram. The middle diagram describes how the tangent cone $\mathcal{T}_{\mathcal{C}}(\mathbf{q})$ is obtained. In order to show that the points $\mathbf{u}$ in the dark grey area do *not* belong to $\mathcal{T}_{\mathcal{C}}(\mathbf{q})$, it is sufficient to find in (4.25) any sequence $\mathbf{q}_n \overset{c}{\to} \mathbf{q}$, $\mu_n \downarrow 0$ which excludes convergence of $\mathbf{u}_n$ to $\mathbf{u}$. One takes, for example, a sequence $\mathbf{q}_n$ of boundary points of $\mathcal{C}$ that converges to $\mathbf{q}$. The resulting cones are depicted in the right diagram of Figure 20.

Both cones are closed subsets of $\mathcal{L}_{\mathcal{M}}(\mathbf{q})$ and, if not empty, they always contain 0. Furthermore, $\mathcal{T}_{\mathcal{C}}(\mathbf{q})$ is convex and is always contained in $\mathcal{K}_{\mathcal{C}}(\mathbf{q})$, which is obvious when setting $\mathbf{q}_1 = \mathbf{q}_2 = \ldots = \mathbf{q}$ in (4.25) and comparing it with (4.24). For $\mathcal{C}$ being convex we

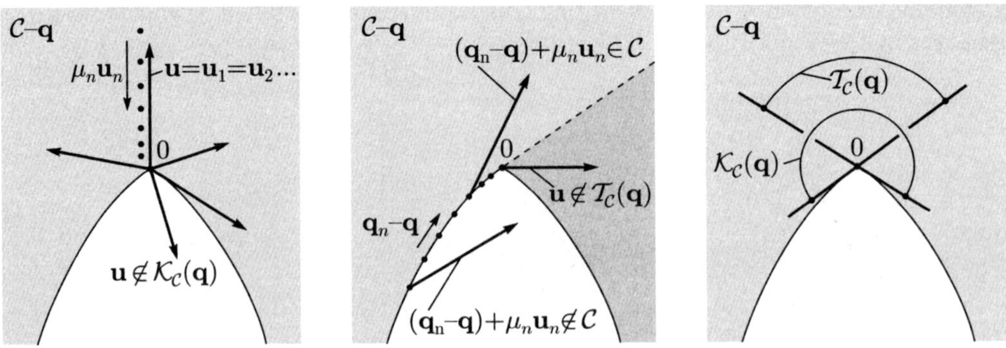

Figure 20. Construction of the cones $\mathcal{K}_{\mathcal{C}}(\mathbf{q})$ and $\mathcal{T}_{\mathcal{C}}(\mathbf{q})$ and at a point $\mathbf{q} \in \mathcal{C}$.

have $\mathcal{T}_\mathcal{C}(\mathbf{q}) = \mathcal{K}_\mathcal{C}(\mathbf{q})$. If $\mathcal{C}$ is a smooth sub-manifold of class C^1, then $\mathcal{T}_\mathcal{C}(\mathbf{q}) = \mathcal{K}_\mathcal{C}(\mathbf{q})$ is the usual tangent space $\mathcal{L}_\mathcal{C}(\mathbf{q})$ to the sub-manifold $\mathcal{C}$ at $\mathbf{q}$. Note also that for $\mathbf{q} \in \text{int}\, \mathcal{C}$ both cones become $\mathcal{T}_\mathcal{C}(\mathbf{q}) = \mathcal{K}_\mathcal{C}(\mathbf{q}) = \mathcal{L}_\mathcal{M}(\mathbf{q})$, which is the tangent space to $\mathcal{M}$ at $\mathbf{q}$. A set $\mathcal{C}$ is called to be *tangentially regular* at a point $\mathbf{q}$ if $\mathcal{T}_\mathcal{C}(\mathbf{q}) = \mathcal{K}_\mathcal{C}(\mathbf{q})$. Tangential regularity holds in particular for the cases mentioned above, i.e. when $\mathcal{C}$ is a convex set or a smooth sub-manifold, or when $\mathbf{q}$ is not a boundary point of $\mathcal{C}$. For non-convex sets we may have $\mathcal{T}_\mathcal{C}(\mathbf{q}) = \mathcal{K}_\mathcal{C}(\mathbf{q})$, but also points $\mathbf{q} \in \mathcal{C}$ at which this equality does not apply. These points are called the *re-entrant corner points* of $\mathcal{C}$.

Since $\mathcal{T}_\mathcal{C}(\mathbf{q})$ is a closed convex cone we may define an object playing the role of normal vectors to $\mathcal{C}$ at $\mathbf{q}$. This object is called the *normal cone* to $\mathcal{C}$ at $\mathbf{q}$ and is defined as the polar of $\mathcal{T}_\mathcal{C}(\mathbf{q})$ via the variational inequality

$$\mathcal{N}_\mathcal{C}(\mathbf{q}) = \{\mathbf{f} \mid \mathbf{f}^\mathsf{T}\mathbf{u} \leq 0 \quad \forall \mathbf{u} \in \mathcal{T}_\mathcal{C}(\mathbf{q})\}. \tag{4.26}$$

We set $\mathcal{N}_\emptyset = \emptyset$ and $\mathcal{N}_\mathcal{C}(\mathbf{q}) = \emptyset$ if $\mathbf{q} \notin \mathcal{C}$. The normal cone is a closed, convex cone in the cotangent space $\mathcal{L}^\star_\mathcal{M}(\mathbf{q})$, always containing 0 if not empty. The cones $\mathcal{T}_\mathcal{C}(\mathbf{q})$ and $\mathcal{N}_\mathcal{C}(\mathbf{q})$ are mutually polar by (4.26). One observes that $\mathcal{N}_\mathcal{C}(\mathbf{q}) = \mathcal{L}^\star_\mathcal{M}(\mathbf{q}) \Leftrightarrow \mathcal{T}_\mathcal{C}(\mathbf{q}) = 0$ and vice versa. If $\mathcal{T}_\mathcal{C}(\mathbf{q})$ is a half-space then $\mathcal{N}_\mathcal{C}(\mathbf{q})$ degenerates to a half-line. If $\mathcal{T}_\mathcal{C}(\mathbf{q})$ is a subspace of $\mathcal{L}_\mathcal{M}(\mathbf{q})$, then $\mathcal{N}_\mathcal{C}(\mathbf{q})$ becomes the associated annihilator as a subspace of $\mathcal{L}^\star_\mathcal{M}(\mathbf{q})$.

For a Riemannian manifold $\mathcal{M}$, we denote by $\mathbf{M}(\mathbf{q})$ the natural isomorphism induced by the metric on $\mathcal{M}$, which takes at $\mathbf{q}$ elements from the tangent to the cotangent space. We introduce, in addition, the image and pre-image of the cones $\mathcal{T}_\mathcal{C}(\mathbf{q})$ and $\mathcal{N}_\mathcal{C}(\mathbf{q})$,

$$\mathcal{N}^\perp_\mathcal{C}(\mathbf{q}) := \mathbf{M}(\mathbf{q})\,\mathcal{T}_\mathcal{C}(\mathbf{q}), \quad \mathcal{T}^\perp_\mathcal{C}(\mathbf{q}) := \mathbf{M}^{-1}(\mathbf{q})\,\mathcal{N}_\mathcal{C}(\mathbf{q}). \tag{4.27}$$

As in (4.22) we have now obtained two pairs of orthogonal cones $\mathcal{T}_\mathcal{C}(\mathbf{q}), \mathcal{T}^\perp_\mathcal{C}(\mathbf{q})$ and $\mathcal{N}_\mathcal{C}(\mathbf{q}), \mathcal{N}^\perp_\mathcal{C}(\mathbf{q})$ which are characterized by

$$\begin{aligned} \mathbf{v} \cdot \mathbf{v}^\perp &\leq 0 \quad \forall \mathbf{v} \in \mathcal{T}_\mathcal{C}(\mathbf{q}),\ \forall \mathbf{v}^\perp \in \mathcal{T}^\perp_\mathcal{C}(\mathbf{q}), \\ \mathbf{r} \cdot \mathbf{r}^\perp &\leq 0 \quad \forall \mathbf{r} \in \mathcal{N}_\mathcal{C}(\mathbf{q}),\ \forall \mathbf{r}^\perp \in \mathcal{N}^\perp_\mathcal{C}(\mathbf{q}), \end{aligned} \tag{4.28}$$

where the dots denote the associated inner products on the tangent and the cotangent space, respectively.

In order to get a better sense how the cones $\mathcal{K}_\mathcal{C}(\mathbf{q})$, $\mathcal{T}_\mathcal{C}(\mathbf{q})$ and $\mathcal{T}^\perp_\mathcal{C}(\mathbf{q})$ are arranged, we refer to Figure 21 where different situations for convex and non-convex sets are shown. For convex sets $\mathcal{C}$ tangential regularity $\mathcal{T}_\mathcal{C} = \mathcal{K}_\mathcal{C}$ holds at every point $\mathbf{q} \in \mathcal{C}$, see configurations (a)–(d). In particular, (a) shows the situation at a smooth boundary, for which $\mathcal{T}_\mathcal{C} = \mathcal{K}_\mathcal{C}$ becomes a half-space and $\mathcal{T}^\perp_\mathcal{C}$ degenerates to a half-line. For $\mathbf{q} \in \text{int}\, \mathcal{C}$ as in (d), one has $\mathcal{K}_\mathcal{C} = \mathcal{T}_\mathcal{C} = \mathcal{L}_\mathcal{M}$ and $\mathcal{T}^\perp_\mathcal{C} = \{0\}$. Non-convex sets $\mathcal{C}$ may be subdivided up into two classes depending on the shape of the boundary. Configurations for which the boundary is tangentially regular at some point $\mathbf{q}$ are depicted in (e)–(h). For these cases we have $\mathcal{T}_\mathcal{C} = \mathcal{K}_\mathcal{C}$, including, for example, smooth sub-manifolds (d) for which $\mathcal{T}_\mathcal{C} = \mathcal{K}_\mathcal{C}$ is recognized to be the classical tangent space of $\mathcal{C}$ at $\mathbf{q}$ and $\mathcal{T}^\perp_\mathcal{C}$ its orthogonal complement. Re-entrant corners ($\mathcal{T}_\mathcal{C} \subset \mathcal{K}_\mathcal{C}$) at $\mathbf{q}$ are depicted in (i)–(l). Note in particular the case (l) of a non-smooth sub-manifold, at which $\mathcal{T}_\mathcal{C}$ may even reduce to 0 and $\mathcal{T}^\perp_\mathcal{C}$ becomes the whole $\mathcal{L}_\mathcal{M}$. This situation also corresponds to a re-entrant corner ($\mathcal{T}_\mathcal{C} \subset \mathcal{K}_\mathcal{C}$).

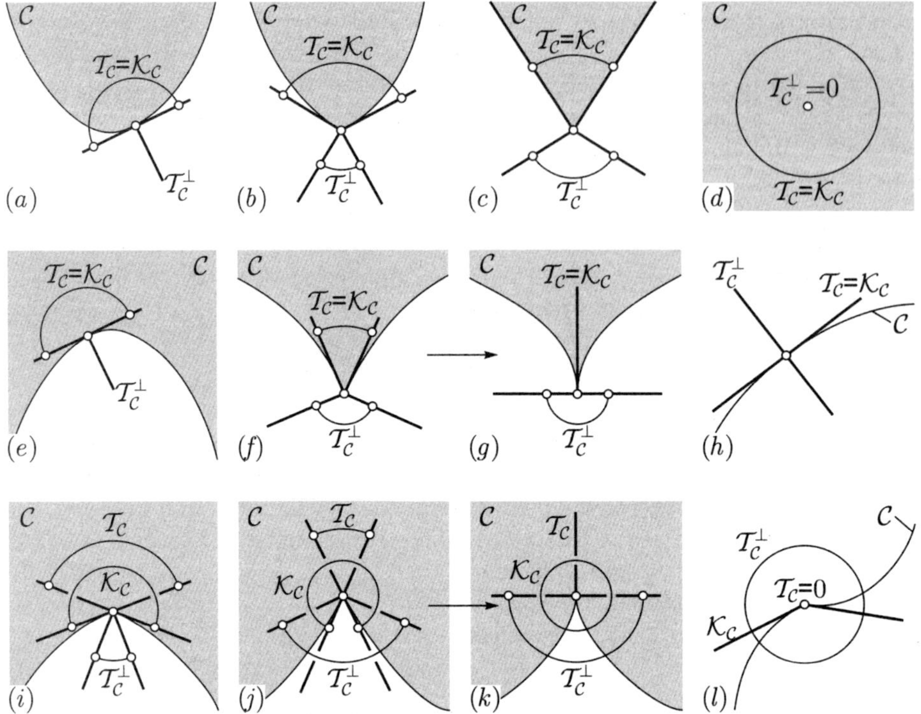

Figure 21. The cones $\mathcal{K}_\mathcal{C}$, $\mathcal{T}_\mathcal{C}$ and $\mathcal{T}_\mathcal{C}^\perp$ at different points of $\mathcal{C}$.

4.5 Cones to Convex Sets

The cones introduced in the preceding section are, in general, quite hard to get. We will therefore discuss only two cases which are of relevance for the geometric impact theory that will be presented in Section 5. The first case relates to convexity of $\mathcal{C} \subset \mathcal{M}$. Convexity as an intrinsic geometric property only makes sense if the associated manifold is flat. We therefore consider $\mathcal{M}$ now as an *affine space* and drop in the notation the point at which the tangent space is evaluated, i.e. we simply write $\mathcal{L}_\mathcal{M}$ for the tangent space of $\mathcal{M}$ without any further specification. For convex sets $\mathcal{C}$ the normal cone $\mathcal{N}_\mathcal{C}(\mathbf{q})$ to $\mathcal{C}$ at $\mathbf{q}$ becomes

$$\mathcal{L}_\mathcal{M}^\star \supset \mathcal{N}_\mathcal{C}(\mathbf{q}) = \{\mathbf{f} \mid \mathbf{f}^T(\mathbf{q}^\star - \mathbf{q}) \leq 0, \quad \mathbf{q} \in \mathcal{C}, \quad \forall \mathbf{q}^\star \in \mathcal{C}\}, \tag{4.29}$$

see e.g. Rockafellar (1972). Its polar, the tangent cone, is then available via the variational inequality

$$\mathcal{L}_\mathcal{M} \supset \mathcal{T}_\mathcal{C}(\mathbf{q}) = \{\mathbf{u} \mid \mathbf{f}^T\mathbf{u} \leq 0, \quad \forall \mathbf{f} \in \mathcal{N}_\mathcal{C}(\mathbf{q})\}, \tag{4.30}$$

cp. (4.26). As soon as a (constant) metric is defined on $\mathcal{M}$, the other two cones are obtained by (4.27), i.e. $\mathcal{N}_\mathcal{C}^\perp(\mathbf{q}) = \mathbf{M}\,\mathcal{T}_\mathcal{C}(\mathbf{q})$, $\mathcal{T}_\mathcal{C}^\perp(\mathbf{q}) = \mathbf{M}^{-1}\,\mathcal{N}_\mathcal{C}(\mathbf{q})$. The left part of Figure 22 shows $\mathcal{T}_\mathcal{C}^\perp(\mathbf{q}_i)$ at different points $\mathbf{q}_i$ of the convex set $\mathcal{C}$. The cone $\mathcal{T}_\mathcal{C}^\perp(\mathbf{q})$ consists of

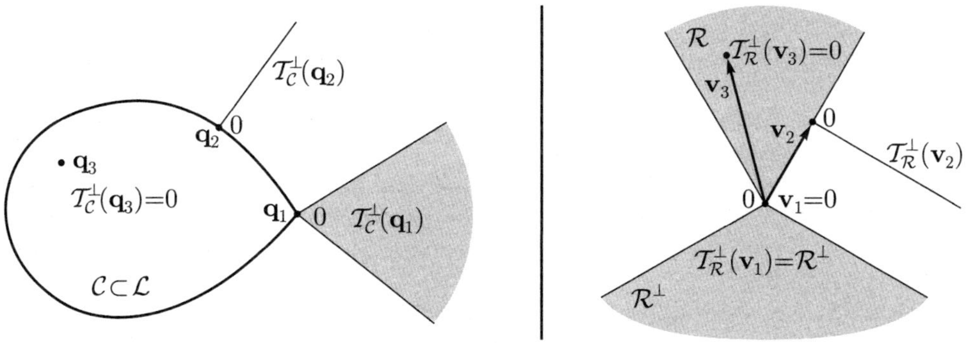

Figure 22. Cones to convex sets, such as convex cones.

all vectors $\mathbf{v} \in \mathcal{L}_{\mathcal{M}}$ which do not make an acute angle with any line segment $\mathbf{q}^{\star} - \mathbf{q}$ in $\mathcal{C}$ with $\mathbf{q}$ as endpoint.

As an example, let us consider the special case that the convex set $\mathcal{C} \subset \mathcal{M}$ is, in addition, a closed cone which we denote by $\mathcal{R}$. This requires to take $\mathcal{M}$ as a *linear space*, for which the associated tangent and the cotangent spaces can simply be regarded as $\mathcal{L}_{\mathcal{M}} = \mathcal{M}$ and $\mathcal{L}_{\mathcal{M}}^{\star} = \mathcal{M}^{\star}$. We want to perform the following task: Let $\mathcal{R}$ be a closed convex cone in an inner product space $\mathcal{M}$ and $\mathbf{v} \in \mathcal{R}$. Determine the cone $\mathcal{T}_{\mathcal{R}}^{\perp}(\mathbf{v}) \subset \mathcal{M}$. The result is depicted in the right diagram of Figure 22, for which we have proceeded in the same way as in the left figure. We observe in particular that $\mathbf{v} \cdot \mathcal{T}_{\mathcal{R}}^{\perp}(\mathbf{v}) = 0$ and $\mathcal{T}_{\mathcal{R}}^{\perp}(\mathbf{v}) \subset \mathcal{R}^{\perp}$ holds for all three cases depicted in this figure, where $\mathcal{R}^{\perp}$ is the cone orthogonal to $\mathcal{R}$. This brings us to the following lemma:

Lemma 4.2. *Let $\mathcal{R}$ and $\mathcal{R}^{\perp}$ be a pair of orthogonal closed convex cones in a finite-dimensional inner product space $\mathcal{M}$. Then the following conditions are equivalent to each other:*

$$\mathbf{v} \in \mathcal{R}, \quad \mathbf{v}^{\perp} \in \mathcal{R}^{\perp}, \quad \mathbf{v} \cdot \mathbf{v}^{\perp} = 0 \tag{4.31}$$

$$\mathbf{v} \in \mathcal{R}, \quad \mathbf{v}^{\perp} \in \mathcal{T}_{\mathcal{R}}^{\perp}(\mathbf{v}) \tag{4.32}$$

$$\mathbf{v}^{\perp} \in \mathcal{R}^{\perp}, \quad \mathbf{v} \in \mathcal{T}_{\mathcal{R}^{\perp}}^{\perp}(\mathbf{v}^{\perp}) \tag{4.33}$$

with $\mathcal{T}_{\mathcal{R}}^{\perp}(\mathbf{v})$ and $\mathcal{T}_{\mathcal{R}^{\perp}}^{\perp}(\mathbf{v}^{\perp})$ the cones orthogonal to the tangent cones to $\mathcal{R}$ at $\mathbf{v}$ and $\mathcal{R}^{\perp}$ at $\mathbf{v}^{\perp}$, respectively.

Proof. In order to show the equivalence of (4.31) and (4.32) we first recall that the cone $\mathcal{T}_{\mathcal{R}}^{\perp}(\mathbf{v})$ can be expressed via the second equation in (4.27) and (4.29) as

$$\mathcal{T}_{\mathcal{R}}^{\perp}(\mathbf{v}) = \{\mathbf{w} \mid \mathbf{w} \cdot (\mathbf{v}^{\star} - \mathbf{v}) \leq 0, \quad \mathbf{v} \in \mathcal{R}, \quad \forall \mathbf{v}^{\star} \in \mathcal{R}\}. \tag{4.34}$$

Further, we note that

$$\mathcal{R}^{\perp} = \{\mathbf{w} \mid \mathbf{w} \cdot \mathbf{v}^{\star} \leq 0, \quad \forall \mathbf{v}^{\star} \in \mathcal{R}\} \tag{4.35}$$

$$\mathcal{T}_{\mathcal{R}}^{\perp}(\mathbf{v}) = \{\mathbf{w} \mid \mathbf{w} \cdot \mathbf{z}^{\star} \leq 0, \quad \mathbf{v} \in \mathcal{R}, \quad \forall \mathbf{z}^{\star} \in \mathcal{R} - \mathbf{v}\} \tag{4.36}$$

$$\mathcal{T}_{\mathcal{R}}^{\perp}(\mathbf{v}) \subset \mathcal{R}^{\perp} \text{ with } \mathcal{T}_{\mathcal{R}}^{\perp}(0) = \mathcal{R}^{\perp} \tag{4.37}$$

where (4.35) is just the definition of $\mathcal{R}^{\perp}$, (4.36) is obtained from (4.34) when setting $\mathbf{z}^{\star} := \mathbf{v}^{\star} - \mathbf{v}$, and (4.37) holds because (4.36) is more restrictive on $\mathbf{w}$ than (4.35): The set $\mathcal{R} - \mathbf{v}$ is the translate of the closed convex cone $\mathcal{R}$ in the direction $-\mathbf{v}$ with $\mathbf{v} \in \mathcal{R}$, hence $\mathcal{R} - \mathbf{v} \supset \mathcal{R}$. Note, however, that $\mathcal{R} - \mathbf{v}$ is no longer a cone!

Suppose now that (4.31) holds. The second condition, $\mathbf{v}^{\perp} \in \mathcal{R}^{\perp}$, means by (4.35) that $\mathbf{v}^{\perp} \cdot \mathbf{v}^{\star} \leq 0 \; \forall \mathbf{v}^{\star} \in \mathcal{R}$. Subtracting from this expression the third condition $\mathbf{v}^{\perp} \cdot \mathbf{v} = 0$ in (4.31) gives $\mathbf{v}^{\perp} \cdot (\mathbf{v}^{\star} - \mathbf{v}) \leq 0 \; \forall \mathbf{v}^{\star} \in \mathcal{R}$. This is by (4.34) together with $\mathbf{v} \in \mathcal{R}$ already $\mathbf{v}^{\perp} \in \mathcal{T}_{\mathcal{R}}^{\perp}(\mathbf{v})$ in (4.32).

To show the converse we rewrite $\mathbf{v}^{\perp} \in \mathcal{T}_{\mathcal{R}}^{\perp}(\mathbf{v})$ in (4.32) with the help of (4.34) as a variational inequality, $\mathbf{v}^{\perp} \cdot (\mathbf{v}^{\star} - \mathbf{v}) \leq 0$, which has to hold for any $\mathbf{v}^{\star} \in \mathcal{R}$ when $\mathbf{v} \in \mathcal{R}$. Choose now $\mathbf{v}^{\star} = 2\mathbf{v} \in \mathcal{R}$ and $\mathbf{v}^{\star} = 0 \in \mathcal{R}$ which is possible because $\mathcal{R}$ is a closed cone and $\mathbf{v} \in \mathcal{R}$, and evaluate the variational expression for both cases. This gives $\mathbf{v}^{\perp} \cdot \mathbf{v} \leq 0$ on the one hand and $\mathbf{v}^{\perp} \cdot \mathbf{v} \geq 0$ on the other hand, hence $\mathbf{v}^{\perp} \cdot \mathbf{v} = 0$ which is the third condition in (4.31). It remains to show that $\mathbf{v}^{\perp} \in \mathcal{R}^{\perp}$. This is immediately obtained from (4.37), because $\mathbf{v}^{\perp} \in \mathcal{T}_{\mathcal{R}}^{\perp}(\mathbf{v})$.

Finally, to show equivalence of (4.31) and (4.33), one proceeds precisely in the same manner. □

By using this lemma, we may now prove the orthogonal decomposition (Theorem 4.1) from Section 4.3:

Proof of Theorem 4.1. We have to show that every $\mathbf{u}$ in an inner product space $\mathcal{M}$ can be uniquely decomposed into a sum $\mathbf{u} = \mathbf{v} + \mathbf{v}^{\perp}$ such that $\mathbf{v} \in \mathcal{R}$, $\mathbf{v}^{\perp} \in \mathcal{R}^{\perp}$ and $\mathbf{v} \cdot \mathbf{v}^{\perp} = 0$. In other words, we have to show that for every $\mathbf{u} \in \mathcal{M}$ there exists a unique $\mathbf{v} \in \mathcal{R}$ such that $\mathbf{u} - \mathbf{v} \in \mathcal{R}^{\perp}$ and $\mathbf{v} \cdot (\mathbf{u} - \mathbf{v}) = 0$. The latter two conditions are by (4.31), (4.32) equivalent to the single condition $\mathbf{u} - \mathbf{v} \in \mathcal{T}_{\mathcal{R}}^{\perp}(\mathbf{v})$, which can be mapped via $\mathbf{M}$ to the dual $\mathcal{M}^{\star}$ to give $\mathbf{M}(\mathbf{u} - \mathbf{v}) \in \mathcal{N}_{\mathcal{R}}(\mathbf{v})$. It is known from convex analysis that the normal cone $\mathcal{N}_{\mathcal{R}}(\mathbf{v})$ to a convex set $\mathcal{R}$ at a point $\mathbf{v} \in \mathcal{R}$ may be equivalently expressed by the sub-differential of the associated indicator function $I_{\mathcal{R}}(\mathbf{v})$. We therefore have to show that for any $\mathbf{u} \in \mathcal{M}$ there exists a unique $\mathbf{v}$ fulfilling the inclusion $0 \in -\mathbf{M}(\mathbf{u} - \mathbf{v}) + \partial I_{\mathcal{R}}(\mathbf{v})$. This inclusion is recognized as the necessary and sufficient optimality condition of the strictly convex function $Z(\mathbf{v}) = \frac{1}{2}\|\mathbf{u} - \mathbf{v}\|^2 + I_{\mathcal{R}}(\mathbf{v})$, which minimizes the distance between $\mathbf{v} \in \mathcal{R}$ and $\mathbf{u}$ and which is known to have a unique optimal solution, denoted by $\mathbf{v} = \mathrm{prox}_{\mathcal{R}}(\mathbf{u})$. □

4.6 Cones to Simple Unilateral Constraints

In this section we will finally discuss the case that a Riemannian manifold $\mathcal{M}$ with metric $\mathbf{M}(\mathbf{q})$ is restricted to a non-empty subset $\mathcal{C}$ by k simple unilateral constraints $g^i(\mathbf{q}) \geq 0$ as depicted in Figure 23,

$$\mathcal{C} := \{\mathbf{q} \mid g^i(\mathbf{q}) \geq 0, \quad i = 1, \ldots, k\}. \tag{4.38}$$

We assume the functions $g^i : \mathcal{M} \to \mathbb{R}$ to be of class C^1 and their level curves $g^i(\mathbf{q}) = 0$ to intersect transversally. We denote by $\mathbf{w}^i(\mathbf{q})$ the differential of g^i and by $\mathcal{H}$ the indices

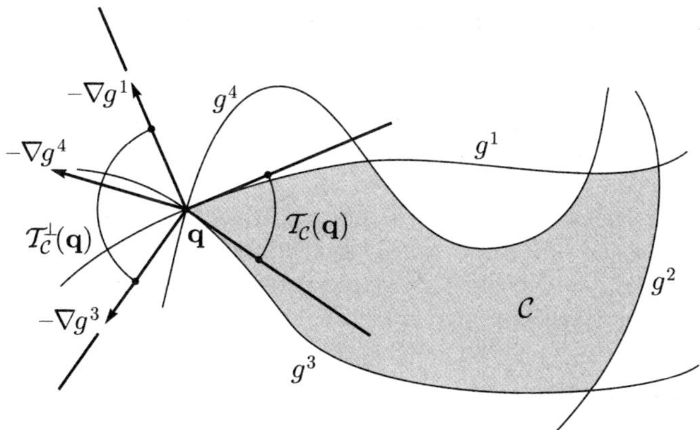

Figure 23. Tangent cone for simple unilateral constraints.

of the active set of constraints at a given point $\mathbf{q}$,

$$\mathcal{L}^\star_{\mathcal{M}}(\mathbf{q}) \ni \mathbf{w}^{i\mathsf{T}}(\mathbf{q}) := \tfrac{\partial g^i}{\partial \mathbf{q}}, \qquad \mathcal{H}(\mathbf{q}) := \{i \mid g^i(\mathbf{q}) = 0\}, \tag{4.39}$$

and assume that $\mathbf{w}^i \neq 0$ when $i \in \mathcal{H}$. The normal cone $\mathcal{N}_\mathcal{C}(\mathbf{q})$ to $\mathcal{C}$ at $\mathbf{q}$ is then finitely generated by the co-vectors $-\mathbf{w}^i$ of the active set,

$$\mathcal{L}^\star_{\mathcal{M}}(\mathbf{q}) \supset \mathcal{N}_\mathcal{C}(\mathbf{q}) = \Big\{\mathbf{f} \,\Big|\, \mathbf{f} = -\sum_{i\in\mathcal{H}} \lambda_i\, \mathbf{w}^i(\mathbf{q}), \quad \lambda_i \geq 0\Big\}. \tag{4.40}$$

The associated cone $\mathcal{T}_\mathcal{C}^\perp(\mathbf{q})$ in $\mathcal{L}_\mathcal{M}(\mathbf{q})$ becomes

$$\mathcal{L}_\mathcal{M}(\mathbf{q}) \supset \mathcal{T}_\mathcal{C}^\perp(\mathbf{q}) = \mathbf{M}^{-1}\mathcal{N}_\mathcal{C}(\mathbf{q}) = \Big\{\mathbf{v} \,\Big|\, \mathbf{v} = -\sum_{i\in\mathcal{H}} \lambda_i\, \nabla g^i(\mathbf{q}), \quad \lambda_i \geq 0\Big\}, \tag{4.41}$$

where $\nabla g^i(\mathbf{q})$ denotes the gradient of g^i at point $\mathbf{q}$, $\nabla g^i(\mathbf{q}) = \mathbf{M}^{-1}(\mathbf{q})\, \mathbf{w}^i(\mathbf{q})$.

The tangent cone $\mathcal{T}_\mathcal{C}(\mathbf{q})$ is now accessible through (4.28). It consists of all vectors $\mathbf{u}$ fulfilling $\mathbf{u}\cdot\mathbf{v} \leq 0$ for all $\mathbf{v} \in \mathcal{T}_\mathcal{C}^\perp(\mathbf{q})$, i.e. fulfilling $\sum_{i\in\mathcal{H}} -\lambda_i\, \nabla g^i(\mathbf{q})\cdot\mathbf{u} \leq 0$ for all $\lambda_i \geq 0$. The latter condition yields

$$\mathcal{L}_\mathcal{M}(\mathbf{q}) \supset \mathcal{T}_\mathcal{C}(\mathbf{q}) = \Big\{\mathbf{u} \,\Big|\, -\nabla g^i(\mathbf{q})\cdot\mathbf{u} \leq 0, \quad \forall i \in \mathcal{H}\Big\}, \tag{4.42}$$

which is obvious when choosing $\lambda_i = 1$ together with $\lambda_j = 0$ $(j \neq i)$ and cycling through all $i \in \mathcal{H}$.

It should be noted that the normal and the tangent cone associated with simple unilateral constraints are already polyhedral. More general situations occur if the set $\mathcal{C}$ of admissible displacements may not be obtained by the intersection of simple smooth inequality constraints, such as the circular cone $\{\,\lambda(\mathbf{e}_3 + \mathcal{D}_{12}),\ \lambda \geq 0\,\}$ in $\mathbb{R}^3$, where $\mathbf{e}_3$ denotes the unit vector in the 3-direction and $\mathcal{D}_{12}$ the closed unit disc centered at 0 in the 1-2-plane.

5 A Geometric Concept for Perfect Impacts

We have learned from Newton's cradle in Section 3.8 that not every impact event can be treated by the standard impact constitutive laws from non-smooth dynamics (3.23), (3.26). We therefore extend our approach by the following strategy: We avoid to specify any particular collision law, but we try to identify the set of all possible post-impact velocities that are in accordance with the kinematic and dynamic restrictions. Only the frictionless case is considered. After having determined such a set, Newton's cradle should be accessible. Impacts as the most dynamic processes imaginable justify a geometric approach based on the kinetic metric, which will be chosen in the following.

5.1 Problem and Settings

In order to formulate our problem, we return in particular to (3.28) to decide which of the equations we want to keep or to modify. Considered for the rest of this report is only the frictionless case $\mu_i = 0$. As a consequence, all tangential impulsive forces Λ_T in (3.28) drop out, and the right column of relations is no longer needed. We further omit the index N, because all terms refer from now to the normal direction. The resulting contact model (Figure 8) is once more depicted in Figure 24.

We briefly recall contact kinematics by starting with the gap functions (3.3), which define k inequality constraints on the coordinates $\mathbf{q}$,

$$g^i(\mathbf{q}) \geq 0, \quad i = 1, \ldots, k. \tag{5.1}$$

According to (3.27), we denote the set of active constraints by

$$\mathcal{H}(\mathbf{q}) = \{i \mid g^i(\mathbf{q}) = 0\}. \tag{5.2}$$

The relative velocity in the normal direction (3.4) is obtained by differentiating the gap functions $g^i(\mathbf{q}(t))$ in (5.1) with respect to time. This yields

$$\gamma^i = \frac{\partial g^i}{\partial \mathbf{q}} \mathbf{u}, \tag{5.3}$$

where $\dot{\mathbf{q}}(t) = \mathbf{u}(t)$ and $\dot{g}^i = \gamma^i$ almost everywhere as introduced in Sections 1.2 and 3.2. Note that we have used the notation $\mathbf{w}^{i\mathsf{T}}(\mathbf{q}) \equiv \partial g^i / \partial \mathbf{q}$ in the second line in (3.28). As

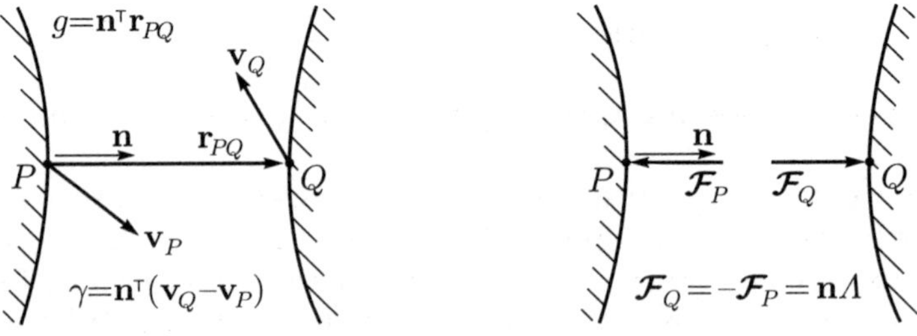

Figure 24. Contact model: Frictionless unilateral contact.

described in the first paragraph of Section 3.4, we still assume that the system arrives at $g^i(\mathbf{q}) = 0$ from an admissible state in the past, leading to non-positive pre-impact relative velocities. *In addition,* we demand that the system leaves the boundary after the collision towards admissible states, i.e. by non-negative post-impact relative velocities,

$$\gamma^{i-} \leq 0, \qquad \gamma^{i+} \geq 0. \tag{5.4}$$

The latter inequality is automatically provided by the impact law from the last line in (3.28), if $\gamma^{i-} \leq 0$ has been presupposed for all $i \in \mathcal{H}$, but is taken here as an *independent* assumption. Both conditions are necessary to keep the solution $\mathbf{q}(t)$ within the admissible set of points.

The impact equations for the frictionless case are taken from the first line in (3.28) with $\mathbf{w}^{i\mathsf{T}}(\mathbf{q}) \equiv \partial g^i/\partial \mathbf{q}$

$$\mathbf{M}(\mathbf{q})(\mathbf{u}^+ - \mathbf{u}^-) = \sum_{i\in\mathcal{H}} \left(\tfrac{\partial g^i}{\partial \mathbf{q}}\right)^{\mathsf{T}} \Lambda_i. \tag{5.5}$$

We keep the compressive character of the scalar impulsive normal forces, which is expressed by the inequality

$$\Lambda_i \geq 0 \tag{5.6}$$

and which has been proposed in the second paragraph of Section 3.4. This inequality is also provided by the Upr-inclusion from the last line of (3.28).

We do, however, *not* keep the impact constitutive law from the last two lines in (3.28). In contrast to equations (5.1)–(5.6) which stem from geometric properties or from fundamental principles in mechanics, the impact law is a much stronger assumption and will therefore be disregarded.

As a new and independent property we demand energetic consistency of the collision. We do not allow for any active behavior of the contacts as, for example, observed in pin ball machines, but assume a dissipative character only. This means that the kinetic energy

$$T = \tfrac{1}{2}\mathbf{u}^{\mathsf{T}}\,\mathbf{M}(\mathbf{q})\,\mathbf{u} \tag{5.7}$$

of the scleronomic system must not increase during the collision,

$$T^+ \leq T^-. \tag{5.8}$$

By this assumption we have also excluded any energy transfer from microscopic internal degrees of freedom to our macroscopic discretization level $\mathbf{q}$. This is guaranteed if, for example, eventual internal oscillations of the collision partners have been faded away at the time of collision, i.e. if the value of T^- is invariant under any spatial discretization of the system, including any possible continuum model.

5.2 General Impact Configurations

In this section, we take the concept of a Riemannian manifold to express impulsive collision behavior within the same mathematical framework as classical mechanics and to develop a geometric picture of impacts. The f-dimensional configuration manifold $\mathcal{M}$ of the mechanical system turns into a Riemannian manifold, if we choose a metric on

$\mathcal{M}$. The most natural choice of such a metric for dynamic collision problems is provided by the mass matrix $\mathbf{M}(\mathbf{q})$, which is the coordinate representation of the kinetic metric with respect to the local coordinates $\mathbf{q}$ on $\mathcal{M}$ in use. We will now need all the concepts introduced in section 4, in particular those of Sections 4.4 and 4.6. The tangent cone to a set $\mathcal{D}$ is denoted by $\mathcal{T}_{\mathcal{D}}(\mathbf{q})$. For $\mathcal{D} = \mathcal{M}$, the tangent cone simplifies to the classical tangent space $\mathcal{L}_M(\mathbf{q})$ with dual $\mathcal{L}^\star_{\mathcal{M}}(\mathbf{q}) = \mathbf{M}\mathcal{T}_{\mathcal{M}}(\mathbf{q})$. Inner products on $\mathcal{L}_{\mathcal{M}}(\mathbf{q})$ are denoted by a dot, $\mathbf{u}\cdot\mathbf{v} := \mathbf{u}^\mathsf{T}\mathbf{M}(\mathbf{q})\,\mathbf{v}$ for $\mathbf{u},\mathbf{v} \in \mathcal{L}_{\mathcal{M}}(\mathbf{q})$, and the associated norm is written as $\|\mathbf{u}\| := \sqrt{\mathbf{u}\cdot\mathbf{u}}$. The gradient $\nabla g^i(\mathbf{q})$ of a function $g^i : \mathcal{M} \to \mathbb{R}$, $\mathbf{q} \to g^i(\mathbf{q})$ is obtained by $\nabla g^i(\mathbf{q}) = \mathbf{M}^{-1}(\partial g^i/\partial\mathbf{q})^\mathsf{T}$ and is an element of $\mathcal{L}_{\mathcal{M}}(\mathbf{q})$.

We are now going to write equations (5.3)–(5.8) exclusively in terms of elements of $\mathcal{L}_{\mathcal{M}}(\mathbf{q})$. This affects in particular the differentials in (5.3) and (5.5) that will be expressed by the associated gradients, and equation (5.5) that will be mapped by $\mathbf{M}^{-1}(\mathbf{q})$ from $\mathcal{L}^\star_{\mathcal{M}}(\mathbf{q})$ to $\mathcal{L}_{\mathcal{M}}(\mathbf{q})$. The kinetic energy (5.7) is finally written by using the norm on $\mathcal{L}_{\mathcal{M}}(\mathbf{q})$. One obtains three equations referring to kinematics (5.3), kinetics (5.5) and energy (5.7) together with associated restrictions (5.4), (5.6) and (5.8) that have to hold for every $i \in \mathcal{H}$:

$$\begin{aligned}
&\text{kinematics:} && \gamma^i(\mathbf{u}) = \nabla g^i(\mathbf{q})\cdot\mathbf{u}, && \gamma^i(\mathbf{u}^+) \geq 0, \quad \gamma^i(\mathbf{u}^-) \leq 0,\\
&\text{kinetics:} && \mathbf{u}^+ - \mathbf{u}^- = \sum_{i\in\mathcal{H}} \nabla g^i(\mathbf{q})\,\Lambda_i, && \Lambda_i \geq 0,\\
&\text{energy:} && T(\mathbf{u}) = \tfrac{1}{2}\|\mathbf{u}\|^2, && T(\mathbf{u}^+) \leq T(\mathbf{u}^-).
\end{aligned} \tag{5.9}$$

This set of equations and restrictions provides the basis for a geometric interpretation of the collision process, because the tangent space $\mathcal{L}_{\mathcal{M}}(\mathbf{q})$ is equipped with an inner product such that all geometric concepts are available.

In order to get now a complete picture of the impact process, we return to the inequality constraints (5.1). They turn $\mathcal{M}$ into a manifold with boundary, characterized by the set of admissible displacements (4.38),

$$\mathcal{C} = \{\mathbf{q} \mid g^i(\mathbf{q}) \geq 0, \quad i = 1,\dots,k\}. \tag{5.10}$$

According to (4.42), the tangent cone $\mathcal{T}_{\mathcal{C}}(\mathbf{q})$ and its orthogonal counterpart $\mathcal{T}_{\mathcal{C}}^\perp(\mathbf{q})$ from (4.41) are now available. We have

$$\begin{aligned}
\mathcal{T}_{\mathcal{C}}(\mathbf{q}) &= \Big\{\mathbf{u} \,\Big|\, -\nabla g^i(\mathbf{q})\cdot\mathbf{u} \leq 0, \quad \forall i \in \mathcal{H}\Big\},\\
\mathcal{T}_{\mathcal{C}}^\perp(\mathbf{q}) &= \Big\{\sum_{i\in\mathcal{H}} -\nabla g^i(\mathbf{q})\,\Lambda_i, \quad \Lambda_i \geq 0\Big\},\\
\mathcal{B}_{\|\mathbf{u}^-\|}(\mathbf{q}) &:= \Big\{\mathbf{u} \,\Big|\, \|\mathbf{u}\| \leq \|\mathbf{u}^-\|\Big\},
\end{aligned} \tag{5.11}$$

where we have also introduced the energy ball $\mathcal{B}_{\|\mathbf{u}^-\|}(\mathbf{q})$ with radius $\|\mathbf{u}^-\|$ and midpoint placed at the origin. By using these sets, (5.9) may now be expressed as

$$\begin{aligned}
&\text{kinematic consistency:} && \mathbf{u}^+ \in \mathcal{T}_{\mathcal{C}}(\mathbf{q}), \quad \mathbf{u}^- \in -\mathcal{T}_{\mathcal{C}}(\mathbf{q}),\\
&\text{kinetic consistency:} && \mathbf{u}^+ \in \mathbf{u}^- - \mathcal{T}_{\mathcal{C}}^\perp(\mathbf{q}),\\
&\text{energetic consistency:} && \mathbf{u}^+ \in \mathcal{B}_{\|\mathbf{u}^-\|}(\mathbf{q}).
\end{aligned} \tag{5.12}$$

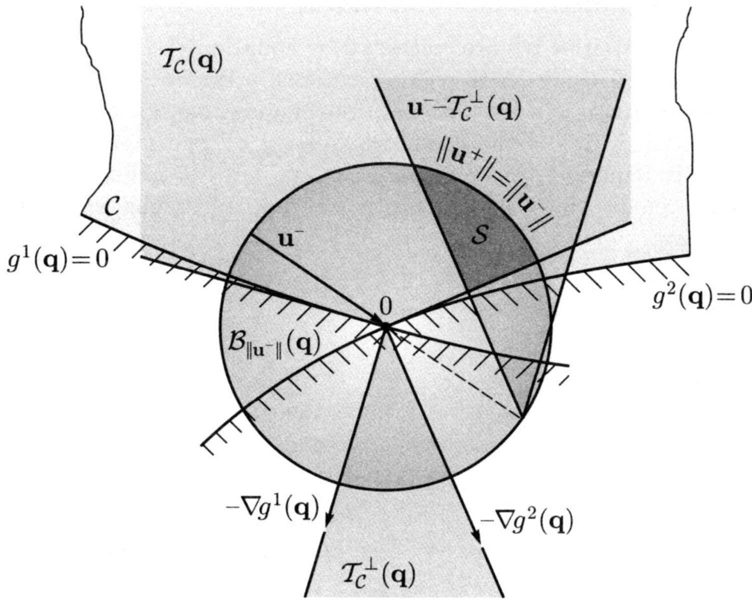

Figure 25. Geometry of multi-contact collisions.

Thus, for a given pre-impact velocity $\mathbf{u}^- \in -\mathcal{T}_\mathcal{C}(\mathbf{q})$, kinematic, kinetic and energetic consistency restrict the set of admissible post-impact velocities to

$$\mathcal{S} := \mathcal{T}_\mathcal{C}(\mathbf{q}) \cap (\mathbf{u}^- - \mathcal{T}_\mathcal{C}^\perp(\mathbf{q})) \cap \mathcal{B}_{\|\mathbf{u}^-\|}(\mathbf{q}). \tag{5.13}$$

This convex set is shown together with the orthogonal pair of cones $\mathcal{T}_\mathcal{C}(\mathbf{q})$, $\mathcal{T}_\mathcal{C}^\perp(\mathbf{q})$ and the energy ball $\mathcal{B}_{\|\mathbf{u}^-\|}(\mathbf{q})$ in Figure 25.

Note in particular three special cases that are contained in (5.13): First of all, assume that $\mathbf{q}$ is not a boundary point of $\mathcal{C}$, i.e. $\mathbf{q} \in \operatorname{int} \mathcal{C}$. For this situation, $\mathcal{T}_\mathcal{C}^\perp(\mathbf{q}) = \{0\}$, and $\mathcal{S}$ reduces to a single element $\mathcal{S} = \{\mathbf{u}^-\}$. As a consequence, $\mathbf{u}^+ = \mathbf{u}^-$ which means that there is no velocity jump. Secondly, suppose that the boundary of $\mathcal{C}$ is smooth at $\mathbf{q}$. In this case, $\mathcal{T}_\mathcal{C}(\mathbf{q})$ becomes a half-space and $\mathcal{T}_\mathcal{C}^\perp(\mathbf{q})$ degenerates to a half-line. This special configuration will be discussed in more detail in Section 5.7. Finally, suppose that $\mathcal{C}$ is a (smooth) sub-manifold of $\mathcal{M}$ as for classical systems under perfect constraints. For this case, $\mathcal{T}_\mathcal{C}(\mathbf{q})$ becomes the tangent space to $\mathcal{C}$ at $\mathbf{q}$ and $\mathcal{T}_\mathcal{C}^\perp(\mathbf{q})$ its orthogonal complement. For $\mathbf{u}^- \in -\mathcal{T}_\mathcal{C}(\mathbf{q}) \equiv \mathcal{T}_\mathcal{C}(\mathbf{q})$, the sets $\mathcal{T}_\mathcal{C}(\mathbf{q})$ and $\mathbf{u}^- - \mathcal{T}_\mathcal{C}^\perp(\mathbf{q})$ intersect precisely at the single element $\mathbf{u}^-$. As a result, we have again $\mathbf{u}^+ = \mathbf{u}^-$, i.e. continuous velocities. Another interesting consequence from cases one and three is that we *must not* demand an energy ball smaller than the one we have proposed. Otherwise, we would immediately get the empty set for the post-impact velocities. The restriction of the kinetic energy is, indeed, a critical concept and should be used with caution. Much better is to speak about energy levels or levels of dissipation that are accessed by an impact event, such as it will be introduced in Section 5.5.

5.3 Impacts with Global Dissipation Index

The post-impact velocities $\mathbf{u}^+$ are restricted by equation (5.13) to the set $\mathcal{S}$, but are still not uniquely defined. In order to pick a particular element out of this set, one needs an impact law in the sense of an additional, independent equation. We present in this section a construction which is the geometric version of the impact law suggested by Moreau (1988a). It requires, according to Theorem 4.1, an orthogonal decomposition of the pre-impact velocity $\mathbf{u}^-$ with respect to the pair of orthogonal cones $\mathcal{T}_C(\mathbf{q})$ and $\mathcal{T}_C^\perp(\mathbf{q})$. Let therefore

$$\mathbf{u}^- = \mathbf{v} + \mathbf{v}^\perp \tag{5.14}$$

such that the vectors $\mathbf{v}$ and $\mathbf{v}^\perp$ satisfy

$$\mathbf{v} \in \mathcal{T}_C(\mathbf{q}), \quad \mathbf{v}^\perp \in \mathcal{T}_C^\perp(\mathbf{q}), \quad \mathbf{v} \cdot \mathbf{v}^\perp = 0. \tag{5.15}$$

This decomposition is unique. The term $\mathbf{v}$ plays the role of the *tangential component* of $\mathbf{u}^-$ which we leave unchanged at the impact. In contrast, we invert the *normal component* $\mathbf{v}^\perp$ by the impact rule

$$\mathbf{v}^\times = -\varepsilon \mathbf{v}^\perp \quad (0 \leq \varepsilon \leq 1), \tag{5.16}$$

where ε is a global dissipation coefficient. The post-impact velocity is then set to be

$$\mathbf{u}^+ = \mathbf{v} + \mathbf{v}^\times \tag{5.17}$$

according to the construction shown in Figure 26, see e.g. Brogliato (1999) for comments on Moreau's sweeping process with soft constraints ($\varepsilon = 0$) in the frictionless case. The impact law provided by equations (5.14)–(5.17) may be summarized in the following way:

Problem 5.1 (Geometric version of Moreau's impact law). *For a given pre-impact velocity* $\mathbf{u}^- \in -\mathcal{T}_C(\mathbf{q})$ *and a given dissipation coefficient* $\varepsilon \geq 0$, *find the post-impact velocity* $\mathbf{u}^+$ *such that*

$$\begin{gathered} \mathbf{u}^- = \mathbf{v} + \mathbf{v}^\perp, \qquad \mathbf{u}^+ = \mathbf{v} - \varepsilon \mathbf{v}^\perp, \\ \mathcal{T}_C(\mathbf{q}) \ni \mathbf{v} \perp \mathbf{v}^\perp \in \mathcal{T}_C^\perp(\mathbf{q}). \end{gathered} \tag{5.18}$$

This post-impact velocity is uniquely determined as a consequence on Theorem 4.1.

From Figure 26 one recognizes immediately a lot of properties of the impact law: A *completely elastic impact* ($\varepsilon = 1$) can be interpreted as a *reflection* on a hyperplane $\mathcal{H}$ with normal $\mathbf{v}^\perp$, whereas a *completely inelastic impact* ($\varepsilon = 0$) corresponds to an *orthogonal projection* of $\mathbf{u}^-$ on $\mathcal{H}$ to give $\mathbf{v}$. In terms of a minimization problem are $\mathbf{v}$ and $\mathbf{v}^\perp$ the nearest points to $\mathbf{u}^-$ in the sets $\mathcal{T}_C(\mathbf{q})$ and $\mathcal{T}_C^\perp(\mathbf{q})$, respectively. The corresponding maps are proximations,

$$\mathbf{v} = \operatorname{prox}_{\mathcal{T}_C(\mathbf{q})}(\mathbf{u}^-), \quad \mathbf{v}^\perp = \operatorname{prox}_{\mathcal{T}_C^\perp(\mathbf{q})}(\mathbf{u}^-). \tag{5.19}$$

For example, the impact law (5.18) might equivalently be stated in terms of proximations

$$\mathbf{u}^+ = (1+\varepsilon) \operatorname{prox}_{\mathcal{T}_C(\mathbf{q})}(\mathbf{u}^-) - \varepsilon \mathbf{u}^- \tag{5.20}$$

when the first equation in (5.19) is used. Further, we recognize that the proximation in (5.20) becomes the identity whenever $\mathbf{u}^- \in \mathcal{T}_C(\mathbf{q})$. In this case $\mathbf{u}^+ \equiv \mathbf{u}^-$, thus no impact occurs.

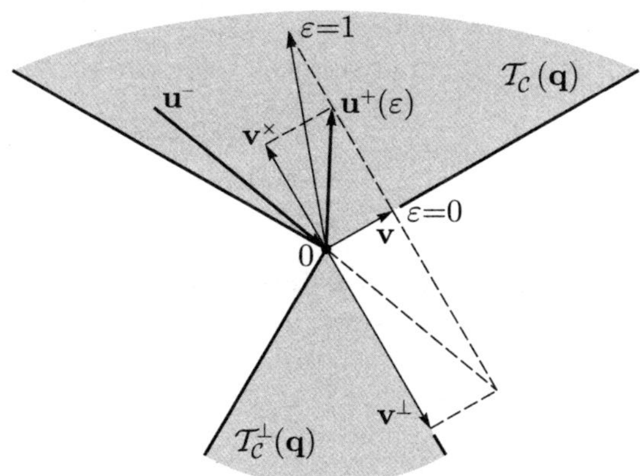

Figure 26. Geometry of impacts with global dissipation index.

Proposition 5.2. *Moreau's impact law* (5.18) *is energetically consistent, as soon as the dissipation coefficient* ε *is restricted to values* $0 \le \varepsilon \le 1$.

Proof. The kinetic energy T satisfies $2T^+ = \|\mathbf{u}^+\|^2 \le \|\mathbf{u}^-\|^2 = 2T^-$, where equality holds for $\varepsilon = 1$ and maximal dissipation is achieved for $\varepsilon = 0$. This is immediately clear from the construction of $\mathbf{u}^+$ and the orthogonality between $\mathbf{v}$ and $\mathbf{v}^\perp$. □

Proposition 5.3. *Moreau's impact law* (5.18) *is kinetically consistent, i.e. it holds that* $\mathbf{u}^- - \mathbf{u}^+ \in \mathcal{T}_C^\perp(\mathbf{q})$.

Proof. By (5.18) we have $\mathbf{u}^- - \mathbf{u}^+ = (1+\varepsilon)\mathbf{v}^\perp$. We further have $\varepsilon \ge 0$ and $\mathbf{v}^\perp \in \mathcal{T}_C^\perp(\mathbf{q})$. Since $\mathcal{T}_C^\perp(\mathbf{q})$ is a cone, it holds that $\alpha\mathbf{v}^\perp \in \mathcal{T}_C^\perp(\mathbf{q})$ whenever $\alpha \ge 0$. □

Proposition 5.4. *Moreau's impact law* (5.18) *is kinematically consistent, i.e. it holds that* $\mathbf{u}^+ \in \mathcal{T}_C(\mathbf{q})$ *whenever* $\mathbf{u}^- \in -\mathcal{T}_C(\mathbf{q})$.

Proof. We first consider the case $\varepsilon = 0$, for which $\mathbf{u}^+ = \mathbf{v} \in \mathcal{T}_C(\mathbf{q})$ by (5.18), independent of the values of $\mathbf{u}^-$.

Let now $\varepsilon > 0$. By assumption, we have $\mathbf{u}^- \in -\mathcal{T}_C(\mathbf{q})$ or, equivalently, $-\mathbf{u}^- \cdot \mathbf{z}^\perp \le 0$ for all $\mathbf{z}^\perp \in \mathcal{T}_C^\perp(\mathbf{q})$. With $\mathbf{u}^- = \mathbf{v} + \mathbf{v}^\perp$ and $\varepsilon > 0$, this variational inequality may be written as $-\varepsilon\mathbf{v} \cdot \mathbf{z}^\perp \le \varepsilon\mathbf{v}^\perp \cdot \mathbf{z}^\perp$, $\forall \mathbf{z}^\perp \in \mathcal{T}_C^\perp(\mathbf{q})$.

We have to show that $\mathbf{u}^+ \in \mathcal{T}_C(\mathbf{q})$ or, equivalently, $\mathbf{u}^+ \cdot \mathbf{z}^\perp \le 0$ for all $\mathbf{z}^\perp \in \mathcal{T}_C^\perp(\mathbf{q})$. With $\mathbf{u}^+ = \mathbf{v} - \varepsilon\mathbf{v}^\perp$, this variational inequality becomes $\mathbf{v} \cdot \mathbf{z}^\perp \le \varepsilon\mathbf{v}^\perp \cdot \mathbf{z}^\perp, \forall \mathbf{z}^\perp \in \mathcal{T}_C^\perp(\mathbf{q})$.

It is therefore sufficient to show that $\mathbf{v} \cdot \mathbf{z}^\perp \le -\varepsilon\mathbf{v} \cdot \mathbf{z}^\perp$ holds true for all $\mathbf{z}^\perp \in \mathcal{T}_C^\perp(\mathbf{q})$. With $\mathbf{v} \in \mathcal{T}_C(\mathbf{q})$ by (5.18), the product $\mathbf{v} \cdot \mathbf{z}^\perp$ is always less than or equal to zero. The inequality is therefore always fulfilled, if $\varepsilon \ge -1$, which is in accordance with our assumption $\varepsilon > 0$. □

Note that the impact law (5.18) *may* produce incompatible post-impact velocities $\mathbf{u}^+$ when the pre-impact velocities $\mathbf{u}^-$ are not from $-\mathcal{T}_C(\mathbf{q})$. Note also that for a *completely inelastic impacts* ($\varepsilon = 0$) compatible post-impact velocities $\mathbf{u}^+ \equiv \mathbf{v} \in \mathcal{T}_C(\mathbf{q})$ are *always* obtained.

With the help of equations (4.31) and (4.32) from Lemma 4.2 we may leave the symmetric formulation of Moreau's impact law (5.18) and may write it as a double inclusion:

Problem 5.5 (Moreau's impact law as a double-inclusion). *For a given pre-impact velocity* $\mathbf{u}^- \in -\mathcal{T}_C(\mathbf{q})$ *and a given dissipation coefficient* $\varepsilon \geq 0$*, find the post-impact velocity* $\mathbf{u}^+$ *such that*

$$\begin{aligned} \mathbf{u}^- &= \mathbf{v} + \mathbf{v}^\perp, & \mathbf{u}^+ &= \mathbf{v} - \varepsilon \mathbf{v}^\perp, \\ \mathbf{v} &\in \mathcal{T}_C(\mathbf{q}), & \mathbf{v}^\perp &\in \mathcal{T}^\perp_{\mathcal{T}_C(\mathbf{q})}(\mathbf{v}). \end{aligned} \tag{5.21}$$

This post-impact velocity is uniquely determined as a consequence on Problem 5.1 and Lemma 4.2.

The expressions in (5.21) look more complicated than those in (5.18), but comply indeed much better with the nature of mechanics. As a fundamental concept in mechanics, forces are regarded as elements of vector spaces. What we actually did by demanding $\mathbf{M}\mathbf{v}^\perp \in \mathbf{M}\mathcal{T}_C^\perp(\mathbf{q})$ is to take the impulsive reactions caused by the unilateral constraints from a closed convex cone as a subset of a vector space, i.e. the cotangent space to $\mathcal{M}$. This approach makes sense in view of the concept of orthogonal complements extended to cones and is still present in (5.21). The second linear space used in mechanics is the tangent space to $\mathcal{M}$, which contains the *virtual* velocities. Only for *scleronomic* systems it coincides with the space of velocities. For rheonomic systems, $\mathcal{M}$ has to be regarded as a breathing manifold, and the space of velocities becomes linear *affine* as a consequence. In such a situation, it doesn't make much sense to propose a convex cone $\mathcal{T}_C(\mathbf{q})$ to address the kinematically admissible velocities. A polyhedral set for simple unilaterally constrained motion seems to be the right choice, instead. We have obtained with (5.21) a representation of the impact law that unambiguously requires the reservoir of impulsive reactions to be a closed convex cone, but does not necessarily prescribe convexity or even any conical shape of $\mathcal{T}_C(\mathbf{q})$. This representation will be used in Section 5.4 to extend Moreau's impact law to the case of re-entrant corners, for which convexity of the admissible velocities is no longer available. We will now use (5.21) to show the relation between Moreau's impact law and the collision problem that we addressed in Section 3.

Proposition 5.6. *The frictionless collision problem from* (3.28),

$$\begin{aligned} \mathbf{M}\left(\mathbf{u}^+ - \mathbf{u}^-\right) &= \sum_{i \in \mathcal{H}} \left(\tfrac{\partial g^i}{\partial \mathbf{q}}\right)^{\mathsf{T}} \Lambda_i, \\ \gamma^i = \tfrac{\partial g^i}{\partial \mathbf{q}} \mathbf{u}, \quad \xi^i &= \gamma^{i+} + \varepsilon^i \gamma^{i-}, \quad -\Lambda_i \in \operatorname{Upr}(\xi^i), \end{aligned} \tag{5.22}$$

and Moreau's impact law (5.21) *are equivalent in the following sense: For any* $\mathbf{u}^+$ *satisfying* (5.21)*, values of* Λ_i *and* γ^{i+} *can be found such that* (5.22) *holds when all restitution coefficients are chosen to be equal to each other with* $\varepsilon = \varepsilon^i = \varepsilon^j$, $\forall i, j \in \mathcal{H}$.

Proof. We will rewrite (5.21) and (5.22) several times to eliminate superfluous variables. In particular, we solve the two equations in the first line of (5.21) for $\mathbf{v}$ and $\mathbf{v}^\perp$, which gives $\mathbf{v} = \frac{1}{1+\varepsilon}(\mathbf{u}^+ + \varepsilon\mathbf{u}^-)$ and $\mathbf{v}^\perp = \frac{-\varepsilon}{1+\varepsilon}(\mathbf{u}^+ - \mathbf{u}^-)$. Since $\mathcal{T}_C(\mathbf{q})$ and $\mathcal{T}^\perp_{\mathcal{T}_C(\mathbf{q})}(\cdot)$ are cones, we may cancel non-negative scalars in $\mathbf{v}$ and $\mathbf{v}^\perp$. Equation (5.21) may therefore be written as

$$\mathbf{u}^+ + \varepsilon\mathbf{u}^- \in \mathcal{T}_C(\mathbf{q}), \qquad -(\mathbf{u}^+ - \mathbf{u}^-) \in \mathcal{T}^\perp_{\mathcal{T}_C(\mathbf{q})}(\mathbf{u}^+ + \varepsilon\mathbf{u}^-). \tag{5.23}$$

We further rewrite equation (5.22) by exclusively using elements of the tangent space and take into account $\varepsilon = \varepsilon^i = \varepsilon^j$, $\forall i, j \in \mathcal{H}$,

$$\begin{gathered} \mathbf{u}^+ - \mathbf{u}^- = \sum_{i\in\mathcal{H}} \nabla g^i(\mathbf{q})\, \Lambda_i, \\ \gamma^i = \nabla g^i(\mathbf{q}) \cdot \mathbf{u}, \quad \xi^i = \gamma^{i+} + \varepsilon\, \gamma^{i-}, \quad -\Lambda_i \in \operatorname{Upr}(\xi^i). \end{gathered} \tag{5.24}$$

The term $\mathbf{u}^+ - \mathbf{u}^-$ in (5.23) is now eliminated by the first equation in (5.24). This gives

$$\mathbf{u}^+ + \varepsilon\mathbf{u}^- \in \mathcal{T}_C(\mathbf{q}), \qquad -\sum_{i\in\mathcal{H}} \nabla g^i(\mathbf{q})\, \Lambda_i \in \mathcal{T}^\perp_{\mathcal{T}_C(\mathbf{q})}(\mathbf{u}^+ + \varepsilon\mathbf{u}^-). \tag{5.25}$$

Further, we may eliminate $\gamma^{i\pm}$ from (5.24) to express ξ^i directly in terms of $\mathbf{u}^\pm$,

$$\xi^i = \nabla g^i(\mathbf{q}) \cdot (\mathbf{u}^+ + \varepsilon\mathbf{u}^-), \quad -\Lambda_i \in \operatorname{Upr}(\xi^i). \tag{5.26}$$

We finally abbreviate in (5.25) and (5.26) the term $\mathbf{u}^+ + \varepsilon\mathbf{u}^-$ by $\mathbf{z}$. Moreau's impact law (5.21) takes now the form

$$\mathbf{z} \in \mathcal{T}_C(\mathbf{q}), \qquad -\sum_{i\in\mathcal{H}} \nabla g^i(\mathbf{q})\, \Lambda_i \in \mathcal{T}^\perp_{\mathcal{T}_C(\mathbf{q})}(\mathbf{z}), \tag{5.27}$$

and the frictionless collision problem (5.22) has reduced to

$$\xi^i = \nabla g^i(\mathbf{q}) \cdot \mathbf{z}, \quad -\Lambda_i \in \operatorname{Upr}(\xi^i). \tag{5.28}$$

According to (2.13), the k Upr-inclusions can be expressed by the inequality-complementarity conditions $\Lambda_i \geq 0$, $\xi^i \geq 0$, $\Lambda_i\, \xi^i = 0$, which in turn might be written thanks to Lemma 2.1 as a variational inequality,

$$-\sum_{i\in\mathcal{H}} \Lambda_i(\xi^{\star i} - \xi^i) \leq 0, \quad \xi^i \geq 0, \quad \forall \xi^{\star i} \geq 0. \tag{5.29}$$

We have to show that (5.29) can be derived from (5.27). Let us first prove the inequality $\xi^i \geq 0$ in (5.29). This result is immediately available from the definition of the tangent cone $\mathcal{T}_C(\mathbf{q})$ in (5.11) and the equation for ξ^i in (5.28), because $\mathbf{z}$ is an element of $\mathcal{T}_C(\mathbf{q})$ by (5.27). We investigate now the second inclusion in (5.27), which can be expressed with the help of (4.34) as

$$-\sum_{i\in\mathcal{H}} \Lambda_i\, \nabla g^i(\mathbf{q}) \cdot (\mathbf{z}^\star - \mathbf{z}) \leq 0, \quad \mathbf{z} \in \mathcal{T}_C(\mathbf{q}), \quad \forall \mathbf{z}^\star \in \mathcal{T}_C(\mathbf{q}). \tag{5.30}$$

Identify now $\xi^{\star i}$ as $\nabla g^i(\mathbf{q}) \cdot \mathbf{z}^\star$, which proves immediately $\xi^{\star i} \geq 0$ by the same reasons as for ξ^i, and take into account the equation for ξ^i in (5.28). As a result, we have obtained the first inequality in (5.29). The last thing to show is that the different $\xi^{\star i}$ in (5.29) can be independently addressed by the values of $\mathbf{z}^\star$ via $\xi^{\star i} = \nabla g^i(\mathbf{q}) \cdot \mathbf{z}^\star$ when $\mathbf{z}^\star$ varies over $\mathcal{T}_\mathcal{C}(\mathbf{q})$. If this is assured, then all contact impulsions Λ_i are uniquely determined. Otherwise, one has an overconstrained system which leads still to a solution of (5.29), but non-uniqueness in the scalar contact impulsions Λ_i will occur. □

Note, however, that (5.21) can be regarded as a generalization of (5.22), because this formulation hold beyond simple unilateral constraints. It is still valid for (tangentially regular) sets $\mathcal{C}$ which can not be expressed by the intersection of a finite number of inequalities $g^i(\mathbf{q}) \geq 0$ with non-vanishing gradients $\nabla g^i(\mathbf{q})$ at their boundaries. An example in $\mathbb{R}^3$ of such a set is already the circular cone $\{\lambda(\mathbf{e}+\mathcal{D}) \mid \lambda \geq 0\}$ with $0 \neq \mathbf{e} \perp \mathcal{D}$ and $\mathcal{D}$ the closed unit disc in $\mathbb{R}^2$.

5.4 Impacts at Re-Entrant Corner Points

We finally consider the case that the boundary of the admissible displacements is not tangentially regular at the point of impact $\mathbf{q}$, but can be approximated well by the contingent cone $\mathcal{K}_\mathcal{C}(\mathbf{q})$. Note that such a configuration can *not* be generated by the intersection of smooth simple unilateral constraints, but occurs in practice when, for example, two rectangular blocks hit each other at their corners. We try to find a reasonable extension of Moreau's impact law which we do by just replacing the tangent cone with the contingent cone in the appropriate equations. The set of admissible velocities is now defined by the contingent cone, whereas the impulsive forces at the impact are still taken from the convex normal cone with the corresponding set $\mathcal{T}_\mathcal{C}^\perp(\mathbf{q})$ in the tangent space. We thus rewrite Problem 5.5 as:

Problem 5.7 (Extension of Moreau's impact law to re-entrant corners). *For a given pre-impact velocity* $\mathbf{u}^- \in -\mathcal{K}_\mathcal{C}(\mathbf{q})$ *and a given dissipation coefficient* $\varepsilon \geq 0$, *find the post-impact velocity* $\mathbf{u}^+$ *such that*

$$\begin{aligned} \mathbf{u}^- &= \mathbf{v} + \mathbf{v}^\perp, & \mathbf{u}^+ &= \mathbf{v} - \varepsilon \mathbf{v}^\perp, \\ \mathbf{v} &\in \mathcal{K}_\mathcal{C}(\mathbf{q}), & \mathbf{v}^\perp &\in \mathcal{T}^\perp_{\mathcal{K}_\mathcal{C}(\mathbf{q})}(\mathbf{v}), \end{aligned} \tag{5.31}$$

where $\mathcal{K}_\mathcal{C}(\mathbf{q})$ *denotes the (non-convex) contingent cone to* $\mathcal{C}$ *at* $\mathbf{q}$.

The geometric construction of the post-impact velocity $\mathbf{u}^+$ is sketched in Figure 27. Note that the inclusion $\mathcal{T}^\perp_{\mathcal{K}_\mathcal{C}(\mathbf{q})}(\mathbf{v}) \subset \mathcal{T}^\perp_\mathcal{C}(\mathbf{q})$ still holds. Of course, the orthogonal decomposition as performed in (5.18) does not longer apply, because convexity of the two participating cones is required there. A representation in terms of proximal points also fails, because due to the lack of convexity of $\mathcal{K}_\mathcal{C}(\mathbf{q})$ one is no longer able to express (5.31) as a *minimization* problem. The *extended* statement of (5.19) is: Find the stationary points of the non-convex function

$$\Phi(\mathbf{v}) = \tfrac{1}{2}\|\mathbf{v} - \mathbf{u}^-\|^2 + I_{\mathcal{K}_\mathcal{C}(\mathbf{q})}(\mathbf{v}), \tag{5.32}$$

where $I_\mathcal{N}(\mathbf{v})$ denotes the value of the indicator function of $\mathcal{N}$ at $\mathbf{v}$, i.e. $I_\mathcal{N}(\mathbf{v}) = 0$ for $\mathbf{v} \in \mathcal{N}$ and $I_\mathcal{N}(\mathbf{v}) = +\infty$ for $\mathbf{v} \notin \mathcal{N}$. The solution set of this problem contains, among

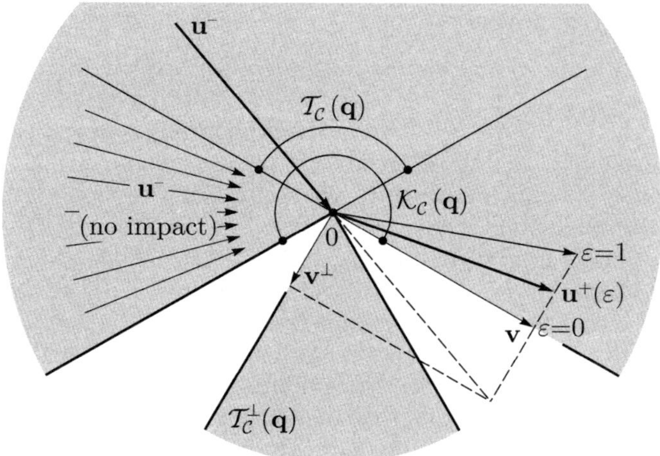

Figure 27. Impact at a re-entrant corner.

others, the proximal points to $\mathbf{u}^-$ in $\mathcal{K}_C(\mathbf{q})$ and is set up by the zeros of the generalized gradient of (5.32). These are the values of $\mathbf{v}$ which satisfy

$$0 \in \bar{\partial}\Phi(\mathbf{v}) = \mathbf{M}(\mathbf{v}-\mathbf{u}^-) + \bar{\partial}I_{\mathcal{K}_C(\mathbf{q})}(\mathbf{v}) = \mathbf{M}(\mathbf{v}-\mathbf{u}^-) + \mathcal{N}_{\mathcal{K}_C(\mathbf{q})}(\mathbf{v}). \tag{5.33}$$

With $\mathbf{v}^{\perp} = \mathbf{u}^- - \mathbf{v}$ and by taking the terms in (5.33) from the cotangent space to the tangent space, we obtain

$$\mathbf{v}^{\perp} \in \mathcal{T}^{\perp}_{\mathcal{K}_C(\mathbf{q})}(\mathbf{v}), \tag{5.34}$$

which is together with the restriction $\mathbf{v} \in \mathcal{K}_C(\mathbf{q})$ again the impact law from the second line in (5.31).

Due to the non-convexity of $\mathcal{K}_C(\mathbf{q})$ one can *not* expect the solution set of (5.31) to consist of one element only. Non-unique post-impact velocities *must* be accepted to occur for certain configurations, such as for the symmetric case depicted in Figure 28 for $\varepsilon = 1$ and $\varepsilon = 0$, where three different solutions are met. An impact law for re-entrant corners was also introduced by Kane et al. (1999) in the sense that (5.32) is taken as a *minimization* problem, leading to a restricted set of solutions.

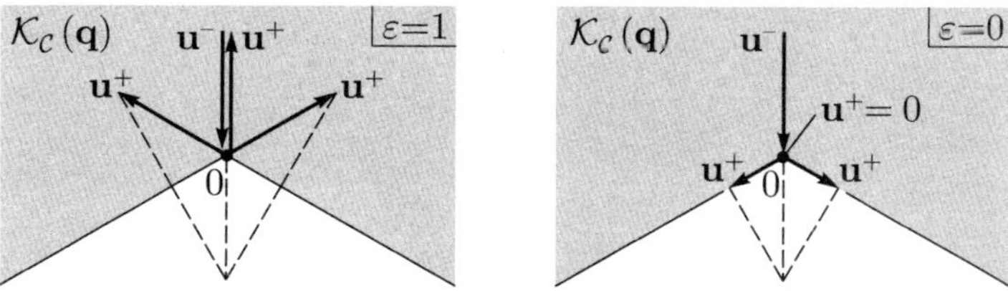

Figure 28. Non-uniqueness of the impact law at re-entrant corners.

5.5 The Geometry of Impacts

Let us return to Figure 25 and discuss once more the geometric properties of perfect collisions, but now by taking into account what we have learned from Moreau's impact law (5.18). For $\varepsilon = 0$, the impact law yields $\mathbf{u}^+ = \mathbf{v}$, which corresponds to the unique post-impact velocity of minimal length under the restrictions (5.13). We call the associated point in $\mathcal{S}$ the *point of maximal dissipation*. Thus, a uniquely determined half-line passing this point and emanating from the tip $\mathbf{u}^-$ of the translated cone $-\mathcal{T}_C^\perp(\mathbf{q})$ is accessible through Moreau's impact law by values $\varepsilon \geq -1$. It is easily seen that $\varepsilon = 0$ addresses the point of maximal dissipation, whereas conservation of kinetic energy is assured for $\varepsilon = 1$. Values of ε in between correspond to different levels of dissipation, defining a family of concentric spheres in the tangent space. As we have learned from Newton's cradle in Section 3.8, Moreau's half-line does not comprise collision events that benefit from impulse transfer between non-neighboring bodies via travelling waves. This behavior is also excluded in classical text books on single impacts by saying that wave effects have to be negligible. The framework achieved by this assumption is usually called the *mechanical impact theory*. However, a precise mathematical formulation of this term has never been given. We therefore suggest to *define* the mechanical impact theory as to consist precisely of those impact events that can be accessed through Moreau's half-line.

Obviously, not every post-impact velocity in $\mathcal{S}$ can be reached by Moreau's law. In view of a future parameterization of $\mathcal{S}$ we want to give already here a characterization of general collisions based on geometric properties. We define the *impact topology* as the entity connected to the different half-lines of $\mathcal{T}_C^\perp(\mathbf{q})$. In other words, all post-impact

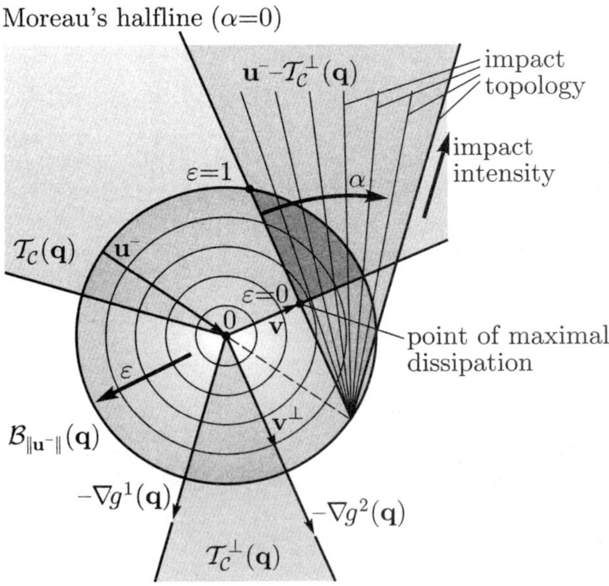

Figure 29. Geometric characterization of impacts.

velocities pointing for some given $\Lambda_i > 0$ to the same half-line $\mathbf{u}^- + \mathbb{R}_0^+ \sum_{i\in\mathcal{H}} \Lambda_i \nabla g^i$ in $\mathbf{u}^- - \mathcal{T}_\mathcal{C}^\perp(\mathbf{q})$ are said to have the same impact topology. This classification corresponds to a constant impact impulse ratio, as called by other authors. We further propose to take the angle α between a certain half-line of $\mathcal{T}_\mathcal{C}^\perp(\mathbf{q})$ and Moreau's half-line as a measure of deviation from the mechanical impact theory, which also quantifies the wave effects at the collision. We further take as *impact intensity* the distance of the post-impact velocity $\mathbf{u}^+$ from the tip of the translated cone $\mathbf{u}^- - \mathcal{T}_\mathcal{C}^\perp(\mathbf{q})$, which is nothing else than $\|\mathbf{u}^+ - \mathbf{u}^-\|$. As soon as the post-impact velocity $\mathbf{u}^+$ has been located in the set $\mathbf{u}^- - \mathcal{T}_\mathcal{C}^\perp(\mathbf{q})$, it can in addition be characterized by the dissipation coefficient ε associated with the energetic sphere on which it is placed.

In Proposition 5.6 a representation of Moreau's impact law in local contact coordinates has been derived, showing that

$$\Lambda_i \geq 0, \quad \gamma^{i+} + \varepsilon\,\gamma^{i-} \geq 0, \quad \Lambda_i\,(\gamma^{i+} + \varepsilon\,\gamma^{i-}) = 0 \tag{5.35}$$

for each contact i in the active set $\mathcal{H}$. The complementarity conditions in (5.35) express that each contact that takes an impulsive force $\Lambda_i > 0$ has to fulfil the classical Newtonian impact law $\gamma^{i+} = -\varepsilon\,\gamma^{i-}$. However, if the contact point is to be regarded as not to participate in the impact ($\Lambda_i = 0$), then the post-impact relative velocity is allowed for having values $\gamma^{i+} > -\varepsilon\gamma^{i-}$. This impact behavior is said to have a *global dissipation index*, because the *same* restitution coefficient ε is taken in (5.35) for each individual contact. In order to extend this concept of local representation to general impacts in $\mathcal{S}$, a matrix of impact coefficients ε_j^i has to be introduced in the inequality impact law (5.35) as proposed in Frémond (1995),

$$\Lambda_i \geq 0, \quad \gamma^{i+} + \varepsilon_j^i\,\gamma^{j-} \geq 0, \quad \Lambda_i\,(\gamma^{i+} + \varepsilon_j^i\,\gamma^{j-}) = 0. \tag{5.36}$$

This matrix of impact coefficients allows to access impact topologies different from Moreau's half-line, such as spatially separated chain-like contacts which directly interact with each other. Restrictions on the coefficients ε_j^i such that $\mathbf{u}^+ \in \mathcal{S}$ is guaranteed have still to be worked out.

5.6 Example: The Geometry of Newton's Cradle

To study the geometry of impacts on a particular example, we return to Newton's cradle with three balls of equal masses $m_1 = m_2 = m_3 =: m$ (Figure 30) that was already presented in Section 3.8. The horizontal absolute displacements of the three masses are

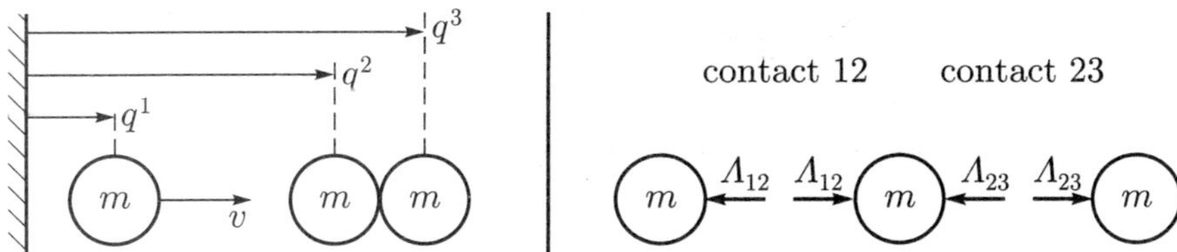

Figure 30. Mechanical model of Newton's cradle with three balls.

denoted by q^1, q^2, q^3, and the associated velocities by u^1, u^2, u^3, respectively. The inequality contact constraints (5.1) are expressed by the two gap functions

$$g^{12} = q^2 - q^1 \geq 0, \qquad g^{23} = q^3 - q^2 \geq 0. \tag{5.37}$$

For the pre-impact state we assume that balls two and three are at rest ($u^{2-} = u^{3-} = 0$) by touching each other ($g^{23} = 0$), and that ball one hits ball two ($g^{12} = 0$) with a velocity $u^{1-} = v > 0$. We therefore have a pre-impact velocity state $\mathbf{u}^- = (v,\, 0,\, 0)^\mathsf{T}$ with both contacts contained in the set of active constraints $\mathcal{H}$ from (5.2). The impact equations (5.5) for this configuration are

$$\begin{pmatrix} m & 0 & 0 \\ 0 & m & 0 \\ 0 & 0 & m \end{pmatrix} \begin{pmatrix} u^{1+} - u^{1-} \\ u^{2+} - u^{2-} \\ u^{3+} - u^{3-} \end{pmatrix} = \begin{pmatrix} -1 \\ 1 \\ 0 \end{pmatrix} \Lambda_{12} + \begin{pmatrix} 0 \\ -1 \\ 1 \end{pmatrix} \Lambda_{23}, \tag{5.38}$$

where Λ_{12} and Λ_{23} denote the impulsive forces between balls 1-2 and 2-3, respectively. From this equation one identifies the gradients of the gap functions as

$$\nabla g^{12} = \frac{1}{m} \begin{pmatrix} -1 \\ 1 \\ 0 \end{pmatrix}, \qquad \nabla g^{23} = \frac{1}{m} \begin{pmatrix} 0 \\ -1 \\ 1 \end{pmatrix} \tag{5.39}$$

which define by (5.11) the edges of $\mathcal{T}_\mathcal{C}^\perp(\mathbf{q})$. As shown in Figure 31, kinetic consistency (5.12) yields a subset $\mathbf{u}^- - \mathcal{T}_\mathcal{C}^\perp(\mathbf{q})$ in the deviatoric plane of the u^1-u^2-u^3 trihedral, which is the plane that intersects the three axes by equal values v. The set $\mathcal{S}$ of admissible post-impact velocities $\mathbf{u}^+$ is obtained according to (5.13) by subsequent intersection with the tangent cone $\mathcal{T}_\mathcal{C}(\mathbf{q})$ and the energy ball $\mathcal{B}_v(\mathbf{q})$.

The point of maximal dissipation in $\mathcal{S}$ as the point in $\mathbf{u}^- - \mathcal{T}_\mathcal{C}^\perp(\mathbf{q})$ with minimal distance from the origin, is recognized as the post-impact motion, for which the three

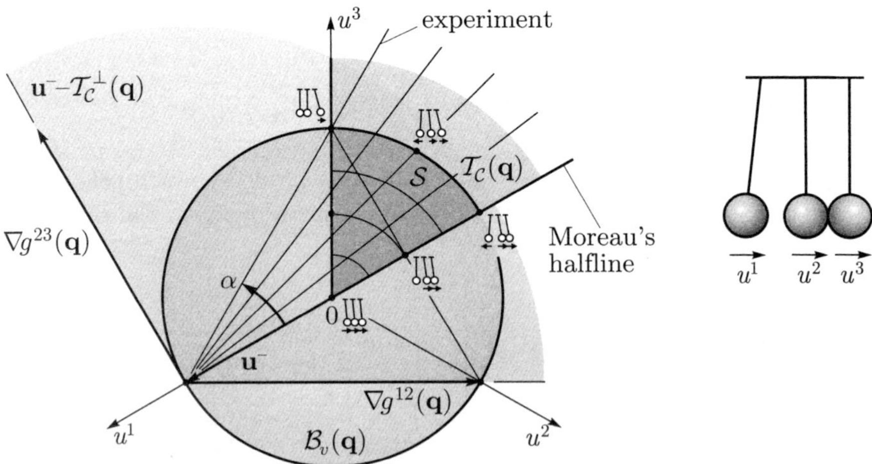

Figure 31. Impact geometry of Newton's cradle with three balls.

balls move together as one rigid body with a common velocity $u^{1,2,3+} = \frac{1}{3}v$. This point defines Moreau's half-line, which emanates from the tip of $\mathbf{u}^- - \mathcal{T}_\mathcal{C}^\perp(\mathbf{q})$ and carries the topology of impacts with global dissipation index. On its intersection with the energetic sphere that assures conservation of kinetic energy, balls two and three remain in contact after the impact and move with a common velocity of $u^{2,3+} = \frac{2}{3}v$ to the right, whereas ball one bounces back to the left with a velocity of $u^{1+} = -\frac{2}{3}v$. In reality, balls one and two stand nearly idle ($u^{1,2+} = 0$) after the impact, and ball three leaves with an approximate velocity $u^{3+} = v$ to the right. This behavior is on the same energetic sphere, but takes place on the topological half-line that deviates by a maximal possible angle ($\alpha = 30°$) from Moreau's impact law. Note that only one point of this half-line belongs to $\mathcal{S}$. As a result, Newton's cradle can be seen as a device that shows experimentally a collision behavior that maximizes wave effects under all imaginable experiments with three arbitrary bodies of equal mass.

5.7 Special Impact Configurations

Let us finally return to Figures 25 and 29 to discuss one particular case which is best known from nearly all text books on dynamics. We suppose that the boundary of $\mathcal{C}$ is smooth at the point of impact $\mathbf{q}$. Such a situation is met, for example, if the system is constrained by only one inequality (5.1), which corresponds at the same time to only one possible contact point in the system. We speak then about a single collision. In this case, the tangent cone $\mathcal{T}_\mathcal{C}(\mathbf{q})$ becomes a half-space and $\mathcal{T}_\mathcal{C}^\perp(\mathbf{q})$ degenerates to Moreau's half-line, as depicted in the left part of Figure 32. As a consequence, the set of admissible post-impact velocities $\mathcal{S}$ from (5.13) reduces to an interval on Moreau's half-line, which extends from the point of maximal dissipation to the point at which the energetic sphere with radius $\mathbf{u}^-$ intersects. For this configuration, only one coefficient is necessary to parameterize $\mathcal{S}$, which is the classical restitution coefficient ε.

Smooth boundaries also allow for very soft impacts that correspond to a nearly tangential touchdown, which is shown in the right part of Figure 32. Suppose that the system arrives with the same speed $\|\mathbf{u}^-\|$ at the point of impact $\mathbf{q}$, but with an angle of

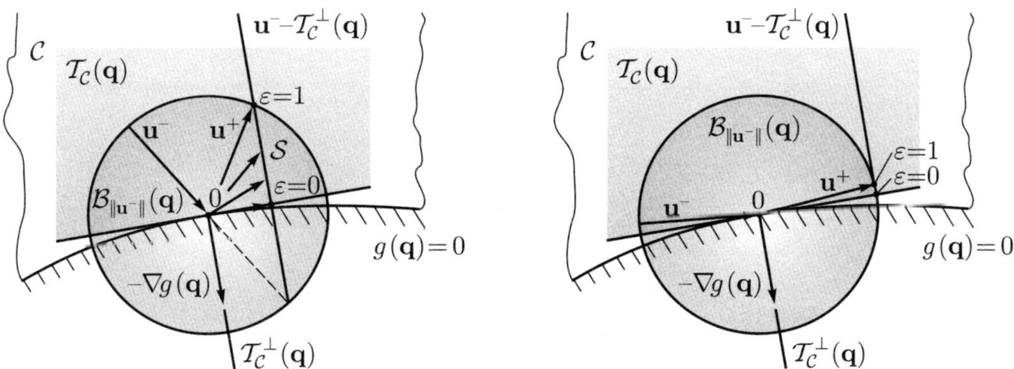

Figure 32. Impact against a smooth boundary and tangential touchdown.

incidence that becomes flatter and flatter. By this process, the interval $\mathcal{S}$ of admissible post-impact velocities shrinks more and more, and so the impulsive force to keep the system within $\mathcal{C}$. In the limit, for which $\mathbf{u}^- \perp -\nabla g(\mathbf{q})$, the points of maximal dissipation and conservation of energy agree, and one obtains $\mathbf{u}^+ = \mathbf{u}^-$ independent of the value that has been chosen for the restitution coefficient. Note however, that orthogonality has still to be understood in the sense of the kinetic metric.

To give at least one illustrative example of a non-trivial tangential touchdown, we discuss an elastic rod of mass m that impacts with velocity v against a rigid half-space. We take a discretized model of the rod, consisting of n masses of equal size m/n with springs in between as shown in Figure 33. We denote the linear displacements of the n masses by $\mathbf{q} = (q^1, \ldots, q^n)^\mathsf{T}$ and assume a pre-impact velocity

$$\mathbf{u}^- = -v\,\mathbf{1}, \tag{5.40}$$

where $\mathbf{1} := (1, 1, \ldots, 1, 1)^\mathsf{T}$. There is only one inequality constraint (5.1)

$$g(\mathbf{q}) = q^1 \geq 0 \tag{5.41}$$

that prevents the first discrete mass to move into the wall. Note that the finite forces resulting from the springs are not needed in the impact equations (5.5). With $\partial g/\partial \mathbf{q} = (1, 0, \ldots, 0, 0) =: \mathbf{e}^{1\mathsf{T}}$, the latter become

$$\tfrac{m}{n}\,(\mathbf{u}^+ - \mathbf{u}^-) = \mathbf{e}^1 \Lambda, \tag{5.42}$$

from which we identify the mass matrix $\mathbf{M}$ to derive the gradient of $g(\mathbf{q})$,

$$\mathbf{M} = \operatorname{diag}\left(\tfrac{m}{n}\right) = \tfrac{m}{n}\,\mathbf{I}^n \quad \Rightarrow \quad \nabla g(\mathbf{q}) = \mathbf{M}^{-1}\left(\tfrac{\partial g}{\partial \mathbf{q}}\right)^\mathsf{T} = \tfrac{n}{m}\,\mathbf{e}^1. \tag{5.43}$$

According to (4.13), we compute the angle α between $-\mathbf{u}^-$ and ∇g,

$$\begin{aligned} \cos\alpha &= \frac{-\mathbf{u}^{-\mathsf{T}}\,\mathbf{M}\,\nabla g}{\sqrt{\mathbf{u}^{-\mathsf{T}}\,\mathbf{M}\,\mathbf{u}^-}\;\sqrt{\nabla g^\mathsf{T}\,\mathbf{M}\,\nabla g}} \\ &= \frac{v\,\frac{m}{n}\,\frac{n}{m}\,\mathbf{1}^\mathsf{T}\,\mathbf{e}^1}{\sqrt{v^2\,\frac{m}{n}\,\mathbf{1}^\mathsf{T}\,\mathbf{1}}\;\sqrt{(\frac{n}{m})^2\,\frac{m}{n}\,\mathbf{e}^{1\mathsf{T}}\,\mathbf{e}^1}} \\ &= \frac{v}{\sqrt{v^2\,\frac{m}{n}\,n}\;\sqrt{\frac{n}{m}}} = \frac{1}{\sqrt{n}}. \end{aligned} \tag{5.44}$$

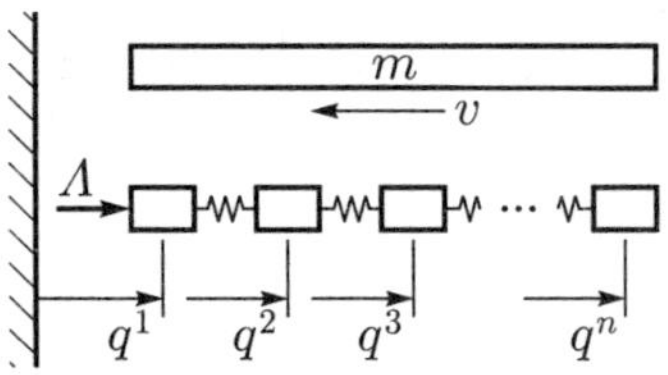

Figure 33. Collision of an elastic rod.

Let us now refine the spatial discretization of the rod by taking more and more discrete masses. We obtain

$$\cos\alpha \xrightarrow{n\to\infty} 0 \quad \Rightarrow \quad \alpha \xrightarrow{n\to\infty} 90^\circ, \tag{5.45}$$

which approaches a tangential touchdown. As smaller the first discrete mass becomes, as smaller is the impulsive force Λ necessary to keep it on the right side of the wall. For the same reason, the kinetic energy of the overall system is less and less affected by the collision of the very first particle.

5.8 Example: Collisions at the Rocking Rod

As a last example, we discuss the rocking motion of a rigid rod that is symmetrically placed on top of two obstacles. This problem belongs to the class at which an impulsive force at a closed contact or even a separation process is induced by a collision on another contact point in the system. We model the rocking under the assumption of completely inelastic impacts, which causes the impact process to be maximally dissipative.

The mechanical model of the rocking rod is depicted in Figure 34. It consists of a homogeneous slender rigid rod of mass m and length $2l$ and two obstacles with a distance $2a$ from each other that are placed below the rod. To describe the linear and angular displacements of the rod, we choose as coordinates $\mathbf{q} = (x, y, \varphi)^\mathsf{T}$ as shown in Figure 34, and denote by $\mathbf{u} = (u, v, \omega)^\mathsf{T}$ the associated velocities. The gap functions (5.1) between the two obstacles and the rod are

$$g^1(\mathbf{q}) = y\cos\varphi - (x+a)\sin\varphi \geq 0, \qquad g^2(\mathbf{q}) = y\cos\varphi - (x-a)\sin\varphi \geq 0. \tag{5.46}$$

We place now the rod symmetrically on top of the two obstacles, lift its left end carefully off such that contact with the right obstacle is still maintained, and let it drop. It then turns around the right contact point, where we assume enough friction to prevent the rod from sliding. After some time, the rod will hit the left obstacle, which is the collision that we will investigate. This collision takes place at values $\mathbf{q}_0 = (0,0,0)^\mathsf{T}$, for which $g^1(\mathbf{q}_0) = g^2(\mathbf{q}_0) = 0$. As a consequence, both inequality constraints have to be considered in the active set (5.2). With

$$\left.\frac{\partial g^1}{\partial \mathbf{q}}\right|_{\mathbf{q}_0} = \begin{pmatrix} 0 & 1 & -a \end{pmatrix}, \qquad \left.\frac{\partial g^2}{\partial \mathbf{q}}\right|_{\mathbf{q}_0} = \begin{pmatrix} 0 & 1 & a \end{pmatrix}, \tag{5.47}$$

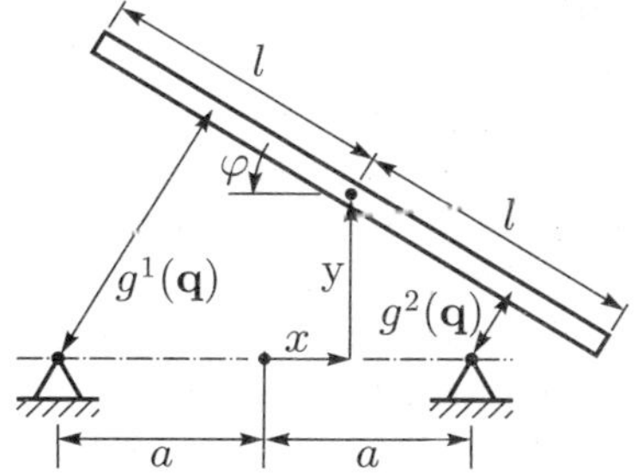

Figure 34. Mechanical model of the rocking rod.

the relative velocities (5.3) at the impact $\mathbf{q}_0$ are obtained as

$$\gamma^1 = u - a\omega, \qquad \gamma^2 = u + a\omega \tag{5.48}$$

with $\gamma^{1-} < 0$ and $\gamma^{2-} = 0$ due to our assumed pre-impact state. The collision itself is modelled as maximally dissipative, i.e. $\varepsilon := \varepsilon^1 = \varepsilon^2 = 0$ applies. The impact equations (5.5) are

$$\begin{pmatrix} m & 0 & 0 \\ 0 & m & 0 \\ 0 & 0 & \frac{1}{3}ml^2 \end{pmatrix} \begin{pmatrix} u^+ - u^- \\ v^+ - v^- \\ \omega^+ - \omega^- \end{pmatrix} = \begin{pmatrix} 0 \\ 1 \\ -a \end{pmatrix} \Lambda_1 + \begin{pmatrix} 0 \\ 1 \\ a \end{pmatrix} \Lambda_2, \tag{5.49}$$

from which we identify the mass matrix $\mathbf{M}$ to compute the gradients of the inequality constraints,

$$\nabla g^1(\mathbf{q}_0) = \frac{1}{m} \begin{pmatrix} 0 \\ 1 \\ -3\frac{a}{l^2} \end{pmatrix}, \qquad \nabla g^2(\mathbf{q}_0) = \frac{1}{m} \begin{pmatrix} 0 \\ 1 \\ 3\frac{a}{l^2} \end{pmatrix}, \tag{5.50}$$

as well as the angle α between them,

$$\cos\alpha = \frac{\nabla g^1(\mathbf{q}_0) \cdot \nabla g^2(\mathbf{q}_0)}{\|\nabla g^1(\mathbf{q}_0)\| \, \|\nabla g^1(\mathbf{q}_0)\|} = \frac{1 - 3\left(\frac{a}{l}\right)^2}{1 + 3\left(\frac{a}{l}\right)^2}. \tag{5.51}$$

We observe that this angle depends on the distance $2a$ between the obstacles and conclude that orthogonality of the gradients $\nabla g^i(\mathbf{q}_0)$ is attained if

$$\alpha = 90° \quad \Rightarrow \quad a = \frac{l}{\sqrt{3}}. \tag{5.52}$$

We may now choose among two methods on how to attack this collision problem. Either, one could solve the associated linear complementarity problem (3.31) to get the post-impact velocities $\mathbf{u}^+$ and transferred impulsive forces Λ_i, or to study the geometric properties of this impact configuration, from which we might get some qualitative answers why and how rocking works. We take here the second approach.

The impact geometry of the rocking rod is depicted in Figure 35, see also (Brogliato, 1999). The tangent cone $\mathcal{T}_C(\mathbf{q}_0)$ at the impact is generated by the intersection of two half-spaces with outward normals $-\nabla g^i(\mathbf{q}_0)$ according to (5.50), which approximate the two inequality constraints $g^i(\mathbf{q}) \geq 0$ (5.46) at the displacements $\mathbf{q}_0 = 0$ at which the collision takes place. The angle α of the cone $\mathcal{T}_C^{\perp}(\mathbf{q}_0)$ is by (5.51) related to the distance $2a$ between the obstacles and increases with a.

For small values of the distance between the obstacles ($a < l/\sqrt{3}$) the rod is rocking, which corresponds to the left diagram with $\alpha < 90°$. Since the right contact is closed just before the impact ($\gamma^{2-} = \nabla g^2(\mathbf{q}_0) \cdot \mathbf{u}^- = 0$), the pre-impact velocity $\mathbf{u}^-$ is at the boundary of the corresponding half-space and is directed towards the corner of the tangent cone formed by the intersection with the second half-space. The orthogonal decomposition (5.18) of $\mathbf{u}^-$ yields in this case $\mathbf{v} \neq 0$, $\mathbf{v}^{\perp} \neq 0$ and thus a post-impact velocity $\mathbf{u}^+ = \mathbf{v}$, because $\varepsilon = 0$. The system therefore leaves the impact configuration with a velocity $\mathbf{u}^+$ which is on the boundary of the other half-space ($\gamma^{1+} = \nabla g^1(\mathbf{q}_0) \cdot \mathbf{u}^+ =$

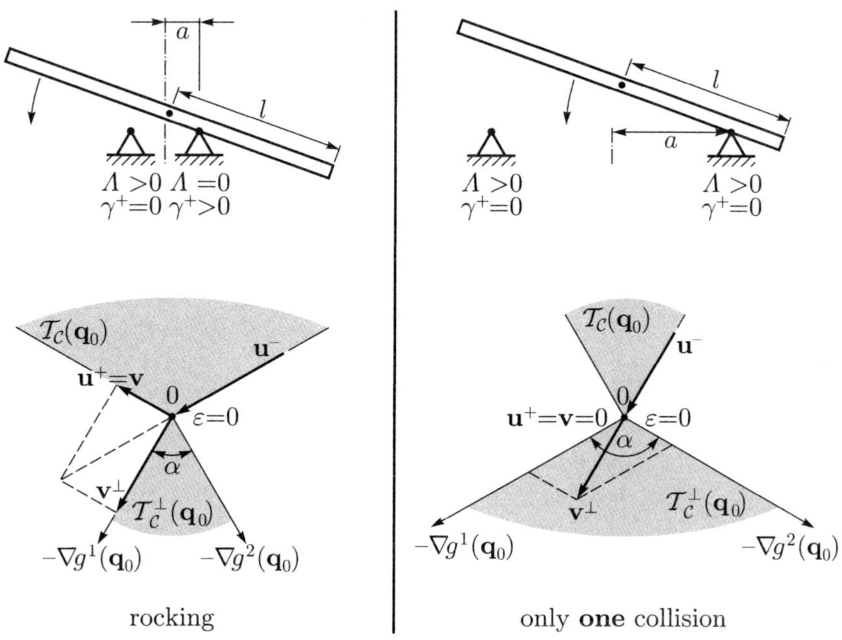

Figure 35. Impact geometry of the rocking rod.

0) and which is reduced in length proportional to the value of the pre-impact velocity, $\|\mathbf{u}^+\| = \nu\|\mathbf{u}^-\|$ for some $0 < \nu < 1$. Further, since $\mathbf{v}^\perp$ points into the same direction as $-\nabla g^1(\mathbf{q}_0)$, scalar impulsive forces $\Lambda_1 > 0$, $\Lambda_2 = 0$ are observed. The system is conservative for the impact-free motion. As a consequence, the post-impact velocity $\mathbf{u}^+$ is reversed to become the pre-impact velocity of the succeeding collision, $\mathbf{u}^-_{n+1} = -\mathbf{u}^+_n$. The rod is therefore rocking by an alternate turning around the two contact points. A collision with one of the obstacles causes an instantaneous detachment at the other contact with no impulsive force being transferred there. There is an infinite number of dissipative impacts (taking place in a finite time interval), leading to a loss of kinetic energy after each collision. At the end, when this infinite impact sequence has been passed, the system comes to a state at which the rod rests on both obstacles.

In the right diagram the distance between the two obstacles has been chosen large enough ($a > l/\sqrt{3}$) to admit an angle $\alpha > 90°$. In this case, the orthogonal decomposition yields $\mathbf{v} = 0$, $\mathbf{v}^\perp \neq 0$. Thus, $\mathbf{u}^+ = 0$, and the rod stops turning immediately after this very first impact and is never moving again. Further, $\mathbf{v}^\perp$ is in the interior of $\mathcal{T}_C^\perp(\mathbf{q}_0)$. This means that we have impulsive forces $\Lambda_{12} > 0$ at *both* contacts, produced by the collision of the rod against the left obstacle only. Note finally that the critical value $a = l/\sqrt{3}$, at which rocking changes to only one impact, corresponds by (5.52) to the situation that the gradients of the inequality constraints become orthogonal in the kinetic metric.

Bibliography

M. Anitescu, F. A. Potra, and D. E. Stewart. Time-stepping for three-dimensional rigid body dynamics. *Comp. Methods Appl. Mech. Eng.*, 177(3):183–197, 1999.

J. Aubin and I. Ekeland. *Applied Nonlinear Analysis.* John Wiley & Sons, New York, 1984.

B. Brogliato. *Nonsmooth Mechanics.* Communications and Control Engineering Series, Springer, London, 1999.

R. W. Cottle, J. S. Pang, and R. E. Stone. *The Linear Complementarity Problem.* Computer Science and Scientific Computing, Academic Press, London, 1992.

J. Elstrodt. *Maß- und Integrationstheorie.* Springer Verlag, Berlin Heidelberg New York, 1996.

M. Frémond. Rigid bodies collisions. *Physics Letters A*, 204:33–41, 1995.

M. Frémond. *Non-Smooth Thermomechanics.* Springer Verlag, Berlin, 2002.

Ch. Glocker. *Dynamik von Starrkörpersystemen mit Reibung und Stößen.* VDI-Fortschrittberichte Mechanik/Bruchmechanik, Reihe 18, Nr. 182, VDI-Verlag, Düsseldorf, 1995.

Ch. Glocker. Velocity jumps induced by C^0-constraints. *Proceedings of the ASME 1999 Design Engineering Technical Conferences, Las Vegas, Nevada, September 12–15,* 8 pages on CD-Rom, 1999.

Ch. Glocker. *Set-Valued Force Laws — Dynamics of Non-Smooth Systems.* Lecture Notes in Applied Mechanics, Vol. 1, Springer Verlag, Berlin, Heidelberg, 2001a.

Ch. Glocker. A geometric interpretation of Newtonian impacts with global dissipation coefficient. In *Cahiers Stephanois de Mathematiques Appliques*, Publ. de l'Equipe d'Analyse Numérique, UPRES EA 3058, no.2, 14 pages online, 2001b. Univ. Jean Monnet, Saint-Etienne.

Ch. Glocker. Concepts for modeling impacts without friction. *Acta Mechanica*, 168:1–19, 2004.

Ch. Glocker and U. Aeberhard. The geometry of Newton's cradle. In *Progress in Non-Smooth Mechanics and Analysis*, Kluwer Acad. Publ., to appear in 2005.

Ch. Glocker and C. Studer. Formulation and preparation for numerical evaluation of linear complementarity systems in dynamics. *Multibody System Dynamics*, to appear in 2005.

M. Jean. The non-smooth contact dynamics method. *Comput. Methods Appl. Mech. Engrg.*, 177:235–257, 1999.

C. Kane, E. A. Repetto, M. Ortiz, and J. E. Marsden. Finite element analysis of nonsmooth contact. *Comp. Methods Appl. Mech. Eng.*, 180:1–26, 1999.

M. D. P. Monteiro Marques. *Differential Inclusions in Nonsmooth Mechanical Problems. Shocks and Dry Friction.* Birkhäuser Verlag, Basel, 1993.

J. J. Moreau. Unilateral contact and dry friction in finite freedom dynamics. In *Non-Smooth Mechanics and Applications*, CISM Courses and Lectures no. 302, pages 1–82, 1988a. Springer Verlag, Wien.

J. J. Moreau. Bounded variation in time. In *Topics in Nonsmooth Mechanics*, pages 1–74, 1988b. Birkhäuser Verlag, Basel.

J. J. Moreau. Some numerical methods in multibody dynamics: Application to granular materials. *European Journal of Mechanics A/Solids,* 21 suppl. 4:93–114, 1994.

K. G. Murty. *Linear Complementarity, Linear and Nonlinear Programming.* Sigma Series in Applied Mathematics, Vol. 3, Heldermann Verlag, Berlin, 1988.

L. Paoli and M. Schatzman. A numerical scheme for impact problems I: The one-dimensional case. *SIAM Journal on Numerical Analysis,* 40(2):702–733, 2002a.

L. Paoli and M. Schatzman. A numerical scheme for impact problems II: The multidimensional case. *SIAM Journal on Numerical Analysis,* 40(2):734–768, 2002b.

J. G. Papastavridis. *Analytical Mechanics.* Oxford University Press, New York, 2002.

M. Payr and Ch. Glocker. Oblique frictional impact of a bar: Analysis and comparison of different impact laws. *Nonlinear Dynamics,* to appear in 2005.

F. Pfeiffer and Ch. Glocker. *Multibody Dynamics with Unilateral Contacts.* John Wiley & Sons, New York, 1996.

R. T. Rockafellar. *Convex Analysis.* Princeton Univ. Press, Princeton, New Jersey, 1972.

W. Rudin. *Real & Complex Analysis.* Tata McGraw-Hill Publ., New Delhi, 1981.

D. E. Stewart. Convergence of a time-stepping scheme for rigid body dynamics and resolution of Painlevé's problem. *Archives of Rational Mechanics and Analysis,* 145:215–260, 1998.

D. E. Stewart and J. C. Trinkle. An implicit time-stepping scheme for rigid body dynamics with inelastic collisions and Coulomb friction. *Int. J. Numer. Methods Engineering,* 39(15):2673–2691, 1996.

Approximation of variational and hemivariational inequalities of elliptic type. Applications to contact problems with friction.

Jaroslav Haslinger*

* Department of Numerical Mathematics, Faculty of Mathematics and Physics, Charles University, Praha, Czech Republic

Abstract This article deals with approximations of differential inclusions of elliptic type. We start with problems which involve only monotone mappings,i.e. with classical variational inequalities of the first and the second kind. Next we show how to approximate a class of inclusion problems called hemivariational inequalities with nonmonotone multivalued mappings. These results are then used for the approximation and the numerical realization of contact problems with different models of friction.

Introduction

The mathematical model of many problems of mathematical physics leads to the following inclusion problem:

$$\left.\begin{array}{l} Find\ u \in V\ such\ that \\ f - Au \in Bu\ , \end{array}\right\} \qquad (P)$$

where A is a *singlevalued* mapping from a Banach space V into a subspace $Y \subseteq V'$ (dual of V), B is a *multivalued* mapping from V into Y and $f \in V'$ is given. Introducing the new variable $\Xi := f - Au$, problem (P) can be written as follows:

$$\left.\begin{array}{l} Find\ (u, \Xi) \in V \times Y\ such\ that \\ Au + \Xi = f\ in\ V' \\ \Xi \in\ Bu\ . \end{array}\right\} \qquad (P)'$$

If the mapping B is *monotone*, problem (P) or $(P)'$ leads to an inequality of elliptic type. The mathematical theory of elliptic inequalities, which goes back to sixties, is now very well established from the theoretical as well as the computational point of view [Lions and Stampacchia 1967], [Lions 1969], [Glowinski and Lions and Trémolière 1981]. The monotonicity assumption on the relation between u and Ξ, expressed by the inclusion $\Xi \in Bu$ is however very restrictive. In practice we meet situations when this relation is not monotone (some of them are discussed in this book). To describe such problems, appropriate tools of nonsmooth analysis have to be used. In eighties, P. D. Panagiotopoulos introduced the so–called *hemivariational inequalities* by means of which many nonmonotone phenomena in solid mechanics such as nonmonotone friction and unilateral conditions can be modeled [Panagiotopoulos 1985].

Since solutions to (P) (or $(P)'$) are hardly available, an approximation of continuous models is necessary. By an approximation of (P) we call the following problem:

$$\left.\begin{array}{l} \textit{Find } u_h \in V_h \textit{ such that} \\ f_h - A_h u_h \in B_h u_h \ , \end{array}\right\} \qquad (P)_h$$

where $V_h \subseteq V$ is a finite–dimensional approximation of V and A_h, B_h, f_h are appropriate discretizations of A, B and f, respectively. In a similar way one can define an approximation of $(P)'$:

$$\left.\begin{array}{l} \textit{Find } (u_h, \Xi_h) \in V_h \times Y_h \textit{ such that} \\ A_h u_h + \Xi_h = f_h \\ \Xi_h \in B_h u_h \ , \end{array}\right\} \qquad (P)'_h$$

where Y_h is a finite–dimensional approximation of Y.

The first part of this article deals with approximations of abstract variational and hemivariational inequalities of elliptic type in Hilbert spaces. In Section 1 we briefly recall primal, mixed and dual formulations of elliptic inequalities of the first and the second kind, their approximations and convergence results. Sections 2 and 3 deal with unconstrained and constrained scalar hemivariational inequalities. We focus on their approximation by finite elements using the scheme $(P)'_h$ and on the convergence analysis. This presentation is based on results from (Haslinger and Miettinen and Panagiotopoulos [1999]), where a reader can find more material including approximations of vector and parabolic hemivariational inequalities. The rest of this article is devoted to the application of the abstract results to the formulation, approximation and the numerical realization of contact problems with friction. Section 4 concerns of the classical Signorini problem with Coulomb's friction in 2 and 3D and a coefficient of friction which eventually may depend on the solution itself. We use a fixed point approach combined with the dual formulation of each iterative step which turns out to be an efficient way of solving these problems. In Section 5 we show how hemivariational inequalities can be used for modeling contact problems with nonmonotone friction. Both sections are completed with model examples illustrating the theory.

The article tries to be self–contained, neverthless some knowledge of elements of nonsmooth analysis [Clarke 1983] and Sobolev spaces [Nečas 1967] is necessary.

Acknowledgement The author of this article wishes to express his gratitude to ing. O. Vlach, dr. R. Kučera, PhD and Mgr. Z. Morávková who computed the examples of Sections 4 and 5. This work was supported by grant IAA1075402 of the Czech Academy of Sciences and by MSM 0021620839. A part of this contribution concerning hemivariational inequalities contains results of the bilateral co–operation between Charles University and Aristotle University in Thessaloniki.

1 Primal, mixed and dual formulations of variational inequalities of elliptic type and their approximations

Let V be a real Hilbert space, V' its dual with a duality pairing $\langle\ ,\ \rangle$. The norm in V, V' will be denoted by $\|\ \|$, $\|\ \|_*$, respectively. Let $K \subset V$ be a non-empty, closed convex set. Finally, by $a : V \times V \mapsto \mathbb{R}^1$ we denote a bilinear form satisfying the following assumptions:

$$\begin{aligned} &(continuity) && \exists M = const. > 0 : && |a(u,v)| \leq M\|u\|\|v\| \quad \forall u,v \in V\ , && (1.1)\\ &(V-ellipticity) && \exists \alpha = const. > 0 : && a(v,v) \geq \alpha\|v\|^2 \quad \forall v \in V\ . && (1.2)\end{aligned}$$

In all applications presented in this article, the form a will be also *symmetric* in V:

$$a(u,v) = a(v,u) \quad \forall u,v \in V\ . \tag{1.3}$$

By *an abstract variational inequality of the first kind* we call the problem

$$\textit{Find } u \in K: \ a(u, v-u) \geq \langle f, v-u\rangle \quad \forall v \in K\ , \tag{1.4}$$

where $f \in V'$ is given. The following theorem holds.

Theorem 1.1. *Let the bilinear form a satisfy (1.1) and (1.2). Then the variational inequality (1.4) has a unique solution u for any $f \in V'$. In addition, if a is symmetric in V, then u can be equivalently characterized as the solution of the following minimization problem:*

$$\textit{Find } u \in K: \quad J(u) = \min_{v\in K} J(v)\ , \tag{1.5}$$

where $J(v) := \frac{1}{2}a(v,v) - \langle f, v\rangle$.

Denote by $A \in L(V, V')$ a lincar, continuous mapping defined by

$$\langle Au, v\rangle = a(u,v) \quad \forall u,v \in V\ .$$

The equivalent expressions of (1.4) and (1.5) are:

$$\begin{aligned} f - Au &\in N_K(u)\ ,\\ 0 &\in \partial J(u) + N_K(u)\ ,\end{aligned}$$

respectively, where $N_K(u)$ is the normal cone to K at u and ∂ stands for the subgradient of a convex function.

Let $j : V \mapsto \mathbb{R}^1 \cup \{-\infty, \infty\}$ be a *convex, weakly lower semicontinuous* and *proper functional.* By *an abstract variational inequality of the second kind* we call the following problem:

$$\textit{Find } u \in V: \ a(u, v-u) + j(v) - j(u) \geq \langle f, v-u\rangle \quad \forall v \in V\ . \tag{1.6}$$

Theorem 1.2. *Let (1.1), (1.2) be satisfied. Then the variational inequality (1.6) has a unique solution for any* $f \in V'$. *In addition, if* a *is symmetric in* V, *(1.6) is equivalent to*

$$\textit{Find } u \in V: \quad J(u) = \min_{v \in V} J(v) , \tag{1.7}$$

where $J(v) := \frac{1}{2}a(v,v) + j(v) - \langle f, v \rangle$.

As before, (1.6), (1.7) are equivalent to the following inclusion problems:

$$f - Au \in \partial j(u) ,$$
$$0 \in \partial J(u) ,$$

respectively.

If the bilinear form a is symmetric, variational inequalities of elliptic type lead to *constrained* minimization problems for generally *non-smooth* functionals. One of possibilities how to treat such a type of problems is to use a duality technique. Next we present the main idea of this approach.

Let Q be another Hilbert space, Q' its dual and $\Lambda \subset Q$ be a *closed, convex cone* containing the zero element of Q. Further, let $[\ ,\ ]$ denote a duality pairing between Q and Q', $\|\ \|_Q$ be the norm in Q, and $b: V \times Q \mapsto \mathbb{R}^1$ be a *continuous* bilinear form. We shall suppose that the following characterization of K holds:

$$v \in K \Leftrightarrow v \in V \textit{ and } b(v,\mu) \le [g,\mu]\ \forall \mu \in \Lambda , \tag{1.8}$$

where $g \in Q'$ is given.

By *the mixed variational formulation* of (1.4) we call the following problem:

$$\left.\begin{array}{rcl} & \textit{Find } (w,\lambda) \in V \times \Lambda \textit{ such that} & \\ a(w,v) + b(v,\lambda) & = & \langle f, v \rangle\ \forall v \in V , \\ b(w, \mu - \lambda) & \le & [g, \mu - \lambda]\ \forall \mu \in \Lambda . \end{array}\right\} \tag{1.9}$$

The relation between (1.4) and (1.9) follows from the next lemma.

Lemma 1.3. *Let* $(w,\lambda) \in V \times \Lambda$ *be a solution to (1.9). Then* w *solves (1.4).*

Proof. Inserting $\mu = 0, 2\lambda$ into $(1.9)_3$ we see that

$$b(w,\lambda) = [g,\lambda] \text{ and } b(w,\mu) \le [g,\mu]\ \forall \mu \in \Lambda , \tag{1.10}$$

so that $w \in K$. Substitution $v := v - w$, $v \in K$ into $(1.9)_2$ leads to

$$a(w, v - w) \ge a(w, v - w) + b(v - w, \lambda) = \langle f, v - w \rangle\ \forall v \in K$$

making use of (1.8) and (1.10). Hence w solves (1.4). □

Let $B \in L(V, Q')$ be a continuous mapping defined by:

$$[Bv, \mu] = b(v, \mu) \quad \forall (v, \mu) \in V \times Q.$$

Then (1.9) is equivalent to:

$$\left.\begin{array}{r} Aw + B^t\lambda = f \ in \ V' \\ Bw - g \in N_\Lambda(\lambda) \ in \ Q' \ , \end{array}\right\} \tag{1.11}$$

where B^t is the transpose of B.

If the bilinear form a is *symmetric* in V, the mixed formulation (1.9) is equivalent to the following *saddle-point* problem:

$$\left.\begin{array}{l} Find \ (w,\lambda) \in V \times \Lambda \ such \ that \\ \mathcal{L}(w,\mu) \le \mathcal{L}(w,\lambda) \le \mathcal{L}(v,\lambda) \ \forall (v,\mu) \in V \times \Lambda \ , \end{array}\right\} \tag{1.12}$$

where $\mathcal{L}(v,\mu) := \frac{1}{2}a(v,v) + b(v,\mu) - [g,\mu] - \langle f,v\rangle$.

Next we prove the existence and uniqueness of a solution to (1.9) provided that the bilinear form a is *symmetric*. To this end we shall need the following assumption:

$$\exists \beta = const. > 0 : \quad \sup_{\substack{v\in V \\ v\neq 0}} \frac{b(v,\mu)}{\|v\|} \ge \beta \|\mu\|_Q \ \forall \mu \in Q \ . \tag{1.13}$$

Theorem 1.4. *Let (1.13) be satisfied. Then there exists a unique saddle-point of* $\mathcal{L}$ *on* $V \times \Lambda$ *or equivalently, the mixed formulation (1.9) has a unique solution.*

Proof. The mapping

$$v \mapsto \mathcal{L}(v,\mu) \ , \quad v \in V$$

is strictly convex and weakly lower semicontinuous in V for any $\mu \in \Lambda$ and

$$\mu \mapsto \mathcal{L}(v,\mu) \ , \quad \mu \in \Lambda$$

is concave and weakly continuous in Q for any $v \in V$. Further

$$\lim_{\|v\|\to\infty} \mathcal{L}(v,\mu) = +\infty \quad \forall \mu \in Q.$$

Thus it is sufficient to show (see [Ekeland and Temam 1976]) that

$$\lim_{\substack{\|\mu\|_Q\to\infty \\ \mu\in\Lambda}} \inf_{v\in V} \mathcal{L}(v,\mu) = -\infty \ . \tag{1.14}$$

This is a consequence of (1.13). Indeed, let $\bar{\mu} \in \Lambda$ be fixed and $u_{\bar{\mu}} \in V$ be the element satisfying

$$\mathcal{L}(u_{\bar{\mu}},\bar{\mu}) = \inf_{v\in V} \mathcal{L}(v,\bar{\mu}) = \frac{1}{2}a(u_{\bar{\mu}},u_{\bar{\mu}}) + b(u_{\bar{\mu}},\bar{\mu}) - [g,\bar{\mu}] - \langle f,u_{\bar{\mu}}\rangle \ . \tag{1.15}$$

Such $u_{\bar{\mu}} \in V$ is unique and solves the problem

$$a(u_{\bar{\mu}},v) + b(v,\bar{\mu}) = \langle f,v\rangle \quad \forall v \in V \ . \tag{1.16}$$

Inserting $v := u_{\bar{\mu}}$ into (1.16) and eliminating the second and the fourth term on the right of (1.15) we obtain:

$$\inf_{v \in V} \mathcal{L}(v, \bar{\mu}) \leq -\frac{\alpha}{2} \|u_{\bar{\mu}}\|^2 + \|g\|_{Q'} \|\bar{\mu}\|_Q . \tag{1.17}$$

From (1.13) and (1.16) it follows that

$$\beta \|\bar{\mu}\|_Q \leq M \|u_{\bar{\mu}}\| + \|f\|_* .$$

This together with (1.17) yields (1.14). Therefore there exists at least one saddle-point (w, λ). Since the mapping $v \mapsto \mathcal{L}(v, \mu)$ is strictly convex in V for any $\mu \in Q$, the first component w is unique. The uniqueness of the second component is a simple consequence of (1.13). □

In a similar way one can introduce the mixed formulation of the variational inequality of the second kind (1.6). Let the convex functional j in (1.6) be such that

$$j(v) = \sup_{\mu \in \Lambda} b(v, \mu) , \tag{1.18}$$

where $b : V \times Q \mapsto \mathbb{R}^1$ is a continuous bilinear form and $\Lambda \subseteq Q$ is a *bounded, closed convex* subset of Q. The mixed variational formulation of (1.6) reads as follows:

$$\left.\begin{array}{lcll} \multicolumn{4}{l}{\textit{Find } (w, \lambda) \in V \times \Lambda \textit{ such that}} \\ a(w, v) + b(v, \lambda) & = & \langle f, v \rangle & \forall v \in V \\ b(w, \mu - \lambda) & \leq & 0 & \forall \mu \in \Lambda . \end{array}\right\} \tag{1.19}$$

It is easy to prove the following lemma.

Lemma 1.5. *Let (w, λ) be a solution to (1.19). Then w solves (1.6).*

Proof. From (1.18) and $(1.19)_3$ it follows that

$$j(w) = b(w, \lambda) .$$

Inserting $v := v - w$, $v \in V$ into $(1.19)_2$ and using that $j(v) \geq b(v, \lambda)$ for any $v \in V$ we obtain:

$$a(w, v - w) + j(v) - j(w) \geq \langle f, v - w \rangle \quad \forall v \in V .$$

□

If the bilinear form a is symmetric, (1.19) is again equivalent to the following saddle-point problem:

$$\left.\begin{array}{l} \textit{Find } (w, \lambda) \in V \times \Lambda \textit{ such that} \\ \mathcal{L}(w, \mu) \leq \mathcal{L}(w, \lambda) \leq \mathcal{L}(v, \lambda) \quad \forall (v, \mu) \in V \times \Lambda , \end{array}\right\} \tag{1.20}$$

where $\mathcal{L}(v, \mu) := \frac{1}{2} a(v, v) + b(v, \mu) - \langle f, v \rangle$. The existence of a saddle-point of $\mathcal{L}$ on $V \times \Lambda$ follows from

Theorem 1.6. *Let j be given by (1.18) with Λ being a closed, bounded convex subset of Q. Then there exists a saddle-point of $\mathcal{L}$ on $V \times \Lambda$, whose first component w is unique.*

Proof. The existence of a saddle-point is a consequence of boundedness of Λ (see [Ekeland and Temam 1976]). □

Let us consider the mixed formulation (1.11) of an elliptic inequality of the first kind. From $(1.11)_1$ one can express the first component w:

$$w = A^{-1}(f - B^t\lambda) \ . \tag{1.21}$$

Inserting (1.21) into $(1.11)_2$ we obtain:

$$BA^{-1}f - BA^{-1}B^t\lambda - g \in N_\Lambda(\lambda) \ .$$

This is a variational inequality of the first kind for the second component λ which is equivalent to:

$$\left.\begin{array}{l} Find\ \lambda \in \Lambda\ such\ that \\ \mathcal{B}(\lambda, \mu - \lambda) \geq [\mathcal{F}, \mu - \lambda] \quad \forall \mu \in \Lambda \ , \end{array}\right\} \tag{1.22}$$

where $\mathcal{B} : Q \times Q \mapsto \mathbb{R}^1$, $\mathcal{F} \in Q'$ is the bilinear, linear form, respectively defined by

$$\mathcal{B}(\mu, \lambda) := [BA^{-1}B^t\mu, \lambda] \tag{1.23}$$

$$\mathcal{F} := BA^{-1}f - g \ . \tag{1.24}$$

Theorem 1.7. *Let (1.13) be satisfied. Then the variational inequality (1.22) has a unique solution λ which is at the same time the second component of the solution to (1.9).*

Proof. From (1.13) it easily follows that $\mathcal{B}$ is a continuous and Q-elliptic bilinear form on Q. □

The variational inequality (1.22) is termed the *dual formulation* of (1.4). In a similar way we can introduce the dual formulation of an elliptic inequality of the second kind (1.6).

Approximations of elliptic variational inequalities

Let $\{V_h\}$, $h \to 0+$ be a family of finite-dimensional subspaces of V, $\dim V_h = n(h) \to \infty$ as $h \to 0+$. Further, let $K_h \subseteq V_h$ be a *closed, convex* subset of V_h not necessarily being a part of K. By an approximation of (1.4) we mean the following problem:

$$Find\ u_h \in K_h : \ a(u_h, v_h - u_h) \geq \langle f, v_h - u_h \rangle \quad \forall v_h \in K_h \ , \tag{1.25}$$

or

$$Find\ u_h \in K_h : \ J(u_h) = \min_{v_h \in K_h} J(v_h) \quad \forall v_h \in K_h \ , \tag{1.26}$$

where $J(v) := \frac{1}{2}a(v, v) - \langle f, v \rangle$ if a is symmetric in V.

To ensure convergence of approximate solutions u_h, $h \to 0+$ to the solution of (1.4), the system $\{K_h\}$, $h \to 0+$ has to satisfy the following assumptions:

$$\forall v \in K \ \exists \{v_h\},\ v_h \in K_h:\ v_h \to v \ in \ V,\ h \to 0+\ ; \tag{1.27}$$

$$if \ v_h \rightharpoonup v \ in \ V \ (weakly),\ v_h \in K_h \Rightarrow v \in K\ . \tag{1.28}$$

The following convergence result holds.

Theorem 1.8. *Let (1.27) and (1.28) be satisfied. Then the sequence $\{u_h\}$ of solutions to (1.25) tends strongly to the solution u of (1.4).*

Remark 1.9. Let us notice that (1.28) is automatically satisfied whenever $K_h \subset K$ for every $h > 0$, i.e when K_h are *inner approximations* of K.

Let $h > 0$ be fixed, $\dim V_h = n := n(h)$, and $\{\varphi_i\}_{i=1}^n$ be a basis of V_h. Then V_h can be identified with the Eucledean space $\mathbb{R}^n$ by means of the isomorphism $\mathcal{T} : V_h \mapsto \mathbb{R}^n$ defined by

$$\mathcal{T} v_h = \mathbf{x} = (x_1, \ldots, x_n) \in \mathbb{R}^n, \quad v_h \in V_h\ ,$$

where $v_h = \sum_{i=1}^n x_i \varphi_i$.

By $\mathcal{K} \subset \mathbb{R}^n$ we denote a closed convex subset of $\mathbb{R}^n$:

$$\mathcal{K} = \{\mathbf{x} \in \mathbb{R}^n |\ \mathcal{T}^{-1}\mathbf{x} \in K_h\}\ ,$$

where $\mathcal{T}^{-1} : \mathbb{R}^n \mapsto V_h$ is the inverse of $\mathcal{T}$. The algebraic form of (1.25) reads as follows:

$$Find \ \mathbf{u} \in \mathcal{K}:\ (\mathbf{A}\mathbf{u}, \mathbf{v} - \mathbf{u})_{\mathbb{R}^n} \geq (\mathbf{f}, \mathbf{v} - \mathbf{u})_{\mathbb{R}^n} \quad \forall \mathbf{v} \in \mathcal{K}\ , \tag{1.29}$$

where $(\ ,\ )_{\mathbb{R}^n}$ stands for the scalar product in $\mathbb{R}^n$, $\mathbf{A} = (a_{ij})_{i,j=1}^n$ is the stiffness matrix and $\mathbf{f} = (f_1, \ldots, f_n) \in \mathbb{R}^n$ is the load vector whose elements are defined by:

$$a_{ij} = a(\varphi_j, \varphi_i)\ , \quad f_i = \langle f, \varphi_i \rangle\ ;\ i, j = 1, \ldots, n\ .$$

In addition, if a is symmetric, problem (1.26) leads to the following non-linear mathematical programming problem:

$$Find \ \mathbf{u} \in \mathcal{K}:\ \mathcal{J}(\mathbf{u}) = \min_{\mathbf{v} \in \mathcal{K}} \mathcal{J}(\mathbf{v})\ , \tag{1.30}$$

where $\mathcal{J}(\mathbf{v}) = \frac{1}{2}(\mathbf{A}\mathbf{v}, \mathbf{v})_{\mathbb{R}^n} - (\mathbf{f}, \mathbf{v})_{\mathbb{R}^n}$.

A similar convergence result can be established for a variational inequality of the second kind.

Problem 1.10 Let $u_h \in V_h$ be a solution of the following discrete variational inequality of the second kind:

$$a(u_h, v_h - u_h) + j(v_h) - j(u_h) \geq \langle f, v_h - u_h \rangle \quad \forall v_h \in V_h\ . \tag{1.31}$$

Suppose that the system $\{V_h\}$, $h \to 0+$ is *dense* in V:

$$\forall v \in V\ \exists \{v_h\}, v_h \in V_h : \quad v_h \to v\ in\ V, h \to 0+$$

and j is *continuous* in V:

$$v_h \to v\ in\ V\ \Rightarrow\ j(v_h) \to j(v),\ h \to 0+\ .$$

Then $u_h \to u$ in V and u solves (1.6). Prove!

We now pass to an approximation of the mixed formulation (1.9). Let $\{V_h\}$, $\{Q_H\}$ be two systems of finite dimensional subspaces of V and Q, respectively, $\dim V_h = n(h) \to \infty$, $h \to 0+$, $\dim Q_H = m(H) \to \infty$, $H \to 0+$. Let us notice that we use two different letters for mesh sizes in order to emphasize that meshes used for the construction of V_h and Q_H are independent, in general. Nevertheless we shall suppose that

$$h \to 0+\ \Leftrightarrow\ H \to 0+\ . \tag{1.32}$$

Finally, let $\{\Lambda_H\}$, $H \to 0+$, $\Lambda_H \subset Q_H$ be a system of closed, convex cones containing the zero element of Q. The approximation of the mixed formulation (1.9) reads as follows:

$$\left.\begin{array}{l} Find\ (w_h, \lambda_H) \in V_h \times \Lambda_H\ such\ that \\ a(w_h, v_h) + b(v_h, \lambda_H) = \langle f, v_h\rangle \quad \forall v_h \in V_h \\ b(w_h, \mu_H - \lambda_H) \le [g, \mu_H - \lambda_H] \quad \forall \mu_H \in \Lambda_H\ . \end{array}\right\} \tag{1.33}$$

To prove the existence and uniqueness of a solution to (1.33) we shall suppose that the following condition is satisfied:

$$Let\ \mu_H \in Q_H\ be\ such\ that\ b(v_h, \mu_H) = 0\ \forall v_h \in V_h\ .\ Then\ \mu_H = 0 \tag{1.34}$$

It holds:

Theorem 1.11. *Let (1.34) be satisfied. Then there exists a unique solution (w_h, λ_H) of (1.33).*

Proof. From (1.34) it follows that there exists a constant $\beta := \beta(h, H) > 0$ depending on h and H, in general and such that

$$\sup_{\substack{v_h \in V_h \\ v_h \neq 0}} \frac{b(v_h, \mu_H)}{\|v_h\|} \ge \beta \|\mu_H\|_Q\ . \tag{1.35}$$

The rest of the proof proceeds in the same way as the one of Theorem 1.4. □

We now give the interpretation of the first component w_h of the solution to (1.33). Denote by

$$K_{hH} = \{v_h \in V_h \mid b(v_h, \mu_H) \le [g, \mu_H]\ \forall \mu_H \in \Lambda_H\} \tag{1.36}$$

a closed, convex subset of V_h. K_{hH} can be viewed to be an approximation of K. From $(1.33)_3$ it easily follows that $w_h \in K_{hH}$ and from $(1.33)_2$ we see that w_h solves the variational inequality of the first kind:

$$w_h \in K_{hH} : a(w_h, v_h - w_h) \geq \langle f, v_h - w_h \rangle \quad \forall v_h \in K_{hH} . \tag{1.37}$$

Next we shall study the relation between solutions to (1.9) and (1.33). To this end we shall need the following assumptions:

$$\forall v \in V \ \exists \{v_h\}, v_h \in V_h : \ v_h \to v \ in \ V \ , \ h \to 0+; \tag{1.38}$$

$$\forall \mu \in \Lambda \ \exists \{\mu_H\}, \mu_H \in \Lambda_H : \ \mu_H \to \mu \ in \ Q \ , \ H \to 0+; \tag{1.39}$$

$$\mu_H \rightharpoonup \mu \ (weakly) \ in \ Q, \mu_H \in \Lambda_H \ \Rightarrow \ \mu \in \Lambda . \tag{1.40}$$

The following convergence result holds:

Lemma 1.12. *Let (1.38)-(1.40) be satisfied and* (w_h, λ_H) *be a solution to (1.33). Then any weak accumulation point of* $\{(w_h, \lambda_H)\}$ *solves (1.9).*

Proof. To simplify notation we shall suppose that the original sequence $\{(w_h, \lambda_H)\}$ tends weakly to some (w, λ):

$$w_h \rightharpoonup w \text{ in } V, \ \lambda_H \rightharpoonup \lambda \text{ in } Q . \tag{1.41}$$

We want to prove that (w, λ) solves (1.9). Let $(\bar{v}, \bar{\mu}) \in V \times \Lambda$ be arbitrary, but fixed. Then there exists a sequence $\{(\bar{v}_h, \bar{\lambda}_H)\}$, $\bar{v}_h \in V_h$, $\bar{\mu}_H \in \Lambda_H$ such that

$$\bar{v}_h \to \bar{v} \text{ in } V, \ \bar{\mu}_H \to \bar{\mu} \text{ in } Q \tag{1.42}$$

as follows from (1.38) and (1.39). Using $(\bar{v}_h, \bar{\mu}_H)$ as test functions in (1.33) and letting $h, H \to 0+$ there we obtain:

$$a(w, \bar{v}) + b(\bar{v}, \lambda) = \langle f, \bar{v} \rangle \tag{1.43}$$

$$\lim_{h,H \to 0+} b(w_h, \bar{\mu}_H - \lambda_H) \leq [g, \bar{\mu} - \lambda] \tag{1.44}$$

provided that the limit on the left of (1.44) exists. We now prove that

$$\lim_{h,H \to 0+} b(w_h, \bar{\mu}_H - \lambda_H) = b(w, \bar{\mu} - \lambda) .$$

From this and (1.40) which implies that $\lambda \in \Lambda$, the assertion of lemma follows. Clearly it will be sufficient to show that

$$\lim_{h,H \to 0+} b(w_h, \lambda_H) = b(w, \lambda) . \tag{1.45}$$

From $(1.33)_3$ it follows:

$$b(w_h, \bar{\mu}_H) - [g, \bar{\mu}_H - \lambda_H] \leq b(w_h, \lambda_H) ,$$

so that

$$b(w, \bar{\mu}) - [g, \bar{\mu} - \lambda] \le \liminf_{h,H \to 0+} b(w_h, \lambda_H) \tag{1.46}$$

holds for any $\bar{\mu} \in \Lambda$. Choosing $\bar{\mu} = \lambda$ in (1.46) we have:

$$b(w, \lambda) \le \liminf_{h,H \to 0+} b(w_h, \lambda_H) \ . \tag{1.47}$$

On the other hand from $(1.33)_2$ we see that

$$b(w_h, \lambda_H) = \langle f, w_h \rangle - a(w_h, w_h) \ .$$

Therefore

$$\limsup_{h,H \to 0+} b(w_h, \lambda_H) \le \langle f, w \rangle - a(w, w) = b(w, \lambda)$$

making use of weak lower semicontinuity of a. From this and (1.47), the equality (1.45) follows. □

A standard way of proving the existence of weakly convergent subsequences is to show that original sequences are bounded in norms of the respective spaces. To ensure the boundedness of $\{w_h\}$ we shall suppose that there exists a bounded sequence $\{v_{hH}\}$ of elements of K_{hH}:

$$\exists c = const. > 0 \ \exists \{v_{hH}\}, v_{hH} \in K_{hH} : \ \|v_{hH}\| \le c \quad \forall h, H > 0 \ . \tag{1.48}$$

The boundedness of $\{\lambda_H\}$ will be guaranteed by the so-called *Ladyzhenskaya-Babuška-Brezzi* (LBB) condition:

$$\exists \beta = const. > 0 : \ \sup_{\substack{v_h \in V_h \\ v_h \ne 0}} \frac{b(v_h, \mu_H)}{\|v_h\|} \ge \beta \|\mu_H\|_Q \tag{1.49}$$

holds for any $\mu_H \in Q_H$ and any $h, H > 0$.

Remark 1.13. Unlike to (1.35), the constant β in (1.49) is now independent of h and H.

Problem 1.14 Let (1.48) and (1.49) be satisfied. Prove that the sequence $\{(w_h, \lambda_H)\}$ is bounded in $V \times Q$.

With the LBB-condition at hand one can improve the convergence result of Lemma 1.12.

Theorem 1.15. *Let (1.38)-(1.40), (1.48) and (1.49) be satisfied. Suppose that (1.9) has a unique solution (w, λ). Then*

$$w_h \to w \ in \ V \ , \ h \to 0+ \ ,$$
$$\lambda_H \to \lambda \ in \ Q \ , \ H \to 0+ \ .$$

For the proof we refer to [Haslinger and Hlaváček and Nečas 1996].

Let the mesh parameters h and H be fixed. The algebraic form of (1.33) reads as follows:

$$\left.\begin{array}{l} \textit{Find } (\mathbf{w}, \boldsymbol{\lambda}) \in \mathbb{R}^n \times \boldsymbol{\Lambda} \textit{ such that} \\ \mathbf{A}\mathbf{w} + \mathbf{B}^t\boldsymbol{\lambda} = \mathbf{f} \\ (\mathbf{B}\mathbf{w}, \boldsymbol{\mu} - \boldsymbol{\lambda})_{\mathbb{R}^m} \le (\mathbf{g}, \boldsymbol{\mu} - \boldsymbol{\lambda})_{\mathbb{R}^m} \quad \forall \boldsymbol{\mu} \in \boldsymbol{\Lambda} \ , \end{array}\right\} \tag{1.50}$$

where $\mathbf{A} = (a_{ij})_{i,j=1}^n$, $a_{ij} = a(\varphi_j, \varphi_i)$ is the stiffness matrix, $\mathbf{B} = (b_{kl})_{k=1,\dots,m}^{l=1,\dots,n}$, $b_{kl} = b(\psi_k, \varphi_l)$ is the kinematic transformation matrix, $\mathbf{f} = (f_i)_{i=1}^n$, $f_i = \langle f, \varphi_i \rangle$ and $\mathbf{g} = (g_j)_{j=1}^m$, $g_j = [g, \psi_j]$. Further, $\{\varphi_i\}_{i=1}^n$, $\{\psi_j\}_{j=1}^m$ are basis functions of V_h and Q_H, respectively and

$$\boldsymbol{\Lambda} = \{\boldsymbol{\mu} = (\mu_1, \dots, \mu_m) \in \mathbb{R}^m | \ \textstyle\sum_{j=1}^m \mu_j \psi_j \in \Lambda_H\} \ . \tag{1.51}$$

The components of $\mathbf{w}$ and $\boldsymbol{\lambda}$ are the coordinates of w_h and λ_H with respect to $\{\varphi_i\}_{i=1}^n$ and $\{\psi_j\}_{j=1}^m$, respectively.

In a similar way one can introduce an approximation of the mixed formulation of elliptic inequalities of the second kind (1.19). Let $\{\Lambda_H\}$, $\Lambda_H \subset Q_H$ be a system of *uniformly bounded*, closed convex subsets of Q_H:

$$\exists c = const. > 0 : \ \|\mu_H\|_Q \le c \quad \forall \mu_H \in \Lambda_H, \ \forall H > 0 \ . \tag{1.52}$$

The approximation of (1.19) is defined as follows:

$$\left.\begin{array}{l} \textit{Find } (w_h, \lambda_H) \in V_h \times \Lambda_H \textit{ such that} \\ a(w_h, v_h) + b(v_h, \lambda_H) = \langle f, v_h \rangle \quad \forall v_h \in V_h \\ b(w_h, \mu_H - \lambda_H) \le 0 \quad \mu_H \in \Lambda_H \ . \end{array}\right\} \tag{1.53}$$

Since the set Λ_H is bounded we immediately have the following existence result.

Theorem 1.16. *There exists a solution (w_h, λ_H) to (1.53) whose first component w_h is unique. In addition, if (1.34) is satisfied then also the second component λ_H is unique.*

Denote

$$j_H(v_h) = \sup_{\mu_H \in \Lambda_H} b(v_h, \mu_H) \ , \quad v_h \in V_h \tag{1.54}$$

the approximation of the functional j defined by (1.18). It is left as an easy exercise to prove:

Lemma 1.17. *Let (w_h, λ_H) be a solution to (1.53). Then $w_h \in V_h$ solves the variational inequality of the second kind:*

$$a(w_h, v_h - w_h) + j_H(v_h) - j_H(w_h) \ge \langle f, v_h - w_h \rangle \quad \forall v_h \in V_h \ . \tag{1.55}$$

Problem 1.18 Prove Lemma 1.17.

Due to the boundedness of the system $\{\Lambda_H\}$, $H \to 0+$, the convergence analysis is simpler compared with the previous case.

Theorem 1.19. *Let (1.38)-(1.40) and (1.52) be satisfied. Suppose that (1.19) has a unique solution (w, λ). Then*

$$w_h \to w \text{ in } V\ , \quad \lambda_H \rightharpoonup \lambda \text{ in } Q\ .$$

In addition, if the LBB-condition (1.35) is satisfied, then $\lambda_H \to \lambda$ in Q.

Problem 1.20 Prove Theorem 1.19.

The rest of this part deals with an approximation of the dual variational formulation (1.22). This time we have to approximate not only the convex set Λ but also the mapping A^{-1} appearing in the definition of $\mathcal{B}$ and $\mathcal{F}$ (see (1.23), (1.24)) since its explicit form is not available, in general.

Let $\{\Lambda_H\}$, $H \to 0+$, $\Lambda_H \subset Q_H$ and $\{A_h^{-1}\}$, $h \to 0+$, $A_h^{-1} \in L(V_h', V_h)$ be systems of approximations of Λ and A^{-1}, respectively. The approximation of the dual variational formulation (1.22) reads as follows:

$$\left.\begin{array}{l} \textit{Find } \lambda_H \in \Lambda_H \textit{ such that} \\ \mathcal{B}_h(\lambda_H, \mu_H - \lambda_H) \geq [\mathcal{F}_h, \mu_H - \lambda_H] \quad \forall \mu_H \in \Lambda_H\ , \end{array}\right\} \tag{1.56}$$

where $\mathcal{B}_h(\mu_H, \lambda_H) := [BA_h^{-1}B^t\mu_H, \lambda_H]$ and $\mathcal{F}_h := BA_h^{-1}f - g$. The mapping A_h^{-1} acting in finite dimensional spaces will be represented by the inverse $\mathbf{A}^{-1}$ of the stiffness matrix $\mathbf{A}$. Analogously to the continuous setting, the approximation (1.56) can be derived from the approximation of the respective mixed formulation by eliminating the first component w_h. Suppose that (1.34) is satisfied. Then problem (1.56) has a unique solution λ_H. Since λ_H is at the same time a part of the solution to (1.33), convergence results follow directly from those ones established for the mixed variational formulation.

2 Unconstrained hemivariational inequalities of elliptic type and their approximations

A common feature of elliptic inequalities of the first and the second kind presented in the previous section is the fact that a multivalued part is given by a *maximal monotone* mapping, here by the subgradient of a convex functional. The aim of this section is to introduce a more general class of problems called *hemivariational inequalities* which involves also *nonmonotone* multivalued relations. We start with *scalar* hemivariational inequalities.

Let $V \subseteq H^1(\Omega, \mathbb{R}^d), d \geq 1$ be a space of generally vector-valued functions which are defined in a bounded domain $\Omega \subseteq \mathbb{R}^n$ with the Lipschitz boundary $\partial\Omega$. Throughout this section we shall suppose that

$$V \cap C^\infty(\overline{\Omega}, \mathbb{R}^d) \text{ is dense in } V\ . \tag{2.1}$$

As before, let $a : V \times V \mapsto \mathbb{R}^1$ denote a bilinear form satisfying (1.1) and (1.2). A nonmonotone relation will be defined in a non-empty open part $\omega \subseteq \partial\Omega$. Further, let Z be a space of *scalar* functions in ω and $\Pi \in L(V, Z)$ be a mapping such that

$$y \in V \cap C^\infty(\overline{\Omega}, \mathbb{R}^d) \Rightarrow \Pi y \in L^\infty(\omega) . \tag{2.2}$$

Finally, by Y we denote another space of scalar functions defined in ω being in duality with Z. The respective duality pairing will be denoted by $\langle\ ,\ \rangle_{Y\times Z}$.

By an *abstract hemivariational* inequality of scalar type we call the following problem:

$$\left.\begin{array}{l} Find\ (u, \Xi) \in V \times Y\ such\ that \\ a(u,v) + \langle \Xi, \Pi v\rangle_{Y\times Z} = \langle f, v\rangle \quad \forall v \in V \\ \Xi(x) \in \hat{b}(\Pi u(x)) \quad for\ a.a.\ x \in \omega\ , \end{array}\right\} \tag{2.3}$$

where $f \in V'$ and $\hat{b} : \mathbb{R}^1 \mapsto 2^{\mathbb{R}^1}$ is a multifunction, whose construction will be now described.

Let $b \in L^\infty_{loc}(\mathbb{R}^1)$ be a locally bounded, measurable function in $\mathbb{R}^1$:

$$\forall r > 0\ \exists c := c(r) = const. > 0 : \ |b(\xi)| \le c\ for\ a.a.\ |\xi| \le r\ . \tag{2.4}$$

For any $\epsilon > 0$ we introduce two auxiliary functions $\underline{b}_\epsilon$, $\overline{b}_\epsilon : \mathbb{R}^1 \mapsto \mathbb{R}^1$:

$$\underline{b}_\epsilon(\xi) = \operatorname*{ess\,inf}_{|\tau-\xi|\le\epsilon} b(\tau)\ , \quad \overline{b}_\epsilon(\xi) = \operatorname*{ess\,sup}_{|\tau-\xi|\le\epsilon} b(\tau)\ . \tag{2.5}$$

Since $\underline{b}_\epsilon$ and $\overline{b}_\epsilon$ are monotone with respect to ϵ, the following limits exist:

$$\underline{b}(\xi) = \lim_{\epsilon\to 0+} \underline{b}_\epsilon(\xi)\ , \quad \overline{b}(\xi) = \lim_{\epsilon\to 0+} \overline{b}_\epsilon(\xi)\ . \tag{2.6}$$

The multifunction $\hat{b}$ associated with $b \in L^\infty_{loc}(\mathbb{R}^1)$ is now defined at any $\xi \in \mathbb{R}^1$ by:

$$\hat{b}(\xi) = [\underline{b}(\xi), \overline{b}(\xi)]\ . \tag{2.7}$$

It results from b by filling in the gaps at points where b is discontinuous. This function is then used in (2.3). Observe that monotonicity of $\hat{b}$ *is not* required.

Remark 2.1. Here we restrict ourselves to the case in which the function b *does not depend* on $x \in \omega$. A more general situation when $b : \omega\times\mathbb{R}^1 \mapsto \mathbb{R}^1$ is analyzed in [Haslinger and Miettinen and Panagiotopoulos 1999]. A monotone relation can be prescribed not only on a part of $\partial\Omega$ but also in a subdomain $\omega \subseteq \Omega$ as in [Haslinger and Miettinen and Panagiotopoulos 1999].

To prove the existence of a solution to (2.3), additional assumptions on b will be needed. Here and in what follows we shall suppose that b satisfies:

$$\exists\overline{\xi} > 0 : \qquad \operatorname*{ess\,sup}_{\xi\in(-\infty,-\overline{\xi})} b(\xi) \le 0 \le \operatorname*{ess\,inf}_{\xi\in(\overline{\xi},\infty)} b(\xi)\ . \tag{2.8}$$

The first existence result is given in the following theorem.

Theorem 2.2. *Let $b \in L^{\infty}_{loc}(\mathbb{R}^1)$ satisfy (2.8). Then there exists $(u, \Xi) \in V \times L^1(\omega) \cap V'$ being a solution to (2.3).*

Remark 2.3. If b satisfies (2.4) and (2.8) then $Y = L^1(\omega) \cap V'$ in (2.3).

It would be possible to prove this theorem directly. Here we use another approach which is based on an appropriate discretization of (2.3) followed by a convergence analysis.

Before we introduce an approximation of (2.3) let us collect auxiliary results which will be used. Let $b \in L^{\infty}_{loc}(\mathbb{R}^1)$ satisfy (2.8). Then there exists $\rho > 0$ such that

$$\left.\begin{array}{lll} b(\xi) \geq 0 & for\ a.\ a. & \xi \in (\overline{\xi}, \infty) \\ |b(\xi)| \leq \rho & for\ a.\ a. & |\xi| \leq \overline{\xi} \\ b(\xi) \leq 0 & for\ a.\ a. & \xi \in (-\infty, -\overline{\xi})\ . \end{array}\right\} \tag{2.9}$$

By $\beta \in C_0^{\infty}([-1,1])$ we denote a nonnegative function such that $\int_{-1}^{1} \beta(\tau)\, d\tau = 1$ and define the regularization $b^{\varkappa}$ of b by

$$b^{\varkappa}(\xi) = \frac{1}{\varkappa} \int_{-\infty}^{\infty} b(\xi - \tau) \beta(\frac{\tau}{\varkappa})\, d\tau\ ,\ \varkappa \to 0+\ .$$

It holds that

$$b^{\varkappa} \in C^{\infty}(\mathbb{R}^1) \quad for\ any\ \varkappa > 0 \tag{2.10}$$

and there exist two numbers $\hat{\rho}, \hat{\xi} > 0$ which do not depend on $\varkappa$ and such that

$$\left.\begin{array}{ll} b^{\varkappa}(\xi) \geq 0 & \xi \geq \hat{\xi} \\ |b^{\varkappa}(\xi)| \leq \hat{\rho} & |\xi| \leq \hat{\xi} \\ b^{\varkappa}(\xi) \leq 0 & \xi \leq -\hat{\xi}\ . \end{array}\right\} \tag{2.11}$$

Problem 2.4 Prove (2.9)-(2.11).

Remark 2.5. A similar result to (2.11) with possibly different $\hat{\rho}$, $\hat{\xi}$ holds also for $\underline{b}_{\varkappa}$, $\overline{b}_{\varkappa}$, $\underline{b}$, $\overline{b}$.

Finally we shall need the following property of $\hat{b}$:

Lemma 2.6. *Let $\{\xi_k\}$, $\{\eta_k\}$ be such that $\eta_k \in \hat{b}(\xi_k)$ for all k and $\xi_k \to \xi$, $\eta_k \to \eta$, $k \to \infty$. Then $\eta \in \hat{b}(\xi)$.*

For the proof we refer to [Haslinger and Miettinen and Panagiotopoulos 1999].

Discretization and convergence analysis

We now describe a discretization of (2.3). Since a solution to (2.3) has two components and both will be approximated, we have to construct two finite element spaces. For simplicity of our presentation we shall suppose that Ω is a *plane polygonal domain*.

Let $\mathcal{D}_h$, $\mathcal{T}_h$ be a triangulation of $\overline{\Omega}$ and a partition of $\overline{\omega} \subseteq \partial\Omega$, respectively. With any $\mathcal{D}_h$ and $\mathcal{T}_h$ the following finite-dimensional spaces will be associated:

$$V_h = \{v_h \in C(\overline{\Omega}; \mathbb{R}^d) | \ v_{h|T} \in (P_1(T))^d \ \ \forall T \in \mathcal{D}_h\} \cap V \ ,$$
$$Y_h = \{\mu_h \in L^\infty(\omega) | \ \mu_{h|K} \in P_0(K) \ \ \forall K \in \mathcal{T}_h\} \ ,$$

i.e. V_h is the space of continuous, piecewise linear functions over $\mathcal{D}_h$ which satisfy the same boundary conditions as functions from V, while Y_h contains all piecewise constant functions over $\mathcal{T}_h$. The spaces V_h, Y_h will serve as the approximations of V and Y, respectively.

Denote by W_h the image of V_h realized by Π:

$$W_h = \Pi(V_h) \tag{2.12}$$

and define a mapping $P_h \in L(W_h, Y_h)$, whose choice will be specified later.

Further we choose one point $x_h^i \in int\, K_i$ for each $K_i \in \mathcal{T}_h$, $i = 1, \ldots, m$. The values of $\mu_h \in Y_h$ at $\{x_h^i\}_{i=1}^m$ will be used as the degree of freedom in Y_h, i.e if $\mu_h \in Y_h$ then $\mu_h = \sum_i \mu_h(x_h^i)\chi_i$, where χ_i is the characteristic function of $int\ K_i$.

The discretization of (2.3) reads as follows:

$$\left.\begin{aligned} &\textit{Find } (u_h, \Xi_h) \in V_h \times Y_h \textit{ such that} \\ &a(u_h, v_h) + \int_\omega \Xi_h P_h(\Pi v_h)\, ds = \langle f, v_h \rangle \quad \forall v_h \in V_h \\ &\Xi_h(x) \in \hat{b}(P_h(\Pi u_h)(x)) \quad \textit{for a.a. } x \in \omega \ . \end{aligned}\right\} \tag{2.13}$$

Remark 2.7. Since both, Ξ_h and $P_h(\Pi u_h)$ are piecewise constant in ω, the inclusion $(2.13)_3$ is equivalent to m inclusions at $\{x_h^i\}$:

$$\Xi_h(x_h^i) \in \hat{b}(P_h(\Pi u_h)(x_h^i)), \ i = 1, \ldots, m \ .$$

To prove the existence of a solution to (2.13) we first introduce the family of regularized problems:

$$\left.\begin{aligned} &\textit{Find } u_h^\varkappa \in V_h \textit{ such that} \\ &a(u_h^\varkappa, v_h) + \int_\omega b^\varkappa(P_h(\Pi u_h^\varkappa))P_h(\Pi v_h)\, ds = \langle f, v_h \rangle \quad \forall v_h \in V_h \ , \end{aligned}\right\} \tag{2.14}$$

where $b^\varkappa$ is the regularization of b and $\varkappa \to 0+$.

Lemma 2.8. *Problem (2.14) has at least one solution $u_h^\varkappa$ for any $h, \varkappa > 0$ which is uniformly bounded:*

$$\exists r > 0 : \ \|u_h^\varkappa\| \le r \quad \forall h, \varkappa > 0 \ . \tag{2.15}$$

Proof. Let $T_h^\varkappa : V_h \mapsto V_h'$ be a mapping defined by

$$\langle T_h^\varkappa(y_h), v_h \rangle_h := a(y_h, v_h) + \int_\omega b^\varkappa(P_h(\Pi y_h))P_h(\Pi v_h)\, ds - \langle f, v_h \rangle \ . \tag{2.16}$$

Clearly $u_h^\varkappa \in V_h$ solves (2.14) if and only if

$$\langle T_h^\varkappa(u_h^\varkappa), v_h\rangle_h = 0 \quad \forall v_h \in V_h \ . \tag{2.17}$$

The existence of a solution to (2.17) follows from a consequence of the Brouwer fixed point theorem. Indeed, it is easy to show that $T_h^\varkappa$ is continuous in V_h:

$$y_h^n \to y_h, \ \ y_h^n, y_h \in V_h \ \Rightarrow \ \|T_h^\varkappa(y_h^n) - T_h^\varkappa(y_h)\|_{*,h} \to 0, \ n \to \infty \ , \tag{2.18}$$

where

$$\|T_h^\varkappa(v_h)\|_{*,h} := \sup_{\substack{\|z_h\|\le 1 \\ z_h \in V_h}} \langle T_h^\varkappa(v_h), z_h\rangle_h \ .$$

Thus it is sufficient to show that

$$\langle T_h^\varkappa(z_h), z_h\rangle_h > 0 \quad \forall \|z_h\| = r \tag{2.19}$$

for some $r > 0$. In addition it will be seen that r does not depend on h and $\varkappa$. The first and the third term in (2.16) can be estimated in a standard way:

$$a(z_h, z_h) - \langle f, z_h\rangle \ge \alpha\|z_h\|^2 - \|f\|_*\|z_h\| \ . \tag{2.20}$$

The estimate of the second term is more involved:

$$\begin{aligned} \int_\omega b^\varkappa(P_h(\Pi z_h))P_h(\Pi z_h)\,ds &= \int_{I_1} b^\varkappa(P_h(\Pi z_h))P_h(\Pi z_h)\,ds + \int_{I_2} b^\varkappa(P_h(\Pi z_h))P_h(\Pi z_h)\,ds \\ &\ge \int_{I_1} b^\varkappa(P_h(\Pi z_h))P_h(\Pi z_h)\,ds \ , \end{aligned} \tag{2.21}$$

where $I_1 = \{x \in \omega |\ |P_h(\Pi z_h)(x)| \le \hat{\xi}\}$, $I_2 = \omega \setminus I_1$ and $\hat{\xi}$ is as in (2.11). The integral over I_2 is nonnegative due to $(2.11)_{1,3}$. The integral over I_1 can be estimated from below:

$$\int_{I_1} b^\varkappa(P_h(\Pi z_h))P_h(\Pi z_h)\,ds \ge -\hat{\rho}\hat{\xi}\, meas\, \omega \ ,$$

making use of $(2.11)_2$. From this, (2.20) and (2.21) we arrive at the existence of r which is independent of h, $\varkappa$ and such that (2.19) holds. □

Letting $\varkappa \to 0+$ in (2.14) we obtain the following existence result for the discrete hemivariational inequality (2.13).

Theorem 2.9. *For any $h > 0$ there exists at least one solution (u_h, Ξ_h) of (2.13) such that*

$$\|u_h\| \le r \quad \forall h > 0 \ , \tag{2.22}$$

where r is the same as in (2.15).

Proof. Let $\{u_h^\varkappa\}$, $\kappa \to 0+$ be a bounded sequence of solutions to (2.14). We may assume that

$$u_h^\varkappa \to u_h\ , \quad \varkappa \to 0+ \tag{2.23}$$

for some element $u_h \in V_h$ satisfying (2.22). The sequences $\{b^\varkappa(P_h(\Pi u_h^\varkappa(x_h^i)))\}$, $i = 1, \dots, m$ are bounded as well as follows from (2.4). Therefore there exists $\Xi_h \in Y_h$ such that

$$b^\varkappa(P_h(\Pi u_h^\varkappa(x_h^i))) \to \Xi_h(x_h^i),\ \varkappa \to 0+,\ \forall i = 1, \dots, m\ . \tag{2.24}$$

Passing to the limit with $\varkappa \to 0+$ in (2.14) and using (2.23), (2.24) we see that the couple (u_h, Ξ_h) satisfies the equation $(2.13)_2$. It remains to prove that Ξ_h satisfies the inclusion $(2.13)_3$. From (2.23) it follows that

$$P_h(\Pi u_h^\varkappa) \to P_h(\Pi u_h) \quad in\ L^\infty(\omega)$$

as $\varkappa \to 0+$. Thus for any $\epsilon > 0$ there exists $\varkappa_0 := \varkappa_0(\epsilon) > 0$ such that

$$|P_h(\Pi u_h^\varkappa)(x) - P_h(\Pi u_h)(x)| < \frac{\epsilon}{2} \tag{2.25}$$

holds for any $x \in \bigcup_{i=1}^m int\, K_i$ and all $\varkappa \le \varkappa_0$. Let $\varkappa \le \min(\varkappa_0, \epsilon/2)$. Then

$$\begin{aligned} b^\varkappa(P_h(\Pi u_h^\varkappa)(x_h^i)) &\le \bar{b}_\varkappa(P_h(\Pi u_h^\varkappa)(x_h^i)) \le \bar{b}_{\epsilon/2}(P_h(\Pi u_h^\varkappa)(x_h^i)) \\ &\overset{(2.25)}{\le} \bar{b}_\epsilon(P_h(\Pi u_h)(x_h^i)) \end{aligned} \tag{2.26}$$

making use that $b^\varkappa(\xi) \le \bar{b}_\varkappa(\xi)$ for all $\xi \in \mathbb{R}^1$ as follows from the definition of $b^\varkappa$ and $\bar{b}_\varkappa$. Passing to the limit first with $\varkappa \to 0+$ and then with $\epsilon \to 0+$ in (2.26) we obtain:

$$\Xi(x_h^i) \le \bar{b}(P_h(\Pi u_h)(x_h^i)) \quad i = 1, \dots, m\ . \tag{2.27}$$

Similary one can show that $\Xi(x_h^i) \ge \underline{b}(P_h(\Pi u_h)(x_h^i))$, $i = 1, \dots, m$. □

The found solution (u_h, Ξ_h) of (2.13) is not unique, in general. Let us suppose that the first component u_h is already known. A natural question arises, namely under which conditions the second component Ξ_h for this particular u_h is unique. The answer is given in the following theorem.

Theorem 2.10. *Let the mapping P_h map W_h onto Y_h. If (u_h, Ξ_h), $(u_h, \overline{\Xi}_h)$ are solutions of (2.13) then $\Xi_h = \overline{\Xi}_h$.*

Proof. The difference $\Xi_h - \overline{\Xi}_h \in Y_h$ satisfies the equation

$$\int_\omega (\Xi_h - \overline{\Xi}_h) P_h(\Pi v_h)\, ds = 0 \quad \forall v_h \in V_h\ . \tag{2.28}$$

Since P_h maps W_h onto Y_h it follows from (2.28) that $\Xi_h = \overline{\Xi}_h$. □

In the subsequent part of this section we shall analyze if there is a relation between solutions of the discretized hemivariational inequality (2.13) and a solution to the original continuous setting (2.3). Convergence properties of a sequence $\{(u_h, \Xi_h)\}$ will be established under the following additional assumptions on $\{V_h\}$ and $\{P_h\}$, $h \to 0+$:

$$\forall v \in V \cap C^\infty(\overline{\Omega}, \mathbb{R}^d)\ \exists \{v_h\},\ v_h \in V_h \textit{ such that } v_h \to v \textit{ in } V \textit{ and } C(\overline{\Omega}, \mathbb{R}^d); \tag{2.29}$$

$$\textit{if } y_h \rightharpoonup y \textit{ in } V,\ y_h \in V_h \Rightarrow \exists \textit{ a subsequence of } \{y_h\} \textit{ (denoted by the same symbol) such that } P_h(\Pi y_h)(x) \to \Pi y(x) \textit{ a.e. in } \omega,\ h \to 0+\ ; \tag{2.30}$$

$$\textit{if } y_h \to y \in V \cap C^\infty(\overline{\Omega}, \mathbb{R}^d) \textit{ in } C(\overline{\Omega}, \mathbb{R}^d),\ y_h \in V_h \Rightarrow P_h(\Pi y_h) \to \Pi y \textit{ in } L^\infty(\omega),\ h \to 0+\ . \tag{2.31}$$

We now prove the main result of this section.

Theorem 2.11. *Let (2.29)-(2.31) be satisfied and $\{(u_h, \Xi_h)\}$, $h \to 0+$ be a sequence of solutions to (2.13) with $\{u_h\}$ being bounded in V. Then there exist: a subsequence of $\{(u_h, \Xi_h)\}$ (denoted by the same symbol) and an element $(u, \Xi) \in V \times L^1(\omega) \cap V'$ such that*

$$\left.\begin{array}{ll} u_h \rightharpoonup u & \textit{in } V\ , \\ \Xi_h \rightharpoonup \Xi & \textit{in } L^1(\omega),\ h \to 0+\ . \end{array}\right\} \tag{2.32}$$

In addition, (u, Ξ) is a solution of (2.3) with $Y = L^1(\omega) \cap V'$. Any accumulation point of $\{(u_h, \Xi_h)\}$ in the sense of (2.32) has this property.

Proof. Since $\{u_h\}$ is bounded, one can find its subsequence and $u \in V$ such that $(2.32)_1$ holds. Next we prove that $\{\Xi_h\}$ is weakly compact in $L^1(\omega)$. To this end it is sufficient to show that for any $\gamma > 0$ there exists $\delta > 0$ such that

$$\int_{\omega_0} |\Xi_h|\, ds < \gamma \tag{2.33}$$

holds for any $\Xi_h \in \{\Xi_h\}$ and any measurable subset $\omega_0 \subseteq \omega$ such that $meas\, \omega_0 \le \delta$.

Let $I_1 = \{x \in \omega |\ |P_h(\Pi u_h)(x)| \le \hat{\xi}\}$, $I_2 = \omega \setminus I_1$, where $\hat{\xi}$ is as in Remark 2.5. Then

$$\begin{aligned} \int_\omega |\Xi_h P_h(\Pi u_h)|\, ds &= \int_{I_1} |\Xi_h P_h(\Pi u_h)|\, ds + \int_{I_2} |\Xi_h P_h(\Pi u_h)|\, ds \\ &\le \int_\omega \Xi_h P_h(\Pi u_h)\, ds + 2 \int_{I_1} |\Xi_h P_h(\Pi u_h)|\, ds \\ &\le -a(u_h, u_h) + \langle f, u_h \rangle + 2\hat{\rho}\hat{\xi}\, meas\, \omega \\ &\le \|f\|_* \|u_h\| + 2\hat{\rho}\hat{\xi}\, meas\, \omega \le C = const. > 0 \end{aligned} \tag{2.34}$$

holds for any $h > 0$. Here we used Remark 2.5 and $(2.13)_2$.

Let $\gamma > 0$ be given. From (2.34) we see that there exists $q_0 > 0$ such that

$$\frac{1}{q_0} \int_\omega |\Xi_h P_h(\Pi u_h)|\, ds \le \frac{\gamma}{2} \quad \forall h > 0 \, . \tag{2.35}$$

On the other hand there exists $\delta > 0$ such that

$$\delta \operatorname*{ess\,sup}_{|P_h(\Pi u_h)(x)| \le q_0} |\Xi_h(x)| \le \frac{\gamma}{2} \quad \forall h > 0 \tag{2.36}$$

as follows from (2.4). Let $\omega_0 \subset \omega$ be an arbitrary measurable set such that $meas\, \omega_0 \le \delta$. For $x \in \omega$ and every $h > 0$ we have:

$$|\Xi_h(x)| \le \frac{1}{q_0} |\Xi_h P_h(\Pi u_h)(x)| + \operatorname*{ess\,sup}_{|P_h(\Pi u_h)(x)| \le q_0} |\Xi_h(x)| \, .$$

From this, (2.35) and (2.36) we arrive at (2.33). Therefore one can pass to a subsequence of $\{\Xi_h\}$ such that $(2.32)_2$ holds for an element $\Xi \in L^1(\omega)$. It remains to show that (u, Ξ) solves (2.3).

Let $\overline{v} \in V \cap C^\infty(\overline{\Omega}, \mathbb{R}^d)$ be given. Then there exists a sequence $\{\overline{v}_h\}$, $\overline{v}_h \in V_h$ such that

$$\left.\begin{array}{ll} \overline{v}_h \to \overline{v} & \text{in } V \text{ and } C(\overline{\Omega}, \mathbb{R}^d) \\ P_h(\Pi \overline{v}_h) \to \Pi \overline{v} & \text{in } L^\infty(\omega),\ h \to 0+ \, , \end{array}\right\} \tag{2.37}$$

as follows from (2.29) and (2.31). Inserting $\overline{v}_h$ as a test function into $(2.13)_2$, letting $h \to 0+$ and using (2.32), (2.37) we have:

$$a(u, \overline{v}) + \int_\omega \Xi \Pi \overline{v}\, ds = \langle f, \overline{v} \rangle \quad \forall \overline{v} \in V \cap C^\infty(\overline{\Omega}, \mathbb{R}^d) \, .$$

The function Ξ can be identified with an element from V'. Indeed,

$$\exists c > 0 : \quad \left| \int_\omega \Xi \Pi \overline{v}\, ds \right| \le c \|\overline{v}\| \tag{2.38}$$

for any $\overline{v} \in V \cap C^\infty(\overline{\Omega}, \mathbb{R}^d)$ and in view of (2.1) the duality pairing $\langle\ ,\ \rangle_{Y \times Z}$ is defined as the extension of the previous integral over ω by continuity.

Next we shall show that Ξ satisfies the inclusion $(2.3)_3$. From (2.30) and Egoroff's theorem we may assume that the sequence $\{P_h(\Pi u_h)\}$, $h \to 0+$ tends to Πu uniformly in ω up to sets of a small measure:

$$\left.\begin{array}{l} \forall \epsilon > 0\ \forall \delta > 0\ \exists \omega_0 \subseteq \omega\, ,\ meas\, \omega_0 < \delta\ \exists \overline{h} > 0 : \\ |P_h(\Pi u_h)(x) - \Pi u(x)| < \epsilon/2 \quad \forall x \in \omega \setminus \omega_0\ \forall h \le \overline{h} \, . \end{array}\right\} \tag{2.39}$$

Let $h \le \min(\overline{h}, \epsilon/2)$. It holds:

$$\Xi_h(x) \le \overline{b}(P_h(\Pi u_h)(x)) \le \overline{b}_{\epsilon/2}(P_h(\Pi u_h)(x)) \le \overline{b}_\epsilon(\Pi u(x)) \tag{2.40}$$

for any $x \in \omega \setminus \omega_0$ making use of $(2.13)_3$, the definition of $\bar{b}$ and (2.39).

Let $\Phi \in L^\infty(\omega)$ be a non-negative function in ω. Then

$$\int\limits_{\omega\setminus\omega_0} \Xi\Phi\, ds = \lim_{h\to 0+} \int\limits_{\omega\setminus\omega_0} \Xi_h\Phi\, ds \le \int\limits_{\omega\setminus\omega_0} \bar{b}_\epsilon(\Pi u)\Phi\, ds \tag{2.41}$$

for any $\epsilon > 0$ sufficiently small as follows from $(2.32)_2$ and (2.40). Passing to the limit with $\epsilon \to 0+$ and using Fatou's lemma in (2.41) we arrive at

$$\int\limits_{\omega\setminus\omega_0} \Xi\Phi\, ds \le \int\limits_{\omega\setminus\omega_0} \bar{b}(\Pi u)\Phi\, ds\ . \tag{2.42}$$

In a similar way one can prove that

$$\int\limits_{\omega\setminus\omega_0} \Xi\Phi\, ds \ge \int\limits_{\omega\setminus\omega_0} \underline{b}(\Pi u)\Phi\, ds\ . \tag{2.43}$$

Since (2.42) and (2.43) are satisfied for arbitrary $\Phi \in L^\infty(\omega)$, $\Phi \ge 0$, we may conclude that $\Xi(x) \in \hat{b}(\Pi u(x))$ for a.a. $x \in \omega$ using that $meas\,\omega_0 < \delta$ and $\delta > 0$ can be chosen arbitrarily small. $\square$

The existence of a solution to (2.3) is now a consequence of the convergence result established in Theorem 2.11. Let us comment, in brief on assumptions (2.29)-(2.31) which play the crucial role in the proof. The assumption (2.29) combined with (2.1) is a standard approximation property of finite element spaces. Instead of (2.30) we usually have at our disposal a stronger result for $\{P_h\}$, $h \to 0+$, namely:

$$y_h \rightharpoonup y \ \ in\ V,\ \ y_h \in V_h \ \ \Rightarrow\ \ \|P_h(\Pi y_h) - \Pi y\|_{L^s(\omega)} \to 0,\ h \to 0+ \tag{2.44}$$

for some $s \ge 1$. Then (2.30) is a consequence of (2.44). The choice of P_h satisfying (2.30) and (2.31) is discussed in details in [Haslinger and Miettinen and Panagiotopoulos 1999]. We return to this point in Section 5 dealing with contact problems with nonmonotone friction.

The convergence result of Theorem 2.11 can be improved if the function b satisfies the following growth condition:

$$\left.\begin{array}{l} \exists\ constants\ c_1, c_2 > 0\ \ such\ that \\ |b(\xi)| \le c_1 + c_2|\xi|^{p/q}\ \ for\ a.a.\ \xi \in \mathbb{R}^1\ , \end{array}\right\} \tag{2.45}$$

where $1/p + 1/q = 1$, $p \in [1, \infty)$ if $n = 2$ or $p \in [1, p^\star)$, $p^\star = 2(n-1)/(n-2)$ if $n > 2$ ($\Omega \subset \mathbb{R}^n$). In addition to (2.45) we shall suppose that Π and P_h satisfy the following assumptions:

$$\Pi \in L(V, L^p(\omega))\ ; \tag{2.46}$$

$$the\ property\ (2.44)\ holds\ with\ some\ s \ge p\ , \tag{2.47}$$

where p is the same as in (2.45).

The following lemma is left as an easy exercise.

Lemma 2.12. *Let (2.45)-(2.47) be satisfied. Then there exists a constant $c > 0$ such that*

$$\|\Xi_h\|_{L^q(\omega)} \leq c \quad \forall h > 0 \,, \tag{2.48}$$

where q is as in (2.45).

Problem 2.13 Prove Lemma 2.12 .

We now prove the following convergence result:

Theorem 2.14. *Let (2.29), (2.45), (2.46) and (2.47) be satisfied. Let $\{(u_h, \Xi_h)\}$, $h \to 0+$ be a sequence of solutions to (2.13) with $\{u_h\}$ being bounded in V. Then there exist: a subsequence of $\{(u_h, \Xi_h)\}$ (denoted by the same symbol) and an element $(u, \Xi) \in V \times L^q(\omega)$ such that*

$$\left.\begin{array}{lll} u_h \to u & \text{in } V \,, & \\ \Xi_h \rightharpoonup \Xi & \text{in } L^q(\omega), \ h \to 0+ \,. & \end{array}\right\} \tag{2.49}$$

In addition, (u, Ξ) solves (2.3) with $Y = L^q(\omega)$. Any accumulation point of $\{(u_h, \Xi_h)\}$ in the sense (2.49) has this property.

Proof. From (2.48) it follows that one can find a subsequence of $\{\Xi_h\}$ and a function $\Xi \in L^q(\omega)$ such that $(2.49)_2$ holds. The fact that (u, Ξ) satisfies $(2.3)_2$ is a consequence of (2.29) and (2.47). The inclusion $\Xi(x) \in \hat{b}(\Pi u(x))$ a.e. in ω has been already proven in Theorem 2.11. It remains to show that $\{u_h\}$ tends strongly to u in V.
In view of (2.29) and (2.47) there exists a sequence $\{\overline{u}_h\}$, $\overline{u}_h \in V_h$ such that

$$\left.\begin{array}{lll} \overline{u}_h \to u & \text{in } V & \\ P_h(\Pi \overline{u}_h) \to \Pi u & \text{in } L^p(\omega), \ h \to 0+ \,. & \end{array}\right\} \tag{2.50}$$

Further

$$\alpha \|u_h - \overline{u}_h\|^2 \leq a(u_h, u_h - \overline{u}_h) - a(\overline{u}_h, u_h - \overline{u}_h) \to 0, \ h \to 0+ \,. \tag{2.51}$$

Indeed, the second term on the right of (2.51) tends to zero since $u_h - \overline{u}_h \rightharpoonup 0$ in V. To prove that also the first term vanishes as $h \to 0+$ we use $(2.13)_2$:

$$\begin{aligned} |a(u_h, u_h - \overline{u}_h)| &\leq \int_\omega |\Xi_h P_h(\Pi u_h - \Pi \overline{u}_h)| \, ds + |\langle f, u_h - \overline{u}_h \rangle| \\ &\leq \|\Xi_h\|_{L^q(\omega)} \|P_h(\Pi u_h - \Pi \overline{u}_h)\|_{L^p(\omega)} + |\langle f, u_h - \overline{u}_h \rangle| \to 0+ \end{aligned}$$

as follows from (2.48) and the fact that $P_h \Pi u_h - P_h \Pi \overline{u}_h \to 0$ in $L^p(\omega)$ which is a consequence of (2.47) and $(2.50)_2$. The triangle inequality and (2.51) yield strong convergence of $\{u_h\}$ to u. □

In the rest of this section we shall show that a large class of hemivariational inequalities is related to a substationary point problem for a locally Lipschitz functional in V. To this end we shall suppose that the bilinear form a is *symmetric* in V, the mapping P and the function b satisfy the following assumptions:

$$\Pi \in L(V, L^2(\omega)) \; ; \tag{2.52}$$

$$\exists c > 0 : \; |b(\xi)| \le c(1 + |\xi|) \; a.e \; in \; \mathbb{R}^1 \; . \tag{2.53}$$

Define a functional $\Phi : V \mapsto \mathbb{R}^1$ by

$$\Phi(v) = \int_\omega \int_0^{\Pi v} b(t) \, dt \, ds \quad \forall v \in V \; . \tag{2.54}$$

Problem 2.15 Let (2.52) and (2.53) be satisfied. Prove that Φ is Lipschitz continuous in $V \subseteq H^1(\Omega, \mathbb{R}^d)$.

By J we denote the superpotential corresponding to (2.3):

$$J(v) = \frac{1}{2} a(v, v) - \langle f, v \rangle + \Phi(v), \quad v \in V \; , \tag{2.55}$$

i.e. J is a *Lipschitz perturbation* of a strictly convex quadratic functional.

Let $u \in V$ be a *substationary point* of J in V:

$$0 \in \overline{\partial} J(u) \; , \tag{2.56}$$

where $\overline{\partial}$ stands for the Clark gradient of J. Taking into account the definition of J, the inclusion (2.56) is equivalent to

$$f - Au \in \overline{\partial} \Phi(u) \; , \tag{2.57}$$

where $A \in L(V, V')$ is defined by $\langle Au, v \rangle = a(u, v) \; \forall u, v \in V$. Introducing the new variable $\Xi := f - Au$, (2.57) can be rewritten as follows:

$$\left. \begin{array}{l} Au + \Xi = f \quad in \; V' \\ \Xi \in \overline{\partial} \Phi(u) \; . \end{array} \right\} \tag{2.58}$$

If from $\Xi \in \overline{\partial}\Phi(u)$ it follows that $\Xi(x) \in \hat{b}(\Pi u(x))$ a.e. in ω, we arrive at a hemivariational inequality.

Algebraic formulation of discrete hemivariational inequalities. Transformation to a substationary point problem

Throughout this part we shall suppose that a is *symmetric* in V and the discretization parameter $h > 0$ is *fixed*. In what follows we derive the algebraic form of (2.13).

Let $dim \; V_h = n$, dim $Y_h = m$ and $dim \; W_h = p$. Then V_h, Y_h and W_h can be identified with $\mathbb{R}^n$, $\mathbb{R}^m$ and $\mathbb{R}^p$, respectively. Since the mappings $\Pi_{|V_h}$ and P_h are linear, they are represented by $(p \times n)$, $(m \times p)$ matrices $\mathbf{\Pi}$ and $\mathbf{P}$, respectively. The integral over ω in $(2.13)_2$ can be evaluated exactly:

$$\int_\omega \Xi_h P_h(\Pi v_h)\, ds = \sum_{i=1}^m c_i \Xi_i (\mathbf{P\Pi v})_i \ ,$$

where $\mathbf{\Xi} = (\Xi_1, \dots, \Xi_m)$, $\Xi_i = \Xi_{h|K_i}$, $c_i = meas\ K_i$, $K_i \in \mathcal{T}_h$ and $\mathbf{v}$ is the vector of the nodal values of $v_h \in V_h$. To simplify notation we introduce the new variables $\Xi_i := c_i \Xi_i$, $i = 1, \dots, m$ and denote $\mathbf{\Lambda} := \mathbf{P\Pi} \in L(\mathbb{R}^n, \mathbb{R}^m)$.

The algebraic form of (2.13) now reads as follows:

$$\left.\begin{array}{l} Find\ (\mathbf{u}, \mathbf{\Xi}) \in \mathbb{R}^n \times \mathbb{R}^m\ such\ that \\ (\mathbf{Au}, \mathbf{v})_{\mathbb{R}^n} + (\mathbf{\Xi}, \mathbf{\Lambda v})_{\mathbb{R}^m} = (\mathbf{f}, \mathbf{v})_{\mathbb{R}^n} \quad \forall \mathbf{v} \in \mathbb{R}^n \\ \Xi_i \in c_i \hat{b}((\mathbf{\Lambda u})_i), \quad i = 1, \dots, m \ , \end{array}\right\} \tag{2.59}$$

where $\mathbf{A}$ is the stiffness matrix and $\mathbf{f}$ is the load vector. Two unknowns appear in (2.59), namely $\mathbf{u}$ and $\mathbf{\Xi}$. Next, we eliminate $\mathbf{\Xi}$ by introducing an appropriate discrete superpotential whose construction will be now described.

Denote by $\Phi_h : V_h \mapsto \mathbb{R}^1$ a locally Lipschitz functional defined by

$$\Phi_h(v_h) = \int_\omega \int_0^{P_h(\Pi v_h)} b(t)\, dt\, ds \ , \quad v_h \in V_h \ . \tag{2.60}$$

We shall approximate Φ_h by applying the rectangular formula to the outer integral over ω:

$$\Phi_h(v_h) \approx \sum_{i=1}^m c_i \int_0^{P_h(\Pi v_h)(x_h^i)} b(t)\, dt := \Psi(\mathbf{v}) \ ,$$

where $c_i = meas\, K_i$, $K_i \in \mathcal{T}_h$. The *discrete superpotential* $\mathcal{J}$ is now defined by

$$\mathcal{J}(\mathbf{v}) = \frac{1}{2}(\mathbf{Av}, \mathbf{v})_{\mathbb{R}^n} - (\mathbf{f}, \mathbf{v})_{\mathbb{R}^n} + \Psi(\mathbf{v}) \ , \quad \mathbf{v} \in \mathbb{R}^n \ . \tag{2.61}$$

Instead of (2.59) we shall consider the following substationary point problem:

$$\left.\begin{array}{l} Find\ \mathbf{u} \in \mathbb{R}^n\ such\ that \\ \mathbf{0} \in \overline{\partial} \mathcal{J}(\mathbf{u}) \ . \end{array}\right\} \tag{2.62}$$

Problem (2.62) can be numerically realized by nonsmooth optimization methods. Some of them are discussed in another article of this book (see [Outrata 2004]). Next we shall analyze the relation between solutions to (2.59) and (2.62).

Theorem 2.16. *Let* $\mathbf{u}$ *be a solution of (2.62). Then there exists* $\mathbf{\Xi} \in \mathbb{R}^m$ *such that* $\mathbf{\Lambda}^t \mathbf{\Xi} \in \overline{\partial} \Psi(\mathbf{u})$ *and* $(\mathbf{u}, \mathbf{\Xi})$ *solves (2.59).*

Sketch of the proof. Let $\widehat{\Psi} : \mathbb{R}^m \mapsto \mathbb{R}^1$ be defined by

$$\widehat{\Psi}(\boldsymbol{\theta}) = \sum_{i=1}^{m} c_i \int_0^{\theta_i} b(t)\,dt := \sum_{i=1}^{m} \widehat{\Psi}_i(\theta_i) \;, \tag{2.63}$$

$\boldsymbol{\theta} = (\theta_1, \dots, \theta_m) \in \mathbb{R}^m$. If $\mathbf{g} \in \overline{\partial}\widehat{\Psi}(\boldsymbol{\theta})$, $\mathbf{g} = (g_1, \dots, g_m)$ then $g_i \in \overline{\partial}\widehat{\Psi}_i(\theta_i)$, $i = 1, \dots, m$. From the definition of $\widehat{\Psi}_i$ it follows that

$$\overline{\partial}\widehat{\Psi}_i(\theta_i) \subseteq c_i \hat{b}(\theta_i) \quad \forall i = 1, \dots, m \;. \tag{2.64}$$

Since $\Psi(\mathbf{v}) = \widehat{\Psi}(\boldsymbol{\Lambda}\mathbf{v})$ it is known that (see [Clarke 1983]):

$$\overline{\partial}\Psi(\mathbf{v}) \subseteq \boldsymbol{\Lambda}^t \overline{\partial}\widehat{\Psi}(\boldsymbol{\Lambda}\mathbf{v}) \;. \tag{2.65}$$

Let $\mathbf{u} \in \mathbb{R}^n$ be a solution of (2.62) or equivalently

$$\mathbf{f} - \mathbf{A}\mathbf{u} \in \overline{\partial}\Psi(\mathbf{u}) \;. \tag{2.66}$$

Setting $\boldsymbol{\Xi}' = \mathbf{f} - \mathbf{A}\mathbf{u}$ we see that $(\mathbf{u}, \boldsymbol{\Xi}') \in \mathbb{R}^n \times \mathbb{R}^n$ satisfies

$$\left.\begin{array}{l} (\mathbf{A}\mathbf{u}, \mathbf{v})_{\mathbb{R}^n} + (\boldsymbol{\Xi}', \mathbf{v})_{\mathbb{R}^n} = (\mathbf{f}, \mathbf{v})_{\mathbb{R}^n} \quad \forall \mathbf{v} \in \mathbb{R}^n \\ \boldsymbol{\Xi}' \in \overline{\partial}\Psi(\mathbf{u}) \;. \end{array}\right\} \tag{2.67}$$

From (2.65) it follows that $\boldsymbol{\Xi}' = \boldsymbol{\Lambda}^t \boldsymbol{\Xi}$ for some $\boldsymbol{\Xi} \in \overline{\partial}\widehat{\Psi}(\boldsymbol{\Lambda}\mathbf{u})$. Using this expression of $\boldsymbol{\Xi}'$ in $(2.67)_1$ we see that $(2.59)_2$ is satisfied. The inclusion $(2.59)_3$ follows from (2.64). □

We proved that any substationary point of $\mathcal{J}$ coincides with the first component of a solution to (2.59). This does not exclude situations when there are solutions of (2.59) whose first components *are not* substationary points of $\mathcal{J}$. In this sense, (2.59) is more general than (2.62). The next theorem gives sufficient conditions under which both formulations are equivalent.

Theorem 2.17. *Let there exist one sided limits $b(\xi\pm)$ for any $\xi \in \mathbb{R}^1$ and $\mathbf{P}$ maps $\mathbb{R}^p$ onto $\mathbb{R}^m$. If $(\mathbf{u}, \boldsymbol{\Xi})$ is a solution of (2.59), then $\mathbf{u}$ is a substationary point of $\mathcal{J}$ and $\boldsymbol{\Lambda}^t\boldsymbol{\Xi} \in \overline{\partial}\Psi(\mathbf{u})$.*

Proof. From (2.63) it follows that

$$(\overline{\partial}\widehat{\Psi}(\boldsymbol{\theta}))_i = \overline{\partial}\widehat{\Psi}_i(\theta_i) \quad \forall i = 1, \dots, m \;. \tag{2.68}$$

The existence of one sided limits $b(\xi\pm)$ for every $\xi \in \mathbb{R}^1$ ensures (see [Chang 1981]) that

$$\overline{\partial}\widehat{\Psi}_i(\theta_i) = c_i \hat{b}(\theta_i) \quad \forall i = 1, \dots, m \;. \tag{2.69}$$

Since $\boldsymbol{\Lambda} = \mathbf{P}\boldsymbol{\Pi}$ maps $\mathbb{R}^n$ onto $\mathbb{R}^m$ we have (see [Clarke 1983]):

$$\overline{\partial}\Psi(\boldsymbol{v}) = \boldsymbol{\Lambda}^t \overline{\partial}\widehat{\Psi}(\boldsymbol{\Lambda}\mathbf{v}) \;. \tag{2.70}$$

Let $(\mathbf{u}, \Xi)$ be a solution of (2.59). Then $\Xi_i \in \overline{\partial}\widehat{\Psi}_i((\mathbf{\Lambda u})_i)$ and $\Xi \in \overline{\partial}\widehat{\Psi}(\mathbf{\Lambda u})$ making use of (2.68) and (2.69). In addition, $\mathbf{\Lambda}^t \Xi \in \overline{\partial}\Psi(\mathbf{u})$ as follows from (2.70). □

Let us suppose that we have already found the first component $\mathbf{u}$ of a solution to (2.59), for example by solving (2.62). Then it is very easy to compute the remaining component Ξ. From $(2.59)_2$ we see that Ξ satisfies the equation

$$\mathbf{\Lambda}^t \Xi = \mathbf{f} - \mathbf{A u} \ . \tag{2.71}$$

If $\mathbf{\Lambda}$ maps $\mathbb{R}^n$ onto $\mathbb{R}^m$ then $ker\, \mathbf{\Lambda}^t = \{\mathbf{0}\}$ and (2.71) has a unique solution. Let us observe that $\mathbf{\Lambda}$ maps $\mathbb{R}^n$ onto $\mathbb{R}^m$ if and only if P_h maps W_h onto Y_h. The uniqueness of the solution to (2.71) is also a consequence of Theorem 2.10.

3 Constrained hemivariational inequalities of scalar type and their approximations

This section deals with another class of hemivariational inequalities when additional constraints are imposed on the first component u of its solution.

By *an abstract constrained hemivariational inequality of scalar type* we call the following problem:

$$\left.\begin{array}{l} Find\ (u,\Xi) \in K \times Y\ such\ that \\ a(u, v-u) + \langle \Xi, \Pi v - \Pi u\rangle_{Y\times Z} \geq \langle f, v-u\rangle \quad \forall v \in K \\ \Xi(x) \in \hat{b}(\Pi u(x)) \quad for\ a.a.\ x \in \omega\ , \end{array}\right\} \tag{3.1}$$

where $K \subseteq V$ is a non empty, closed convex subset of V. The meaning of the remaining symbols appearing in (3.1) is exactly the same as in Section 2.

To prove the existence of at least one solution to (3.1) we proceed as in Section 2: we first discretize (3.1) and prove the existence of solutions to the discrete hemivariational inequality. The existence result for (3.1) then will follow from the convergence analysis. We start with a discretization of (3.1).

Let $\{V_h\}$, $\{Y_h\}$, $h \to 0+$ be two systems of finite element spaces constructed in Section 2. We introduce a system $\{K_h\}$, $h \to 0+$ of *closed, convex subsets* of V_h which define an approximation of K. The *discretization* of (3.1) reads as follows:

$$\left.\begin{array}{l} Find\ (u_h, \Xi_h) \in K_h \times Y_h\ such\ that \\ a(u_h, v_h - u_h) + \int\limits_\omega \Xi_h P_h(\Pi v_h - \Pi u_h)\, ds \geq \langle f, v_h - u_h\rangle \quad \forall v_h \in K_h \\ \Xi_h(x) \in \hat{b}(P_h(\Pi u_h(x))) \quad for\ a.a.\ x \in \omega\ , \end{array}\right\} \tag{3.2}$$

where P_h is the same as in Section 2.

Instead of (3.2) we shall consider its *penalized* and *regularized* form:

$$\left.\begin{array}{l} Find\ u_h^{\varkappa\epsilon} \in V_h\ such\ that \\ a(u_h^{\varkappa\epsilon}, v_h) + \int\limits_\omega b^\varkappa(P_h(\Pi u_h^{\varkappa\epsilon}))P_h(\Pi v_h)\, ds + \frac{1}{\epsilon}\langle \beta_h'(u_h^{\varkappa\epsilon}), v_h\rangle_h = \langle f, v_h\rangle\ \forall v_h \in V_h\ , \end{array}\right\} \tag{3.3}$$

where $\langle\ ,\ \rangle_h$ is a duality pairing between V_h and V_h', $\varkappa$ and ϵ are two positive parameters destined to tend to zero, $b^\varkappa$ is the regularization of b, $\beta_h : V_h \mapsto \mathbb{R}^1$ is a penalty functional releasing the constraint $v_h \in K_h$ and satisfying the following assumptions:

$$\beta_h \textit{ is convex and continuously Fréchet differentiable in } V_h\ ; \tag{3.4}$$

$$\textit{the Fréchet derivative } \beta_h'(v_h) = 0 \ \textit{ iff } v_h \in K_h\ . \tag{3.5}$$

Let h, $\varkappa$ and ϵ be fixed. The existence of a solution to (3.3) follows from

Lemma 3.1. *Let the function $b \in L^\infty_{loc}(\mathbb{R}^1)$ satisfy the sign condition (2.8) and let $0 \in K_h$ for every $h > 0$. Then (3.3) admits at least one solution $u_h^{\varkappa\epsilon}$ for any h, $\varkappa$ and ϵ which is uniformly bounded:*

$$\exists r > 0 : \ \|u_h^{\varkappa\epsilon}\| \le r \quad \forall h, \varkappa, \epsilon > 0\ . \tag{3.6}$$

Proof. We shall proceed as in the proof of Lemma 2.8. Let $T_h^{\varkappa\epsilon} : V_h \mapsto V_h'$ be a mapping defined by

$$\begin{aligned}\langle T_h^{\varkappa\epsilon}(y_h), v_h\rangle_h :=& a(y_h, v_h) + \int_\omega b^\varkappa(P_h(\Pi y_h))P_h(\Pi v_h)\, ds \\ &+ \frac{1}{\epsilon}\langle \beta_h'(y_h), v_h\rangle_h - \langle f, v_h\rangle\ .\end{aligned}$$

A function $u_h^{\varkappa\epsilon} \in V_h$ solves (3.3) if and only if $T_h^{\varkappa\epsilon}(u_h^{\varkappa\epsilon}) = 0$. The existence of such $u_h^{\varkappa\epsilon}$ follows again from a consequence of Brouwer's fixed point theorem. From the assumptions of our lemma one can easily prove that there exists $r > 0$ which does not depend on h, $\varkappa$, ϵ and such that

$$\langle T_h^{\varkappa\epsilon}(z_h), z_h\rangle_h > 0 \quad \forall \|z_h\| = r\ .$$

The only difference between $T_h^{\varkappa\epsilon}$ and $T_h^{\varkappa}$ defined by (2.16) is the presence of the penalty term which however can be neglected. Indeed:

$$\frac{1}{\epsilon}\langle \beta_h'(z_h), z_h\rangle_h = \frac{1}{\epsilon}\langle \beta_h'(z_h) - \beta_h'(0), z_h - 0\rangle_h \ge 0$$

using that $0 \in K_h\ \forall h > 0$, (3.5) and monotonicity of the mapping $z_h \mapsto \beta_h'(z_h)$.

Let $h, \epsilon > 0$ be fixed whereas the regularization parameter $\varkappa$ tends to zero. Since $\{u_h^{\varkappa\epsilon}\}$, $\{b^\varkappa(P_h(\Pi u_h^{\varkappa\epsilon}(x_h^i)))\}$, $i = 1, \dots, m$ are bounded, one can pass to their subsequences (denoted by the same symbol) such that

$$u_h^{\varkappa\epsilon} \to u_h^\epsilon,\ b^\varkappa(P_h(\Pi u_h^{\varkappa\epsilon}(x_h^i))) \to \Xi_h^\epsilon(x_h^i),\ i - 1, \dots, m \tag{3.7}$$

as $\varkappa \to 0+$ for some $u_h^\epsilon \in V_h$ and $\Xi_h^\epsilon \in Y_h$. From (3.7) it easily follows that $(u_h^\epsilon, \Xi_h^\epsilon)$ satisfies

$$a(u_h^\epsilon, v_h) + \int_\omega \Xi_h^\epsilon P_h(\Pi v_h)\, ds + \frac{1}{\epsilon}\langle \beta_h'(u_h^\epsilon), v_h\rangle_h = \langle f, v_h\rangle \quad \forall v_h \in V_h\ . \tag{3.8}$$

Arguing exactly as in the proof of Theorem 2.9 (see (2.25), (2.26)) one can show that

$$\Xi_h^\epsilon(x_h^i) \in \hat{b}(P_h(\Pi u_h^\epsilon)(x_h^i)), \; i = 1, \dots, m \, . \tag{3.9}$$

□

We now prove the following existence result.

Theorem 3.2. *Let all the assumptions of Lemma 3.1 be satisfied. Then (3.2) has at least one solution for any $h > 0$.*

Proof. We keep $h > 0$ fixed and pass to the limit with $\epsilon \to 0+$. Let $\{(u_h^\epsilon, \Xi_h^\epsilon)\}$ be a bounded sequence of solutions to (3.8) and (3.9). Therefore one can pass to a subsequence (still denoted by the same symbol) such that

$$\left.\begin{array}{l} u_h^\epsilon \to u_h \\ \Xi_h^\epsilon \to \Xi_h \, , \quad \epsilon \to 0+ \end{array}\right\} \tag{3.10}$$

for some $(u_h, \Xi_h) \in V_h \times Y_h$. Next we prove that (u_h, Ξ_h) solves (3.2).

From (3.8) and (3.10) we have:

$$\exists C = const. > 0 : \quad |\langle \beta_h'(u_h^\epsilon), v_h \rangle_h| \le C\epsilon \|v_h\| \quad \forall v_h \in V_h \tag{3.11}$$

so that

$$\langle \beta_h'(u_h), v_h \rangle_h = \lim_{\epsilon \to 0+} \langle \beta_h'(u_h^\epsilon), v_h \rangle_h = 0 \quad \forall v_h \in V_h \, .$$

From this and (3.5) we conclude that $u_h \in K_h$. Let $v_h \in K_h$ be an arbitrary element and insert $v_h := v_h - u_h^\epsilon$ into (3.8):

$$a(u_h^\epsilon, v_h - u_h^\epsilon) + \int_\omega \Xi_h^\epsilon P_h(\Pi v_h - \Pi u_h^\epsilon)\, ds + \frac{1}{\epsilon}\langle \beta_h'(u_h^\epsilon) - \beta_h'(v_h), v_h - u_h^\epsilon \rangle_h = \langle f, v_h - u_h^\epsilon \rangle$$

using that $\beta_h'(v_h) = 0$. The penalty term is nonpositive because of monotonicity of the mapping $v_h \mapsto \beta_h'(v_h)$. Hence

$$a(u_h^\epsilon, v_h - u_h^\epsilon) + \int_\omega \Xi_h^\epsilon P_h(\Pi v_h - \Pi u_h^\epsilon)\, ds \ge \langle f, v_h - u_h^\epsilon \rangle \, .$$

Passing here to the limit with $\epsilon \to 0+$ we obtain $(3.2)_2$. The inclusion $(3.2)_3$ is now a consequence of (3.9), (3.10) and Lemma 2.6. □

We close this section with the convergence result.

Theorem 3.3. *Let the following assumptions be satisfied:*

- *the function b satisfies the sign condition (2.8) and the growth condition (2.45);*
- *the mappings P_h and Π satisfy (2.46) and (2.47);*
- *the system $\{K_h\}$ is such that $0 \in K_h$ for any $h > 0$ and (1.27), (1.28) are satisfied.*

Let $\{(u_h, \Xi_h)\}$, $h \to 0+$ *be a sequence of solutions to (3.2) such that* $\{u_h\}$ *is bounded in* V. *Then there exist: a subsequence of* $\{(u_h, \Xi_h)\}$ *(denoted by the same symbol), and an element* $(u, \Xi) \in K \times L^q(\omega)$, *where* q *is as in (2.45) such that*

$$\left.\begin{array}{lll} u_h \to u & \text{in } V & \\ \Xi_h \rightharpoonup \Xi & \text{in } L^q(\omega), & h \to 0+ \ . \end{array}\right\} \tag{3.12}$$

In addition, (u, Ξ) *solves (3.1) with* $Y = L^q(\omega)$. *Any accumulation point of* $\{(u_h, \Xi_h)\}$ *in the sense of (3.12) possesses this property.*

Proof. From (2.45)-(2.47) it follows that $\{\Xi_h\}$ is bounded in $L^q(\omega)$ (see also Lemma 2.12). Therefore one can suppose that

$$\left.\begin{array}{lll} u_h \rightharpoonup u & \text{in } V & \\ \Xi_h \rightharpoonup \Xi & \text{in } L^q(\omega), & h \to 0+ \ . \end{array}\right\} \tag{3.13}$$

From (1.28) and $(3.13)_1$ it follows that $u \in K$. Let $\overline{v} \in K$ be given. There exists a sequence $\{\overline{v}_h\}$, $\overline{v}_h \in K_h$ such that $\overline{v}_h \to \overline{v}$ in V and also

$$\left.\begin{array}{ll} P_h(\Pi u_h) \to \Pi u & \text{in } L^p(\omega) \\ P_h(\Pi \overline{v}_h) \to \Pi \overline{v}_h & \text{in } L^p(\omega) \end{array}\right\} \tag{3.14}$$

making use of (1.27), (2.47) and $(3.13)_1$. Using $\overline{v}_h$ as a test function and letting $h \to 0+$ in $(3.2)_2$ we easily obtain:

$$a(u, \overline{v} - u) + \int_\omega \Xi(\Pi\overline{v} - \Pi u)\, ds \geq \langle f, \overline{v} - u\rangle \tag{3.15}$$

as follows from (3.13), (3.14) and weak lower semicontinuity of the bilinear form a. The inclusion $(3.1)_3$ can be proven in the same way as in Theorem 2.11.

Next we show that $\{u_h\}$ tends strongly to u in V. From (1.27) it follows that there exists $\{\overline{u}_h\}$, $\overline{u}_h \in K_h$ such that

$$\overline{u}_h \to u \quad \text{in } V, \ h \to 0+ \ . \tag{3.16}$$

Further

$$\alpha\|u_h - \overline{u}_h\|^2 \leq a(u_h - \overline{u}_h, u_h - \overline{u}_h) = a(u_h, u_h - \overline{u}_h) - a(\overline{u}_h, u_h - \overline{u}_h) \ . \tag{3.17}$$

It is easy to see that

$$a(\overline{u}_h, u_h - \overline{u}_h) \to 0\ , \quad h \to 0+ \tag{3.18}$$

in view of $u_h - \overline{u}_h \rightharpoonup 0$ in V and (3.16). To estimate the first term on the right of (3.17) we use $(3.2)_2$:

$$\begin{aligned} a(u_h, u_h - \overline{u}_h) &\leq -\int_\omega \Xi_h(P_h(\Pi u_h - \Pi\overline{u}_h))\, ds + \langle f, u_h - \overline{u}_h\rangle \\ &\leq \|\Xi_h\|_{L^q(\omega)}\|P_h(\Pi u_h - \Pi\overline{u}_h)\|_{L^p(\omega)} + \langle f, u_h - \overline{u}_h\rangle \ . \end{aligned}$$

Therefore

$$\limsup_{h\to 0+} a(u_h, u_h - \overline{u}_h) \le \limsup_{h\to 0+} \{ \|\Xi_h\|_{L^q(\omega)} \|P_h(\Pi u_h - \Pi \overline{u}_h)\|_{L^p(\omega)} + \langle f, u_h - \overline{u}_h \rangle \} = 0 \ .$$

Here we used (2.47), (3.13) and the fact that $u_h - \overline{u}_h \rightharpoonup 0$ in V. From this and (3.18) we obtain:

$$\limsup_{h\to 0+} \|u_h - \overline{u}_h\|^2 = 0$$

implying that $\|u_h - \overline{u}_h\| \to 0$, $h \to 0+$. This and the triangle inequality yield $(3.12)_1$. □

To derive the algebraic form of (3.2) we proceed as in the unconstrained case. Let $h > 0$ be given. Then V_h, W_h and Y_h are isomorphic with $\mathbb{R}^n$, $\mathbb{R}^p$, and $\mathbb{R}^m$, respectively. The set $K_h \subset V_h$ can be identified with a closed convex subset $\mathcal{K} \subseteq \mathbb{R}^n$. The algebraic form of (3.2) now reads as follows:

$$\left.\begin{array}{l} \textit{Find } (\mathbf{u}, \Xi) \in \mathcal{K} \times \mathbb{R}^m \textit{ such that} \\ (\mathbf{A}\mathbf{u}, \mathbf{v} - \mathbf{u})_{\mathbb{R}^n} + (\Xi, \mathbf{\Lambda}(\mathbf{v} - \mathbf{u}))_{\mathbb{R}^m} \ge (\mathbf{f}, \mathbf{v} - \mathbf{u})_{\mathbb{R}^n} \quad \forall \mathbf{v} \in \mathcal{K} \\ \Xi_i \in c_i \hat{b}((\mathbf{\Lambda}\mathbf{u})_i) \ , \quad \forall i = 1, \ldots, m \ . \end{array}\right\} \tag{3.19}$$

The meaning of all symbols used in (3.19) is exactly the same as in Section 2.

Analogous to the unconstrained case we shall eliminate the second component Ξ by introducing the discrete superpotentional $\mathcal{J}$ given by (2.61). Instead of (3.19) we shall seek for substationary points $\mathbf{u}$ of $\mathcal{J}$ on $\mathcal{K}$:

$$\mathbf{0} \in \overline{\partial} \mathcal{J}(\mathbf{u}) + N_{\mathcal{K}}(\mathbf{u}) \ . \tag{3.20}$$

This problem can be numericaly realized again by methods of non-smooth optimization. The relation between both formulations (3.19) and (3.20) is the same as in the unconstrained case, see Theorems 2.16 and 2.17.

4 Contact problems with Coulomb's friction in elastostatics

This section deals with different variational formulations of contact problems with friction, their approximations and numerical realization. We shall use theoretical results of Section 1. To simplify our presentation we restrict ourselves to the case of *one* deformable body which is supported by a *perfectly rigid* half–plane in 2D or a half–space in 3D. We shall distinguish two cases of coefficients of friction $\mathcal{F}$, accordingly wheather or not $\mathcal{F}$ depends on a solution. We first start with two–dimensional contact problems with friction (see [Haslinger and Hlaváček and Nečas and Lovíšek 1988]).

4.1 Two-dimensional contact problems with friction

Let a deformable body be represented by a bounded domain $\Omega \subset \mathbb{R}^2$ with the Lipschitz boundary $\partial\Omega$ which is split into three non–empty, non–overlapping open parts Γ_u, Γ_p and Γ_c, i.e. $\partial\Omega = \overline{\Gamma}_u \cup \overline{\Gamma}_p \cup \overline{\Gamma}_c$. The body is subject to body forces of density $F \in (L^2(\Omega))^2$

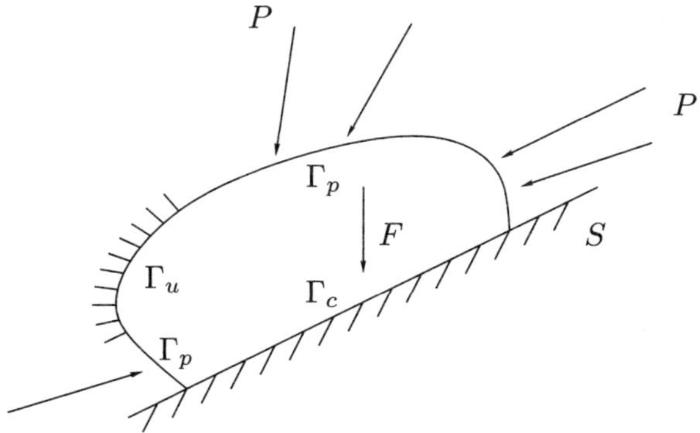

Figure 1.

and surface tractions of density $P \in (L^2(\Gamma_p))^2$ acting on Γ_p. Further the body is fixed on Γ_u and unilaterally supported by a rigid half–plane S along Γ_c (see Fig 1.). Γ_c itself will be represented by *a straight segment.*

Our aim is to find an equilibrium state of Ω taking into account the influence of friction on the contact part Γ_c. This state is characterized by a *symmetric stress tensor* $\tau = (\tau_{ij})_{i,j=1}^2$, a *linearized strain tensor* $\epsilon = (\epsilon_{ij})_{i,j=1}^2$ and a *displacement vector* $u = (u_1, u_2)$ mutually linked by the following equations and the boundary conditions (here and in what follows we shall use the summation convention):

(equilibrium equations)

$$\frac{\partial \tau_{ij}}{\partial x_j} + F_i = 0 \quad in\ \Omega,\ i = 1,2\ ; \tag{4.1}$$

(a stress-strain relation)

$$\tau_{ij} = c_{ijkl}\epsilon_{kl}\ , \quad c_{ijkl} \in L^\infty(\Omega) \quad i,j,k,l = 1,2\ ; \tag{4.2}$$

(a compatibility condition)

$$\epsilon := (\epsilon_{ij}(u))_{i,j=1}^2 = \frac{1}{2}(\frac{\partial u_i}{\partial x_j} + \frac{\partial u_j}{\partial x_i})_{i,j=1}^2\ ; \tag{4.3}$$

(kinematical boundary conditions)

$$u_i = 0 \quad on\ \Gamma_u,\ i = 1,2\ ; \tag{4.4}$$

(static boundary conditions)

$$\tau_{ij}n_j = P_i \quad on\ \Gamma_p,\ i = 1,2\ ; \tag{4.5}$$

(non-penetration conditions)

$$u_n := u \cdot n \le 0, \quad T_n := \tau_{ij} n_i n_j \le 0, \quad u_n T_n = 0 \quad on\ \Gamma_c\ ; \tag{4.6}$$

(Coulomb's law of friction)

$$\left.\begin{array}{rcl} u_t := u \cdot t = 0 & \Rightarrow & |T_t| \le -\mathcal{F} T_n \\ u_t \ne 0 & \Rightarrow & T_t = \mathcal{F} T_n\, sign(u_t) \end{array}\right\} \quad on\ \Gamma_c\ , \tag{4.7}$$

if $\mathcal{F} := \mathcal{F}(x)$ does not depend on u, where $T_t := \tau_{ij} n_j t_i$. The symbols n, t denote the unit outward normal, tangential vector, respectively to $\partial\Omega$. If $\mathcal{F}$ depends on u, more precisely on $|u_t|$, the Coulomb law of friction reads as follows:

$$\left.\begin{array}{rcl} u_t = 0 & \Rightarrow & |T_t| \le -\mathcal{F}(0) T_n \\ u_t \ne 0 & \Rightarrow & T_t = \mathcal{F}(|u_t|) T_n\, sign(u_t) \end{array}\right\} \quad on\ \Gamma_c\ . \tag{4.8}$$

The elasticity coefficients c_{ijkl} in (4.2) satisfy the following symmetry and ellipticity conditions:

$$c_{ijkl} = c_{jikl} = c_{klij} \quad a.e.\ in\ \Omega\ ;$$
$$\exists \alpha > 0 : c_{ijkl}(x) \xi_{ij} \xi_{kl} \ge \alpha \xi_{ij} \xi_{ij}$$

for all $\xi_{ij} = \xi_{ji} \in \mathbb{R}^1$ and a.a $x \in \Omega$. Finally, the coefficient $\mathcal{F}$ of Coulomb's friction satisfies:

$$\mathcal{F} \in L^\infty(\Gamma_c)\ , \quad \mathcal{F} \ge 0\ , \quad supp\, \mathcal{F} \subset \Gamma_c \tag{4.9}$$

if $\mathcal{F}$ does not depend on u and

$$\left.\begin{array}{l} \mathcal{F} \in L^\infty(\Gamma_c \times \mathbb{R}^1_+)\ , \ \mathcal{F} \ge 0\ , \ supp\, \mathcal{F}(\cdot, \xi) \subset \Gamma_c \ \ \forall \xi \in \mathbb{R}^1_+\ ; \\ \exists C_L > 0 : \ |\mathcal{F}(x,\xi) - \mathcal{F}(x,\bar{\xi})| \le C_L |\xi - \bar{\xi}| \ \ \forall \xi, \bar{\xi} \in \mathbb{R}^1_+ \\ \ and\ a.a.\ x \in \Gamma_c \end{array}\right\} \quad on\ \Gamma_c \tag{4.10}$$

in the opposite case.

By a *classical solution* of the contact problem with Coulomb's friction we call any displacement field u satisfying (4.1)-(4.7) ((4.8)) with $\tau := \tau(u) = (\tau_{ij}(u))^2_{i,j=1}$, where $\tau_{ij}(u) = c_{ijkl} \epsilon_{kl}(u)$.

To give the weak formulation of this problem we introduce the following function space:

$$V = \{v = (v_1, v_2) \in (H^1(\Omega))^2 |\ v_i = 0\ on\ \Gamma_u\ ;\ i = 1, 2\} \tag{4.11}$$

and its closed convex subset

$$K = \{v \in V |\ v_n \le 0\ on\ \Gamma_c\}\ . \tag{4.12}$$

By a *weak solution* of the contact problem with Coulomb's friction we mean any displacement field $u \in K$ satisfying the following *implicit* variational inequality of elliptic type:

$$\left.\begin{array}{l} \displaystyle\int_\Omega \tau_{ij}(u) \epsilon_{ij}(v - u)\, dx - \int_{\Gamma_c} \mathcal{F} T_n(u)(|v_t| - |u_t|)\, ds \\ \quad \ge \ell(v - u) \quad \forall v \in K\ , \end{array}\right\} \quad on\ \Gamma_c \tag{4.13}$$

where

$$\ell(v) := \int_{\Omega} F_i v_i \, dx + \int_{\Gamma_p} P_i v_i \, ds \ .$$

The coefficient $\mathcal{F}$ appearing in $(4.13)_2$ reads either as $\mathcal{F} := \mathcal{F}(x)$ if $\mathcal{F}$ does not depend on u or $\mathcal{F} := \mathcal{F}(x, |u_t|)$ in the oposite case. Next we shall consider both cases separately.

CASE $\mathcal{F} := \mathcal{F}(x)$

Two different approaches were used for proving the existence of a solution to (4.13), namely

a) *fixed point approach;*

b) *approximation of* (4.13) *by a simultaneous penalization and regularization.*

The numerical realization of contact problems presented in this article will be based *solely* on *a*). Nevertheless let us say a few words on *b*) because from the theoretical point of view this approach gives stronger results compared with *a*). The basic idea is simple: the unilateral constraint $u_n \le 0$ on Γ_c in (4.13) is replaced by adding the penalty functional

$$\frac{1}{\epsilon} \int_{\Gamma_c} u_n^+ v_n \, ds \ , \quad \epsilon \to 0+ \ ,$$

where a^+ stands for the positive part of a real number a, while the absolute value $| \ |$ in the friction term is approximated by appropriate smooth convex functions Φ_η, $\eta \to 0+$. The regularized and penalized version of (4.13) is given by the following variational equation:

$$\left.\begin{array}{l} \textit{Find } u_{\epsilon\eta} \in V \textit{ such that} \\ \displaystyle\int_{\Omega} \tau_{ij}(u_{\epsilon\eta}) \epsilon_{ij}(v) \, dx + \frac{1}{\epsilon} \int_{\Gamma_c} (u_{\epsilon\eta})_n^+ v_n \, ds + \\ \displaystyle\frac{1}{\epsilon} \int_{\Gamma_c} (u_{\epsilon\eta})_n^+ \mathcal{F} \nabla \Phi_\eta((u_{\epsilon\eta})_t) v_t \, ds = \ell(v) \ \forall v \in V \end{array}\right\} \qquad (4.14)$$

In [Eck and Jarušek 1998] it was shown that (4.14) has a solution for any $\epsilon, \eta > 0$. Passing to the limit first with $\eta \to 0+$, then with $\epsilon \to 0+$ in (4.14), the authors proved the existence of a function $u \in K$ satisfying (4.13) provided that the coefficient $\mathcal{F}$ is sufficiently small. In addition, they derived an upper bound for $\mathcal{F}$ guaranteeing the existence of a solution.

As we have already mentioned, the approach *b*) is powerfull from the theoretical point of view but not very convenient for practical computations. The reason is that a discrete version of (4.13) leads to a system of non–linear equations depending on two small parameters ϵ and η. It turns out (see [Eck and Steinbach and Wendland 1999]) that the numerical process is very sensitive to their choice. Therefore *a*) is prefered. Before we present this approach let us introduce several notation.

Denote

$$W = \{v \in H^1(\Omega)|\ v = 0 \ on \ \Gamma_u\} \tag{4.15}$$

and

$$H^{1/2}(\Gamma_c) = \{\varphi \in L^2(\Gamma_c)|\ \exists v \in W :\ v = \varphi \ on \ \Gamma_c\} \ , \tag{4.16}$$

i.e. $H^{1/2}(\Gamma_c)$ is the trace space on Γ_c of functions which belong to W. The norm in $H^{1/2}(\Gamma_c)$ can be introduced as follows:

$$\|\varphi\|_{1/2,\Gamma_c} = \inf_{\substack{v \in W \\ v=\varphi \ on \ \Gamma_c}} |v|_{1,\Omega} \ . \tag{4.17}$$

Problem 4.1 Prove that $H^{1/2}(\Gamma_c)$ endowed with the norm (4.17) is a Banach space.

The dual of $H^{1/2}(\Gamma_c)$ will be denoted by $H^{-1/2}(\Gamma_c)$ and $\langle\ ,\ \rangle$ stands for a duality pairing between $H^{-1/2}(\Gamma_c)$ and $H^{1/2}(\Gamma_c)$. Finally, we shall need the following cones of all non–negative elements of $H^{1/2}(\Gamma_c)$ and $H^{-1/2}(\Gamma_c)$:

$$H_+^{1/2}(\Gamma_c) = \{\varphi \in H^{1/2}(\Gamma_c) \mid \varphi \geq 0 \ a.e. \ on \ \Gamma_c\} \tag{4.18}$$

$$H_+^{-1/2}(\Gamma_c) = \{\mu \in H^{-1/2}(\Gamma_c) \mid \langle \mu, \varphi \rangle \geq 0 \ \forall \varphi \in H_+^{1/2}(\Gamma_c)\} \ . \tag{4.19}$$

Let $g \in H_+^{-1/2}(\Gamma_c)$ be given. Instead of Coulomb's law of friction we shall consider the following simplified model:

$$\left.\begin{array}{l} u_t = 0 \Rightarrow |T_t| \leq \mathcal{F}g \\ u_t \neq 0 \Rightarrow T_t = -\mathcal{F}g \, sign(u_t) \end{array}\right\} \ on \ \Gamma_c \ , \tag{4.20}$$

i.e. the *unknown* normal stress T_n in (4.7) is now replaced by a known slip bound g. We use this model of friction in order to introduce an auxiliary contact problem with *given* friction. Its variational formulation leads to the following elliptic inequality of the second kind:

$$\left.\begin{array}{l} Find \ u := u(g) \in K \ such \ that \\ \displaystyle\int_\Omega \tau_{ij}(u)\epsilon_{ij}(v-u)\,dx + \langle \mathcal{F}g, |v_t| - |u_t| \rangle \geq \ell(v-u) \quad \forall v \in K \ . \end{array}\right\} \tag{4.21}$$

From Theorem 1.2 it easily follows that (4.21) has a unique solution for every $g \in H_+^{-1/2}(\Gamma_c)$ (we write $u(g)$ in order to emphasize that u depends on a particular choice of g). This makes it possible to define the mapping Φ from $H_+^{-1/2}(\Gamma_c)$ into itself by

$$\Phi(g) = -T_n(u(g)) \ , \tag{4.22}$$

where $T_n(u(g))$ is the normal component of the contact stress which corresponds to the solution $u(g)$ of (4.21). From (4.13) and (4.22) we see that a function $u \in K$ is a weak solution of a contact problem with Coulomb's friction if and only if $-T_n(u)$ is *a fixed point* of Φ:

$$\Phi(-T_n(u)) = -T_n(u) \ . \tag{4.23}$$

The existence of at least one fixed point of Φ has been shown in [Nečas and Jarušek and Haslinger 1980] for an infinity long strip and in [Jarušek 1983] for a bounded domain provided that the coefficient $\mathcal{F}$ of Coulomb's friction is small enough.

The fixed point formulation of contact problems with Coulomb's friction is convenient for their numerical realization. A natural way how to find fixed points of Φ is to use the method of *successive approximations*:

$$\left.\begin{array}{l} g_0 \in H_+^{-1/2}(\Gamma_c) \ given \ ; \\ for \ k \in \mathbb{N} \ set \ g_{k+1} = \Phi(g_k) \ . \end{array}\right\} \tag{4.24}$$

The advantage of this approach is readily seen: each iterative step in (4.24) leads to a contact problem with given friction, characterized by the slip bound g_k known from the previous iteration. Let us mention however that the mapping Φ *is not* contractive. Therefore convergence of the sequence $\{g_k\}$ to a fixed point of Φ is not guaranteed. As we shall see later on, the situation is different for discretized problems: for an appropriate discretization of Φ the method of successive approximations *is* convergent provided that $\mathcal{F}$ is sufficiently small. Unfortunately the respective bounds for $\mathcal{F}$ are *mesh dependent*. The numerical efficiency of (4.24) depends among others, how efficiently each iterative step, represented by a contact problem with given friction can be realized.

Primal, mixed and dual formulations of contact problems with given friction

Since a contact problem with given friction is the main step in (4.24), we focus on it. We present its different variational formulations. Because of the symmetry of the elasticity coefficients c_{ijkl}, problem (4.21) is equivalent to the following minimization problem over K:

$$\left.\begin{array}{l} Find \ u \in K \ such \ that \\ J(u) \le J(v) \ \forall v \in K \ , \end{array}\right\} \tag{4.25}$$

where $J(v) = \frac{1}{2}a(v,v) + j(v) - \ell(v)$, with

$$\left.\begin{array}{l} a(u,v) = \displaystyle\int_\Omega c_{ijkl}\epsilon_{ij}(u)\epsilon_{kl}(v)\,dx \ , \\ j(v) = \langle \mathcal{F}g, |v_t|\rangle \ , \\ \ell(v) = \displaystyle\int_\Omega F_i v_i\,dx + \int_{\Gamma_p} P_i v_i\,ds \ . \end{array}\right\} \tag{4.26}$$

(4.25) is called the *primal* variational formulation of a contact problem with given friction. This is a constrained minimization problem for the non–differentiable functional J.

In order to derive the mixed formulation, we shall suppose in what follows that the slip bound g is represented by a non–negative function from $L^2(\Gamma_c)$ (this assumption *is not* necessary but it simplifies the presentation). We use Lagrange multipliers to release the unilateral constraint $u_n \le 0$ on Γ_c and to regularize the non–differentiable term j. Denote

$$\left.\begin{array}{l}\Lambda_n = H_+^{-1/2}(\Gamma_c)\\ \Lambda_t(g) = \{\mu_t \in L^2(\Gamma_c) \mid |\mu_t| \le \mathcal{F}g \textit{ a.e. on } \Gamma_c\}\ .\end{array}\right\} \tag{4.27}$$

It can be proven that

$$v \in K \textit{ if and only if } v \in V \textit{ and } \langle \mu, v_n\rangle \le 0\ \forall \mu \in \Lambda_n$$

and

$$j(v) = \sup_{\mu_t \in \Lambda_t(g)} \int_{\Gamma_c} \mu_t v_t\, ds\ .$$

Therefore

$$\min_{v \in K} J(v) = \min_{v \in V} \sup_{\substack{\mu_n \in \Lambda_n\\ \mu_t \in \Lambda_t(g)}} \mathcal{L}(v, \mu_n, \mu_t)\ ,$$

where

$$\mathcal{L}(v, \mu_n, \mu_t) = J(v) + \langle \mu_n, v_n\rangle + \langle \mu_t, v_t\rangle \tag{4.28}$$

is the Lagrange function.

According to Section 1, by a *mixed* variational formulation of a contact problem with given friction we call a problem of finding a saddle–point of $\mathcal{L}$ on $V \times \Lambda_n \times \Lambda_t(g)$:

$$\left.\begin{array}{l}\textit{Find } (w, \lambda_n, \lambda_t) \in V \times \Lambda_n \times \Lambda_t(g) \textit{ such that}\\ \mathcal{L}(w, \mu_n, \mu_t) \le \mathcal{L}(w, \lambda_n, \lambda_t) \le \mathcal{L}(v, \lambda_n, \lambda_t)\\ \quad \forall v \in V,\ \forall(\mu_n, \mu_t) \in \Lambda_n \times \Lambda_t(g)\ ,\end{array}\right\} \tag{4.29}$$

or equivalently

$$\left.\begin{array}{l}\textit{Find } (w, \lambda_n, \lambda_t) \in V \times \Lambda_n \times \Lambda_t(g) \textit{ such that}\\ a(w, v) = \ell(v) - \langle \lambda_n, v_n\rangle - \langle \lambda_t, v_t\rangle \quad \forall v \in V\\ \langle \mu_n - \lambda_n, w_n\rangle + \langle \mu_t - \lambda_t, w_t\rangle \le 0 \quad \forall(\mu_n, \mu_t) \in \Lambda_n \times \Lambda_t(g)\ .\end{array}\right\} \tag{4.30}$$

It holds:

Theorem 4.2. *There exists a unique solution $(w, \lambda_n, \lambda_t)$ of (4.30). In addition,*

$$w = u\ ,\ \lambda_n = -T_n(u)\ ,\ \lambda_t = -T_t(u)\ ,$$

where $u \in K$ solves (4.25).

Problem 4.3 Prove Theorem 4.2.

These are the main advantages of the mixed formulation (4.30):

- the constrained minimization problem for the non–smooth functional J is now replaced by a smooth problem for the Lagrange function $\mathcal{L}$ with constraints imposed only on the Lagrange multipliers. In addition, these constraints are *simple* (box–constraints);

- mixed formulation (4.30) enables us to approximate not only the displacement field u but also the Lagrange multipliers λ_n, λ_t which have an important mechanical interpretation (contact stresses). This fact will be used for numerical realization of contact problems with Coulomb's friction.

Let $B_n, B_t : V \mapsto H^{1/2}(\Gamma_c)$ be the following trace mappings:

$$B_n v = v_n \,, \quad B_t v = v_t \quad on\ \Gamma_c \,. \tag{4.31}$$

Problem 4.4 Prove that B_n and B_t are linear continuous mappings from V *onto* $H^{1/2}(\Gamma_c)$.

The mixed formulation (4.30) is equivalent to the following system of generalized equations:

$$\left.\begin{array}{l} Find\ (w, \lambda_n, \lambda_t) \in V \times \Lambda_n \times \Lambda_t(g)\ such\ that \\ Aw = \ell - B_n^t \lambda_n - B_t^t \lambda_t \quad in\ V' \\ B_n w \in N_{\Lambda_n}(\lambda_n) \\ B_t w \in N_{\Lambda_t(g)}(\lambda_t) \,, \end{array}\right\} \tag{4.32}$$

where $A \in L(V, V')$ is associated with the bilinear form a defined by $(4.26)_1$ and B_n^t, B_t^t denote the transposes of B_n and B_t, respectively.

From the well-known properties of saddle-points (see [Ekeland and Temam 1976]) we know that

$$\mathcal{L}(w, \lambda_n, \lambda_t) = \min_{v \in V} \sup_{\substack{\mu_n \in \Lambda_n \\ \mu_t \in \Lambda_t(g)}} \mathcal{L}(v, \mu_n, \mu_t) = \max_{\substack{\mu_n \in \Lambda_n \\ \mu_t \in \Lambda_t(g)}} \inf_{v \in V} \mathcal{L}(v, \mu_n, \mu_t) = - \min_{\substack{\mu_n \in \Lambda_n \\ \mu_t \in \Lambda_t(g)}} \mathcal{S}(\mu_n, \mu_t) \,,$$

where

$$\mathcal{S}(\mu_n, \mu_t) := - \inf_{v \in V} \mathcal{L}(v, \mu_n, \mu_t)$$

is the *dual* functional. It is easy to show that $\mathcal{S}$ is a quadratic functional:

$$\mathcal{S}(\mu) = \frac{1}{2} \mathcal{B}(\mu, \mu) - f(\mu) \,, \quad \mu = (\mu_n, \mu_t) \in (H^{-1/2}(\Gamma_c))^2 \,,$$

where the bilinear form $\mathcal{B} : (H^{-1/2}(\Gamma_c))^2 \times (H^{-1/2}(\Gamma_c))^2 \mapsto \mathbb{R}^1$, the linear form $f : (H^{-1/2}(\Gamma_c))^2 \mapsto \mathbb{R}^1$ are defined by

$$\mathcal{B}(\mu, \nu) = \langle \mu_n, A^{-1}(\nu_n, \nu_t) n \rangle + \langle \mu_t, A^{-1}(\nu_n, \nu_t) t \rangle \,, \tag{4.33}$$

$$f(\mu) = -\langle \mu_n, A^{-1}(\ell) n \rangle - \langle \mu_t, A^{-1}(\ell) t \rangle \,, \tag{4.34}$$

respectively, $\mu = (\mu_n, \mu_t)$, $\nu = (\nu_n, \nu_t) \in (H^{-1/2}(\Gamma_c))^2$. Here $A^{-1}(\nu_n, \nu_t) := z_1$, $A^{-1}(\ell) := z_2$, where z_1, z_2 are solutions to the following linear elasticity problems:

$$z_1 \in V : \quad a(z_1, v) = -\langle \nu_n, v_n \rangle - \langle \nu_t, v_t \rangle \quad \forall v \in V \tag{4.35}$$

$$z_2 \in V : \quad a(z_2, v) = \ell(v) \quad \forall v \in V \,. \tag{4.36}$$

Problem 4.5 Prove that $\mathcal{B}$ and f which define the dual functional $\mathcal{S}$ are given by (4.33), (4.34), respectively.

The *dual* variational formulation of a contact problem with given friction reads as follows:

$$\left.\begin{array}{l} Find\ \lambda = (\lambda_n, \lambda_t) \in \Lambda_n \times \Lambda_t(g)\ such\ that \\ \mathcal{S}(\lambda) \leq \mathcal{S}(\mu) \quad \forall \mu = (\mu_n, \mu_t) \in \Lambda_n \times \Lambda_t(g)\ , \end{array}\right\} \tag{4.37}$$

or equivalently

$$\left.\begin{array}{l} Find\ \lambda \in \Lambda_n \times \Lambda_t(g)\ such\ that \\ \mathcal{B}(\lambda, \mu - \lambda) \geq f(\mu - \lambda) \quad \forall \mu \in \Lambda_n \times \Lambda_t(g)\ . \end{array}\right\} \tag{4.38}$$

The variational inequality (4.38) can be derived directly from the mixed formulation (4.32): we first express w from $(4.32)_2$ and then eliminate it from $(4.32)_{3,4}$ (see also (1.21)-(1.24)).

From Theorem 1.7 we obtain the following existence result:

Theorem 4.6. *Problem (4.37) (or (4.38)) has a unique solution $\lambda = (\lambda_n, \lambda_t)$. In addition,*

$$\lambda_n = -T_n(u)\ , \quad \lambda_t = -T_t(u)\ on\ \Gamma_c\ ,$$

where $T_n(u)$, $T_t(u)$ is the normal, tangential contact stress, respectively corresponding to the solution u of (4.25).

Remark 4.7. Let us observe that the formulation of a contact problem with Coulomb's friction in terms of contact stresses leads to the following *quasivariational inequality*:

$$\left.\begin{array}{l} Find\ \lambda = (\lambda_n, \lambda_t) \in \Lambda_n \times \Lambda_t(\lambda_n)\ such\ that \\ \mathcal{B}(\lambda, \mu - \lambda) \geq f(\mu - \lambda) \quad \forall \mu \in \Lambda_n \times \Lambda_t(\lambda_n)\ . \end{array}\right\} \tag{4.39}$$

Approximations of contact problems with given friction

This subsection deals with approximations of the previous variational formulations of contact problems with given friction. To avoid the use of curved elements we shall suppose that $\Omega \subset \mathbb{R}^2$ is a *polygonal domain*. We start with the approximation of the primal formulation (4.25).

Let $\{\mathcal{T}_h\}$, $h \to 0+$ be a *regular* family of triangulations of $\overline{\Omega}$ (see [Ciarlet 1978]). With any $\mathcal{T}_h$ we associate the space V_h of continuous, piecewise linear vector functions over $\mathcal{T}_h$:

$$V_h = \{v_h = (v_{h1}, v_{h2}) \in (C(\overline{\Omega}))^2 \mid v_{h|_T} \in (P_1(T))^2 \quad \forall T \in \mathcal{T}_h,\ v_h = 0\ on\ \Gamma_u\} \tag{4.40}$$

and its closed convex subset K_h defined by

$$K_h = \{v_h \in V_h \mid v_{hn}(a_i) \leq 0 \quad \forall a_i \in \mathcal{N}_h\}\ , \tag{4.41}$$

where $\mathcal{N}_h$ is the set of all contact nodes, i.e. the nodes of $\mathcal{T}_h$ lying on $\overline{\Gamma}_c \setminus \overline{\Gamma}_u$. Since Γ_c is supposed to be represented by a straight segment, it holds that $K_h \subset K \ \forall h \to 0+$. The approximation of the primal formulation is defined by (1.26) with K_h given by (4.41) and J as in (4.25). The sequence $\{u_h\}$ of approximate solutions tends to a unique solution u of (4.25) provided that (1.27) is satisfied. A simple criterion for the satisfaction of (1.27) is given in the following theorem.

Theorem 4.8. *Let $K \cap (C^\infty(\overline{\Omega}))^2$ be dense in K. Then (1.27) is satisfied.*

Some cases when the density assumption of Theorem 4.8 is satisfied are studied in [Haslinger and Hlaváček and Nečas and Lovíšek 1988].

For numerical realization of contact problems with given friction *we shall not* use the primal formulation. Besides non–smoothness of J there is yet another reason: the discrete unilateral conditions lead to linear inequality constraints which *are not simple* (i.e. box–type constraints), in general.

We now pass to a discretization of the mixed formulation (4.30). Let L^1_H, L^2_H be two finite dimensional spaces which are in duality with $C(\overline{\Gamma}_c)$ and denote by $\langle \ , \ \rangle_l$, $l = 1, 2$ the respective duality pairings. The subscript H stands for a discretization parameter characterizing the dimension of L^l_H, $l = 1, 2$. By $\Lambda_{nH} \subset L^r_H$, $\Lambda_{tH}(g) \subset L^s_H$, $r, s \in \{1, 2\}$ we denote two convex sets which will serve as approximations of Λ_n and $\Lambda_t(g)$, respectively. The discretization of the mixed formulation (4.30) reads as follows:

$$\left.\begin{array}{l} \textit{Find } (w_h, \lambda^n_H, \lambda^t_H) \in V_h \times \Lambda_{nH} \times \Lambda_{tH}(g) \textit{ such that} \\ a(w_h, v_h) = \ell(v_h) - \langle \lambda^n_H, v_{hn}\rangle_r - \langle \lambda^t_H, v_{ht}\rangle_s \quad \forall v_h \in V_h \\ \langle \mu^n_H - \lambda^n_H, w_{hn}\rangle_r + \langle \mu^t_H - \lambda^t_H, w_{ht}\rangle_s \le 0 \quad \forall (\mu^n_H, \mu^t_H) \in \Lambda_{nH} \times \Lambda_{tH}(g)\ , \end{array}\right\} \tag{4.42}$$

where V_h is defined by (4.40), $r, s \in \{1, 2\}$, $v_{hn} := v_h \cdot n$ and $v_{ht} := v_h \cdot t$.

To ensure the existence and uniqueness of a solution to (4.42) we shall suppose that the following assumption is satisfied:

$$\left.\begin{array}{l} \textit{if } (\mu^n_H, \mu^t_H) \in \Lambda_{nH} \times \Lambda_{tH}(g) \textit{ is such that} \\ \langle \mu^n_H, v_{hn}\rangle_r + \langle \mu^t_H, v_{ht}\rangle_s = 0 \quad \forall v_h \in V_h\ , \\ \textit{then } \mu^n_H = \mu^t_H = 0\ . \end{array}\right\} \tag{4.43}$$

The algebraic form of (4.43) reads as follows:

$$\left.\begin{array}{l} \textit{Find } (\mathbf{w}, \boldsymbol{\lambda}_n, \boldsymbol{\lambda}_t) \in \mathbb{R}^p \times \boldsymbol{\Lambda}_n \times \boldsymbol{\Lambda}_t(g) \textit{ such that} \\ \mathbf{A}\mathbf{w} + \mathbf{B}^t_n \boldsymbol{\lambda}_n + \mathbf{B}^t_t \boldsymbol{\lambda}_t = \mathbf{f} \\ (\mathbf{B}_n\mathbf{w}; \boldsymbol{\mu}_n - \boldsymbol{\lambda}_n)_{\mathbb{R}^{m_r}} + (\mathbf{B}_t\mathbf{w}; \boldsymbol{\mu}_t - \boldsymbol{\lambda}_t)_{\mathbb{R}^{m_s}} \le 0 \quad \forall (\boldsymbol{\mu}_n, \boldsymbol{\mu}_t) \in \boldsymbol{\Lambda}_n \times \boldsymbol{\Lambda}_t(g)\ , \end{array}\right\} \tag{4.44}$$

where $\boldsymbol{\Lambda}_n \subseteq \mathbb{R}^{m_r}$, $\boldsymbol{\Lambda}_t(g) \subseteq \mathbb{R}^{m_s}$, $r, s \in \{1, 2\}$ are convex sets which are isometrically isomorphic with Λ_{nH}, $\Lambda_{tH}(g)$, respectively, $m_l = dim\, L^l_H$, $l = 1, 2$ and $p = dim\, V_h$. Further $\mathbf{w}$, $\boldsymbol{\lambda}_n$, $\boldsymbol{\lambda}_t$ are the coordinates of w_h, λ^n_H, λ^t_H with respect to chosen basis in V_h, L^r_H and L^s_H, respectively. Finally, $\mathbf{A}$ is the stiffness matrix, $\mathbf{f}$ is the force vector, $\mathbf{B}_n$, $\mathbf{B}_t$ are kinematic transformation matrices linking the primal variable $\mathbf{w}$ with the Lagrange multipliers $\boldsymbol{\lambda}_n$ and $\boldsymbol{\lambda}_t$, respectively.

Problem 4.9 Write the explicit form of $\mathbf{B}_n$ and $\mathbf{B}_t$.

From Section 1 we know that the first component w_h of the solution to (4.42) solves the following variational inequality:

$$w_h \in K_{hH} : \ a(w_h, v_h - w_h) + j_H(v_h) - j_H(w_h) \geq \ell(v_h - w_h) \quad \forall v_h \in K_{hH} \ , \tag{4.45}$$

where

$$K_{hH} = \{v_h \in V_h \mid \langle \mu_H^n, v_{hn} \rangle_r \leq 0 \quad \forall \mu_H^n \in \Lambda_{nH}\} \ , \tag{4.46}$$

$$j_H(v_h) = \sup_{\mu_H^t \in \Lambda_{tH}(g)} \langle \mu_H^t, v_{ht} \rangle_s \quad r, s \in \{1, 2\} \tag{4.47}$$

is an approximation of K and j, respectively. Next, we shall specify two possible constructions of L_H^1 and L_H^2 which are frequently used in computations.

Let $\{\delta_i\}_{i=1}^{m_1}$ be a family of the Dirac distributions concentrated at the contact nodes, i.e. $\delta_i(\varphi) = \varphi(a_i)$, $\varphi \in C(\overline{\Gamma}_c)$, $a_i \in \mathcal{N}_h$, $i = 1, \dots, m_1 = card\,\mathcal{N}_h$. By L_H^1 we denote the space spanned by $\{\delta_i\}_{i=1}^{m_1}$:

$$L_H^1 = \{\mu_H \mid \mu_H = \sum_{i=1}^{m_1} \mu_{Hi}\delta_i \, , \ \mu_{Hi} \in \mathbb{R}^1\} \ . \tag{4.48}$$

The duality pairing $\langle \ , \ \rangle_1$ is defined by

$$\langle \mu_H, \varphi \rangle_1 := \sum_{i=1}^{m_1} \mu_{Hi}\varphi(a_i) \, , \ \mu_H \in L_H^1, \ \varphi \in C(\overline{\Gamma}_c) \ . \tag{4.49}$$

If $\Lambda_{nH} \subset L_H^1$ then we set

$$\Lambda_{nH} = \{\mu_H \in L_H^1 \mid \mu_H = \sum_{i=1}^{m_1} \mu_{Hi}\delta_i \, , \ \mu_{Hi} \geq 0 \ \forall i = 1, \dots, m_1\} \tag{4.50}$$

so that Λ_{nH} is isometrically isomorphic with $\mathbf{\Lambda}_n = \mathbb{R}_+^{m_1}$. It is readily seen that the set K_{hH} coincides with K_h defined in (4.41).

If $\Lambda_{tH}(g) \subset L_H^1$ then

$$\Lambda_{tH}(g) = \{\mu_H \in L_H^1 \mid \mu_H = \sum_{i=1}^{m_1} \mu_{Hi}\delta_i \, , \ |\mu_{Hi}| \leq (\mathcal{F}g)(a_i) \ \forall i = 1, \dots, m_1\} \tag{4.51}$$

provided that $\mathcal{F}g \in C(\overline{\Gamma}_c)$. The set $\Lambda_{tH}(g)$ is isometrically isomorphic with $\mathbb{R}_{[\,]}^{m_1}$, where

$$\mathbb{R}_{[\,]}^{m_1} = \{\boldsymbol{\alpha} \in \mathbb{R}^{m_1} \mid |\alpha_i| \leq (\mathcal{F}g)(a_i) \quad i = 1, \dots, m_1\} \ . \tag{4.52}$$

It is easy to show that

$$j_H(v_h) = \sum_{i=1}^{m_1} (\mathcal{F}g)(a_i)|v_{ht}(a_i)| \tag{4.53}$$

for this choice of $\Lambda_{tH}(g)$.

Problem 4.10 Prove (4.53).

Instead of Dirac distributions one can use *piecewise polynomial* approximations of Lagrange multipliers. Below we shall present the simplest case, namely the approximation by *piecewise constant* functions.

Let $\{\mathcal{T}_H\}$, $H \to 0+$ be a family of regular partitions of $\overline{\Gamma}_c$. With any $\mathcal{T}_H$ we associate the space

$$L_H^2 = \{\mu_H \in L^2(\Gamma_c) \mid \mu_{H|S_i} \in P_0(S_i) \quad \forall S_i \in \mathcal{T}_H\} , \tag{4.54}$$

where S_i, $i = 1, \ldots, m_2$ are segments of $\mathcal{T}_H$ and $P_0(S_i)$ denotes the space of all constants on S_i. The duality pairing $\langle\ ,\ \rangle_2$ is defined by

$$\langle \mu_H, \varphi \rangle_2 := \int_{\Gamma_c} \mu_H \varphi \, ds \ , \quad \mu_H \in L_H^2 \ , \quad \varphi \in L^2(\Gamma_c) \tag{4.55}$$

(observe that this duality is defined for a larger class of φ unlike to the previous case).

If $\Lambda_{nH}, \Lambda_{tH}(g) \subset L_H^2$, then

$$\Lambda_{nH} = \{\mu_H \in L_H^2 \mid \mu_H = \sum_{i=1}^{m_2} \mu_{Hi}\chi_i, \quad \mu_{Hi} \geq 0, \quad i = 1, \ldots, m_2\}$$

$$\Lambda_{tH}(g) = \{\mu_H \in L_H^2 \mid \mu_H = \sum_{i=1}^{m_2} \mu_{Hi}\chi_i, \quad |\mu_{Hi}| \leq \int_{S_i} \mathcal{F} g \, ds, \quad i = 1, \ldots, m_2\} ,$$

where χ_i stands for the characteristic function of $int\, S_i$, $i = 1, \ldots, m_2$. The convex set K_{hH} is now an *external* approximation of K:

$$K_{hH} = \{v_h \subset V_h \mid \int_{S_i} v_{hn} \, ds \leq 0 \quad \forall i = 1, \ldots, m_2\} . \tag{4.56}$$

It is worth mentioning that Λ_{nH} and $\Lambda_{tH}(g)$ for both L_H^l, $l = 1, 2$ are convex sets with simple (box) constraints.

Problem 4.11 Let $\Lambda_{nH}, \Lambda_{tH}(g) \subset L_H^1$. Prove that (4.43) is satisfied with $r = s = 1$. If $\Lambda_{nH}, \Lambda_{tH} \subset L_H^2$, prove that (4.43) is satisfied provided that the ratio H/h is sufficiently large, i.e. the partition $\mathcal{T}_H$ of $\overline{\Gamma}_c$ used for the construction of L_H^2 is *coarser* than the partition of $\overline{\Gamma}_c$ generated by $\mathcal{T}_{h|\overline{\Gamma}_c}$.

Let $\Lambda_{nH}, \Lambda_{tH}(g) \subset L_H^2$. To ensure convergence of approximate solutions of (4.42) we shall need the LBB–condition (1.49). Since L_H^2 is a subspace of $H^{-1/2}(\Gamma_c)$, the LBB–condition for this particular case reads as follows:

$$\exists \beta = const. > 0 : \sup_{\substack{v_h \in V_h \\ v_h \neq 0}} \frac{\langle \mu_H^n, v_{hn} \rangle_2 + \langle \mu_H^t, v_{ht} \rangle_2}{\|v_h\|_{1,\Omega}} \geq \beta[\|\mu_H^n\|_{-1/2,\Gamma_c} + \|\mu_H^t\|_{-1/2,\Gamma_c}] \tag{4.57}$$

holds for every $(\mu_H^n, \mu_H^t) \in (L_H^2)^2$ and every $h, H > 0$, where $\| \ \|_{-1/2,\Gamma_c}$ denotes the norm in $H^{-1/2}(\Gamma_c)$. To satisfy (4.57) we need, roughly speaking, to use again coarser partitions $\mathcal{T}_H$ of $\overline{\Gamma}_c$ compared with $\mathcal{T}_{h|\Gamma_c}$. For more details we refer to [Haslinger and Panagiotopoulos 1984] and also [Girault and Glowinski 1995]. The following convergence result is a direct consequence of Theorems 1.15 and 1.19.

Theorem 4.12. *Let $\Lambda_{nH}, \Lambda_{tH}(g) \subset L_H^2$ and (4.57) be satisfied. Further, let $h \to 0+$ if and only if $H \to 0+$. Then the sequence of approximate solutions $\{(w_h, \lambda_H^n, \lambda_H^t)\}$ to (4.42) tends to a unique solution $(w, \lambda_n, \lambda_t)$ of (4.30):*

$$\left.\begin{array}{l} w_h \to w \quad \text{in } V \\ \lambda_H^n \to \lambda_n \ , \ \lambda_H^t \to \lambda_t \quad \text{in } H^{-1/2}(\Gamma_c) \ , \ h, H \to 0+ \ . \end{array}\right\} \tag{4.58}$$

Problem 4.13 Prove Theorem 4.12.

One of important features of this mixed formulation is the fact that also the contact stresses $T_n(u)$ and $T_t(u)$ on Γ_c are approximated as follows from (4.58).

Next we use the mixed formulation (4.44) to derive the discrete dual formulation of contact problems with given friction. From $(4.44)_2$ one can express $\mathbf{w}$:

$$\mathbf{w} = \mathbf{A}^{-1}(\mathbf{f} - \mathbf{B}_n^t \boldsymbol{\lambda}_n - \mathbf{B}_t^t \boldsymbol{\lambda}_t) \ . \tag{4.59}$$

Inserting (4.59) into $(4.44)_3$ and rearranging terms we arrive at the variational inequality for $\boldsymbol{\lambda}_n$ and $\boldsymbol{\lambda}_t$:

$$\left.\begin{array}{l} \text{Find } \boldsymbol{\lambda} = (\boldsymbol{\lambda}_n, \boldsymbol{\lambda}_t) \in \boldsymbol{\Lambda}_n \times \boldsymbol{\Lambda}_t(g) \text{ such that} \\ (\mathbf{Q}\boldsymbol{\lambda}, \boldsymbol{\mu} - \boldsymbol{\lambda}) \geq (\mathbf{h}, \boldsymbol{\mu} - \boldsymbol{\lambda}) \ \forall \boldsymbol{\mu} = (\boldsymbol{\mu}_n, \boldsymbol{\mu}_t) \in \boldsymbol{\Lambda}_n \times \boldsymbol{\Lambda}_t(g) \ , \end{array}\right\} \tag{4.60}$$

where

$$\mathbf{Q} = \begin{pmatrix} \mathbf{Q}_{nn} & \mathbf{Q}_{nt} \\ \mathbf{Q}_{tn} & \mathbf{Q}_{tt} \end{pmatrix} , \qquad \mathbf{h} = (\mathbf{h}_n, \mathbf{h}_t)$$

with $\mathbf{Q}_{ij} = \mathbf{B}_i \mathbf{A}^{-1} \mathbf{B}_j^t$, $\mathbf{h}_i = \mathbf{B}_i \mathbf{A}^{-1} \mathbf{f}$, $i, j \in \{n, t\}$.

Since the matrix $\mathbf{Q}$ is symmetric and positive definite, problem (4.60) is equivalent to the following quadratic programming problem:

$$\left.\begin{array}{l} \text{Find } \boldsymbol{\lambda} = (\boldsymbol{\lambda}_n, \boldsymbol{\lambda}_t) \in \boldsymbol{\Lambda}_n \times \boldsymbol{\Lambda}_t(g) \text{ such that} \\ \boldsymbol{\mathcal{S}}(\boldsymbol{\lambda}) \leq \boldsymbol{\mathcal{S}}(\boldsymbol{\mu}) \quad \forall \boldsymbol{\mu} \in \boldsymbol{\Lambda}_n \times \boldsymbol{\Lambda}_t(g) \ , \end{array}\right\} \tag{4.61}$$

where

$$\boldsymbol{\mathcal{S}}(\boldsymbol{\lambda}) = \frac{1}{2}(\boldsymbol{\mu}_n, \boldsymbol{\mu}_t) \begin{pmatrix} \mathbf{Q}_{nn} & \mathbf{Q}_{nt} \\ \mathbf{Q}_{tn} & \mathbf{Q}_{tt} \end{pmatrix} \begin{pmatrix} \boldsymbol{\mu}_n \\ \boldsymbol{\mu}_t \end{pmatrix} - (\mathbf{h}_n, \mathbf{h}_t) \begin{pmatrix} \boldsymbol{\mu}_n \\ \boldsymbol{\mu}_t \end{pmatrix} \ . \tag{4.62}$$

These are the main advantages of the dual formulation (4.60) (or (4.61)):

- the convex sets $\boldsymbol{\Lambda}_n$ and $\boldsymbol{\Lambda}_t(g)$ are defined by simple (box) constraints;

- the number of the dual variables is much less than the number of the primal variables. This number still can be reduced by using coarser partitions of $\overline{\Gamma}_c$ if piecewise polynomial approximations of the Lagrange multipliers are used;
- spectral properties of the matrix $\mathbf{Q}$ are favorable for the use of conjugate gradient methods;
- it is an efficient method for the numerical realization of contact problems with other models of friction as will be seen from the next part of this section.

Approximation and the numerical realization of contact problems with Coulomb's friction

As we have already mentioned at the beginning of this section, numerical realization will be based on the fixed point formulation and the method of successive approximations. We have to define an appropriate discretization of the mapping Φ from (4.23).

To this end we use the mixed formulation (4.42) with L_H^1 and L_H^2 defined by (4.48) and (4.54), respectively. We shall suppose that both Λ_{nH} and $\Lambda_{tH}(g)$ are subsets of the *same* L_H^l, $l = 1, 2$. If it is so one can associate with any $g_H \in \Lambda_{nH}$ a closed, convex set $\Lambda_{tH}(g_H) \subset L_H^l$, $l = 1, 2$.

Indeed, if $\Lambda_{nH} \subset L_H^1$ and $g_H \in \Lambda_{nH}$, $g_H = \sum_{i=1}^{m_1} g_{Hi}\delta_i$, $g_{Hi} \geq 0$ $\forall i = 1, \ldots, m_1$, then

$$\Lambda_{tH}(g_H) = \{\mu_H \in L_H^1 \mid \sum_{i=1}^{m_1} |\mu_{Hi}|\delta_i \leq \mathcal{F} g_H\} , \tag{4.63}$$

where $\mathcal{F} g_H \in L_H^1$ and $\langle \mathcal{F} g_H, \varphi\rangle_1 := \langle g_H, \mathcal{F}\varphi\rangle_1 = \sum_{i=1}^{m_1} g_{Hi}\mathcal{F}(a_i)\varphi(a_i)$. It is readily seen that $\mu_H \in \Lambda_{tH}(g_H)$ iff $|\mu_{Hi}| \leq \mathcal{F}(a_i) g_{Hi}$ $\forall i = 1, \ldots, m_1$.

Similarly, if $\Lambda_{nH} \subset L_H^2$ and $g_H \in \Lambda_{nH}$, $g_H = \sum_{i=1}^{m_2} g_{Hi}\chi_i$, $g_{Hi} \geq 0$ $\forall i = 1, \ldots, m_2$ then

$$\Lambda_{tH}(g_H) = \{\mu_H \in L_H^2 \mid \sum_{i=1}^{m_2} |\mu_{Hi}|\chi_i \leq \sum_{i=1}^{m_2} (\int_{S_i} \mathcal{F} g_H \, ds)\chi_i\} . \tag{4.64}$$

Let $g_H \in \Lambda_{nH}$ be fixed and consider the mixed formulation (4.42) on $V_h \times \Lambda_{nH} \times \Lambda_{tH}(g_H)$ with $\Lambda_{tH}(g_H)$ given by (4.63) or (4.64). We shall suppose that (4.43) is satisfied. Therefore (4.42) has a unique solution $(w_h(g_H), \lambda_H^n(g_H), \lambda_H^t(g_H))$ which depends on a particular choice of $g_H \in \Lambda_{nH}$. This makes it possible to define the mapping Φ_H from Λ_{nH} into itself by

$$\Phi_H(g_H) = \lambda_H^n(g_H) , \tag{4.65}$$

where $\lambda_H^n(g_H)$ is a part of the solution to (4.42) which approximates the normal contact stress $-T_n(u(g))$. The mapping Φ_H will serve as a discretization of Φ. Analogously to the continuous setting we say that $w_h(\overline{g}_H) \in V_h$ is a solution of a discrete contact problem with Coulomb's friction if and only if $\overline{g}_H$ is a fixed point of Φ_H in Λ_{nH}:

$$\Phi_H(\overline{g}_H) = \overline{g}_H . \tag{4.66}$$

The proof of the existence of a fixed point of Φ_H is much simpler than the one for Φ. It can be shown that Φ_H has at least one fixed point for *any* coefficient of friction and the mapping Φ_H is *contractive* (see [Haslinger 1983]) for $\mathcal{F}$ sufficiently small, implying the uniqueness of the solution. Unfortunately, the bound for $\|\mathcal{F}\|_{C(\overline{\Gamma}_c)}$ ensuring that Φ_H is a contraction *is mesh dependent.* It decays as $H \to 0+$.

To find fixed points of Φ_H we use the method of successive approximations. We now present two strategies which are based on the algebraic form of the dual formulation of contact problems with given friction (4.61):

$$\left.\begin{array}{l} \textit{Find } \boldsymbol{\lambda} = (\boldsymbol{\lambda}_n, \boldsymbol{\lambda}_t) \in \boldsymbol{\Lambda}_n \times \boldsymbol{\Lambda}_t(\mathbf{g}) \textit{ such that} \\ \boldsymbol{\mathcal{S}}(\boldsymbol{\lambda}) \leq \boldsymbol{\mathcal{S}}(\boldsymbol{\mu}) \quad \forall \boldsymbol{\mu} = (\boldsymbol{\mu}_n, \boldsymbol{\mu}_t) \in \boldsymbol{\Lambda}_n \times \boldsymbol{\Lambda}_t(\mathbf{g}) \ , \end{array}\right\} \tag{4.67}$$

where $\boldsymbol{\mathcal{S}}$ is defined by (4.62) and $\boldsymbol{\Lambda}_n$, $\boldsymbol{\Lambda}_t(\mathbf{g})$ are isometrically isomorphic with Λ_{nH}, $\Lambda_{tH}(g_H)$, respectively. Here $\mathbf{g} \in \mathbb{R}^{m_l}$, $l = 1, 2$ denotes the vector of the coordinates of $g_H \in \Lambda_{nH}$ with respect to $\{\delta_i\}_{i=1}^{m_1}$ or $\{\chi_i\}_{i=1}^{m_2}$.

VARIANT I. (*standard successive approximations*).

Initialize $\boldsymbol{\lambda}_n^{(0)} \in \boldsymbol{\Lambda}_n$; $i := 0$;
repeat
  $i := i + 1$;
  $\boldsymbol{\lambda}^{(i)} = \ \arg\min \boldsymbol{\mathcal{S}}(\boldsymbol{\mu})$
    subject to $\boldsymbol{\mu}_n \in \boldsymbol{\Lambda}_n$, $\boldsymbol{\mu}_t \in \boldsymbol{\Lambda}_t(\boldsymbol{\lambda}_n^{(i-1)})$;
until stopping criterion.

This is a standard successive approximation method. The quadratic functional $\boldsymbol{\mathcal{S}}$ is minimized on $\boldsymbol{\Lambda}_n \times \boldsymbol{\Lambda}_t(\boldsymbol{\lambda}_n^{(i-1)})$. The set $\boldsymbol{\Lambda}_t(\boldsymbol{\lambda}_n^{(i-1)})$ is updated only if the solution $\boldsymbol{\lambda}^{(i)} = (\boldsymbol{\lambda}_n^{(i)}, \boldsymbol{\lambda}_t^{(i)})$ of the i-th iterative step is known. Each iterative step is a quadratic programming problem with simple constraints which can be solved by conjugate gradient methods [Dostál 1997].

VARIANT II. (*the Gauss-Seidel like approach*).

This time $\boldsymbol{\mathcal{S}}$ is minimized on $\boldsymbol{\Lambda}_n$ and $\boldsymbol{\Lambda}_t(\boldsymbol{\lambda}_n^{(i-1)})$ *separately.* Unlike the classical Gauss-Seidel method, the set $\boldsymbol{\Lambda}_t(\cdot)$ is updated after *each* partial minimization of $\boldsymbol{\mathcal{S}}$ on $\boldsymbol{\Lambda}_n$. This variant reads as follows:

Initialize $\boldsymbol{\lambda}_n^{(0)} \in \boldsymbol{\Lambda}_n$; $i := 0$;
repeat
  $i := i + 1$;
  $\boldsymbol{\lambda}_t^{(i)} = \ \arg\min \boldsymbol{\mathcal{S}}(\boldsymbol{\lambda}_n^{(i-1)}, \boldsymbol{\mu}_t)$
    subject to $\boldsymbol{\mu}_t \in \boldsymbol{\Lambda}_t(\boldsymbol{\lambda}_n^{(i-1)})$;
  $\boldsymbol{\lambda}_n^{(i)} = \ \arg\min \boldsymbol{\mathcal{S}}(\boldsymbol{\mu}_n, \boldsymbol{\lambda}_t^{(i)})$
    subject to $\boldsymbol{\mu}_n \in \boldsymbol{\Lambda}_n$;
until stopping criterion.

Each iterative step is split into two quadratic programming subproblems for finding the subvectors $\boldsymbol{\lambda}_t^{(i)}$ and $\boldsymbol{\lambda}_n^{(i)}$ separately. VARIANT II is the dual version of the famous Panagiotopoulos splitting algorithm (see [Panagiotopoulos 1975]). Convergence of this algorithm was recently studied in [Bisegna and Lebon and Maceri 2001] and [Haslinger and Dostál and Kučera 2002]. It turns out that if VARIANT II converges then it is much more efficient than VARIANT I (see [Haslinger and Dostál and Kučera 2002]).

CASE $\mathcal{F} := \mathcal{F}(x, |u_t|)$

The mathematical analysis of contact problems with solution dependent coefficients of friction has been done in [Eck and Jarušek 1998] using a penalization and dualization approach. For reasons that have been mentioned at the beginning of this section, the fixed point approach is preferred from the computational point of view. We now describe it in more details.

Denote $X = H_+^{1/2}(\Gamma_c) \times H_+^{-1/2}(\Gamma_c)$. For any $(\varphi, g) \in X$ we shall consider the following variational inequality of the second kind:

$$\left.\begin{array}{l} \textit{Find } u := u(\varphi, g) \in K \textit{ such that} \\ a(u, v-u) + \langle \mathcal{F}(\varphi) g, |v_t| - |u_t| \rangle \geq \ell(v-u) \quad \forall v \in K\ , \end{array}\right\} \tag{4.68}$$

where a and ℓ are defined in (4.26). The unique solution u of (4.68) depends on a particular choice of the couple $(\varphi, g) \in X$. Let $\Psi : X \mapsto X$ be a mapping defined by

$$\Psi(\varphi, g) = (|u_t|; -T_n(u))\ , \tag{4.69}$$

where $|u_t|$ is the trace of the absolute value of the tangential component of u on Γ_c and $T_n(u)$ is the normal contact stress corresponding to the solution u of (4.68).

We say that $u \in K$ is *a weak solution* of a contact problem with Coulomb's friction and a solution dependent coefficient of friction $\mathcal{F}$ if and only if the couple $(|u_t|; -T_n(u))$ is a fixed point of Ψ:

$$\Psi(|u_t|; -T_n(u)) = (|u_t|; -T_n(u))\ . \tag{4.70}$$

As before we shall use the method of successive approximations for finding fixed points of Ψ in X. Each iterative step leads to a contact problem with given friction and a coefficient of friction which does not depend on the solution. Therefore it can be realized by methods discussed in the previous parts of this section.

To simplify our presentation we restrict ourselves to problems with *given* friction. In addition, we shall suppose that the slip bound g is represented by a non–negative square integrable function in Γ_c. The weak formulation of this problem reads as follows:

$$\left.\begin{array}{l} \textit{Find } u \in K \textit{ such that} \\ a(u, v-u) + \displaystyle\int_{\Gamma_c} \mathcal{F}(|u_t|) g (|v_t| - |u_t|)\, ds \geq \ell(v-u) \quad \forall v \in K\ . \end{array}\right\} \tag{4.71}$$

We outline how to prove that (4.71) has a solution. For more details we refer to [Haslinger and Vlach 2005]. It is readily seen that (4.71) can be again formulated as a fixed point problem. Indeed, for $\varphi \in H^{1/2}_+(\Gamma_c)$ let $u := u(\varphi) \in K$ be a solution of (4.68) keeping $g \in L^2_+(\Gamma_c)$ fixed and define the mapping $\widetilde{\Psi} : H^{1/2}_+(\Gamma_c) \mapsto H^{1/2}_+(\Gamma_c)$ by

$$\widetilde{\Psi}(\varphi) = |u_t| \,. \tag{4.72}$$

Clearly, $u \in K$ is a solution of (4.71) if and only if $|u_t|$ is a fixed point of $\widetilde{\Psi}$:

$$\widetilde{\Psi}(|u_t|) = |u_t| \,. \tag{4.73}$$

First, let us suppose that $\mathcal{F}$ satisfies (4.10). If the constant C_L of Lipschitz continuity is small enough, problem (4.71) has a *unique solution* as follows from the next theorem.

Theorem 4.14. *Let $\widetilde{\Psi}$ be defined by (4.72) and the constant C_L in (4.10) be small enough. Then $\widetilde{\Psi}$ is contractive in $L^2_+(\Gamma_c)$:*

$$\exists q \in (0,1) \quad \|\widetilde{\Psi}(\varphi) - \widetilde{\Psi}(\overline{\varphi})\|_{0,\Gamma_c} \le q\|\varphi - \overline{\varphi}\|_{0,\Gamma_c} \tag{4.74}$$

holds for any $\varphi, \overline{\varphi} \in L^2_+(\Gamma_c)$.

From this theorem it also follows that the method of successive approximations:

$$\left.\begin{array}{l} \varphi_0 \in H^{1/2}_+(\Gamma_c) \ \textit{given} \,; \\ \textit{for } k \in \mathbb{N} \textit{ set } \varphi_{k+1} = \widetilde{\Psi}(\varphi_k) \end{array}\right\} \tag{4.75}$$

is convergent.

The assumption (4.10) on $\mathcal{F}$ is too strong. The existence of a solution to (4.71) (but not uniqueness) can be established under the following weaker assumptions:

$$\mathcal{F} \in C(\Gamma_c \times \mathbb{R}^1_+); \quad 0 \le \mathcal{F}(x,t) \le \overline{\mathcal{F}} \quad \forall (x,t) \in \Gamma_c \times \mathbb{R}^1_+ \,, \tag{4.76}$$

where $\overline{\mathcal{F}} > 0$ is given. The existence of a fixed point of $\widetilde{\Psi}$ can be proven directly by using the weak version of the Schauder theorem. Here we use another approach which will be based on a suitable approximation of (4.68).

Let $\Omega \subset \mathbb{R}^2$ be a *polygonal* domain and $\{\mathcal{T}_h\}$, $h \to 0+$ be a family of *regular* triangulations of $\overline{\Omega}$. With any $\mathcal{T}_h$ we associate the convex set K_h defined by (4.41) and the convex cone Λ_h, where

$$\begin{aligned} \Lambda_h = \{\varphi_h \in C(\overline{\Gamma}_c) \mid \ & \varphi_h \textit{ is piecewise linear on } \mathcal{T}_{h|_{\Gamma_c}}, \quad \varphi_h \ge 0 \textit{ on } \Gamma_c \\ & \textit{and } \varphi_h(x) = 0 \textit{ if } x \in \overline{\Gamma}_c \cap \overline{\Gamma}_u \ne \emptyset\} \,. \end{aligned} \tag{4.77}$$

For every $\varphi_h \in \Lambda_h$ we consider the following discrete problem:

$$\left.\begin{array}{l} \textit{Find } u_h := u_h(\varphi_h) \in K_h \textit{ such that} \\ a(u_h, v_h - u_h) + \displaystyle\int_{\Gamma_c} \mathcal{F}(\varphi_h) g(|v_{ht}| - |u_{ht}|)\, ds \ge \ell(v_h - u_h) \quad \forall v_h \in K_h \,. \end{array}\right\} \tag{4.78}$$

Let $\widetilde{\Psi}_h : \Lambda_h \mapsto \Lambda_h$ be the discretization of $\widetilde{\Psi}$ defined by

$$\widetilde{\Psi}_h(\varphi_h) = r_h(|u_{ht}|) , \tag{4.79}$$

where $u_h \in K_h$ solves (4.78) and r_h is the piecewise linear Lagrange interpolation operator on $\mathcal{T}_{h|\Gamma_c}$.

We say that $\overline{u}_h \in K_h$ is an approximate solution of (4.71) if and only if $r_h(|\overline{u}_{ht}|)$ is a fixed point of $\widetilde{\Psi}_h$ in Λ_h, i.e.:

$$\begin{aligned} \overline{u}_h \in K_h : \quad a(\overline{u}_h, v_h - \overline{u}_h) + \int_{\Gamma_c} \mathcal{F}(r_h|\overline{u}_{ht}|) g(|v_{ht}| - |\overline{u}_{ht}|)\, ds \\ \geq \ell(v_h - \overline{u}_h) \quad \forall v_h \in K_h . \end{aligned} \tag{4.80}$$

The existence of a fixed point of $\widetilde{\Psi}_h$ follows from Brower's fixed point theorem.

Problem 4.15 Prove that the mapping $\widetilde{\Psi}_h$ defined by (4.79) is continuous in Λ_h and maps $\Lambda_h \cap B_r$ into itself for some $r > 0$ which does not depend on h, where $B_r = \{\varphi_h \in \Lambda_h \mid \ \|\varphi_h\|_{1/2,\Gamma_c} \leq r\}$.

One can show that discrete problems (4.80) are close to the continuous setting (4.71) as follows from the next theorem.

Theorem 4.16. *Let $\{\overline{\varphi}_h\}$ be a sequence of fixed points of $\widetilde{\Psi}_h$ in Λ_h, $h \to 0+$. Then one can pass to a subsequence $\{\overline{\varphi}_{h'}\} \subset \{\overline{\varphi}_h\}$ such that*

$$\overline{\varphi}_{h'} \rightharpoonup \overline{\varphi} \quad \text{in } H^{1/2}(\Gamma_c) , \quad h' \to 0+ .$$

In addition, $\overline{\varphi}$ is a fixed point of $\widetilde{\Psi}$.

For numerical realization of each iterative step in (4.75) we use the dual variational formulation (4.37) (or (4.38)) with the following minor change: the set $\Lambda_t(g)$ has to be now replaced by

$$\Lambda_t(|u_t^{(k)}|) = \{\mu_t \in L^2(\Gamma_c) \mid \ |\mu_t| \leq \mathcal{F}(|u_t^{(k)}|)g \quad on\ \Gamma_c\} . \tag{4.81}$$

The dual formulation enables us to approximate directly the contact stresses on Γ_c. On the other hand, the algorithm (4.75) requires the update of $|u_t|$ on Γ_c at each iterative step. A natural question arises, namely is it possible to get this information from the dual formulation? The answer is positive. To see that we use again the duality approach in order to release the constraints $\mu_n \in \Lambda_n$, $\mu_t \in \Lambda_t(\varphi)$.

Let $Y := H^{-1/2}(\Gamma_c) \times L^2(\Gamma_c) \times (L^2_+(\Gamma_c))^2 \times H^{1/2}_+(\Gamma_c) \mapsto \mathbb{R}^1$ and $\mathcal{L} : Y \mapsto \mathbb{R}^1$ be the Lagrangian defined by

$$\begin{aligned} \mathcal{L}(\mu, v_t^1, v_t^2, v_n) = \mathcal{S}(\mu) - \int_{\Gamma_c} (\mathcal{F}(\varphi)g - \mu_t) v_t^1\, ds \\ - \int_{\Gamma_c} (\mathcal{F}(\varphi)g + \mu_t) v_t^2\, ds - \langle \mu_n, v_n \rangle , \end{aligned}$$

where $\varphi \in H_+^{1/2}(\Gamma_c)$ is fixed and $\mu = (\mu_n, \mu_t) \in H^{-1/2}(\Gamma_c) \times L^2(\Gamma_c)$. It is easy to see that

$$\min_{\substack{\mu_n \in \Lambda_n \\ \mu_t \in \Lambda_t(\varphi)}} \mathcal{S}(\mu) = \min_{\substack{\mu_n \in H^{-1/2}(\Gamma_c) \\ \mu_t \in L^2(\Gamma_c)}} \sup_{\substack{v_t^1, v_t^2 \in L_+^2(\Gamma_c) \\ v_n \in H_+^{1/2}(\Gamma_c)}} \mathcal{L}(\mu, v_t^1, v_t^2, v_n) \ . \tag{4.82}$$

Problem 4.17 Prove that $\mathcal{L}$ has a unique saddle–point $(\lambda_n, \lambda_t, w_t^1, w_t^2, w_n)$ in Y. In addition,

$$(\lambda_n, \lambda_t, w_t^1, w_t^2, w_n) = (-T_n(u), -T_t(u), u_t^-, u_t^+, u_n) \ , \tag{4.83}$$

where u_t^-, u_t^+ denotes the negative, positive part, respectively of u_t on Γ_C and $u \in K$ is a solution of (4.68).

From (4.83) we see that required information on $|u_t|$ is hidden in the Lagrange multipliers w_t^1, w_t^2 releasing the constraint $\mu_t \in \Lambda_t(\varphi)$.

Remark 4.18. Contact problems with Coulomb's friction and $\mathcal{F}$ depending on a solution can be treated in a similar way. Using the dual formulation of (4.68) at each step of the method of successive approximations, one has to update not only $|u_t^{(k)}|$ on Γ_c but also the slip bound $T_n(u^{(k)})$. The admissible set for μ_t now depends on two parameters $\varphi \in H_+^{1/2}(\Gamma_c)$ and $g \in H_+^{-1/2}(\Gamma_c)$:

$$\Lambda_t(\varphi, g) = \{\mu_t \in L^2(\Gamma_c) \mid \ |\mu_t| \leq \mathcal{F}(\varphi) g \quad on\ \Gamma_c\}$$

(here again we consider $g \in L_+^2(\Gamma_c)$ for simplicity). Thus the same approach can be used for a large class of contact problems involving friction.

Next we shall solve two model examples of contact problems with Coulomb's friction. The elastic body is represented by the rectangle $\Omega = (0,5) \times (0,1)$. A homogeneous and isotropic material is characterized by the Young modulus $E = 21.19e10$ [Pa] and Poisson's ratio $\sigma = 0.277$. The partition of $\partial\Omega$ into Γ_c, Γ_u and $\Gamma_p = \Gamma_p^1 \cup \Gamma_p^2$ is seen from Fig. 2. The top Γ_p^2 of the body is subject to the linearly distributed surface

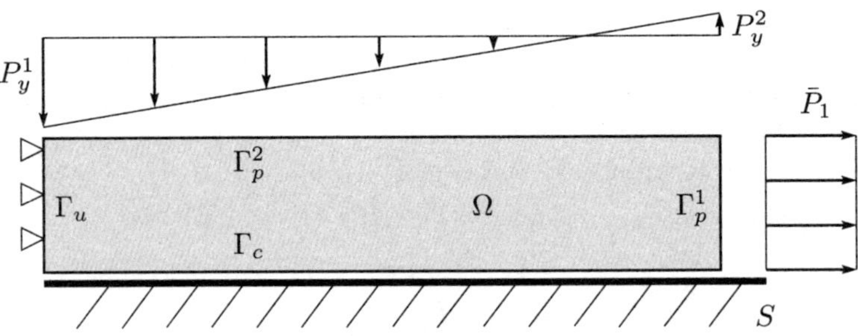

Figure 2.

traction $P = (0, P_y)$, where $P_y = (1-\lambda)P_y^1 + \lambda P_y^2$, $\lambda \in (0,1)$, $P_y^1 = -8.0\mathrm{e}6\ [Nm^{-1}]$ and $P_y^2 = 1.0\mathrm{e}6\ [Nm^{-1}]$. On the vertical side Γ_p^1 the uniform tensile force $P = (\bar{P}_1, 0)$ is applied, where $\bar{P}_1 = 6.0\mathrm{e}6\ [Nm^{-1}]$. Two cases of $\mathcal{F}$ are considered:

a) $\mathcal{F}$ does not depend on the solution: $\mathcal{F} = 0.3$ on Γ_c;

b) $\mathcal{F}$ depends on the solution:

$$\mathcal{F}(t) = \begin{cases} 0.3 & t \in (0; 10^{-5}) \\ 0.3 - 0.05(t - 10^{-5})^2 param^2 & t \in (10^{-5}; 1/param) \\ 0.2 + 0.05/(1 + 2param(t - 10^{-5} - 1/param)) & t \in (1/param; \infty)\ . \end{cases}$$

In our example we choose $param = 60.0\mathrm{e}3$ (see Fig. 3).

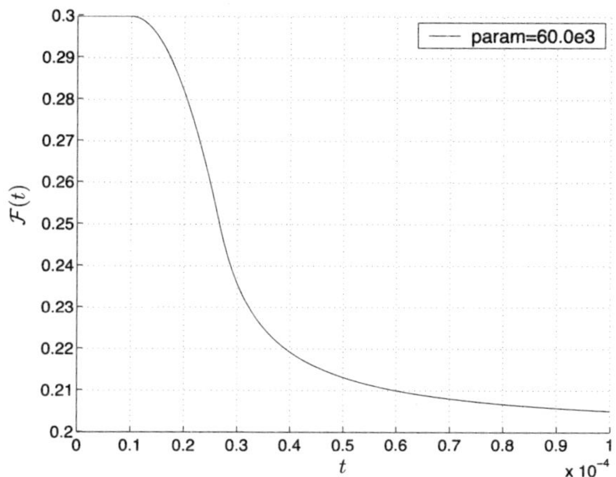

Figure 3.

For the numerical realization we use the method of successive approximations (4.24) in the case *a)* and

$$\begin{cases} (\varphi_0, g_0) \in H_+^{1/2}(\Gamma_c) \times H_-^{-1/2}(\Gamma_c)\ given\ ; \\ for\ k \in \mathbb{N}\ set\ (\varphi_{k+1}, g_{k+1}) = \Psi(\varphi_k, g_k) \end{cases} \tag{4.84}$$

in the case *b)*, where Ψ is the mapping defined by (4.69). As we already know, each iterative step leads to a contact problem (4.68) which will be solved using its dual formulation in terms of contact stresses.

The triangulation $\mathcal{T}_h$ of $\overline{\Omega}$ consists of 4500 triangles. The Lagrange multipliers are discretized by Dirac distributions concentrated at the contact nodes. The total number of the dual variables is 300. If $\mathcal{F}$ does not depend on u (case *a)*), the fixed point formulation (4.24) is solved by VARIANT II. If $\mathcal{F}$ depends on u (case *b)*), then the k-th iteration in (4.84) is realized by the classical Gauss-Seidel method, i.e. the separate minimization of the dual functional $\mathcal{S}$ on Λ_n and $\Lambda_t(\varphi_{k-1}, g_{k-1})$, where the later set is

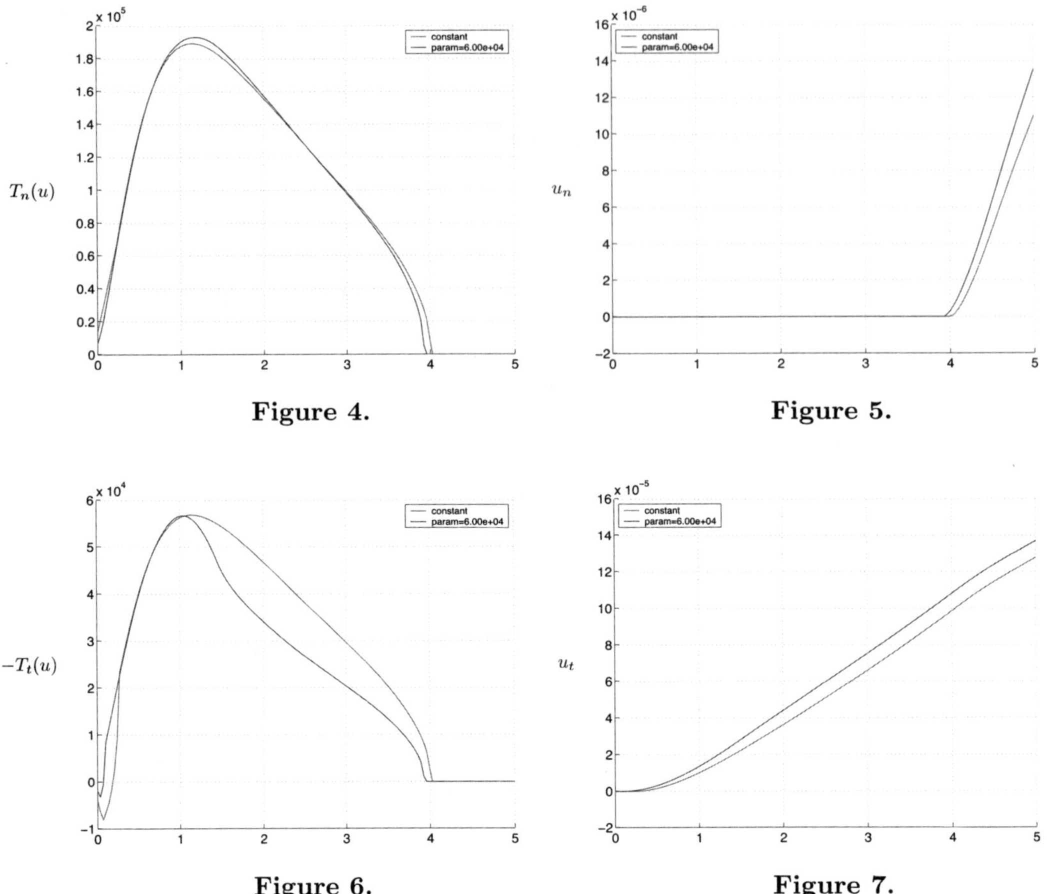

Figure 4. **Figure 5.**

Figure 6. **Figure 7.**

defined in Remark 4.18. The distribution of the normal contact stresses and the graph of the normal component of the displacement vector along Γ_c are depicted in Figs. 4 and 5, respectively, (blue for a) and red for b)).

The similar graphs for the tangential stresses and the tangential component of displacements on Γ_c are depicted in Figs. 6 and 7, respectively.

Finally, Fig. 8 shows the graph of the products $\mathcal{F}|T_n(u)|$ and $\mathcal{F}(|u_t|)|T_n(u)|$, respectively. A number of iterative steps of the method of successive approximations needed to get the solution with a prescribed accuracy is 15 for a) and 13 for b).

A typical number of floating point operations was 6.0e7 for a) and 4.6e7 for b). Computations were done in Matlab on 1 GHz Athlon (256 MB RAM).

4.2 3D contact problems with friction

Variational formulations of contact problems in 3D, their derivation and mutual relations are *the same* as in the plane case with appropriate modifications of notations

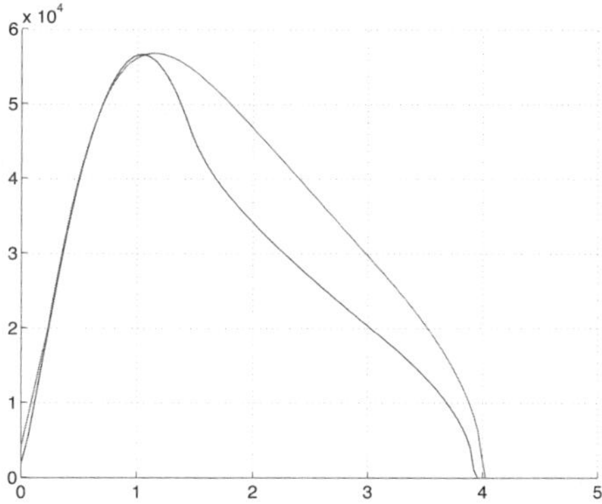

Figure 8.

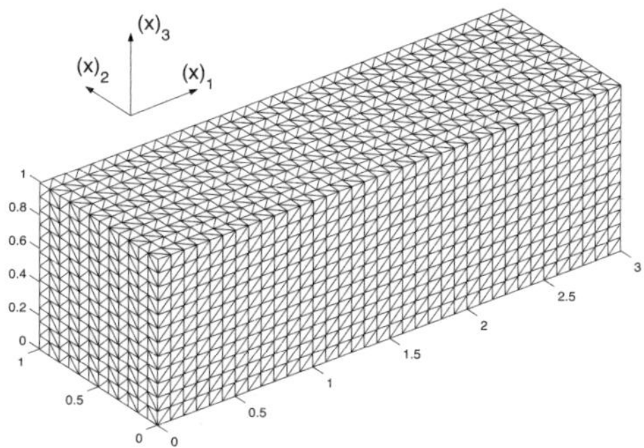

Figure 9.

reflecting a 3D situation.

Let $\Omega \subset \mathbb{R}^3$ be a bounded domain with the Lipschitz boundary $\partial\Omega$ unilaterally supported along $\Gamma_c \subseteq \partial\Omega$ by a rigid half–space S. We shall suppose that Γ_c is planar, i.e. there is no gap between S and Γ_c for the undeformed configuration. If $\xi : \Gamma_c \mapsto \mathbb{R}^3$ is a vector field defined on Γ_c, we denote by $\xi_t(x)$, $x \in \Gamma_c$ its projection onto the tangent plane to Γ_c at x, i.e. onto S in our particular case:

$$\xi_t = \xi - \xi_n n \ , \quad \xi_n = \xi \cdot n \ . \tag{4.85}$$

The unilateral and friction conditions are expressed as in 2D problems with u_t and T_t defined by (4.85). In 3D the symbol $|\ |$ means the *Euclidean norm* of the respective quantity. In particular, the convex set $\Lambda_t(g)$ in (4.27) used in the mixed and dual formulation is now defined by *quadratic* constraints in contrast to box constraints in the plane case.

This fact makes computations more difficult.

We now describe in brief the approximation of the mixed formulation of 3D contact problems with given friction. To this end we shall suppose that Ω is a *polyhedron* and $\{\mathcal{T}_h\}$, $h \to 0+$ is a regular system of partitions of $\overline{\Omega}$ into *tetrahedrons*. With any $\mathcal{T}_h$ we associate the space V_h of piecewise linear vector functions over $\mathcal{T}_h$:

$$V_h = \{v_h \in (C(\overline{\Omega}))^3 \mid \quad v_{h|T} \in (P_1(T))^3 \ \forall T \in \mathcal{T}_h \quad v_h = 0 \ on \ \Gamma_u\} \ . \tag{4.86}$$

Further, let Λ_{nH} and $\Lambda_{tH}(g)$ be appropriate discretizations of $\Lambda_n = H_+^{-1/2}(\Gamma_c)$ and $\Lambda_t(g)$. As in 2D problems, Λ_{nH} and $\Lambda_{tH}(g)$ can be realized or by Dirac distributions concentrated at contact nodes ($l = 1$) or by piecewise constant functions over a partition $\mathcal{T}_H$ of $\overline{\Gamma}_c$ ($l = 2$). The mixed formulation is defined by (4.42). Eliminating w_h from $(4.42)_2$ we obtain the dual formulation in terms of λ_H^n, λ_H^t, whose matrix form is given by (4.60) or (4.61). The convex cone $\mathbf{\Lambda}_n$ is again represented by $\mathbb{R}_+^{m_l}$, $l = 1, 2$ as in the plane case whereas $\mathbf{\Lambda}_t(g)$ is now determined by quadratic constraints and it has the following form:

$$\mathbf{\Lambda}_t(g) = \{\boldsymbol{\mu} = (\boldsymbol{\mu}_1, \ldots, \boldsymbol{\mu}_{m_l}) \in (\mathbb{R}^2)^{m_l} \mid \quad |\boldsymbol{\mu}_i|^2 \le r_i^2 \ , \ i = 1, \ldots, m_l\} \ , \ l = 1, 2 \ . \tag{4.87}$$

For an appropriate choice of r_i, each subvector $\boldsymbol{\mu}_i \in \mathbb{R}^2$ represents a discretization of $T_t(u)$ at the i–th contact node or in the i–th element of a partition $\mathcal{T}_H$ of $\overline{\Gamma}_c$. Problem (4.61) can be solved by the Gauss-Seidel method. The partial minimization over $\mathbf{\Lambda}_n$ is "standard" since $\mathbf{\Lambda}_n$ is defined by box constraints. The minimization over $\mathbf{\Lambda}_t(g)$ seems to be more involved. Fortunately, the quadratic constraints in (4.87) are *separated*. This fact enables us to propose an efficient algorithm for solving such a type of problems. For more details we refer to [Kučera 2004].

We terminate this section by the following model example of a 3D contact problem with Coulomb's friction. The brick $\Omega = (0, 3) \times (0, 1) \times (0, 1)$ is made of an elastic, isotropic and homogenous material which is characterized by Young's modulus $E = 21.19\text{e}10$ and Poisson's ratio $\sigma = 0.277$. The brick is supported by the rigid foundation $S = \mathbb{R}_-^3$. The partition of $\partial\Omega$ into Γ_u, Γ_c and Γ_p is defined as follows (see Fig. 9):

$$\begin{aligned} \Gamma_u &= \{x \in \partial\Omega \mid x_1 = 0\} \\ \Gamma_c &= \{x \in \partial\Omega \mid x_3 = 0\} \\ \Gamma_p &= \partial\Omega \setminus (\overline{\Gamma}_u \cup \overline{\Gamma}_c) \end{aligned}$$

On Γ_p the following surface tractions of density P are applied:

$$\begin{aligned} P &= (0, 0, 10^6) & on \ \Gamma_p^1 &= \{x \in \Gamma_p \mid x_1 = 3\} \ , \\ P &= (0, 0, -0.66\text{e}6(3 - x_1)) & on \ \Gamma_p^2 &= \{x \in \Gamma_p \mid x_3 = 1\} \ . \end{aligned}$$

On the rest of Γ_p the tractions are equal to zero. Fig. 10 shows the two–dimensional cut of Ω by the plane $x_2 = 0.5$ with marked surface tractions. The coefficient of Coulomb's friction is $\mathcal{F} = 0.35$.

The body is cut into small cubes and each cube is divided into five tetrahedrons. The used partition of $\overline{\Omega}$ is depicted in Fig. 9. The total number of the primal variables is

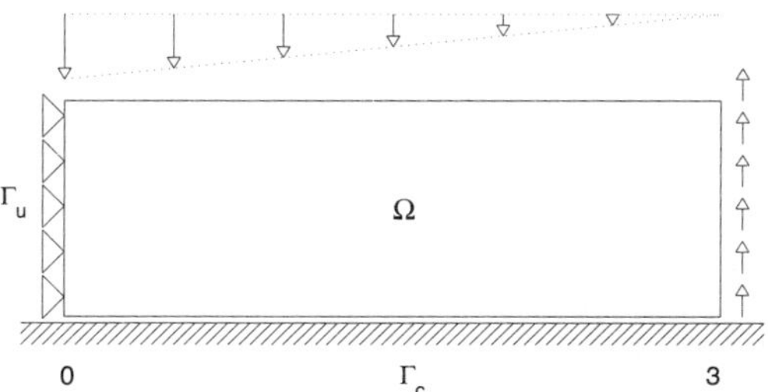

Figure 10.

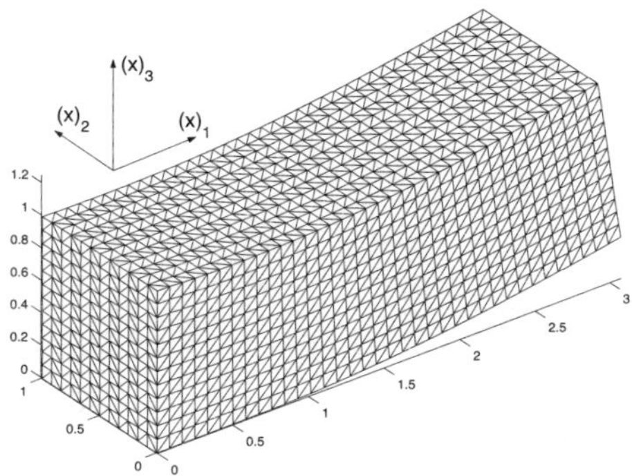

Figure 11.

18 759 . The Lagrange multipliers were approximated by Dirac distributions concentrated at the contact nodes. The total number of the dual variables was 1404. The problem was solved by the method of successive approximations (4.24) using the dual formulation of each iterative step and VARIANT II. The partial minimization of the dual functional $\mathcal{S}$ over the set $\mathbf{\Lambda}_t(g)$ defined by (4.87) was done by the method described in [Kučera 2004]. The number of iterations in (4.24) was 10. The total CPU time needed was 5 min. Computations were done on 3 GHz Pentium (512 MB RAM).

The deformed body is shown in Fig. 11. Further, Figs. 12 and 13 illustrate the distribution of the normal contact stress and the norm of the tangential contact stress along Γ_c. Finally, Fig. 14 shows in more details the distribution of the tangential contact stress at selected contact nodes. The radius of each circle centered at a contact node is equal to $\mathcal{F}|T_n(u)|$, while the segment emanating from the centre represents the vector $T_t(u)$ at this node.

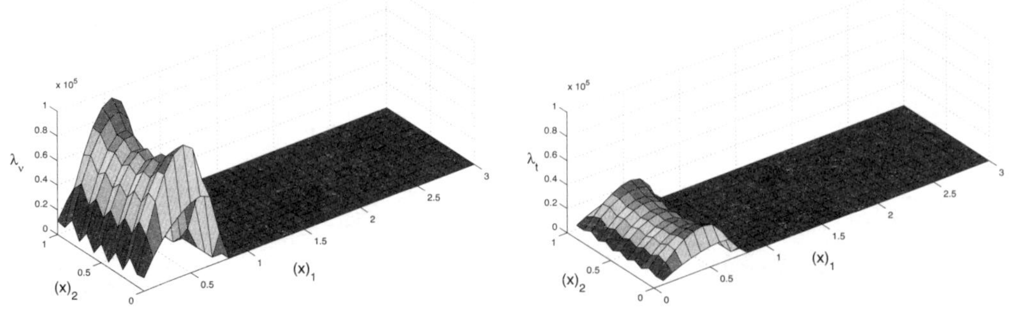

Figure 12. **Figure 13.**

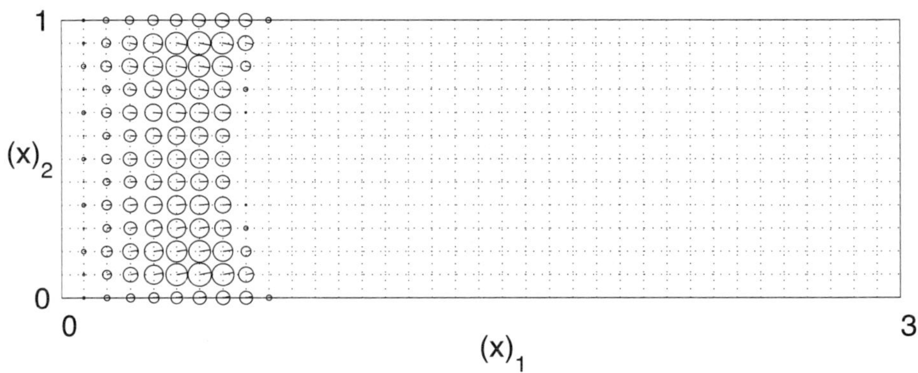

Figure 14.

5 Contact problems with nonmonotone friction

From the previous section we know that the mathematical model of contact problems with given friction leads to an elliptic inequality of the second kind (4.21). Let us suppose that the slip bound g is a positive constant. Inequality (4.21) is equivalent to the following inclusion problem:

$$\left.\begin{array}{l} \textit{Find } (u,\Xi) \in K \times L^2(\Gamma_c) \textit{ such that} \\ a(u, v-u) + \displaystyle\int_{\Gamma_c} \Xi(v_t - u_t)\, ds \ge \ell(v-u) \quad \forall v \in K \\ \Xi \in \partial j(u) \text{ on } \Gamma_c\,, \end{array}\right\} \tag{5.1}$$

where a, j, ℓ are the same as in (4.26) and K is defined by (4.12).

Problem 5.1 Prove that (4.21) and (5.1) are equivalent and $\Xi = -T_t(u)$ on Γ_c.

Since the functional j is convex, the multivalued mapping $v \mapsto \partial j(v)$ is maximal monotone in V. In addition, it is known (see [Brézis 1973]) that

$$\Xi \in \partial j(u) \; iff \; \Xi(x) \in \hat{b}(u_t(x)) \; a.e. \; in \; \Gamma_c \; , \tag{5.2}$$

where $\hat{b} : \mathbb{R}^1 \mapsto 2^{\mathbb{R}^1}$ is the maximal monotone function in $\mathbb{R}^1$ defined by

$$\hat{b}(\xi) = \begin{cases} -g & \xi \in (-\infty, 0) \\ [-g, g] & \xi = 0 \\ g & \xi \in (0, \infty) \; , \end{cases} \tag{5.3}$$

whose graph is depicted in Fig. 15.

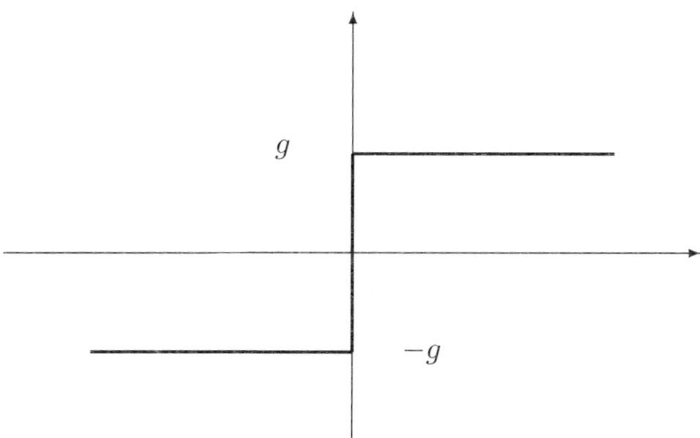

Figure 15.

The multifunction $\hat{b}$ in (5.3) defines a friction law, i.e. a relation between $T_t(u)$ and u_t on Γ_c. In practice we meet problems in which this relation is nonmonotone (see [Panagiotopoulos 1985]).

For this reason we shall consider a friction law of the following form:

$$-T_t(u(x)) \in \hat{b}(u_t(x)) \quad a.e. \; in \; \Gamma_c \; , \tag{5.4}$$

where $\hat{b} : \mathbb{R}^1 \mapsto 2^{\mathbb{R}^1}$ is a multifunction, *not necessarily monotone* and which is defined by (2.7). This multifunction will be constructed by means of a function b which will satisfy the sign condition (2.8) and the growth condition (2.53). The mathematical model of contact problems with nonmonotone friction is given by the following constrained hemivariational inequality:

$$\left. \begin{array}{l} Find \; (u, \Xi) \in K \times L^2(\Gamma_c) \; such \; that \\ a(u, v - u) + \displaystyle\int_{\Gamma_c} \Xi(v_t - u_t) \, ds \geq \ell(v - u) \; \forall v \in K \\ \Xi(x) \in \hat{b}(u_t(x)) \; for \; a.a. \; x \in \Gamma_c \; , \end{array} \right\} \tag{5.5}$$

where a and ℓ are defined in (4.26) and $\hat{b}$ is as in (5.4). Comparing (5.5) with the abstract setting (3.1) we see that $\omega = \Gamma_c$, $Y = L^2(\Gamma_c)$ and $\Pi v = v_t$ on Γ_c.

Next we show how to discretize and numerically realize this type of problems using results of Section 3. To fix ideas let us consider a model example for a rectangular domain Ω unilaterally supported by a rigid halfplane $S = \mathbb{R}^2_-$ with a nonmonotone friction law on Γ_c, see Fig. 16.

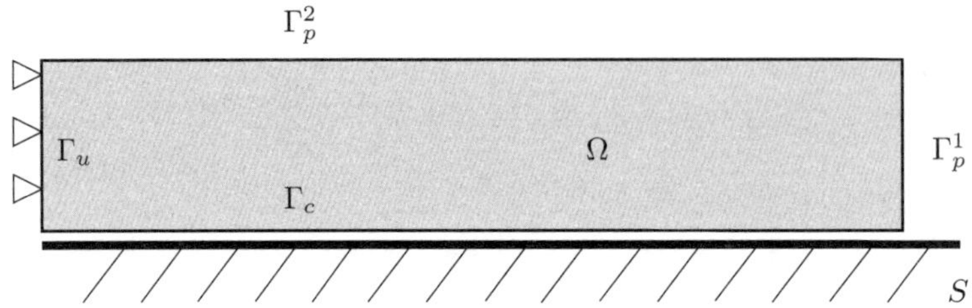

Figure 16.

The partition of $\partial\Omega$ into Γ_u, Γ_c and $\Gamma_p = \Gamma_p^1 \cup \Gamma_p^2$ is seen from Fig. 16. In view of this particular geometry, the convex set K is defined by

$$K = \{v = (v_1, v_2) \in (H^1(\Omega))^2 \mid v = 0 \ on \ \Gamma_u \,,\ v_2 \geq 0 \ on \ \Gamma_c\} \,, \tag{5.6}$$

while the friction law (5.4) reads as follows:

$$-T_1(u(x)) \in \hat{b}(u_1(x)) \quad for \ a.a. \ x \in \Gamma_c \,. \tag{5.7}$$

The existence of a solution to (5.5) will be a consequence of the subsequent convergence analysis. We start with a discretization of K and Y.

Let $\{\mathcal{D}_h\}$, $h \to 0+$ be a system of regular triangulations of $\overline{\Omega}$. With any $\mathcal{D}_h$ we associate the space V_h of piecewise linear vector functions on $\mathcal{D}_h$ vanishing on Γ_u and denote by $K_h = V_h \cap K$ the discretization of K. It is easy to see that

$$K_h = \{v_h = (v_{h1}, v_{h2}) \in V_h \mid v_{h2}(x_h^i) \geq 0 \quad i = 1, \ldots, m\} \,, \tag{5.8}$$

where $\{x_h^i\}_{i=1}^m$ is the set of all contact nodes, i.e. the nodes of $\mathcal{D}_h$ lying on $\overline{\Gamma}_c \setminus \overline{\Gamma}_u$. The intersection of $\overline{\Gamma}_u$ and $\overline{\Gamma}_c$ is a node of $\mathcal{D}_h$ which will be denoted by x_h^0 in what follows. Further, let $x_h^{i+1/2}$ be the midpoint of the interval $[x_h^i, x_h^{i+1}]$, $i = 0, \ldots, m-1$. The partition $\mathcal{T}_h$ which will be used for constructing a discretization Y_h of Y consists of segments S_i, joining the midpoints $x_h^{i-1/2}$ and $x_h^{i+1/2}$, $i = 2, \ldots, m-1$ with the following modifications for the first and the last segment (see Fig. 17):

$$S_1 = [x_h^0, x_h^{3/2}] \,, \quad S_m = [x_h^{m-1/2}, x_h^m] \,.$$

With any such $\mathcal{T}_h$ we associate the space Y_h of all piecewise constant functions on $\mathcal{T}_h$. Their values at $\{x_h^i\}_{i=1}^m$ will serve as the degrees of freedom in Y_h. Clearly, $\dim Y_h =$

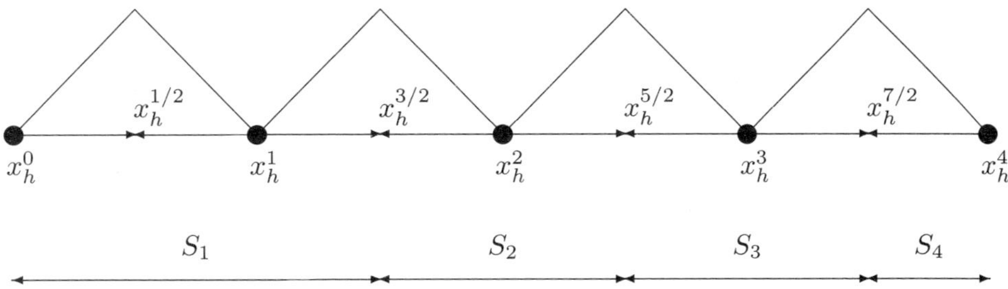

Figure 17.

$\dim W_h = m$, where W_h is the space defined in (2.12). Indeed, since $\Pi v = v_1$, $v \in V$, where

$$V = \{v = (v_1, v_2) \in (H^1(\Omega))^2 \mid v = 0 \textit{ on } \Gamma_u\} , \tag{5.9}$$

we see that W_h consists of all continuous, piecewise linear (scalar) functions over the partition of $\overline{\Gamma}_c$ defined by $\{x_h^i\}_{i=0}^m$ and vanishing at x_h^0.

To discretize (5.5) it remains to specify the mapping $P_h : W_h \mapsto Y_h$ which appears in (3.2). It will be defined as follows:

$$P_h(w_h) = \sum_{i=1}^{m} w_h(x_h^i)\chi_i , \quad w_h \in W_h , \tag{5.10}$$

where χ_i is the characteristic function of the interior of $S_i \subset \mathcal{T}_h$, i.e. P_h associates with any $w_h \in W_h$ its piecewise constant Lagrange interpolate over $\mathcal{T}_h$.

Taking into account the special geometry of Ω, the discretization of (5.5) reads as follows:

$$\left.\begin{array}{l} \textit{Find } (u_h, \Xi_h) \in K_h \times Y_h \textit{ such that} \\ a(u_h, v_h - u_h) + \displaystyle\int_{\Gamma_c} \Xi_h P_h(v_{h1} - u_{h1})\, dx_1 \ge \ell(v_h - u_h) \quad \forall v_h \in K_h \\ \Xi_h(x_h^i) \in \hat{b}(u_{h1}(x_h^i)) \quad i = 1, \ldots, m , \end{array}\right\} \tag{5.11}$$

where K_h, Y_h and P_h are as above.

Since the function b is supposed to satisfy the sign condition (2.8), problem (5.11) has at least one solution (u_h, Ξ_h) for every $h > 0$ as follows from Theorem 3.2.

To obtain a convergence result, one has to verify the assumptions of Theorem 3.3, namely those ones concerning the mappings Π, P_h and the system $\{K_h\}$, $h \to 0+$. From the trace theorem it follows that

$$\Pi \in L((H^1(\Omega))^2, L^2(\Gamma_c)) . \tag{5.12}$$

The assumption (2.47) for P_h defined by (5.10) holds with $s = p = 2$ as proved in [Glowinski and Lions and Trémolière 1981]:

$$y_h \rightharpoonup y \ in \ (H^1(\Omega))^2 \,,\ y_h \in V_h \Rightarrow \|P_h(\Pi y_h) - \Pi y\|_{L^2(\Gamma_c)} \to 0 \,,\ h \to 0+ \ . \tag{5.13}$$

Since K_h, $h > 0$ are inner approximations of K, it is sufficient to verify (1.27). This assumption is a direct consequence of the following density result (for the proof see [Baniotopoulos and Haslinger and Morávková 2005]):

$$\forall v \in K \quad \exists \{v_n\},\ v_n \in K \cap (C^\infty(\overline{\Omega}))^2 : \quad v_n \to v \ in \ (H^1(\Omega))^2 \ . \tag{5.14}$$

On the basis of (5.12)-(5.14) we obtain the following convergence result.

Theorem 5.2. *Let the function b satisfy the sign condition (2.8), the growth condition (2.45) with $p = q = 2$ and P_h be defined by (5.10). Further let $\{(u_h, \Xi_h)\}$ be a sequence of solutions to (5.11) such that $\{u_h\}$ is bounded in $(H^1(\Omega))^2$. Then there exist: a subsequence of $\{(u_h, \Xi_h)\}$ (denoted by the same symbol), and an element $(u, \Xi) \in K \times L^2(\Gamma_c)$ such that*

$$\left.\begin{array}{ll} u_h \to u & in \ (H^1(\Omega))^2 \,, \\ \Xi_h \rightharpoonup \Xi & in \ L^2(\Gamma_c) \,,\ h \to 0+ \ . \end{array}\right\} \tag{5.15}$$

In addition, the pair (u, Ξ) is a solution of (5.5). Any accumulation point of $\{(u_h, \Xi_h)\}$ in the sense of (5.15) possesses this property.

The existence of at least one solution of (5.5) is a consequence of the previous convergence result.

We now describe how to solve numerically the discrete hemivariational inequality (5.11). Let the discretization parameter h be fixed, $\dim V_h = n := n(h)$, $\dim W_h = \dim Y_h = m := m(h)$. Any displacement vector $\mathbf{x} \in \mathbb{R}^n$ will be decomposed as follows: $\mathbf{x} = (\mathbf{x}_i, \mathbf{x}_n, \mathbf{x}_t)$, where $\mathbf{x}_i \in \mathbb{R}^{n-2m}$ is the subvector whose components are the nodal displacements at all interior nodes of $\mathcal{D}_h$, while $\mathbf{x}_n, \mathbf{x}_t \in \mathbb{R}^m$ contain the normal, tangential nodal displacements, respectively at the contact nodes of $\mathcal{T}_h$. The convex set K_h can be identified with the closed, convex subset $\mathcal{K}$ of $\mathbb{R}^n$:

$$\mathcal{K} = \{\mathbf{x} \in \mathbb{R}^n \mid \mathbf{x}_n \geq \mathbf{0}\} \ .$$

Taking into account the definition of Π and P_h we see that

$$\int_{\Gamma_c} \Xi_h P_h(v_{h1} - u_{h1})\, dx_1 = \sum_{i=1}^m c_i \Xi_i (v_{h1}(x_h^i) - u_{h1}(x_h^i)) = \sum_{i=1}^m c_i \Xi_i (\mathbf{v}_t - \mathbf{u}_t)_i \ ,$$

where $c_i = meas\, S_i$, $S_i \in \mathcal{T}_h$, $\boldsymbol{\Xi} = (\Xi_1, \dots, \Xi_m)$, $\Xi_i := \Xi_{h|S_i}$ and $\mathbf{v}, \mathbf{u} \in \mathbb{R}^n$ is the nodal displacement vector corresponding to $v_h, u_h \in V_h$, respectively.

Introducing the new variables $\Xi_i := c_i \Xi_i$, $i = 1, \dots, m$, the algebraic form of the discrete hemivariational inequality (5.11) reads as follows (see also (3.19)):

$$\left.\begin{array}{l} Find \ (\mathbf{u}, \boldsymbol{\Xi}) \in \mathcal{K} \times \mathbb{R}^m \ such \ that \\ (\mathbf{A}\mathbf{u}, \mathbf{v} - \mathbf{u})_{\mathbb{R}^n} + (\boldsymbol{\Xi}, \mathbf{v}_t - \mathbf{u}_t)_{\mathbb{R}^m} \geq (\mathbf{f}, \mathbf{v} - \mathbf{u})_{\mathbb{R}^n} \quad \forall \mathbf{v} \in \mathcal{K} \\ \Xi_i \in c_i \hat{b}((\mathbf{u}_t)_i) \quad \forall i = 1, \dots, m \ , \end{array}\right\} \tag{5.16}$$

where $\mathbf{A} \in L(\mathbb{R}^n, \mathbb{R}^n)$, $\mathbf{f} \in \mathbb{R}^n$ is the stiffness matrix, the load vector, respectively.

Using the fact that all nonlinearities (geometrical and physical) are concentrated on Γ_c, one can eliminate the displacements $\mathbf{u}_i$ at the interior nodes of $\mathcal{D}_h$. The resulting hemivariational inequality on Γ_c has considerably less unknowns as the original one. We describe in brief the main idea of this condensation approach.

The decomposition of $\mathbf{u}$ into $\mathbf{u}_i$, $\mathbf{u}_n$ and $\mathbf{u}_t$ gives rise to the following block structure of the stiffness matrix $\mathbf{A}$:

$$\mathbf{A} = \begin{pmatrix} \mathbf{A}_{ii}\ , & \mathbf{A}_{in}\ , & \mathbf{A}_{it} \\ \mathbf{A}_{ni}\ , & \mathbf{A}_{nn}\ , & \mathbf{A}_{nt} \\ \mathbf{A}_{ti}\ , & \mathbf{A}_{tn}\ , & \mathbf{A}_{tt} \end{pmatrix} .$$

In a similar way one can decompose the load vector $\mathbf{f}$:

$$\mathbf{f} = (\mathbf{f}_i\ ,\ \mathbf{f}_n\ ,\ \mathbf{f}_t) \in \mathbb{R}^{n-2m} \times \mathbb{R}^m \times \mathbb{R}^m\ .$$

Clearly, (5.16) is equivalent to the following problem:

$$\left.\begin{array}{l} \textit{Find } (\mathbf{u}_i, \mathbf{u}_n, \mathbf{u}_t, \Xi) \in \mathbb{R}^{n-2m} \times \mathbb{R}^m_+ \times \mathbb{R}^m \times \mathbb{R}^m \textit{ such that} \\ \mathbf{A}_{ii}\mathbf{u}_i + \mathbf{A}_{in}\mathbf{u}_n + \mathbf{A}_{it}\mathbf{u}_t = \mathbf{f}_i \\ (\mathbf{A}_{ni}\mathbf{u}_i + \mathbf{A}_{nn}\mathbf{u}_n + \mathbf{A}_{nt}\mathbf{u}_t, \mathbf{v}_n - \mathbf{u}_n)_{\mathbb{R}^m} \geq (\mathbf{f}_n, \mathbf{v}_n - \mathbf{u}_n)_{\mathbb{R}^m} \quad \forall \mathbf{v}_n \in \mathbb{R}^m_+ \\ \mathbf{A}_{ti}\mathbf{u}_i + \mathbf{A}_{tn}\mathbf{u}_n + \mathbf{A}_{tt}\mathbf{u}_t + \Xi = \mathbf{f}_t \\ \Xi_i \in c_i\hat{b}((\mathbf{u}_t)_i) \quad \forall i = 1, \ldots, m\ , \end{array}\right\} \quad (5.17)$$

using that $\mathcal{K} = \mathbb{R}^{n-2m} \times \mathbb{R}^m_+ \times \mathbb{R}^m$, i.e. only the normal components of the displacement vector are constrained. From the first equation in (5.17) we can express $\mathbf{u}_i$:

$$\mathbf{u}_i = \mathbf{A}_{ii}^{-1}(\mathbf{f}_i - \mathbf{A}_{in}\mathbf{u}_n - \mathbf{A}_{it}\mathbf{u}_t)\ . \quad (5.18)$$

Substituting (5.18) into $(5.17)_{3,4}$ we obtain the following constrained hemivariational inequality for the quantities defined only on Γ_c:

$$\left.\begin{array}{l} \textit{Find } (\mathbf{u}_c, \Xi) := (\mathbf{u}_n, \mathbf{u}_t, \Xi) \in \mathbb{R}^m_+ \times \mathbb{R}^m \times \mathbb{R}^m \textit{ such that} \\ (\widetilde{\mathbf{A}}\mathbf{u}_c, \mathbf{v}_c - \mathbf{u}_c)_{\mathbb{R}^{2m}} + (\Xi, \mathbf{v}_t - \mathbf{u}_t)_{\mathbb{R}^m} \geq (\widetilde{\mathbf{f}}, \mathbf{v}_c - \mathbf{u}_c)_{\mathbb{R}^{2m}} \quad \forall \mathbf{v}_c \in \mathbb{R}^m_+ \times \mathbb{R}^m \\ \Xi_i \in c_i\hat{b}((\mathbf{u}_t)_i)\ , \quad i = 1, \ldots, m\ , \end{array}\right\} \quad (5.19)$$

where

$$\widetilde{\mathbf{A}} = \begin{pmatrix} \widetilde{\mathbf{A}}_{nn}\ , & \widetilde{\mathbf{A}}_{nt} \\ \widetilde{\mathbf{A}}_{tn}\ , & \widetilde{\mathbf{A}}_{tt} \end{pmatrix}, \quad \widetilde{\mathbf{f}} = (\widetilde{\mathbf{f}}_n\ ,\ \widetilde{\mathbf{f}}_t)\ ,$$

with

$$\begin{aligned} \widetilde{\mathbf{A}}_{ss} &= \mathbf{A}_{ss} - \mathbf{A}_{si}\mathbf{A}_{ii}^{-1}\mathbf{A}_{is}\ , \quad s = n, t \\ \widetilde{\mathbf{A}}_{rs} &= \mathbf{A}_{rs} - \mathbf{A}_{ri}\mathbf{A}_{ii}^{-1}\mathbf{A}_{is}\ . \quad r \neq s\ , \quad r, s \in \{n, t\} \\ \widetilde{\mathbf{f}}_s &= \mathbf{f}_s - \mathbf{A}_{si}\mathbf{A}_{ii}^{-1}\mathbf{f}_i\ , \quad s = n, t\ . \end{aligned}$$

Problem 5.3 Prove that the substitution (5.18) into $(5.17)_{3,4}$ leads to the hemivariational inequality (5.19).

Once $\mathbf{u}_n$ and $\mathbf{u}_t$ are known, the remaining subvector $\mathbf{u}_i$ can be computed from (5.18).

As we have already mentioned in Sections 2 and 3, instead of (5.19) which contains two unknowns $\mathbf{u}_c$ and Ξ we shall solve the following substationary point problem (see (3.20)):

$$\left.\begin{array}{l} \textit{Find } \mathbf{u}_c \in \mathbb{R}^{2m} \textit{ such that} \\ \mathbf{0} \in \overline{\partial}\mathcal{J}(\mathbf{u}_c) + N_{\mathbb{R}^m_+}(\mathbf{u}_n) \ , \end{array}\right\} \tag{5.20}$$

where $\mathcal{J} : \mathbb{R}^{2m} \mapsto \mathbb{R}^1$ is the superpotential (see (2.61)) defined by

$$\mathcal{J}(\mathbf{x}_c) = \frac{1}{2}(\mathbf{x}_c, \widetilde{\mathbf{A}}\mathbf{x}_c)_{\mathbb{R}^{2m}} - (\widetilde{\mathbf{f}}, \mathbf{x}_c)_{\mathbb{R}^{2m}} + \Psi(\mathbf{x}_t) \ , \quad \mathbf{x}_c = (\mathbf{x}_n, \mathbf{x}_t) \in \mathbb{R}^m \times \mathbb{R}^m$$

and

$$\Psi(\mathbf{x}_t) := \sum_{i=1}^{m} c_i \int_0^{(\mathbf{x}_t)_i} b(t)\, dt$$

is the Lipschitz continuous functional. The relation between solutions of (5.19) and (5.20) follows from Theorems 2.16 and 2.17.

Once $\mathbf{u}_i$, $\mathbf{u}_n$ and $\mathbf{u}_t$ are known, the vector $\Xi \in \mathbb{R}^m$ can be computed from $(5.17)_4$.

We close this article by a simple contact problem with a nonmonotone friction law (5.4), where the function $\hat{b} : \mathbb{R}^1 \mapsto 2^{\mathbb{R}^1}$ is defined as follows:

$$\hat{b}(\xi) = \left\{ \begin{array}{ll} [-0.1, 0.1] & \xi = 0 \\ 0.1 & \xi \in (0, 2.\text{e–}5) \\ {[0.08, 0.1]} & \xi = 2.\text{e–}5 \\ 0.08 & \xi \in (2.\text{e–}5, \infty) \end{array} \right.$$

and $\hat{b}(\xi) = -\hat{b}(-\xi)$ for $\xi < 0$ (see Fig. 18)

The body is represented by the rectangle $\Omega = (0, 100) \times (0, 10)$ (in mm). It is made of an elastic, homogenous, isotropic material obeying the plane stress model characterized by Young's modulus $E = 2.1\text{e}5$ $[N/mm^2]$, Poisson's ratio $\sigma = 0.3$ and the element thickness $t = 5\ mm$. The partition of $\partial\Omega$ into Γ_u, Γ_c and $\Gamma_p = \Gamma_p^1 \cup \Gamma_p^2$ is the same as in Fig. 16. On the right vertical side Γ_p^1 the uniform tensile force $P = (P_1, 0)$, $P_1 = 0.06\ N/mm^2$ is applied. On the rest of Γ_p surface tractions as well as body forces are equal to zero.

The used triangulation $\mathcal{D}_h$ of $\overline{\Omega}$ is shown in Fig. 19: the body is carved into 40×4 squares and each square is divided by its diagonal into two triangles. This figure also illustrates the deformed configuration zoomed $10^6\times$.

The numerical values of the horizontal and vertical displacements at four selected nodes $N_1 = (25, 0)$, $N_2 = (50, 0)$, $N_3 = (75, 0)$ and $N_4 = (100, 0)$ lying on Γ_c are shown in Tables 1 and 2. The obtained results are compared with the frictionless case ($g = 0$) and with the monotone friction law defined by $\hat{b}$ depicted in Fig. 15 for $g = 0.08$.

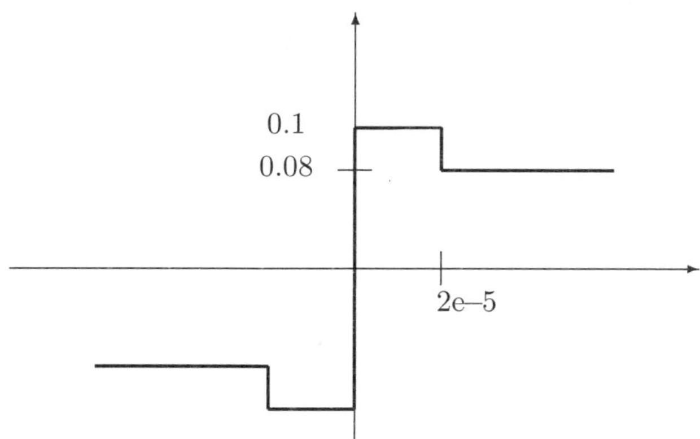

Figure 18.

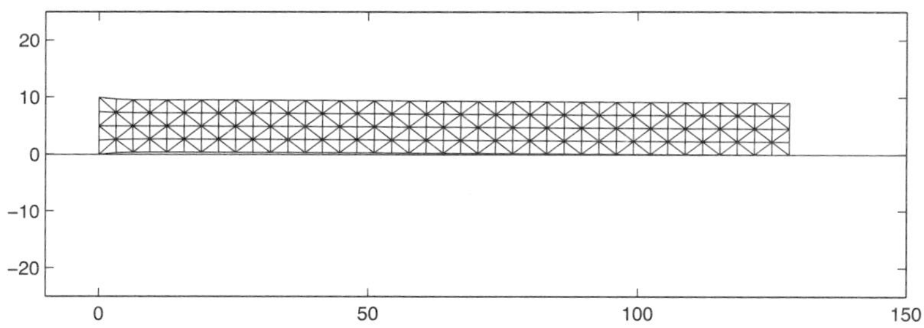

Figure 19.

	$u_1(N_1)$	$u_1(N_2)$	$u_1(N_3)$	$u_1(N_4)$
$g = 0$	0.070918 e–04	0.142347 e–04	0.213776 e–04	0.285204 e–04
monotone	0.069074 e–04	0.139190 e–04	0.209832 e–04	0.281013 e–04
nonmonotone	0.068688 e–04	0.138673 e–04	0.209432 e–04	0.280705 e–04

Table 1.

Computations were realized by using a nonsmooth variant of Newton's method developed in [Lukšan and Vlček 1993]. Once the displacement vector $\mathbf{u}$ is known one can get out the second component of the solution Ξ from $(5.17)_4$ and to verify if Ξ really satisfies the required inclusion which is a part of the definition of hemivariational inequalities. If not, the displacement vector $\mathbf{u}$ is not still good enough and it has to be found with a higher accuracy.

The distribution of the tangential component of the contact stress is shown in Fig. 20. Denote by $\mathbf{r}$ the residual vector in $(5.17)_3$, i.e. $\mathbf{r} = \mathbf{A}_{ni}\mathbf{u}_i + \mathbf{A}_{nn}\mathbf{u}_n + \mathbf{A}_{nt}\mathbf{u}_t - \mathbf{f}_n$. The coordinates of $\mathbf{r}$ represent the discrete normal contact stresses. Their distribution is

	$u_2(N_1)$	$u_2(N_2)$	$u_2(N_3)$	$u_2(N_4)$
$g = 0$	0.428558 e–06	0.428571 e–06	0.4286571 e–06	0.428571 e–06
monotone	0.3963 e–06	0.302824 e–06	0.160708 e–06	−4.235164 e–21
nonmonotone	0.366665 e–06	0.227080 e–06	0.085659 e–06	2.966762 e–21

Table 2.

shown in Fig. 21. It is worth mentioning that in the examples involving friction the deformed body Ω lies above the support S except the last node N_4.

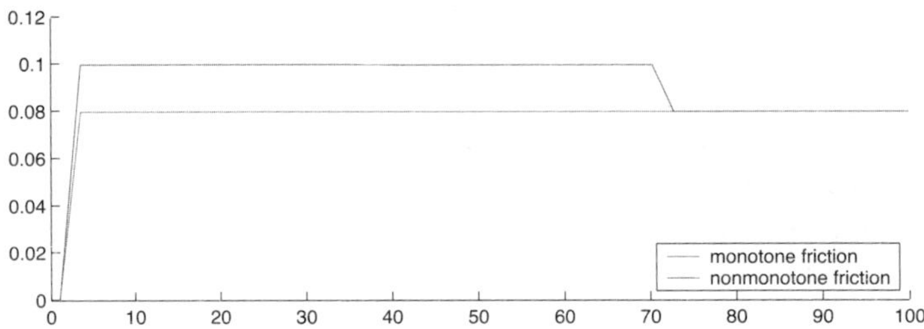

Figure 20.

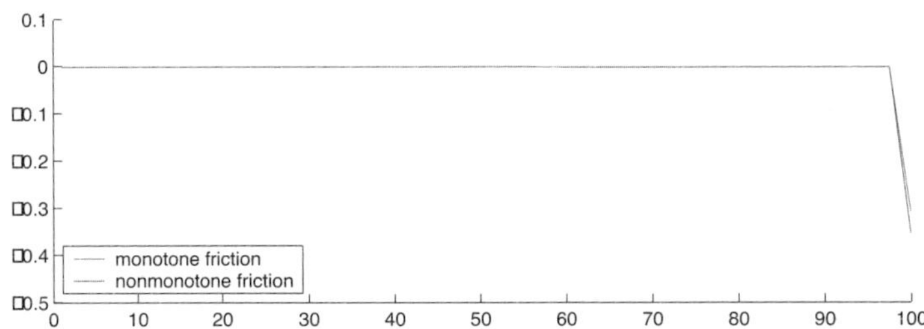

Figure 21.

Bibliography

H. Brézis, *Opérateurs Maximaux Monotones et Semigroupes de Contraction dans les Espaces de Hilbert.* North-Holland, Amsterdam 1973

C. C. Baniotopoulos and J. Haslinger and Z. Morávková, *Mathematical modeling of delamination and nonmonotone friction problems by hemivariational inequalities.* Appl. Math. 50(2005), 1-25.

P. Bisegna and F. Lebon and F. Maceri, *D-PANA: a convergent block-relaxation solution method for the discretized dual formulation of the Signorini-Coulomb contact problem.* C. R. Acad. Sci. Paris, t. 333, Série I (2001), *1053-1058*

K. C. Chang, *Variational methods for nondifferentiable functionals and their applications to partial differential equations.* J. Math. Anal. Appl. 80 (1981), *102-129*

P. G. Ciarlet, *The Finite Element Method for Elliptic Problems.* Studies in Mathematics and its Applications 4, North-Holland, Amsterdam, 1978

F. H. Clarke, *Optimization and Nonsmooth Analysis.* Wiley, New York, 1983

Z. Dostál, *Box constrained quadratic programming with proportioning and projections.* SIAM J. Optim. 7 (1997), *871-887*

Ch. Eck, and J. Jarušek, *Existence result for the static contact problem with Coulomb friction.* Math. Models Methods Appl. Sci 8 (1998), *445-468*

C. Eck and O. Steinbach and W. L. Wendland, *A symmetry boundary element method for contact problems with friction.* Mathematics and Computers in Simulations 50 (1999), *43-61*

I. Ekeland and R. Temam, *Convex Analysis and Variational Problems.* North-Holland, Amsterdam, 1976

R. Glowinski, *Numerical Methods for Nonlinear Variational Problems.* Springer Series in Computational Physics, Springer Verlag, 1984

V. Girault and R. Glowinski, *Error analysis of a fictitious domain method applied to a Dirichlet problem.* Japan J. Indust. Appl. Math. 12 (1995), *487-514*

R. Glowinski and J. L. Lions and R. Trémolière, *Numerical Analysis of Variational Inequalities.* Studies in Math. and its Appl., North-Holland, Amsterdam, 1981

J. Haslinger, *Approximation of the Signorini problem with friction, obeying Coulomb's law.* Math. Methods Appl. Sci. 5 (1983), *422-437*

J. Haslinger and Z. Dostál and R. Kučera, *On a splitting type algorithm for the numerical realization of contact problems with Coulomb friction.* Comput. Methods Appl. Mech. Engrg. 191 (2002), *2261-2281*

J. Haslinger and I. Hlaváček and J. Nečas and J. Lovíšek, *Numerical Solution of Variational Inequalities.* Springer Series in Appl. Math. Sci. 66, Springer-Verlag, New York, 1988

J. Haslinger and I. Hlaváček and J. Nečas, *Numerical Methods for Unilateral Problems in Solid Mechanics.* Volume IV, P. G. Ciarlet and J. L. Lions, eds., North-Holland, 1996

J. Haslinger and M. Miettinen and P. D. Panagiotopoulos, *Finite Element Method for Hemivariational Inequalities. Theory, Methods and Applications.* Kluwer Academic Publishers, Nonconvex Optimization and its Applications, Vol. 35, Dordrecht 1999

J. Haslinger and P. D. Panagiotopoulos, *The reciprocal variational approach to the Signorini problem with friction. Approximation results.* Proc. Roy. Soc. Edinburgh 98A (1984), *365-383*

J. Haslinger and O. Vlach, *Signorini problem with a solution dependent coefficient of friction (model with given friction): Approximation and the numerical realization.* To appear in Appl. Math.

J. Jarušek, *Contact problems with bounded friction. Coercive Case.* Czech. Math. J. 33 (1983), *237-249*

R. Kučera, *Quadratic programming with separated quadratic constraints.* In preparation, 2004

J. L. Lions, *Quelques Méthodes de résolution des problèmes aux limites non linéaries.* Dunod-Gauthier Villar,Paris, 1969

L. Lukšan and J. Vlček, *A bundle-Newton method for nonsmooth unconstrained minimization.* Math. Programming 83 (1993), *373-391*

J. L. Lions and G. Stampacchia, *Variational inequalities.* Commun. Pure Appl. Math. 20 (1967), *493-519*

J. Nečas, *Les méthodes directes en théorie des équations elliptiques.* Academia, Prague, 1967

J. Nečas and J. Jarušek and J. Haslinger, *On the solution of the variational inequality to the Signorini problem with small friction.* Boll. Unione Mat. Ital. 17-B (5A) (1980), *796-811*

J. Outrata, *an article in this book*

P. D. Panagiotopoulos, *A nonlinear programming approach to the unilateral contact and friction boundary value problem in the theory of elasticity.* Ing. Arch. 44 (1975), *421-432*

P. D. Panagiotopoulos, *Inequality Problems in Mechanics and Applications. Convex and Nonconvex energy functions.* Birkhäuser, Basel, 1985

Semicoercive Hemivariational Inequalities, Regularization Methods, Applications on Mechanics

Zdzisław Naniewicz

Cardinal Stefan Wyszyński University
Faculty of Mathematics and Science
Dewajtis 5, 01-815 Warsaw, Poland
E-mail: naniewicz@uksw.edu.pl; naniewicz.z@acn.waw.pl

Abstract. An approach to semicoercive variational-hemivariational or hemivariational inequalities based on a recession technique introduced in (Naniewicz 2003), is developed. First, problems defined on vector-valued function spaces are considered under unilateral growth conditions imposed on nonlinear parts by making use of the Galerkin method. Second, a minimax method relying on Chang's version of Mountain Pass Theorem for locally Lipschitz functionals (Chang 1981) is applied to study semicoercive hemivariational inequalities on vector valued function spaces. Third, the resonant problem governed by the p-Laplacian involving the unilateral growth condition is discussed. Some mechanical problems as exemplifications of the presented approach are shown.

1 Introduction

The theory of hemivariational inequalities begun in the early eighties with the works of P. D. Panagiotopoulos (Panagiotopoulos 1981), (Panagiotopoulos 1983), and has as a main reason for its birth the need for the description of important problems in physics and engineering, where nonmonotone, multivalued boundary or interface conditions occur, or where some nonmonotone, multivalued relations between stress and strain, or reaction and displacement have to be taken into account. The theory of hemivariational inequalities (as the generalization of variational inequalities, cf. (Duvaut & Lions 1972)) has been proved to be very useful in understanding of many problems of mechanics and engineering involving nonconvex, nonsmooth energy functionals. For the study of existence problems related to hemivariational and variational-hemivariational inequalities the reader is referred to (Motreanu & Naniewicz 1996), (Motreanu & Naniewicz 2001), (Motreanu & Naniewicz 2002), (Motreanu & Panagiotopoulos 1995), (Motreanu & Panagiotopoulos 1996), (Naniewicz 1994*b*), (Naniewicz 1995*a*), (Naniewicz & Panagiotopoulos 1995), (Naniewicz 1995*b*), (Naniewicz 1997), (Panagiotopoulos 1985), (Panagiotopoulos 1993), (Motreanu & Panagiotopoulos 1999), (Pop, Panagiotopoulos & Naniewicz 1997), (Gasiński & Papageorgiou 2000),(Adly & Goeleven 2000), (Haslinger, Miettinen & Panagiotopoulos 1999), (Goeleven, Motreanu & Panagiotopoulos 1997), (Dem'yanov, Stavroulakis, Polyakova & Panagiotopoulos 1996) where the theory of pseudomonotone mappings, the critical point theory, the Galerkin method, the finite element method have been used among others to establish solutions.

These lectures concern the study of variational-hemivariational and hemivariational inequalities in the case of noncoercivity. The recession technique introduced in (Naniewicz 2003) will be presented by three types of the aforementioned problems.

In Section 1 we consider a class of semicoercive variational-hemivariational inequalities on vector-valued function spaces under unilateral growth hypotheses imposed on nonconvex parts. We shall deal with the problem of finding $u \in V$ such as to satisfy a variational-hemivariational inequality

$$\langle Au - f, v - u\rangle_V + \Phi(v) - \Phi(u) + \int_\Omega j^0(u; v-u) d\Omega \geq 0, \quad \forall\, v \in V,$$

where $A : V \to V^\star$ is a pseudo-monotone semicoercive operator, $\Phi : V \to \mathbb{R} \cup \{+\infty\}$ a convex, lower semicontinuous, proper function, $j : \mathbb{R}^s \to \mathbb{R}$ a locally Lipschitz function fulfilling the unilateral growth condition (Naniewicz 1994*a*):

$$j^0(\xi; \eta - \xi) \leq \alpha(\rho)(1 + |\xi|^r), \;\; \forall\, \xi, \eta \in \mathbb{R}^s, \; |\eta| \leq \rho, \; \rho \geq 0.$$

Here $j^0(\cdot\,;\cdot)$ stands for the generalized Clarke differential given by (Clarke 1983)

$$j^0(\xi; \eta) = \limsup_{\substack{h \to 0 \\ \lambda \to 0_+}} \frac{j(\xi + h + \lambda\eta) - j(\xi + h)}{\lambda}.$$

To prove the main result the theory of pseudomonotone multivalued mappings combined with the Lipschitz regularization of ∂j and some compactness arguments involving Dunford-Pettis criterion will be applied.

Section 2 is devoted to semicoercive hemivariational inequalities of the form

$$\langle Au - f, v - u\rangle_V + \int_\Omega j^0(u; v-u) d\Omega \geq 0, \quad \forall\, v \in V,$$

treated by making use of the minimax methods relying on Chang's version (Chang 1981) of Mountain Pass Theorem (see (Ambrosetti & Rabinowitz 1973), (Rabinowitz 1986)) for locally Lipschitz functionals and the recession techique developed in (Naniewicz 2003).

In Section 3 a nonlinear hemivariational inequality at ressonance driven by the p-Laplacian will be examined. The existence of solutions for the problem

$$\int_\Omega |Du|^{p-2} \langle Du, Dv - Du\rangle_{\mathbb{R}^N} \, d\Omega - \lambda_1 \int_\Omega |u|^{p-2}\, u(v-u)\, d\Omega + \int_\Omega j^0(u; v-u)\, d\Omega \geq 0, \quad \forall\, v \in W_0^{1,p}(\Omega).$$

where λ_1 is the first eigenvalue of $\left(-\Delta_p, W_0^{1,p}(\Omega)\right)$, is disscused.

In Section 4 some applications related to the Signorini problem will be presented.

2 Variational-hemivariational inequalities with unilateral growth condition

2.1 Statement of the problem and some preliminarities

Let V be a real Hilbert space endowed with the scalar product $(\cdot,\cdot)_V$ and the associated norm $\|\cdot\|_V$. Let Ω be a bounded domain in $\mathbb{R}^m$, $m > 2$, with sufficiently smooth boundary Γ. Assume that V is densly and compactly imbedded into $L^p(\Omega;\mathbb{R}^s)$, $s \geq 1$. We write $\|\cdot\|_{L^p(\Omega;\mathbb{R}^s)}$ for the norm in $L^p(\Omega;\mathbb{R}^s)$. For the pairing over $V^\star \times V$ the symbol $\langle\cdot,\cdot\rangle_V$ will be used, $V^\star$ being the topological dual of V. Moreover, we assume that V admits the orthogonal decomposition $V = \widehat{V} \oplus V_0$ for a finite dimensional linear subspace $V_0 \subset V$. Therefore each element $u \in V$ can be uniquely expressed as $u = \widehat{u} + \theta$ with $\widehat{u} \in \widehat{V}$ and $\theta \in V_0$ and $(\widehat{u},\theta)_V = 0$.

Let $A : V \to V^\star$ be a bounded, pseudo-monotone operator. It means that A maps bounded sets into bounded sets and that the following conditions hold (Brézis 1968), (Browder & Hess 1972):

(i) The effective domain of A, $\mathrm{Dom}(A)$, coincides with the whole V, i.e. $\mathrm{Dom}(A) = V$;

(ii) If $u_n \to u$ weakly in V and $\limsup\limits_{n\to\infty}\langle Au_n, u_n - u\rangle_V \leq 0$ then $\liminf\limits_{n\to\infty}\langle Au_n, u_n - v\rangle_V \geq \langle Au, u - v\rangle_V$ for any $v \in V$.

Note that from (i) and (ii) it follows that A is demicontinuous, i.e.

(iii) If $u_n \to u$ strongly in V then $Au_n \to Au$ weakly in $V^\star$.

Moreover we assume that

$$\langle Au, u\rangle_V \geq c(\|\widehat{u}\|_V)\|\widehat{u}\|_V, \quad \forall u = \widehat{u} + \theta,\ \widehat{u} \in \widehat{V},\ \theta \in V_0, \tag{2.1}$$

$c : \mathbb{R}^+ \to \mathbb{R}$ stands for a coercivity function with $c(r) \to \infty$ as $r \to \infty$. Further, let $j : \mathbb{R}^s \to \mathbb{R}$ be a locally Lipschitz function fulfilling the unilateral growth conditions (Naniewicz 1994*a*):

$$j^0(\xi;\eta-\xi) \leq \alpha(\rho)(1+|\xi|^r), \ \forall\, \xi,\eta \in \mathbb{R}^s,\ |\eta| \leq \rho,\ \rho \geq 0, \tag{2.2}$$

$$j^0(\xi;-\xi) \leq \widehat{k}\,|\xi|, \quad \forall\, \xi \in \mathbb{R}^s, \tag{2.3}$$

where $1 \leq r < p$, $\widehat{k}$ is a nonnegative constant and $\alpha : \mathbb{R}^+ \to \mathbb{R}^+$ is assumed to be a nondecreasing function from $\mathbb{R}^+$ into $\mathbb{R}^+$. Here $j^0(\cdot\,;\cdot)$ stands for the generalized Clarke differential (Clarke 1983), i.e.

$$j^0(\xi;\eta) = \limsup_{\substack{h\to 0\\ \lambda\to 0_+}} \frac{j(\xi+h+\lambda\eta) - j(\xi+h)}{\lambda}. \tag{2.4}$$

Let $\Phi : V \to R \cup \{+\infty\}$ be a convex, lower semicontinuous and proper function, $g \in V^\star$ be an element of $V^\star$. Consider the problem (P) of finding $u \in V$ such as to satisfy the variational-hemivariational inequality

$$\langle Au - f, v-u\rangle_V + \Phi(v) - \Phi(u) + \int_\Omega j^0(u; v-u)d\Omega \geq 0, \quad \forall v \in V, \tag{2.5}$$

where the convention $+\infty - \infty = +\infty$ is applied.

In fact, we establish a stronger result, namely, we show the existence of $u \in V$, $\chi \in L^1(\Omega;\mathbb{R}^s)$ and $\psi \in V^\star$ with the properties that

$$\langle Au - f, v-u\rangle_V + \langle \psi, v-u\rangle_V + \int_\Omega \chi\cdot(v-u)d\Omega = 0,$$
$$\forall v \in V \cap L^\infty(\Omega;\mathbb{R}^s) \tag{2.6}$$

$$\chi \in \partial j(u) \text{ a.e. in } \Omega, \quad \chi\cdot u \in L^1(\Omega), \quad \psi \in \partial\Phi(u), \tag{2.7}$$

where the dot " $\cdot$ " stands for the inner product in $\mathbb{R}^s$. From (2.6) and (2.7) we then derive the validity of variational-hemivariational inequality in the form (2.5).

For $R > 0$ define

$$\widetilde{j^0_R}(\xi;\eta) := \begin{cases} j^0(\xi;\eta) & \text{if } |\xi| \leq R \\ j^0(R\frac{\xi}{|\xi|};\eta) & \text{if } |\xi| > R. \end{cases} \tag{2.8}$$

Lemma 2.1. *The function* $\mathbb{R}^s \times \mathbb{R}^s \ni (\xi,\eta) \mapsto \widetilde{j^0_R}(\xi;\eta)$ *is upper semicontinuous and if* (2.2) – (2.3) *hold, then*

$$j^0_R(\xi;\eta-\xi) \leq \widetilde{\alpha}(r)(1+|\xi|^r), \quad \forall\, \xi \in \mathbb{R}^s, \forall\, \eta \in \mathbb{R}^s, \ |\eta| \leq r, \ r \geq 0. \tag{2.9}$$

$$\widetilde{j^0_R}(\xi;-\xi) \leq \widehat{k}\,|\xi|, \quad \forall\, \xi \in \mathbb{R}^s, \tag{2.10}$$

for some nondecreasing, independent of R *function* $\widetilde{\alpha} : \mathbb{R}^+ \to \mathbb{R}^+$.

Proof. The upper semicontinuity of $\widetilde{j^0_R}(\cdot\,;\cdot)$ follows directly from the upper semicontinuity of $j^0(\cdot\,;\cdot)$. To establish (2.9) and (2.10) it is enough to consider the case $|\xi| > R$ and invoke the estimates

$$\begin{aligned}
&\widetilde{j^0_R}(\xi;\eta-\xi) = j^0(R\tfrac{\xi}{|\xi|};\eta-\xi) \leq j^0(R\tfrac{\xi}{|\xi|};\eta - R\tfrac{\xi}{|\xi|}) + \tfrac{|\xi|-R}{R} j^0(R\tfrac{\xi}{|\xi|}; -R\tfrac{\xi}{|\xi|}) \\
&\leq \alpha(|\eta|)(1+R^r) + \tfrac{|\xi|-R}{R}\widehat{k}R \leq \alpha(\rho)(1+|\xi|^r) + \widehat{k}\,|\xi|\,, \\
&\forall\, \xi,\eta \in \mathbb{R}^s, \ |\eta| \leq \rho, \ \rho \geq 0,
\end{aligned}$$

and

$$\widetilde{j^0_R}(\xi;-\xi) = j^0(R\tfrac{\xi}{|\xi|};-\xi) \leq \tfrac{|\xi|}{R} j^0(R\tfrac{\xi}{|\xi|};-R\tfrac{\xi}{|\xi|}) \leq \tfrac{|\xi|}{R}\widehat{k}R = \widehat{k}\,|\xi|\,,$$

from which the desired estimates easily follow. □

Define a recession function $j^\infty : \mathbb{R}^s \to \mathbb{R} \cup \{+\infty\}$ by (cf. (Goeleven & Théra 1995), (Brézis & Nirenberg 1978), (Baiocchi, Buttazzo, Gastaldi & Tomarelli 1988), (Goeleven 1996))

$$j^\infty(\xi) = \liminf_{\substack{\eta\to\xi \\ t\to+\infty}} [-j^0(t\eta;-\eta)], \quad \xi \in \mathbb{R}^s. \tag{2.11}$$

From now on we assume that $g \in V^\star$ fulfills the compatibility condition

$$\langle f, \theta \rangle_V < \Phi^\infty(\theta) + \int_\Omega j^\infty(\theta) d\Omega, \quad \forall\, \theta \in V_0 \setminus \{0\}, \tag{2.12}$$

where $\Phi^\infty : V \to \mathbb{R} \cup \{+\infty\}$ defined by

$$\Phi^\infty(u) = \lim_{\lambda \to +\infty} \frac{\Phi(u_0 + \lambda u) - \Phi(u_0)}{\lambda}, \quad u_0 \in \operatorname{Dom} \Phi,$$

stands for the recession functional of Φ (cf. (Baiocchi et al. 1988), (Goeleven 1996)).

Lemma 2.2. *Suppose that* (2.1) – (2.3) *and* (2.12) *hold. Then there exists $R_0 > 0$ such that for any $R > R_0$ the set of all $u \in V$ with the property that*

$$\langle Au - f, u \rangle_V + \Phi(u) - \int_\Omega \widetilde{j^0_R}(u; -u)\, d\Omega \leq \mathfrak{C} \tag{2.13}$$

is bounded in V, i.e. there exists $M_R > 0$ such that (2.13) *implies*

$$\|u\|_V \leq M_R. \tag{2.14}$$

Proof. Suppose on the contrary that this claim is not true, i.e. there exists a sequence $\{u_n\}_{n=1}^\infty \subset V$ with the property that

$$\langle Au_n - f, u_n \rangle_V + \Phi(u_n) - \int_\Omega \widetilde{j^0_R}(u_n; -u_n)\, d\Omega \leq \mathfrak{C}, \tag{2.15}$$

where $\|u_n\|_V \to \infty$ as $n \to \infty$. By the hypothesis each element u_n can be represented as

$$u_n = \hat{u}_n + e_n \theta_n, \tag{2.16}$$

where $\widehat{u}_n \in \widehat{V}$, $e_n \geq 0$, $\theta_n \in V_0$, $\|\theta_n\|_V = 1$ and $\langle Au_n, u_n \rangle_V \geq c(\|\widehat{u}_n\|_V)\|\widehat{u}_n\|_V$. Since Φ is lower semicontinuous,

$$\Phi(u) \geq -a\|u\|_V - b, \quad \forall\, u \in V, \tag{2.17}$$

for some $a, b \in \mathbb{R}$. Therefore taking into account (2.3) it follows that

$$\begin{aligned}
\mathfrak{C} &\geq \langle Au_n - f, u_n \rangle_V + \Phi(u_n) - \int_\Omega \widetilde{j^0_R}(u_n; -u_n) d\Omega \\
&\geq c(\|\widehat{u}_n\|_V)\|\widehat{u}_n\|_V - \|f\|_{V^\star}\|\widehat{u}_n\|_V - e_n\langle f, \theta_n \rangle_V \\
&\quad + \Phi(\widehat{u}_n + e_n\theta_n) - \widehat{k}\int_\Omega |\widehat{u}_n + e_n\theta_n|\, d\Omega \\
&\geq c(\|\widehat{u}_n\|_V)\|\widehat{u}_n\|_V - \|f\|_{V^\star}\|\widehat{u}_n\|_V - e_n\langle f, \theta_n \rangle_V \\
&\quad + \Phi(\widehat{u}_n + e_n\theta_n) - k\|\widehat{u}_n\|_V - e_n k\|\theta_n\|_V \\
&\geq \|\widehat{u}_n\|_V\big(c(\|\widehat{u}_n\|_V) - \|f\|_{V^\star} - k\big) - e_n\big(\langle f, \theta_n \rangle_V + k\big) + \Phi(\widehat{u}_n + e_n\theta_n) \\
&\geq \|\widehat{u}_n\|_V\big(c(\|\widehat{u}_n\|_V) - \|f\|_{V^\star} - k - a\big) - e_n\big(\langle f, \theta_n \rangle_V + k + a\big) - b,
\end{aligned} \tag{2.18}$$

where $k = \text{const}$. The obtained estimates imply that $\{e_n\}$ is unbounded. Indeed, if it would not be so, then due to the behavior of $c(\cdot)$ at infinity $\{\widehat{u}_n\}$ had to be bounded. In such a case the contradiction with $\|u_n\|_V \to \infty$ as $n \to \infty$ results. Therefore one can suppose without loss of generality that $e_n \to +\infty$ as $n \to \infty$. The next claim is that

$$\frac{1}{e_n}\widehat{u}_n \to 0 \text{ strongly in } V. \tag{2.19}$$

Indeed, if $\{\|\widehat{u}_n\|_V\}$ is bounded then (2.19) follows immediately. If $\|\widehat{u}_n\|_V \to \infty$ then $c(\|\widehat{u}_n\|_V) \to +\infty$. From (2.18) one has

$$k + a + \|f\|_{V^*} + \frac{\mathfrak{C} + b}{e_n} \geq \left(c(\|\widehat{u}_n\|_V) - \|f\|_{V^*} - k - a\right)\frac{\|\widehat{u}_n\|_V}{e_n}.$$

Thus the boundedness of the sequence

$$\left\{\left(c(\|\widehat{u}_n\|_V) - \|f\|_{V^*} - k - a\right)\frac{\|\widehat{u}_n\|_V}{e_n}\right\}_{n=1}^{\infty}$$

results which in view of

$$c(\|\widehat{u}_n\|_V) - \|f\|_{V^*} - k - a \to +\infty \ \text{ as } n \to \infty,$$

implies the assertion (2.19). The obtained results give rise to the following representation of u_n:

$$u_n = e_n\left(\frac{1}{e_n}\widehat{u}_n + \theta_n\right),$$

where $\frac{1}{e_n}\widehat{u}_n \to 0$ strongly in V and $\theta_n \to \theta$ in V_0 as $n \to \infty$ for some $\theta \in V_0$ with $\|\theta\|_V = 1$ (recall that V_0 has been assumed to be finite dimensional). Moreover, the compact imbedding $V \subset L^p(\Omega; \mathbb{R}^s)$ permits to suppose that

$$\tfrac{1}{e_n}\widehat{u}_n \to 0 \text{ and } \theta_n \to \theta \text{ a.e. in } \Omega \tag{2.20}$$

Using (2.15) together with the semicoercivity of A leads to

$$\begin{aligned}\mathfrak{C} &\geq \langle Au_n - f, u_n\rangle_V + \Phi(u_n) - \int_\Omega \widetilde{j^0_R}(u_n; -u_n)d\Omega \\ &\geq \left(c(\|\widehat{u}_n\|_V) - \|f\|_{V^*}\right)\|\widehat{u}_n\|_V - e_n\langle f, \theta_n\rangle_V \\ &\quad + \Phi(\widehat{u}_n + e_n\theta_n) + e_n\int_\Omega -\widetilde{j^0_R}\Big(e_n(\tfrac{1}{e_n}\widehat{u}_n + \theta_n); -\tfrac{1}{e_n}\widehat{u}_n - \theta_n\Big)d\Omega.\end{aligned}$$

Hence

$$\begin{aligned}\langle f, \theta_n\rangle_V \geq \left(c(\|\widehat{u}_n\|_V) - \|f\|_{V^*}\right)\frac{1}{e_n}\|\widehat{u}_n\|_V + \frac{1}{e_n}\Phi\big(e_n(\tfrac{\widehat{u}_n}{e_n} + \theta_n)\big) + \\ + \int_\Omega -\widetilde{j^0_R}\Big(e_n(\tfrac{\widehat{u}_n}{e_n} + \theta_n); -\tfrac{\widehat{u}_n}{e_n} - \theta_n\Big)d\Omega - \frac{\mathfrak{C}}{e_n}.\end{aligned} \tag{2.21}$$

Now observe that either

$$\left(c(\|\widehat{u}_n\|_V) - \|f\|_{V^\star}\right)\frac{1}{e_n}\|\widehat{u}_n\|_V \to 0 \quad \text{as } n \to \infty,$$

if $\{\|\widehat{u}_n\|_V\}$ is bounded, or

$$\left(c(\|\widehat{u}_n\|_V) - \|f\|_{V^\star}\right)\frac{1}{e_n}\|\widehat{u}_n\|_V \geq 0$$

for sufficiently large n, if $\|\widehat{u}_n\|_V \to \infty$ as $n \to \infty$. Therefore, for each case

$$\liminf_{n\to\infty}\left(c(\|\widehat{u}_n\|_V) - \|f\|_{V^\star}\right)\frac{1}{e_n}\|\widehat{u}_n\|_V \geq 0.$$

Moreover, by (2.10) the estimate follows

$$-\widetilde{j^0_R}\left(e_n(\tfrac{\widehat{u}_n}{e_n} + \theta_n); -\tfrac{\widehat{u}_n}{e_n} - \theta_n\right) \geq -\widehat{k}\left|\frac{\widehat{u}_n}{e_n} + \theta_n\right|. \tag{2.22}$$

This allows the application of Fatou's lemma in (2.21) from which one is led to

$$\begin{aligned}\langle f,\theta\rangle_V &\geq \liminf_{n\to\infty}\frac{1}{e_n}\Phi\left(e_n(\tfrac{\widehat{u}_n}{e_n} + \theta_n)\right)\\ &\quad + \liminf_{n\to\infty}\int_\Omega\left[-\widetilde{j^0_R}\left(e_n(\tfrac{\widehat{u}_n}{e_n} + \theta_n); -\tfrac{\widehat{u}_n}{e_n} - \theta_n\right)\right]d\Omega\\ &\geq \liminf_{n\to\infty}\frac{1}{e_n}\Phi\left(e_n(\tfrac{\widehat{u}_n}{e_n} + \theta_n)\right)\\ &\quad + \int_\Omega \liminf_{n\to\infty}\left[-\widetilde{j^0_R}\left(e_n(\tfrac{\widehat{u}_n}{e_n} + \theta_n); -\tfrac{\widehat{u}_n}{e_n} - \theta_n\right)\right]d\Omega. \end{aligned}\tag{2.23}$$

Taking into account (2.20) and upper semicontinuity of $j^0(\cdot\,,\cdot)$ one can easily verify that

$$\liminf_{n\to\infty}\left[-\widetilde{j^0_R}\left(e_n(\tfrac{\widehat{u}_n}{e_n} + \theta_n); -\tfrac{\widehat{u}_n}{e_n} - \theta_n\right)\right] \geq -j^0_{R+}(\theta; -\theta),$$

where

$$j^0_{R+}(\theta;-\theta) := \begin{cases} j^0(\frac{R}{|\theta|}\theta; -\theta) & \text{a.e. in } \{x \in \Omega\colon \theta(x) \neq 0\}\\ 0 & \text{a.e. in } \{x \in \Omega\colon \theta(x) = 0\}.\end{cases}$$

Further, for a convex, lower semicontinuous $\Phi : V \to \mathbb{R} \cup \{+\infty\}$ we have (Baiocchi et al. 1988)

$$\liminf_{n\to\infty}\frac{1}{e_n}\Phi\left(e_n(\tfrac{\widehat{u}_n}{e_n} + \theta_n)\right) \geq \Phi^\infty(\theta).$$

Thus we arrive at

$$\langle f,\theta\rangle_V \geq \Phi^\infty(\theta) + \int_\Omega -j^0_{R+}(\theta;-\theta)\,d\Omega. \tag{2.24}$$

Since $j^\infty(\cdot)$ is lower semicontinuous and V_0 is finite dimensional, from (2.12) it follows that a $\delta > 0$ can be found such that for any $\theta \in V_0$ with $\|\theta\|_V = 1$,

$$\langle f,\theta\rangle_V + \delta < \Phi^\infty(\theta) + \int_\Omega j^\infty(\theta)d\Omega. \tag{2.25}$$

With the help of Fatou's lemma (permitted by (2.22)) we arrive at

$$\liminf_{R\to\infty} \int_\Omega -j^0_{R+}(\theta;-\theta)d\Omega \geq \int_\Omega j^\infty(\theta)d\Omega. \tag{2.26}$$

The upper semicontinuity of $j^0(\cdot\ \cdot)$ allows us to conclude the existence of $R_\theta > 0$ and $\varepsilon_\theta > 0$ such that

$$\int_\Omega -j^0_{R+}(\theta';-\theta')d\Omega \geq \int_\Omega j^\infty(\theta)d\Omega - \frac{\delta}{2}$$

for each $R > R_\theta$ and $\theta' \in V_0$ with $\|\theta-\theta'\|_V < \varepsilon_\theta$. As the sphere $\{v \in V_0 \colon \|v\|_V = 1\}$ is compact in V_0, there exists $R_0 > 0$ such that

$$\int_\Omega -j^0_{R+}(\theta;-\theta)d\Omega \geq \int_\Omega j^\infty(\theta)d\Omega - \frac{\delta}{2}, \tag{2.27}$$

for any $\theta \in V_0$ with $\|\theta\|_V = 1$, $R > R_0$. This combined with (2.24) contradicts to (2.12). Accordingly, the existence of a constant $M_R > 0$ has been established such that (2.13) implies (2.14), provided $R > R_0$. The proof of Lemma 2.2 is complete. □

2.2 Regularized semicoercive variational-hemivariational inequality

Define a mapping $\Gamma_R : L^p(\Omega;\mathbb{R}^s) \to 2^{L^q(\Omega;\mathbb{R}^s)}$ by the following formula

$$\Gamma_R(v) = \{\psi \in L^q(\Omega;\mathbb{R}^s) : \int_\Omega \psi\cdot w\, d\Omega\ \leq \int_\Omega \widetilde{j^0_R}(v;w)\, d\Omega,\ \ \forall\, w \in L^p(\Omega;\mathbb{R}^s)\}, \tag{2.28}$$

where $\frac{1}{p}+\frac{1}{q} = 1$. It is easy to verify that the restriction of Γ_R to V, $\Gamma_R|_V : V \to 2^{V^\star}$, is pseudo monotone.

For any $R > 0$ the following regularization of the primal problem (P) can be formulated:

Problem (P_R). Find a triple $(u_R, \chi_R, \psi_R) \in V \times L^q(\Omega;\mathbb{R}^s) \times V^\star$, such that

$$\langle Au_R - f, v-u_R\rangle_V + \langle \psi_R, v-u_R\rangle_V + \int_\Omega \chi_R\cdot(v-u_R)d\Omega = 0, \quad \forall\, v \in V \tag{2.29}$$

$$\chi_R \in \Gamma_R(u_R), \quad \psi_R \in \partial\Phi(u_R). \tag{2.30}$$

Proposition 2.1. *Assume that* (2.1)–(2.3), (2.12) *hold. Moreover, suppose*

$$0 \in \mathrm{Dom}\,\Phi. \tag{2.31}$$

Then for any $R > R_0$ (with R_0 as in Lemma 2.2) there exist $u_R \in V$, $\chi_R \in L^q(\Omega;\mathbb{R}^s)$ and $\psi_R \in V^\star$ such that (2.29) *and* (2.30) *hold, i.e. the problem (P_R) has solutions. Moreover, $\{u_R\}_{R>R_0}$ is bounded in V, i.e. there exists $M > 0$ independent of $R > R_0$, such that*

$$\|u_R\|_V \leq M, \quad \forall\, R > R_0. \tag{2.32}$$

Proof. Fix $R > R_0$. The sum of two pseudo-monotone mappings $A + \Gamma_R|_V$ is pseudo-monotone (Browder & Hess 1972). Thus the variational inequality

$$\langle Au + \Gamma_R|_V(u_R) - f, v - u_R \rangle + \Phi(v) - \Phi(u_R) \geq 0, \quad \forall v \in B_{2M_R}, \tag{2.33}$$

with M_R as in Lemma 2.2, $B_{2M_R} = \{v \in V\colon \|v\|_V \leq 2M_R\}$ being a ball in V with the radius $2R$, admits a solution $u_R \in B_{2M_R}$ (cf. (Browder & Hess 1972)). By substituting $v = 0$ into (2.33) we arrive at the inequality

$$\langle Au_R - f, u_R \rangle_V + \Phi(u_R) - \int_\Omega \widetilde{j_R^0}(u_R; -u_R)\, d\Omega \leq \Phi(0)$$

which by Lemma 2.2 with $\mathfrak{C} = \Phi(0)$ allows the conclusion that $\|u_R\|_V \leq M_R$. This together with the validity of (2.33) for any $v \in B_{2M_R}$ implies that, in fact, it holds for any $v \in V$. Thus we easily deduce the existence of a triple (u_R, χ_R, ψ_R) which is a solution of (P_R).

Let us proceed to the boundedness of $\{u_R\}_{R>R_0}$. To begin with let us note that from (2.29) and (2.30) the inequality follows

$$\langle Au_R - f, v - u_R \rangle_V + \Phi(v) - \Phi(u_R) + \int_\Omega \widetilde{j_R^0}(u_R; v - u_R) d\Omega \geq 0, \quad \forall v \in V, \tag{2.34}$$

which by substitution $v = 0$ leads to

$$\langle Au_R - f, u_R \rangle_V + \Phi(u_R) - \int_\Omega \widetilde{j_R^0}(u_R; -u_R) d\Omega \leq \Phi(0).$$

Now suppose on the contrary that the claim is not true. Then there would exist a sequence $R_n \to \infty$ such that $\|u_{R_n}\|_V \to \infty$ as $n \to \infty$, and

$$\langle Au_{R_n} - f, u_{R_n} \rangle_V + \Phi(u_{R_n}) - \int_\Omega \widetilde{j_{R_n}^0}(u_{R_n}; -u_{R_n}) d\Omega \leq \Phi(0).$$

For simplicity of notations, instead of subscript "R_n" we write "n". Thus we have

$$\langle Au_n - f, u_n \rangle_V + \Phi(u_n) - \int_\Omega \widetilde{j_n^0}(u_n; -u_n) d\Omega \leq \Phi(0). \tag{2.35}$$

Now we follow the lines of the proof of Lemma 2.2. Analogously, we arrive at the representation

$$u_n = e_n \left(\frac{1}{e_n} \widehat{u}_n + \theta_n \right),$$

with $e_n \to \infty$, $\frac{1}{e_n} \widehat{u}_n \to 0$ strongly in V and $\theta_n \to \theta$ in V_0 as $n \to \infty$ for some $\theta \in V_0$ with $\|\theta\|_V = 1$, and then at the inequality

$$\langle f, \theta \rangle_V \geq \Phi^\infty(\theta) + \liminf_{n\to\infty} \int_\Omega -\widetilde{j_n^0}\Big(e_n(\tfrac{1}{e_n}\widehat{u}_n + \theta_n); -\tfrac{1}{e_n}\widehat{u}_n - \theta_n\Big) d\Omega. \tag{2.36}$$

But

$$\widetilde{j_n^0}\Big(e_n(\tfrac{1}{e_n}\widehat{u}_n + \theta_n); -\tfrac{1}{e_n}\widehat{u}_n - \theta_n\Big)$$
$$= \begin{cases} j^0\Big(e_n(\frac{1}{e_n}\widehat{u}_n + \theta_n); -\frac{1}{e_n}\widehat{u}_n - \theta_n\Big) & \text{if } |\widehat{u}_n + e_n\theta_n| \le R_n \\ j^0\Big(\frac{R_n}{\left|\frac{1}{e_n}\widehat{u}_n + \theta_n\right|}(\frac{1}{e_n}\widehat{u}_n + \theta_n); -\frac{1}{e_n}\widehat{u}_n - \theta_n\Big) & \text{if } |\widehat{u}_n + e_n\theta_n|\, R_n, \end{cases}$$

with

$$e_n \to \infty \quad \text{and} \quad \frac{R_n}{\left|\frac{1}{e_n}\widehat{u}_n + \theta_n\right|} \to \infty \text{ a.e. in } \Omega,$$

because $\frac{1}{e_n}\widehat{u}_n + \theta_n \to \theta$ a.e. in Ω. Therefore we easily conclude using (2.11) that

$$\liminf_{n\to\infty} \widetilde{j_n^0}\Big(e_n(\tfrac{1}{e_n}\widehat{u}_n + \theta_n); -\tfrac{1}{e_n}\widehat{u}_n - \theta_n\Big) \ge j^\infty(\theta) \quad \text{a.e. in } \Omega.$$

Finally, by Fatou's lemma, permitted due to (2.3), we are led to the conclusion

$$\langle f, \theta\rangle_V \ge \Phi^\infty(\theta) + \int_\Omega j^\infty(\theta)\, d\Omega,$$

contrary to (2.12). This contradiction yields the boundedness of $\{u_R\}_{R>R_0}$. The proof of Proposition 2.1 is complete. □

The next result is related to the compactness property of $\{\chi_R\}_{R>R_0}$ in $L^1(\Omega;\mathbb{R}^s)$.

Proposition 2.2. *Let a triple $(u_R, \chi_R, \psi_R) \in V \times L^q(\Omega;\mathbb{R}^s) \times V^\star$ be a solution of (P_R). Then the set $\{\chi_R\}_{R>R_0}$ is weakly precompact in $L^1(\Omega;\mathbb{R}^s)$.*

Proof. According to the Dunford-Pettis theorem (Ekeland & Temam 1976) it suffices to show that for each $\varepsilon > 0$ a $\delta > 0$ can be determined such that for any $\omega \subset \Omega$ with $|\omega| < \delta$,

$$\int_\omega |\chi_R|\, d\Omega < \varepsilon, \qquad R > R_0. \tag{2.37}$$

Fix $\rho > 0$ and let $\eta \in \mathbb{R}^s$ be such that $|\eta| \le \rho$. Then one has $\chi_R\cdot(\eta - u_R) \le \widetilde{j_R^0}(u_R; \eta - u_R)$ from which, by virtue of (2.9) it results that

$$\chi_R \cdot \eta \le \chi_R \cdot u_R + \widetilde{\alpha}(\rho)(1 + |u_R|^r) \tag{2.38}$$

a.e. in Ω. Let us set $\eta \equiv \frac{\rho}{\sqrt{s}}(\operatorname{sgn}\chi_{R_1}, \ldots, \operatorname{sgn}\chi_{R_s})$, where χ_{R_i}, $i = 1, 2, \ldots, s$, are the components of χ_R and where $\operatorname{sgn} y = 1$ if $y > 0$, $\operatorname{sgn} y = 0$ if $y = 0$, and $\operatorname{sgn} y = -1$ if $y < 0$. It is not difficult to verify that $|\eta| \le \rho$ for almost all $x \in \Omega$ and that $\chi_R \cdot \eta \ge \frac{\rho}{\sqrt{s}}|\chi_R|$. Therefore by virtue of (2.38) one is led to the estimate

$$\frac{\rho}{\sqrt{s}}|\chi_R| \le \chi_R \cdot u_R + \widetilde{\alpha}(\rho)(1 + |u_R|^r).$$

Integrating this inequality over $\omega \subset \Omega$ yields

$$\int_\omega |\chi_R| \, d\Omega \leq \frac{\sqrt{s}}{\rho} \int_\omega \chi_R \cdot u_R \, d\Omega + \frac{\sqrt{s}}{\rho} \widetilde{\alpha}(\rho) \, |\omega| + \frac{\sqrt{s}}{\rho} \widetilde{\alpha}(\rho)(|\omega|)^{\frac{p-r}{p}} \|u_R\|^r_{L^p(\Omega)}. \tag{2.39}$$

Thus, from (2.32) one obtains

$$\begin{aligned} \int_\omega |\chi_R| \, d\Omega &\leq \frac{\sqrt{s}}{\rho} \int_\omega \chi_R \cdot u_R \, d\Omega + \frac{\sqrt{s}}{\rho} \widetilde{\alpha}(\rho) \, |\omega| + \frac{\sqrt{s}}{\rho} \widetilde{\alpha}(\rho)(|\omega|)^{\frac{p-r}{p}} \gamma^r \|u_R\|^r_V \\ &\leq \frac{\sqrt{s}}{\rho} \int_\omega \chi_R \cdot u_R \, d\Omega + \frac{\sqrt{s}}{\rho} \widetilde{\alpha}(\rho) \, |\omega| + \frac{\sqrt{s}}{\rho} \widetilde{\alpha}(\rho)(|\omega|)^{\frac{p-r}{p}} \gamma^r M^r \end{aligned} \tag{2.40}$$

($\|\cdot\|_{L^p(\Omega;\mathbb{R}^s)} \leq \gamma \|\cdot\|_V$). The next claim is that

$$\int_\omega \chi_R \cdot u_R \, d\Omega \leq C \tag{2.41}$$

for some positive constant C not depending on $\omega \subset \Omega$ and $R > R_0$. Indeed, from (2.10) one can easily deduce that

$$\chi_R \cdot u_R + \widehat{k} \, |u_R| \geq 0 \ \text{ a.e. in } \ \Omega.$$

Thus it follows

$$\int_\omega (\chi_R \cdot u_R + \widehat{k} \, |u_R|) \, d\Omega \leq \int_\Omega (\chi_R \cdot u_R + \widehat{k} \, |u_R|) \, d\Omega,$$

and consequently

$$\int_\omega \chi_R \cdot u_R \, d\Omega \leq \int_\Omega \chi_R \cdot u_R \, d\Omega + \widehat{k}_1 \|u_R\|_V.$$

But A maps bounded sets into bounded sets. Therefore, by means of (2.17), (2.29), (2.30) and (2.32),

$$\begin{aligned} &\int_\Omega \chi_R \cdot u_R \, d\Omega \leq \left\langle Au_R - f, -u_R \right\rangle_V + \Phi(0) - \Phi(u_R) \\ &\leq \|Au_R - f\|_{V^\star} \|u_R\|_V + \Phi(0) + a\|u_R\|_V + b \leq C, \quad C = \text{const}, \end{aligned} \tag{2.42}$$

and consequently, (2.41) easily follows. Further, from (2.40) and (2.41), for $\rho > 0$,

$$\int_\omega |\chi_R| \, d\Omega \leq \frac{\sqrt{s}}{\rho} C + \frac{\sqrt{s}}{\rho} \widetilde{\alpha}(\rho) \, |\omega| \ + \frac{\sqrt{s}}{\rho} \widetilde{\alpha}(\rho)(|\omega|)^{\frac{p-r}{p}} \gamma^r M^r. \tag{2.43}$$

This estimate is crucial for (2.37) to be obtained. Namely, let $\varepsilon > 0$. Fix $\rho > 0$ with

$$\frac{\sqrt{s}}{\rho} C < \frac{\varepsilon}{2} \tag{2.44}$$

and determine $\delta > 0$ small enough to fulfill

$$\frac{\sqrt{s}}{\rho} \widetilde{\alpha}(\rho) \, |\omega| + \frac{\sqrt{s}}{\rho} \widetilde{\alpha}(\rho)(|\omega|)^{\frac{p-r}{p}} \gamma^r M^r \leq \frac{\varepsilon}{2},$$

provided that $|\omega| < \delta$. Thus from (2.43) it follows that for any $\omega \subset \Omega$,

$$\int_\omega |\chi_R| \, d\Omega \le \varepsilon, \quad R > R_0, \tag{2.45}$$

whenever $|\omega| < \delta$. Finally, $\{\chi_R\}_{R>R_0}$ is equi-integrable and its precompactness in $L^1(\Omega; \mathbb{R}^s)$ has been proved (Ekeland & Temam 1976). □

2.3 Existence result

Now the main result will be formulated.

Theorem 2.1. *Let $A : V \to V^\star$ be a pseudo-monotone, bounded operator, $j : \mathbb{R}^s \to \mathbb{R}$ a locally Lipschitz function. Let $\Phi : V \to R \cup \{+\infty\}$ be a convex, lower semicontinuous function with $0 \in \operatorname{Dom}\Phi$, $g \in V^\star$ an element of $V^\star$. Suppose that (2.1)–(2.3) and (2.12) hold. Moreover, assume that the following hypothesis holds:*

(H) For any $v \in \operatorname{Dom}\Phi$ there exists a sequence of functions $\{\varepsilon_k\} \subset L^\infty(\Omega)$ with $0 \le \varepsilon_k \le 1$, $\{(1-\varepsilon_k)v\} \subset V \cap L^\infty(\Omega; \mathbb{R}^s)$ and $\widetilde{v}_k := (1-\varepsilon_k)v \to v$ strongly in V such that

$$\lim_{k\to\infty} \Phi(\widetilde{v}_k) = \Phi(v).$$

Then the variational-hemivariational inequality

$$\left\langle Au - f, v - u \right\rangle_V + \Phi(v) - \Phi(u) + \int_\Omega j^0(u; v-u) \, d\Omega \ge 0, \quad \forall\, v \in V \tag{2.46}$$

admits solutions.

Proof. The proof is divided into a sequence of steps.
Step 1. By Proposition 2.1 and Proposition 2.2 it follows that from the set $\{u_R, \chi_R, \psi_R\}_{R>R_0}$ of solutions of (P_R) a sequence $\{u_{R_n}, \chi_{R_n}, \psi_{R_n}\}$ can be extracted with $R_n \to \infty$ as $n \to \infty$ (for simplicity of notations it will be denoted by $\{u_n, \chi_n, \psi_n\}$), such that

$$\left\langle Au_n - f, v - u_n \right\rangle_V + \left\langle \psi_n, v - u_n \right\rangle_V + \int_\Omega \chi_n \cdot (v - u_n) \, d\Omega = 0, \quad \forall\, v \in V, \tag{2.47}$$

and

$$\left.\begin{array}{l} \chi_n \in \Gamma_n(u_n) \quad (\Gamma_n := \Gamma_{R_n}) \\ \psi_n \in \partial\Phi(u_n) \\ u_n \to u \quad \text{weakly in } V \\ \chi_n \to \chi \quad \text{weakly in } L^1(\Omega; \mathbb{R}^s). \end{array}\right\} \tag{2.48}$$

From (2.47) and (2.48) we get

$$\left\langle Au_n - f, v - u_n \right\rangle_V + \Phi(v) - \Phi(u_n) + \int_\Omega \widetilde{j}_n^0(u_n; v - u_n) \, d\Omega \ge 0, \quad \forall\, v \in V. \tag{2.49}$$

Hence by setting $v = 0$ we obtain

$$\Phi(u_n) \le \|Au_n - f\|_{V^\star} \|u_n\|_V + \Phi(0) + k\|u_n\|_V \le C, \quad C = \text{const},$$

which by (2.32), the boundedness of A and lower semicontinuity of Φ yields $\Phi(u) \in \mathbb{R}$, i.e. $u \in \operatorname{Dom}\Phi$.

Step 2. Now we prove that $\chi \in \partial j(u)$ a.e in Ω. Since V is compactly imbedded into $L^p(\Omega;\mathbb{R}^s)$, due to (2.48) one may suppose that

$$u_n \to u \quad \text{strongly in } L^p(\Omega;\mathbb{R}^s). \tag{2.50}$$

This implies that for a subsequence of $\{u_n\}$ (again denoted by the same symbol) one gets $u_n \to u$ a.e. in Ω. Thus Egoroff's theorem can be applied from which it follows that for any $\varepsilon > 0$ a subset $\omega \subset \Omega$ with $\operatorname{mes}\omega < \varepsilon$ can be determined such that $u_n \to u$ uniformly in $\Omega \setminus \omega$ with $u \in L^\infty(\Omega \setminus \omega;\mathbb{R}^s)$. Let $v \in L^\infty(\Omega \setminus \omega;\mathbb{R}^s)$ be an arbitrary function. From the estimate

$$\int_{\Omega\setminus\omega} \chi_n \cdot v \, d\Omega \leq \int_{\Omega\setminus\omega} \widetilde{j_n^0}(u_n;v)\, d\Omega = \int_{\Omega\setminus\omega} j^0(u_n;v)\, d\Omega, \quad \text{(for large } n\text{)}$$

(u_n remains bounded in $\Omega \setminus \omega$ as $n \to \infty$) combined with the weak convergence in $L^1(\Omega;\mathbb{R}^s)$ of χ_n to χ, (2.50) and with the upper semicontinuity of

$$L^\infty(\Omega \setminus \omega;\mathbb{R}^s) \ni u_n \longmapsto \int_{\Omega\setminus\omega} j^0(u_n;v)\, d\Omega$$

it follows

$$\int_{\Omega\setminus\omega} \chi \cdot v \, d\Omega \leq \int_{\Omega\setminus\omega} j^0(u;v)\, d\Omega, \quad \forall\, v \in L^\infty(\Omega \setminus \omega;\mathbb{R}^s).$$

But the last inequality amounts to saying that $\chi \in \partial j(u)$ a.e. in $\Omega \setminus \omega$. Since $|\omega| < \varepsilon$ and ε was chosen arbitrarily,

$$\chi \in \partial j(u) \quad \text{a.e. in } \Omega, \tag{2.51}$$

as claimed.

Step 3. Now it will be shown that

$$\limsup_{n\to\infty} \int_\Omega \widetilde{j_n^0}(u_n; v - u_n)\, d\Omega \leq \int_\Omega j^0(u; v-u)\, d\Omega \tag{2.52}$$

holds for any $v \in V \cap L^\infty(\Omega;\mathbb{R}^s)$. It can be supposed that $u_n \to u$ a.e. in Ω, since $u_n \to u$ in $L^p(\Omega;\mathbb{R}^s)$. Fix $v \in L^\infty(\Omega;\mathbb{R}^s)$ arbitrarily. In view of $\chi_n \in \Gamma_n(u_n)$, Eq. (2.9) implies

$$\widetilde{j_n^0}(u_n; v-u_n) \leq \widetilde{\alpha}(\|v\|_{L^\infty(\Omega;\mathbb{R}^s)})(1 + |u_n|^r). \tag{2.53}$$

From Egoroff's theorem it follows that for any $\varepsilon > 0$ a subset $\omega \subset \Omega$ with $\operatorname{mes}\omega < \varepsilon$ can be determined such that $u_n \to u$ uniformly in $\Omega \setminus \omega$. One can also suppose that ω is small enough to fulfill $\int_\omega \widetilde{\alpha}(\|v\|_{L^\infty(\Omega;\mathbb{R}^s)})(1+|u_n|^r)\, d\Omega \leq \varepsilon$, $n = 1,2,\dots$, and $\int_\omega \widetilde{\alpha}(\|v\|_{L^\infty(\Omega;\mathbb{R}^s)})(1+|u|^r)\, d\Omega \leq \varepsilon$. Hence

$$\begin{aligned} \int_\Omega \widetilde{j_n^0}(u_n; v-u_n)\, d\Omega &\leq \int_{\Omega\setminus\omega} \widetilde{j_n^0}(u_n; v-u_n)\, d\Omega + \varepsilon \\ &= \int_{\Omega\setminus\omega} j^0(u_n; v-u_n)\, d\Omega + \varepsilon \quad \text{(for large } n\text{)}, \end{aligned}$$

which by Fatou's lemma and upper semicontinuity of $j^0(\cdot\,;\cdot)$ yields

$$\limsup_{n\to\infty}\int_\Omega \widetilde{j^0_n}(u_n; v-u_n)\,d\Omega \le \int_\Omega j^0(u; v-u)\,d\Omega + 2\varepsilon.$$

By arbitrariness of $\varepsilon > 0$ one obtains (2.52), as required.
Step 4. Now we show that

$$\chi\cdot u \in L^1(\Omega) \tag{2.54}$$

$$\liminf_{n\to\infty}\int_\Omega \chi_n\cdot u_n\,d\Omega \ge \int_\Omega \chi\cdot u\,d\Omega. \tag{2.55}$$

For this purpose let $0\le\varepsilon_k\le 1$ be as stated in (H). Without loss of generality it can be assumed that $\widetilde{u}_k \to u$ a.e. in Ω. Since it is already known that $\chi\in\partial j(u)$, one can apply (2.3) to obtain $\chi\cdot(-u)\le j^0(u;-u)\le \widehat{k}\,|u|$. Hence

$$\chi\cdot\widetilde{u}_k = (1-\varepsilon_k)\chi\cdot u \ge -\widehat{k}\,|u|\,. \tag{2.56}$$

This implies that the sequence $\{\chi\cdot\widetilde{u}_k\}$ is bounded from below by integrable function and $\chi\cdot\widetilde{u}_k\to\chi\cdot u$ a.e. in Ω. On the other hand, one gets

$$\int_\Omega \chi_n\cdot(\widetilde{u}_k-u_n)\,d\Omega \le \int_\Omega \widetilde{j^0_n}(u_n;\widetilde{u}_k-u_n)\,d\Omega.$$

Thus

$$\int_\Omega \chi\cdot\widetilde{u}_k\,d\Omega - \liminf_{n\to\infty}\int_\Omega \chi_n\cdot u_n\,d\Omega \le \limsup_{n\to\infty}\int_\Omega \widetilde{j^0_n}(u_n;\widetilde{u}_k-u_n)\,d\Omega,$$

and due to (2.52) we are led to the estimate

$$\begin{aligned}\int_\Omega \chi\cdot\widetilde{u}_k\,d\Omega &\le \liminf_{n\to\infty}\int_\Omega \chi_n\cdot u_n\,d\Omega + \int_\Omega j^0(u;\widetilde{u}_k-u)\,d\Omega\\ &\le \liminf_{n\to\infty}\int_\Omega \chi_n\cdot u_n\,d\Omega + \int_\Omega j^0(u;-\varepsilon_k u)\,d\Omega\\ \le \liminf_{n\to\infty}&\int_\Omega \chi_n\cdot u_n\,d\Omega + \int_\Omega \varepsilon_k\widehat{k}\,|u|\,d\Omega \le C,\quad C=\text{const}\,.\end{aligned}$$

Thus by Fatou's lemma we are allowed to conclude that $\chi\cdot u\in L^1(\Omega)$, i.e. (2.54) holds. Taking into account that $\varepsilon_k\to 0$ as $k\to\infty$ a.e. in Ω (passing to a subsequence if necessary) we establish (2.55), as required.
Step 5. In this step it will be shown that

$$\limsup_{n\to\infty}\langle Au_n, u_n-u\rangle_V \le 0. \tag{2.57}$$

Taking into account (2.49) one has

$$\langle Au_n - f, u_n-\widetilde{u}_k\rangle_V \le \Phi(\widetilde{u}_k)-\Phi(u_n) + \int_\Omega \widetilde{j^0_n}(u_n;\widetilde{u}_k-u_n)\,d\Omega.$$

Since

$$\limsup_{n\to\infty}\langle Au_n - f, u_n - \widetilde{u}_k\rangle_V = \limsup_{n\to\infty}\langle Au_n, u_n - u\rangle_V + \langle B - f, u - \widetilde{u}_k\rangle_V,$$

we obtain

$$\limsup_{n\to\infty}\langle Au_n, u_n - u\rangle_V \le \Phi(\widetilde{u}_k) - \Phi(u) + \int_\Omega j^0(u;\widetilde{u}_k - u)\, d\Omega + \langle B - f, \widetilde{u}_k - u\rangle_V. \quad (2.58)$$

Here, because of the boundedness of A, we have supposed without loss of generality that for some $B \in V^\star$, $Au_n \to B$ weakly in $V^\star$. Now observe that $\widetilde{u}_k \to u$ a.e. in Ω, and

$$j^0(u;\widetilde{u}_k - u) = \varepsilon_k j^0(u;-u) \le \varepsilon_k \widehat{k}\,|u| \le \widehat{k}\,|u|,$$

thus Fatou's lemma ensures

$$\limsup_{k\to\infty}\int_\Omega j^0(u;\widetilde{u}_k - u)\, d\Omega \le 0. \quad (2.59)$$

On passing to the limit in (2.58) as $k \to \infty$ and taking into account (H), (2.52) and (2.59) allows to conclude (2.57), as claimed.
Step 6. Now it will be shown that (2.46) holds for any $v \in V$ with $v \in \operatorname{Dom}\Phi$ and $j^0(u; v - u) \in L^1(\Omega)$.
In view of (2.57) the pseudo-monotonicity of A yields

$$Au_n \to Au \text{ weakly in } V \text{ (i.e. } B = Au) \quad (2.60)$$

$$\langle Au_n, u_n\rangle_V \to \langle Au, u\rangle_V. \quad (2.61)$$

Let $v \in V \cap L^\infty(\Omega;\mathbb{R}^s)$ with $\Phi(v) < +\infty$ be chosen arbitrarily. Thus (2.49) combined with (2.52) and lower semicontinuity of Φ yields the assertion. Now let $j^0(u; v - u) \in L^1(\Omega)$ with $v \notin V \cap L^\infty(\Omega;\mathbb{R}^s)$ and $\Phi(v) < +\infty$. According to (H) there exists a sequence $\widetilde{v}_k = (1-\varepsilon_k)v$ such that $\{\widetilde{v}_k\} \subset V \cap L^\infty(\Omega;\mathbb{R}^s)$, $\widetilde{v}_k \to v$ strongly in V and $\Phi(\widetilde{v}_k) \to \Phi(v)$. Since, as already has been established,

$$\langle Au - f, \widetilde{v}_k - u\rangle_V + \Phi(\widetilde{v}_k) - \Phi(u) + \int_\Omega j^0(u;\widetilde{v}_k - u)\, d\Omega \ge 0,$$

so in order to show (2.46) it remains to deduce that

$$\limsup_{k\to\infty}\int_\Omega j^0(u;\widetilde{v}_k - u)\, d\Omega \le \int_\Omega j^0(u; v - u)\, d\Omega.$$

For this purpose let us observe that $\widetilde{v}_k - u = (1-\varepsilon_k)(v - u) + \varepsilon_k(-u)$ which combined with the convexity of $j^0(u;\cdot)$ yields the estimate

$$j^0(u;\widetilde{v}_k - u) \le (1-\varepsilon_k) j^0(u; v - u) + \varepsilon_k j^0(u;-u) \le \left|j^0(u; v - u)\right| + \widehat{k}\,|u|.$$

Thus Fatou's lemma implies the assertion.

Step 7. In the final step of the proof we are to consider the case in which $v \notin \operatorname{Dom}\Phi$ or $j^0(u;v-u) \notin L^1(\Omega)$. Recall that if $j^0(u;v-u) \notin L^1(\Omega)$ then according to the convention that $+\infty - \infty = +\infty$ we have

$$\int_\Omega j^0(u;v-u)\,d\Omega =$$

$$= \begin{cases} +\infty & \text{if } \int_\Omega [j^0(u;v-u)]^+\,d\Omega = +\infty \\ -\infty & \text{if } \int_\Omega [j^0(u;v-u)]^+\,d\Omega < +\infty \ \text{ and } \ \int_\Omega [j^0(u;v-u)]^-\,d\Omega = +\infty, \end{cases}$$

where the notation has been used: $r^+ := \max\{r,0\}$ and $r^- := \max\{-r,0\}$ for any $r \in \mathbb{R}$. Thus, if $v \notin \operatorname{Dom}\Phi$ or $\int_\Omega j^0(u;v-u)\,d\Omega = +\infty$ then $\Phi(v) + \int_\Omega j^0(u;v-u)\,d\Omega = +\infty$ and (2.46) holds immediately.

Finally, it remains to consider the case $v \in \operatorname{Dom}\Phi$ and $\int_\Omega j^0(u;v-u)\,d\Omega = -\infty$. It will be proved that in such a case we are led to the contradiction which means that whenever $v \in \operatorname{Dom}\Phi$, we get either $\int_\Omega j^0(u;v-u)\,d\Omega = +\infty$ or $j^0(u;v-u) \in L^1(\Omega)$ showing in view of the previous results that (2.46) holds.

According to the hypothesis (H) there exists a sequence $\widetilde{v}_k = (1-\varepsilon_k)v$ such that $\{\widetilde{v}_k\} \subset V \cap L^\infty(\Omega;\mathbb{R}^s)$, $\widetilde{v}_k \to v$ strongly in V and $\Phi(\widetilde{v}_k) \to \Phi(v)$. Since, as already has been established,

$$\langle Au - f, \widetilde{v}_k - u\rangle_V + \Phi(\widetilde{v}_k) - \Phi(u) + \int_\Omega j^0(u;\widetilde{v}_k - u)\,d\Omega \geq 0,$$

we get

$$\int_\Omega j^0(u;\widetilde{v}_k - u)\,d\Omega \geq \langle Au - f, -\widetilde{v}_k + u\rangle_V - \Phi(\widetilde{v}_k) + \Phi(u) \geq -C, \quad C = \text{const},$$

and consequently

$$\int_\Omega [j^0(u;\widetilde{v}_k - u)]^+\,d\Omega \geq \int_\Omega [j^0(u;\widetilde{v}_k - u)]^-\,d\Omega - C. \tag{2.62}$$

By the hypothesis, $\int_\Omega [j^0(u;v-u)]^-\,d\Omega = +\infty$ and $\int_\Omega [j^0(u;v-u)]^+\,d\Omega < +\infty$. Since

$$j^0(u;\widetilde{v}_k - u) \leq (1-\varepsilon_k)j^0(u;v-u) \ + \varepsilon_k j^0(u;-u) \leq (1-\varepsilon_k)j^0(u;v-u) + \widehat{k}\,|u|\,,$$

so we obtain

$$\int_\Omega [j^0(u;\widetilde{v}_k - u)]^+\,d\Omega \leq \int_\Omega [j^0(u;v-u)]^+\,d\Omega + \int_\Omega \widehat{k}\,|u|\,d\Omega \leq D, \quad D = \text{const},$$

which combined with (2.62) yields

$$\int_\Omega [j^0(u;\widetilde{v}_k - u)]^-\,d\Omega \leq C + D.$$

Thus the application of Fatou's lemma concludes

$$\int_\Omega [j^0(u;v-u)]^-\,d\Omega \leq C + D.$$

contrary to the assumption that $\int_\Omega j^0(u;v-u)\,d\Omega = -\infty$. This contradiction completes the proof of Theorem 2.1. □

2.4 Some remarks and comments

As it is well known, cf. (Naniewicz & Panagiotopoulos 1995), in case of the classical growth condition of the form

$$|\partial j(\xi)| \leq c(1 + |\xi|^{p-1}), \quad \xi \in \mathbb{R}^s, \tag{2.63}$$

the problem described by hemivariational inequality (2.46) admits a solution $u \in V$ and, moreover, there exist $\chi \in L^q(\Omega; \mathbb{R}^s)$, $1/p + 1/q = 1$, and $\psi \in V^\star$ such that

$$\begin{gathered} \chi \in \partial j(u) \text{ a.e. in } \Omega \quad \text{and} \quad \psi \in \partial\Phi(u), \\ f = Au + \psi + l_\chi, \end{gathered}$$

where $l_\chi \in V^\star$ is a linear continuous functional defined by

$$\langle l_\chi, v\rangle := \int_\Omega \chi \cdot v \, d\Omega, \quad v \in V. \tag{2.64}$$

Recall that the subdifferential $\partial\Phi(u) \subset V^\star$ in the sense of Convex Analysis (Ekeland & Temam 1976) is defined for $u \in \operatorname{Dom}\Phi$ by means of the formula

$$\Phi(v) - \Phi(u) \geq \langle \psi, v - u\rangle, \quad \forall v \in V.$$

Thus in case of (2.63) a statement that $u \in V$ is a solution of hemivariational inequality (2.46) is equivalent to

$$f - Au - l_\chi \in \partial\Phi(u). \tag{2.65}$$

The situation changes essentially when (2.63) is replaced by the unilateral growth condition (2.2). In such a case we have only ensured the $L^1(\Omega)$-regularity of χ and consequently, the corresponding functional l_χ (given by the formula (2.64)) is linear on its domain $\operatorname{Dom} l_\chi \supset L^\infty(\Omega; \mathbb{R}^s) \cap V$, but not necessarily continuous. It may happen that l_χ does not have the continuous extension onto the whole space V (l_χ is discontinuous). If it is the case, l_χ can be extended onto the whole space V as a function from V into $\mathbb{R} \cup \{+\infty, -\infty\}$ by setting

$$l_\chi(v) := \begin{cases} \int_\Omega \chi \cdot v \, d\Omega & \text{if } \chi \cdot v \in L^1(\Omega) \\ +\infty & \text{if } \int_\Omega [\chi \cdot v]^+ \, d\Omega = +\infty \\ -\infty & \text{if } \int_\Omega [\chi \cdot v]^+ \, d\Omega < +\infty \text{ and } \int_\Omega [\chi \cdot v]^- \, d\Omega = +\infty, \end{cases} \tag{2.66}$$

for each $v \in V$. Thus we deal with a functional $l_\chi : V \to \mathbb{R} \cup \{+\infty, -\infty\}$ which is discontinuous whenever $l_\chi(v) = +\infty$ or $l_\chi(v) = -\infty$ for at least one point of V. Notice that l_χ can be expressed as a difference of two convex lower semicontinuous proper functions $l_\chi^+(v) := \int_\Omega [\chi \cdot v]^+ \, d\Omega$ and $l_\chi^-(v) := \int_\Omega [\chi \cdot v]^- \, d\Omega$, $v \in V$, i.e.

$$l_\chi(v) = l_\chi^+(v) - l_\chi^-(v), \quad \forall v \in V. \tag{2.67}$$

Denote by $\mathcal{L}(V)$ the class of all linear densly defined functions $l : V \to \mathbb{R} \cup \{+\infty, -\infty\}$ which can be represented by a difference of two convex lower semicontinuous proper functions $l^+ : V \to \mathbb{R} \cup \{+\infty\}$ and $l^- : V \to \mathbb{R} \cup \{+\infty\}$, i.e. $l = l^+ - l^-$, with the convention that

$$l(v) := \begin{cases} l^+(v) - l^-(v) & \text{if } v \in \operatorname{Dom} l^+ \cap \operatorname{Dom} l^- \\ +\infty & \text{if } v \notin \operatorname{Dom} l^+ \\ -\infty & \text{if } v \in \operatorname{Dom} l^+ \text{ and } v \notin \operatorname{Dom} l^-. \end{cases} \tag{2.68}$$

For a convex, lower semicontinuous, proper function $\varphi : V \to \mathbb{R} \cup \{+\infty\}$ we introduce $\widehat{\partial}\varphi(u) \subset \mathcal{L}(V)$ as follows: if $u \notin \operatorname{Dom}\varphi$ then $\widehat{\partial}\varphi(u) = \emptyset$ while if $u \in \operatorname{Dom}\varphi$ then we set

$$l \in \widehat{\partial}\varphi(u) \quad \Leftrightarrow \quad l(u) \in \mathbb{R} \text{ and } \varphi(v) - \varphi(u) \geq l(v-u), \quad \forall v \in V. \tag{2.69}$$

The formal definition of $\widehat{\partial}\varphi(u)$ coincides with that of $\partial\varphi(u) \subset V^\star$ in the sense of convex analysis. However, $\widehat{\partial}\varphi(u)$ apart from containing elements of $\partial\varphi(u)$, may contain also some discontinuous linear functionals which will be called here discontinuous subgradients. Notice that if $u \in \operatorname{Int}(\operatorname{Dom}\Phi)$, where "Int" means "the interior", then by the Banach-Steinhaus theorem it follows that $\widehat{\partial}\varphi(u)$ and $\partial\varphi(u)$ coincide.

In the terminology of (Pallaschke & Rolewicz 1997) a function $l \in \widehat{\partial}\varphi(u)$ fulfilling (2.69) is said to be a $\mathcal{L}(V)$-subgradient of Φ at $u \in V$. We refer the reader to (Pallaschke & Rolewicz 1997) for the general abstract subdifferential theory.

As we shall see below, the notion of discontinuous subgradient is specially useful in describing some particular aspects of Theorem 2.1.

Theorem 2.2. *Assume all the hypotheses of Theorem 2.1. Then there exists $u \in V$ such that*

$$\langle Au - f, v - u \rangle_V + \Phi(v) - \Phi(u) + \int_\Omega j^0(u; v-u)\, d\Omega \geq 0, \quad \forall v \in V. \tag{2.70}$$

Moreover, there exists $\chi \in L^1(\Omega; \mathbb{R}^s)$, $\chi \in \partial j(u)$ a.e. in Ω, such that for l_χ defined by (2.66) *it follows that*

$$f - Au - l_\chi \in \widehat{\partial}\Phi(u). \tag{2.71}$$

Proof. Following the lines of the proof of Theorem 2.1 we can deduce that the inequality

$$\langle Au - f, v - u \rangle_V + \Phi(v) - \Phi(u) + \int_\Omega \chi \cdot (v - u)\, d\Omega \geq 0$$

holds for any $v \in V \cap L^\infty(\Omega; \mathbb{R}^s)$. It can be written equivalently as

$$\Phi(v) - \Phi(u) \geq \langle -Au + f, v - u \rangle_V - l_\chi(v - u), \quad \forall v \in V \cap L^\infty(\Omega; \mathbb{R}^s), \tag{2.72}$$

where $l_\chi(v-u) = \int_\Omega \chi \cdot (v-u)\, d\Omega$. It must be shown that this inequality holds for any $v \in V$. If $v \notin \operatorname{Dom}\Phi$ then there is nothing to prove because $\Phi(v) = +\infty$.

Let us consider the case $v \in \operatorname{Dom}\Phi$. If $\chi \cdot v \in L^1(\Omega)$ then $-\chi \cdot \widetilde{v}_k \geq -|\chi \cdot v|$ which by Fatou's lemma yields $\liminf_{k\to\infty} -l_\chi(\widetilde{v}_k) \geq -l_\chi(v)$. Thus in view of (H) the assertion follows ($\widetilde{v}_k$ has been taken as in the hypothesis (H)). If $\chi \cdot v \notin L^1(\Omega)$ then there is nothing to prove if $l_\chi(v) = +\infty$ while, as it will be shown, the case $l_\chi(v) = -\infty$ cannot happen.

Indeed, suppose that $l_\chi(v) = -\infty$, i.e. $\int_\Omega [\chi \cdot v]^+ \, d\Omega < +\infty$ and $\int_\Omega [\chi \cdot v]^- \, \Omega = +\infty$. Taking into account (2.72) we are led to $l_\chi(\widetilde{v}_k) \geq -C$ for a constant C. Hence

$$D \geq \int_\Omega [\chi \cdot \widetilde{v}_k]^+ \, d\Omega \geq \int_\Omega [\chi \cdot \widetilde{v}_k]^- \, d\Omega - C$$

for some $D = \text{const}$. But due to Fatou's lemma it yields

$$\int_\Omega [\chi \cdot v]^- \, d\Omega \leq C + D$$

contrary to $\int_\Omega [\chi \cdot v]^- \, d\Omega = +\infty$. This contradiction completes the proof. □

Remark 2.1. The hypothesis (H) of Theorem 2.1 involves the truncation result for vector-valued Sobolev spaces $V = W^{1,r}(\Omega; \mathbb{R}^s)$ formulated as

Theorem 2.3 *((Naniewicz 1997)). For each $v \in W^{1,r}(\Omega; \mathbb{R}^s)$, $r \geq 1$, there exists a sequence of functions $\{\varepsilon_n\} \subset L^\infty(\Omega)$ with $0 \leq \varepsilon_n \leq 1$ such that*

$$\begin{aligned} &\{(1-\varepsilon_n)v\} \subset W^{1,r}(\Omega; \mathbb{R}^s) \cap L^\infty(\Omega; \mathbb{R}^s) \\ &(1-\varepsilon_n)v \to v \quad \textit{strongly in } W^{1,r}(\Omega; \mathbb{R}^s). \end{aligned} \tag{2.73}$$

In the case of scalar valued Sobolev spaces $W^{m,r}(\Omega)$, $r \geq 1$, $m \geq 1$, such truncation procedure is admissible due to famous result of (Hedberg 1978).

3 Semicoercive hemivariational inequalities: A minimax approach

3.1 Statement of the problem and formulation of the main result

Let us consider the case in which $A : V \to V^\star$ is generated by a bilinear, bounded, symmetric form $a : V \times V \to \mathbb{R}$, i.e. $\langle Au, v \rangle = a(u, v)\ \forall\, u, v \in V$ and $\Psi \equiv 0$ on V.

Now the problem is to find $u \in V$ such that

$$(P) \qquad \langle Au - f, v - u \rangle_V + \int_\Omega j^0(x, u(x); v(x))dx \geq 0, \quad \forall v \in V.$$

Our hypotheses on the data in (P) are the following:

$(H1)$ $a : V \times V \to \mathbb{R}$ is a bilinear, bounded, symmetric form satisfying the semicoercivity condition

$$a(v, v) \geq \alpha \|\widehat{v}\|_V^2, \quad \forall v = \widehat{v} + \theta \in V,\ \widehat{v} \in \widehat{V},\ \theta \in V_0,$$

with a constant $\alpha > 0$;

$(H2)$ $j : \Omega \times \mathbb{R}^s \to \mathbb{R}$ is a Carathéodory function which satisfies:

(i) $j(x, \cdot) : \mathbb{R}^s \to \mathbb{R}$ is locally Lipschitz, a.e. $x \in \Omega$, and $j(\cdot, 0) \in L^1(\Omega)$;

(*ii*) there exists a constant $c > 0$ such that

$$|w| \leq c(1 + |\xi|^{p-1}), \ \forall w \in \partial j(x, \xi), \quad \text{a.e } x \in \Omega, \ \forall \xi \in \mathbb{R}^s;$$

(*iii*) there exist constants $\mu > 2$, $a_1 \geq 0$, $a_2 \geq 0$ and $1 \leq \sigma < 2$ such that

$$\mu j(x, \xi) - j^0(x, \xi; \xi) \geq -a_1|\xi|^\sigma - a_2, \quad \text{a.e. } x \in \Omega, \ \forall \xi \in \mathbb{R}^s;$$

(*iv*) $\int_\Omega j(x, 0)dx \leq 0$;

(*v*) there exist constants $c_0 > 0$, $c_1 > 0$ such that

$$j(x, \xi) \geq c_0|\xi|^2 - c_1|\xi|^p, \quad \forall \xi \in \mathbb{R}^s, \ \text{a.e. } x \in \Omega;$$

(*vi*) there exists $v_0 \in V \setminus \{0\}$ such that

$$\liminf_{s \to +\infty} s^{-\sigma} \int_\Omega j(x, sv_0(x))\, dx < \frac{a_1}{\sigma - \mu} \int_\Omega |v_0(x)|^\sigma\, dx;$$

(H_3) $f \in V^*$ fulfills the compatibility condition

$$\langle f, \theta \rangle_V < \int_\Omega j^\infty(x, \theta(x))\, dx, \quad \forall \theta \in V_0, \ \|\theta\|_V = 1,$$

where

$$j^\infty(x, \xi) = \liminf_{\substack{z \to \xi \\ t \to +\infty}} [-j^0(x, tz; -z)], \quad x \in \Omega, \ \xi \in \mathbb{R}^s,$$

is the recession function related to j, and the value $+\infty$ for the integral is admitted.

The notation $j^0(x, \cdot; \cdot)$ (in (P) and (H_2)) stands for the generalized directional derivative in the sense of Clarke ((Clarke 1983), p. 25), namely

$$j^0(x, \xi; z) = \limsup_{\substack{w \in \mathbb{R}^s, \ w \to \xi \\ t \in \mathbb{R}, \ t \downarrow 0}} \frac{1}{t}[j(x, w + tz) - j(x, w)], \quad \text{a.e. } x \in \Omega, \ \forall \xi, z \in \mathbb{R}^s,$$

while $\partial j(x, \cdot) \subset \mathbb{R}^s$ denotes the generalized gradient in the sense of Clarke ((Clarke 1983), p. 27), that is

$$\partial j(x, \xi) = \{w \in \mathbb{R}^s : \ w \cdot z \leq j^0(x, \xi; z), \ \forall z \in \mathbb{R}^s\}, \ \text{a.e. } x \in \Omega, \ \forall \xi \in \mathbb{R}^s.$$

Our nonsmooth variational approach will be applied to the locally Lipschitz functional functional $\Psi : V \to \mathbb{R}$ given by

$$\Psi(v) = \frac{1}{2}a(v, v) + \int_\Omega j(x, v(x))dx - \langle f, v \rangle_V \quad \forall v \in V, \tag{3.1}$$

which is associated to problem (P). The functional Ψ in (3.1) is well defined because $j(\cdot, v(\cdot)) \in L^1(\Omega)$ for every $v \in V$ as can be seen from Lebourg's mean value theorem and assumption (H_2) (*i*), (*ii*) as follows

$$|j(x, v(x))| \leq c(|v(x)| + |v(x)|^p) + |j(x, 0)|, \quad \forall v \in V, \ \text{a.e. } x \in \Omega. \tag{3.2}$$

Since $V \subset L^p(\Omega;\mathbb{R}^s)$, from (3.2) and (H_2) (i) we obtain $j(\cdot, v(\cdot)) \in L^1(\Omega)$. Moreover, the function Ψ defined in (3.1) is Lipschitz continuous on the bounded subsets of V. This property is a direct consequence of the growth condition (H_2) (iii). Therefore it makes sense to consider the generalized gradient $\partial\Psi(u)$ at an arbitrary $u \in V$. Using the calculus with generalized gradients, including Aubin-Clarke's theorem (Clarke 1983), we find that

$$\partial\Psi(u) \subset a(u,\cdot) + \int_\Omega \partial j(x, v(x))\, dx - f, \quad \forall u \in V, \tag{3.3}$$

where the integral term in (3.3) has the meaning in (Clarke 1983), p. 83.

For a later use we recall the Mountain Pass Theorem for locally Lipschitz functionals of (Chang 1981).

Theorem 3.1 ((Chang 1981)). *Let X be a reflexive Banach space and let $F : X \to \mathbb{R}$ be a locally Lipschitz functional satisfying the Palais-Smale condition, i.e. every sequence $\{u_n\}$ in X such that $F(u_n)$ is bounded and there exists a sequence $\{w_n\}$ in X^* with $w_n \in \partial F(u_n)$ converging to 0, contains a strongly convergent subsequence. Assume that*

(a) $F(0) \le 0$ and there exist positive constants γ, ρ with $F(v) \ge \gamma$ if $\|v\| = \rho$,

(b) there is an element $e \in X$ such that $\|e\| \ge \rho$ and $F(e) \le 0$.

Then the number

$$c := \inf\{\max_{0\le t\le 1} F(g(t)) : \ g \in C([0,1],V), \ g(0) = 0, \ g(3.1) = e\}$$

is a critical value of F, which means that there exits a critical point $u \in X$ of F (i.e., $0 \in \partial F(u)$) with $F(u) = c$. In addition, one has $c \ge \gamma$.

Our main result in studying problem (P) is the following theorem.

Theorem 3.2. *Assume that conditions (H_1)-(H_3) are fulfilled. Then there exists a constant $B > 0$ such that if $f \in V^*$ is so that $\|f\|_{V^*} \le B$, problem (P) admits a nontrivial solution $u_f \in V$. Moreover, there exist constants $\beta > 0$ and $t_0 > 0$ independent on $f \in V^*$ such that the following a priori estimate for the solution u_f holds*

$$\beta \le \frac{1}{2}a(u_f, u_f) + \int_\Omega j(x, u_f(x))\, dx - \langle f, u\rangle_V$$

$$\le \max_{0\le t\le t_0} \left(\frac{1}{2}t^2 a(v_0, v_0) + \int_\Omega j(x, tv_0(x))\, dx - t\langle f, v\rangle_V\right), \tag{3.4}$$

whenever $f \in V^*$ satisfies $\|f\|_{V^*} \le B$.

Remark 3.1. Theorem 3.2 provides a minimax characterization of the energy $\Psi(u_f)$ corresponding to the solution u_f, precisely

$$\frac{1}{2}a(u_f, u_f) + \int_\Omega j(x, u_f(x))\, dx - \langle f, u\rangle_V$$

$$= \inf\{\max_{0\le t\le 1} \Psi(g(t)) \ : \ g \in C([0,1],V),\ g(0)=0,\ g(3.1)=t_0 v_0\},$$

for $v_0 \in V \setminus \{0\}$ in (H_2) (vi) and any $t_0 > 0$ sufficiently large (see the proof of Theorem 3.2).

The rest of the paper is organized as follows. Section 2 is devoted to the verification of Palais-Smale condition for the energy functional Ψ in (3.1). Section 3 contains the proof of Theorem 3.2. Section 4 deals with an application of Theorem 3.2 to an elliptic Dirichlet problem at resonance and with discontinuous nonlinearities.

3.2 Proof of Palais-Smale condition

The Palais-Smale condition in the sense of (Chang 1981) for the functional Ψ in (3.1) is proved in the following result.

Proposition 3.1. *Under assumptions (H_1), (H_2) (i)-(iii) and (H_3), the locally Lipschitz functional $\Psi : V \to \mathbb{R}$ in* (3.1) *satisfies the Palais-Smale condition in the sense of (Chang 1981).*

Proof. To check the Palais-Smale condition in the sense of (Chang 1981) for the function Ψ in (3.1), we have to show that every sequence $\{u_n\} \subset V$ satisfying the properties

$$|\Psi(u_n)| \le M, \quad \forall n \ge 1, \tag{3.5}$$

with a constant $M > 0$, and

$$\lambda(u_n) := \min_{w \in \partial\Psi(u_n)} \|w\| \to 0 \quad \text{as } n \to \infty, \tag{3.6}$$

contains a (strongly) convergent subsequence in V. Relation (3.6) yields the existence of $w_n \in \partial I(u_n)$ such that $w_n \to 0$ in V as $n \to \infty$. Taking into account inclusion (3.3) leads to

$$\begin{aligned} a(u_n, v-u_n) + \int_\Omega j^0(x, u_n(x); v(x)-u_n(x))\,dx - \langle f, u_n\rangle_V \\ \ge -\varepsilon_n \|v-u_n\|_V, \quad \forall v \in V, \end{aligned} \tag{3.7}$$

for some sequence $\{\varepsilon_n\} \subset \mathbb{R}^+$ with $\varepsilon_n \to 0$ as $n \to \infty$. Setting $v = 2u_n$ in (3.7) ensures that

$$\varepsilon_n \|u_n\|_V \ge -a(u_n,u_n) - \int_\Omega j^0(x,u_n(x);u_n(x))\,dx + \langle f, u_n\rangle_V \quad \forall n \ge 1. \tag{3.8}$$

Combining (3.5) and (3.8) we infer that

$$M + \frac{1}{\mu}\varepsilon_n \|u_n\|_V \ge \left(\frac{1}{2} - \frac{1}{\mu}\right) a(u_n,u_n)$$

$$+\int_\Omega \left(j(x,u_n(x)) - \frac{1}{\mu} j^0(x,u_n(x);u_n(x))\right) dx + \left(\frac{1}{\mu} - 1\right) \langle f, u_n\rangle_V,$$

where $\mu > 2$ is the constant introduced by assumption (H_2) (iii). On the basis of hypothesis (H_2) (iii) we derive that

$$M + \frac{1}{\mu}\varepsilon_n \|u_n\|_V \geq \left(\frac{1}{2} - \frac{1}{\mu}\right) a(u_n, u_n)$$

$$+\frac{1}{\mu}\left(-a_1\|u\|^\sigma_{L^\sigma(\Omega;\mathbb{R}^s)} - a_2|\Omega|\right) + \left(\frac{1}{\mu} - 1\right)\|f\|_{V^*}\|u_n\|_V. \tag{3.9}$$

The notation $|\Omega|$ stands for the (Lebesgue) measure of Ω in $\mathbb{R}^m$. According to the continuous embeddings $V \subset L^p(\Omega;\mathbb{R}^s) \subset L^\sigma(\Omega;\mathbb{R}^s)$, there is a constant $C_\sigma(\Omega) > 0$ such that

$$\|v\|_{L^\sigma(\Omega;\mathbb{R}^s)} \leq C_\sigma(\Omega)\|v\|_V, \quad \forall v \in V.$$

Due to the direct sum decomposition $V = \widehat{V} \oplus V_0$, the element $u_n \in V$ is expressed uniquely as follows

$$u_n = \widehat{u}_n + e_n\theta_n, \quad \text{with } \widehat{u}_n \in \widehat{V},\ \theta_n \in V_0,\ \|\theta_n\|_V = 1,\ e_n \in \mathbb{R},\ e_n \geq 0. \tag{3.10}$$

Relation (10) and assumption (H_1) enable us to deduce from (3.9) the estimate

$$M + \frac{1}{\mu}\varepsilon_n\|\widehat{u}_n\|_V + \frac{1}{\mu}\varepsilon_n e_n \geq \left(\frac{1}{2} - \frac{1}{\mu}\right)\alpha\|\widehat{u}_n\|^2_V - b(\|\widehat{u}_n\|^\sigma_V + e^\sigma_n + 1)$$

$$+\left(\frac{1}{\mu} - 1\right)\|f\|_{V^*}(\|\widehat{u}_n\|_V + e_n), \quad \forall n \geq 1, \tag{3.11}$$

for a constant $b > 0$.

First, we show that $\{u_n\}$ is bounded in V. Arguing by contradiction, suppose that for a relabelled subsequence one has

$$\|u_n\|_V \to +\infty \quad \text{as} \quad n \to +\infty. \tag{3.12}$$

By (3.11), (3.12) and $\sigma < 2$ we see that $e_n \neq 0$ for n sufficiently large. Dividing (3.11) by e_n we obtain

$$\left(1 - \frac{1}{\mu}\right)\|f\|_{V^*} + \frac{1}{\mu}\varepsilon_n + be^{\sigma-1}_n$$

$$\geq \left[\left(\frac{1}{2} - \frac{1}{\mu}\right)\alpha\|\widehat{u}_n\|_V - b\|\widehat{u}_n\|^{\sigma-1}_V - \frac{M+b}{\|\widehat{u}_n\|_V} + \left(\frac{1}{\mu} - 1\right)\|f\|_{V^*} - \frac{1}{\mu}\varepsilon_n\right]\frac{\|\widehat{u}_n\|_V}{e_n}. \tag{3.13}$$

Inequality (3.13) implies that $\{e_n\}$ is unbounded, so

$$e_n \to +\infty \quad \text{as} \quad n \to +\infty, \tag{3.14}$$

passing eventually to a renamed subsequence. Note that (3.14) holds because otherwise, in view of (3.12), $\|\widehat{u}_n\|_V \to +\infty$ and inequality (3.13) would lead to a contradiction.

We claim that

$$\frac{1}{e_n}\,\widehat{u}_n \to 0 \ \text{ in } V \text{ as } n \to +\infty. \tag{3.15}$$

If the sequence $\{\widehat{u}_n\}$ is bounded, this follows from (3.14). If $\|\widehat{u}_n\|_V \to +\infty$ as $n \to +\infty$ along a subsequence, suppose that (3.15) does not hold. Then one finds a constant $k_0 > 0$ such that $e_n \le k_0\|\widehat{u}_n\|_V$, $\forall n \ge 1$. It turns out from (3.13) that a constant $k > 0$ exists such that

$$\left(1-\frac{1}{\mu}\right)\|f\|_{V^*} + \frac{1}{\mu}\,\varepsilon_n$$

$$\ge \left[\left(\frac{1}{2}-\frac{1}{\mu}\right)\alpha\|\widehat{u}_n\|_V - \frac{M+b}{\|\widehat{u}_n\|_V} - k\|\widehat{u}_n\|_V^{\sigma-1} + \left(\frac{1}{\mu}-1\right)\|f\|_{V^*} - \frac{1}{\mu}\,\varepsilon_n\right]\frac{\|\widehat{u}_n\|_V}{e_n}.$$

Pass to the limit as $n \to +\infty$. Using $\|\widehat{u}_n\|_V \to +\infty$ and the fact that $\frac{\|\widehat{u}_n\|_V}{e_n}$ was supposed not tending to 0, the inequality above in conjunction with $\sigma < 2$ leads to a contradiction. Consequently, the claim in (3.15) holds true.

Setting $v = 0$ in (3.7) yields

$$\varepsilon_n\|u_n\|_V \ge a(u_n, u_n) - \int_\Omega j^0(x, u_n(x); -u_n(x))\,dx - \langle f, v\rangle_V, \quad \forall n \ge 1. \tag{3.16}$$

Knowing from (3.10) that

$$u_n = e_n\left(\frac{1}{e_n}\,\widehat{u}_n + \theta_n\right),$$

by (H_1) we can write (3.16) as follows

$$\langle f, \theta_n\rangle_V \ge (\alpha\|\widehat{u}_n\|_V - \|f\|_{V^*} - \varepsilon_n)\frac{1}{e_n}\|\widehat{u}_n\|_V$$

$$- \int_\Omega j^0(x, e_n\left(\frac{\widehat{u}_n}{e_n} + \theta_n\right); -\frac{\widehat{u}_n}{e_n} - \theta_n)\,dx - \varepsilon_n. \tag{3.17}$$

We show that

$$\liminf_{n\to+\infty}(\alpha\|\widehat{u}_n\|_V - \|f\|_{V^*} - \varepsilon_n)\frac{1}{e_n}\|\widehat{u}_n\|_V \ge 0. \tag{3.18}$$

If $\{\widehat{u}_n\}$ is bounded, then (3.14) implies directly (3.18). If $\{\widehat{u}_n\}$ is unbounded, then considering eventually a subsequence, we have that $\alpha\|\widehat{u}_n\|_V - \|f\|_{V^*} - \varepsilon_n$ is positive, thus (3.18) is satisfied.

By (10) and since the space V_0 is finite dimensional, we may assume that

$$\theta_n \to \theta \ \text{ in } V_0 \text{ as } n \to +\infty, \text{ with } \theta \in V_0,\ \|\theta\| = 1. \tag{3.19}$$

Passing to lim inf in (3.17) and using (3.18), (3.19) and $\varepsilon_n \to 0$, it follows that

$$\langle f, \theta\rangle_V \ge \liminf_{n\to+\infty}\int_\Omega -j^0(x, e_n\left(\frac{\widehat{u}_n}{e_n} + \theta_n\right); -\frac{\widehat{u}_n}{e_n} - \theta_n)\,dx. \tag{3.20}$$

Notice that, thanks to the growth condition (H_2) (iii), Fatou's lemma can be applied to the integral in (3.20). Then (3.20) implies

$$\langle f, \theta \rangle_V \geq \int_\Omega \liminf_{n \to +\infty} \left[-j^0\left(x, e_n \left(\frac{\widehat{u}_n}{e_n} + \theta_n \right); -\frac{\widehat{u}_n}{e_n} - \theta_n \right) \right] dx. \tag{3.21}$$

The compactness of the embedding $V \subset L^p(\Omega; \mathbb{R}^s)$ and (3.15) permit to extract a relabelled subsequence such that

$$\frac{1}{e_n} \widehat{u}_n(x) \to 0 \text{ in } \mathbb{R}^s \text{ for a.e. } x \in \Omega$$

and

$$\theta_n(x) \to \theta(x) \text{ in } \mathbb{R}^s \text{ for a.e. } x \in \Omega.$$

Then (3.21) in conjunction with (3.14) and the definition of the recession function j^∞ given in (H_3) shows that

$$\langle f, \theta \rangle_V \geq \int_\Omega j^\infty(x, \theta(x))\, dx.$$

Since this contradicts assumption (H_3), the boundedness of the sequence $\{u_n\}$ in V is established.

Taking into account the reflexivity of V and the compact embedding $V \subset L^p(\Omega; \mathbb{R}^s)$, we may assume that the sequence $\{u_n\}$ contains a subsequence, denoted again by $\{u_n\}$, such that

$$\begin{aligned} u_n &\rightharpoonup u \text{ weakly in } V, \\ u_n &\to u \text{ strongly in } L^p(\Omega; \mathbb{R}^s), \\ u_n(x) &\to u(x) \text{ for a.e. } x \in \Omega, \end{aligned}$$

for some $u \in V$. Putting $v = u$ in (3.7) yields

$$\begin{aligned} a(u_n, u_n) \leq a(u_n, u) &+ \int_\Omega j^0(x, u_n(x); u(x) - u_n(x))\, dx \\ &- \langle f, u - u_n \rangle_V + \varepsilon_n \|u - u_n\|_V, \quad \forall n \geq 1. \end{aligned} \tag{3.22}$$

Passing to lim sup in (3.22) it results

$$\limsup_{n \to +\infty} a(u_n, u_n) \leq a(u, u). \tag{3.23}$$

According to (3.10) we have

$$u_n = \widehat{u}_n + y_n, \quad u = \widehat{u} + y, \quad \text{with } \widehat{u}_n, \widehat{u} \in \widehat{V},\ y_n, y \in V_0.$$

Since V_0 is finite dimensional it follows that, up to a subsequence, $y_n \to y$ strongly in V. Using this and $\widehat{u}_n \rightharpoonup \widehat{u}$ weakly in V, from (3.23) we deduce that

$$\limsup_{n \to +\infty} a(\widehat{u}_n, \widehat{u}_n) \leq a(\widehat{u}, \widehat{u}).$$

The semicoercivity assumption (H_1) guarantees that $a(\cdot, \cdot)^{\frac{1}{2}}$ is a norm on $\widehat{V}$ equivalent to $\| \cdot \|_V$ on $\widehat{V}$. Consequently, we obtain that $\widehat{u}_n \to \widehat{u}$ strongly in V, thereby $u_n \to u$ strongly in V as $n \to +\infty$. The proof of Palais-Smale condition for the functional Ψ in (3.1) is complete. □

3.3 Proof of Theorem 3.2

We apply Theorem 3.1 for the space $X = V$ and the function $I = \Psi$ introduced in (3.1). The Palais-Smale condition for the locally Lipschitz functional $\Psi : V \to \mathbb{R}$ has been proven in Proposition 3.1. By assumption (H_2) (iv) we have that $I(0) \leq 0$.

Let us check that there exist constants $B > 0$, $\beta > 0$ and $\rho > 0$ such that

$$\Psi(v) \geq \beta, \quad \forall v \in V, \ \|v\|_V = \rho, \tag{3.24}$$

whenever $f \in V^*$ in (3.1) satisfies $\|f\|_{V^*} \leq B$.

Towards this, by assumptions (H_1) and (H_2) (v) we obtain

$$\Psi(v) \geq \frac{1}{2}\alpha\|\widehat{v}\|_V^2 + c_0\|\widehat{v}+\theta\|^2_{L^2(\Omega;\mathbb{R}^s)} - c_1\|\widehat{v}+\theta\|^p_{L^p(\Omega;\mathbb{R}^s)} - \|f\|_{V^*}(\|\widehat{v}\|_V + \|\theta\|_V),$$

for all $v = \widehat{v} + \theta \in V$, with $\widehat{v} \in \widehat{V}$, $\theta \in V_0$. The convexity of the function $t \mapsto t^p$ $(t \in \mathbb{R}_+)$ and Young's inequality show that

$$\Psi(v) \geq \frac{1}{2}\alpha\|\widehat{v}\|_V^2 + c_0\|\widehat{v}+\theta\|^2_{L^2(\Omega;\mathbb{R}^s)} - 2^{p-1}c_1\left(\|\widehat{v}\|^p_{L^p(\Omega;\mathbb{R}^s)} + \|\theta\|^p_{L^p(\Omega;\mathbb{R}^s)}\right)$$
$$-\frac{1}{p}(\|\widehat{v}\|_V^p + \|\theta\|_V^p) - \frac{2(p-1)}{p}\|f\|_{V^*}^{\frac{p}{p-1}}, \quad \forall v = \widehat{v}+\theta \in V. \tag{3.25}$$

The continuous embeddings $V \subset L^2(\Omega;\mathbb{R}^s)$ and $V \subset L^p(\Omega;\mathbb{R}^s)$ ensure the existence of constants (depending on the domain Ω) $C_2(\Omega) > 0$ and $C_p(\Omega) > 0$ such that

$$\|v\|_V \leq C_2(\Omega)\|v\|_{L^2(\Omega;\mathbb{R}^s)} \text{ and } \|v\|_V \leq C_p(\Omega)\|v\|_{L^p(\Omega;\mathbb{R}^s)}, \quad \forall v \in V. \tag{3.26}$$

Notice that for every $\varepsilon > 0$ we have

$$\|\widehat{v}+\theta\|^2_{L^2(\Omega;\mathbb{R}^s)} \geq -\varepsilon\|\widehat{v}\|^2_{L^2(\Omega;\mathbb{R}^s)} + \frac{\varepsilon}{1+\varepsilon}\|\theta\|^2_{L^2(\Omega;\mathbb{R}^s)}.$$

Choosing

$$0 < \varepsilon < \frac{\alpha}{2c_0(C_2(\Omega))^2} \tag{3.27}$$

inequality (3.25) yields

$$\Psi(v) \geq \frac{1}{2}\alpha\|\widehat{v}\|_V^2 + c_0\left(-\varepsilon\|\widehat{v}\|^2_{L^2(\Omega;\mathbb{R}^s)} + \frac{\varepsilon}{1+\varepsilon}\|\theta\|^2_{L^2(\Omega;\mathbb{R}^s)}\right)$$
$$-2^{p-1}c_1\left(\|\widehat{v}\|^p_{L^p(\Omega;\mathbb{R}^s)} + \|\theta\|^p_{L^p(\Omega;\mathbb{R}^s)}\right)$$
$$-\frac{1}{p}(\|\widehat{v}\|_V^p + \|\theta\|_V^p) - \frac{2(p-1)}{p}\|f\|_{V^*}^{\frac{p}{p-1}}, \quad \forall v = \widehat{v}+\theta \in V. \tag{3.28}$$

Combining (3.26) and (3.28) we derive the estimate

$$\Psi(v) \geq \frac{1}{2}\alpha\|\widehat{v}\|_V^2 - c_0\varepsilon(C_2(\Omega))^2\|\widehat{v}\|_V^2 - \left[2^{p-1}c_1(C_p(\Omega))^p + \frac{1}{p}\right]\|\widehat{v}\|_V^p$$

$$-\left[2^{p-1}c_1(C_p(\Omega))^p + \frac{1}{p}\right]\|\theta\|_V^p + \frac{c_0\varepsilon}{1+\varepsilon}\|\theta\|^2_{L^2(\Omega;\mathbb{R}^s)}$$
$$-\frac{2(p-1)}{p}\|f\|_{V^*}^{\frac{p}{p-1}}, \quad \forall v = \widehat{v} + \theta \in V. \tag{3.29}$$

Taking into account that V_0 is a finite dimensional subspace of V, there exists a constant $\overline{c} > 0$ such that

$$\|\theta\|^2_{L^2(\Omega;\mathbb{R}^s)} \geq \overline{c}\|\theta\|_V, \quad \forall \theta \in V_0.$$

This enables us to deduce from (3.29) that

$$\Psi(v) \geq \left[\frac{1}{2}\alpha - c_0\varepsilon(C_2(\Omega))^2 - \left(2^{p-1}c_1(C_p(\Omega))^p + \frac{1}{p}\right)\|\widehat{v}\|_V^{p-2}\right]\|\widehat{v}\|_V^2$$
$$+\left[\frac{c_0\overline{c}^2\varepsilon}{1+\varepsilon} - \left(2^{p-1}c_1(C_p(\Omega))^p + \frac{1}{p}\right)\|\theta\|_V^{p-2}\right]\|\theta\|_V^2$$
$$-\frac{2(p-1)}{p}\|f\|_{V^*}^{\frac{p}{p-1}}, \quad \forall v = \widehat{v} + \theta \in V. \tag{3.30}$$

We are now in a position to choose the constants $B > 0$, $\beta > 0$ and $\rho > 0$ for achieving (3.24). First, on the basis of (3.27) and since $p > 2$, we may fix a number $\rho > 0$ such that

$$A_1 := \frac{1}{2}\alpha - c_0\varepsilon(C_2(\Omega))^2 - \left(2^{p-1}c_1(C_p(\Omega))^p + \frac{1}{p}\right)\rho^{p-2} > 0$$

and

$$A_2 := \frac{c_0\overline{c}^2\varepsilon}{1+\varepsilon} - \left(2^{p-1}c_1(C_p(\Omega))^p + \frac{1}{p}\right)\rho^{p-2} > 0.$$

Then we choose $B > 0$ to satisfy

$$\beta := \min\{A_1, A_2\}\rho^2 - \frac{2(p-1)}{p}B^{\frac{p}{p-1}} > 0.$$

With these choices, inequality (3.30) leads to

$$\Psi(v) \geq \min\{A_1, A_2\}(\|\widehat{v}\|_V^2 + \|\theta\|_V^2) - \frac{2(p-1)}{p}B^{\frac{p}{p-1}} = \beta,$$

for all $v = \widehat{v} + \theta \in V$, with $\|v\|_V = (\|\widehat{v}\|_V^2 + \|\theta\|_V^2)^{\frac{1}{2}} = \rho$, which is just (3.24). Thus it is shown that the functional Ψ in (3.1) verifies the assumption (a) of Theorem 3.1 whenever $\|f\|_{V^*} \leq B$.

We prove now the following property

$$\lim_{t\to+\infty} \Psi(tv_0) = -\infty, \tag{3.31}$$

where the element $v_0 \in V$ is given in assumption (H_2) (vi). To this end, we start from the formula

$$\frac{d}{dt}(t^{-\mu}j(x,ty)) = -t^{-\mu-1}[\mu j(x,ty) - j_y'(x,ty)\cdot(ty)] \text{ a.e. } y \in \mathbb{R}^s, \ \forall t > 0. \tag{3.32}$$

Here the notation j'_y designates the partial derivative of j with respect to the second variable $y \in \mathbb{R}^s$, while $\mu > 2$ is the constant in (H_2) (iii). Equality (3.31) is valid because the function $j(x,\cdot): \mathbb{R}^s \to \mathbb{R}$ is a.e. differentiable. By integration of (3.32) over the interval $[0,t]$, for every $t > 1$, we obtain

$$t^{-\mu} j(x,ty) - j(x,y) = \int_1^t -\tau^{-\mu-1}[\mu j(x,\tau y) - j'_y(x,\tau y)\cdot(\tau y)]\,d\tau$$

$$\leq \int_1^t -\tau^{-\mu-1}[\mu j(x,\tau y) - j^0(x,\tau y;\tau y)]\,d\tau \quad \text{a.e. } y \in \mathbb{R}^s,\ \forall t > 1$$

since always the derivative belongs to the generalized gradient. Using hypothesis (H_2) (iii) and the inequality above we infer that

$$j(x,ty) \leq t^\mu j(x,y) + \frac{a_1|y|^\sigma}{\mu-\sigma}(t^\mu - t^\sigma) + \frac{a_2}{\mu}(t^\mu - 1)$$

$$\leq t^\mu j(x,y) + \frac{a_1|y|^\sigma}{\mu-\sigma}t^\mu + \frac{a_2}{\mu}t^\mu \quad \text{a.e. } y \in \mathbb{R}^s,\ \forall t > 1. \tag{3.33}$$

Setting in (3.33) $y = sv_0(x)$, with $s > 0$ and the element $v_0 \in V$ prescribed by (H_2) (vi), it follows that

$$j(x,tsv_0(x)) \leq \left(j(x,sv_0(x)) + \frac{a_1 s\sigma|v_0(x)|^\sigma}{\mu-\sigma} + \frac{a_2}{\mu}\right)t^\mu \tag{3.34}$$

for a.e. $x \in \Omega$ and for all $t > 1$, $s > 0$.

In virtue of (3.1) and (3.34) we get

$$\Psi(tsv_0) = \frac{1}{2}t^2 s^2 a(v_0,v_0)$$

$$+t^\mu s^\sigma\left[s^{-\sigma}\left(\int_\Omega j(x,sv_0(x))\,dx + \frac{a_2}{\mu}|\Omega|\right) + \frac{a_1}{\mu-\sigma}\|v_0\|^\sigma_{L^\sigma(\Omega;\mathbb{R}^s)}\right]$$

$$-ts\langle f,v_0\rangle_V,\quad \forall t > 1,\ s > 0. \tag{3.35}$$

By assumption (H_2) (vi) we can find a number $s > 0$ such that

$$s^{-\sigma}\left(\int_\Omega j(x,sv_0(x))\,dx + \frac{a_2}{\mu}|\Omega|\right) + \frac{a_1}{\mu-\sigma}\|v_0\|^\sigma_{L^\sigma(\Omega;\mathbb{R}^s)} < 0.$$

With such a fixed number s, passing to the limit in (3.35) as $t \to \infty$ and taking into account that $\mu > 2$, we see that property (3.31) holds true. Now, hypothesis (b) of Theorem 3.1 follows readily from (3.31), uniformly with respect to $\|f\|_{V^*} \leq B$. For this it suffices to take $e = t_0 v_0$ for $t_0 > 0$ sufficiently large.

All the assumptions of Theorem 3.1 are fulfilled. Then Theorem 3.1 ensures that

$$c := \inf\{\max_{0\leq t\leq 1} F(g(t)) :\ g \in C([0,1],V),\ g(0) = 0,\ g(3.1) = t_0 v_0\} \tag{3.36}$$

is a critical value of the functional Ψ introduced in (3.1). Consequently, there exists $u \in V$ and $z \in L^{\frac{p}{p-1}}(\Omega; \mathbb{R}^s)$ such that

$$\frac{1}{2} a(u,u) + \int_\Omega j(x,u(x))dx - \langle f,u \rangle_V = c, \tag{3.37}$$

$$a(u,v) + \int_\Omega z(x)v(x)\,dx = \langle f,v \rangle_V, \quad \forall v \in V, \tag{3.38}$$

and

$$z(x) \in \partial j(x,u(x)) \ \text{ a.e. } x \in \Omega. \tag{3.39}$$

In writing (3.38), (3.39) we made use of the inclusion determining the generalized gradient $\partial\Psi(u)$ in (3.3). By relations (3.38), (3.39) and the maximality property of j^0 (see Clarke (Clarke 1983), Proposition 2.1.2), we obtain that $u \in V$ is a solution of the hemivariational inequality (P).

The first inequality in (3.4) follows from (3.37) and the a priori estimate $c \geq \gamma = \beta$ as stated in Theorem 3.1. The second inequality in (3.4) is obtained by considering in (3.36) the path $g(t) = t\,t_0 v_0$, $t \in [0,1]$, i.e. the segment joining 0 and $t_0 v_0$. The proof is thus complete. □

4 Hemivariational inequalities at resonance

4.1 Formulation of the problem and preliminaries

Let $\Omega \subseteq \mathbb{R}^N$ be a bounded domain with a compact connected C^2-boundary $\partial\Omega$. The problem under consideration is as follows: Find $u \in W^{1,p}(\Omega)$ such that

$$\begin{cases} \operatorname{div}(|Du(x)|^{p-2}\,Du(x)) + \lambda_1\,|u(x)|^{p-2}\,u(x) \in \partial j(x,u(x)) \text{ a.e. on } \Omega \\ u|_{\partial\Omega} = 0, \quad 2 \leq p < \infty. \end{cases} \tag{4.1}$$

By $\partial j(x,u)$ we denote the generalized gradient for locally Lipschitz functionals due to Clarke (Clarke 1983). For the right hand side of (4.1) we suppose only that it satisfies the unilateral growth condition due to Naniewicz (Naniewicz 1994a). Thus the functions $j^0(\cdot,u;v)$ and $j(\cdot,u)$ are not in general summable for every $u, v \in W_0^{1,p}(\Omega)$. Therefore the energy functional has no longer as its effective domain the whole space $W_0^{1,p}(\Omega)$, so we cannot use directly the Mountain-Pass theorem but we have to study the problem in finite dimensional spaces (subspaces of $W_0^{1,p}(\Omega) \cap L^\infty(\Omega)$), in which we can use the Mountain-Pass theorem and then pass to the limit using the Dunford - Pettis criterion.

In order to prove that our energy functional satisfies the (PS) condition we use an extended Poincaré inequality which appears very recently in the paper of Fleckinger-Pellé and Takáč (Fleckinger-Pellé & Tkáč 2002). So for this purpose we assume that our boundary is a compact connected C^2-manifold.

Let us mention some facts about the first eigenvalue of the p-Laplacian. Consider the first eigenvalue λ_1 of $(-\Delta_p, W_0^{1,p}(\Omega))$. From Lindqvist (Lindqvist 1990) we know that

$\lambda_1 > 0$ is isolated and simple, that any two solutions u, v of

$$\left\{ \begin{array}{l} -\Delta_p u := -\operatorname{div}(|Du|^{p-2} Du) = \lambda_1 |u|^{p-2} u \ \text{ a.e. on } \Omega \\ u\,|_{\partial\Omega} = 0, \quad 2 \le p < \infty \end{array} \right\} \tag{4.2}$$

satisfy $u = cv$ for some $c \in \mathbb{R}$. In addition, the λ_1-eigenfunctions do not change sign in Ω. Finally we have the following variational characterization of λ_1 (Rayleigh quotient):

$$\lambda_1 = \inf \left[\frac{\|Du\|_p^p}{\|u\|_p^p} : u \in W_0^{1,p}(\Omega), u \neq 0 \right].$$

We are going to use the mountain - pass theorem of (Chang 1981) and the generalization of the Poincaré inequality of Fleckinger-Pellé-Tkáč (Fleckinger-Pellé & Tkáč 2002): There exists a positive constant $c > 0$ such that:

$$\int_\Omega |Du|^p \, dx - \lambda_1 \int_\Omega |u|^p \, dx \ge$$

$$c\left(|e|^{p-2} \int_\Omega |D\theta|^{p-2} |D\widehat{u}|^2 \, dx + \int_\Omega |D\widehat{u}|^p \, dx \right), \forall\, u \in W_0^{1,p}(\Omega), \tag{4.3}$$

where λ_1 is the first eigenvalue of $\left(-\Delta_p, W_0^{1,p}(\Omega)\right)$, θ is the λ_1-eigenfunction and $u = e\theta + \widehat{u}$ is an orthogonal decomposition of u in $L^2(\Omega)$, $e = \|\theta\|_{L^2(\Omega)}^{-2} \langle u, \theta \rangle_{L^2(\Omega)}$, $\langle \widehat{u}, \theta \rangle_{L^2(\Omega)} = 0$.

Let us denote by $V_0 = \{s\theta\}_{s \in \mathbb{R}}$ the one-dimensional eigenspace spanned by the eigenfunction θ corresponding to the first eigenvalue λ_1 of $\left(-\Delta_p, W_0^{1,p}(\Omega)\right)$, normalized by $\theta > 0$ in Ω and $\|\theta\|_{W_0^{1,p}(\Omega)} = 1$. Due to Anane (Anane 1988) we have $\theta \in L^\infty(\Omega)$. By $V^\perp$ we denote the orthogonal complement in $L^2(\Omega)$ of V_0. Thus for any $u \in W_0^{1,p}(\Omega)$ the decomposition follows

$$u = e\overline{\theta} + \widehat{u} \quad \text{with} \quad e \ge 0,\ \overline{\theta} \in \{\pm\theta\} \subset V_0,\ \widehat{u} \in \widehat{V}, \tag{4.4}$$

where $\widehat{V} := V^\perp \cap W_0^{1,p}(\Omega)$.

Lemma 4.1. *Assume that*

(H0) $j(\cdot, 0) \in L^1(\Omega)$ *and* $j(x, \cdot)$ *is Lipschitz continuous on the bounded subsets of* $\mathbb{R}$ *uniformly with respect to* $x \in \Omega$, *i.e.,* $\forall\, r > 0\ \exists\, K_r > 0$ *such that* $\forall\, |y_1|, |y_2| \le r$,

$$|j(x, y_1) - j(x, y_2)| \le K_r |y_1 - y_2|, \quad \text{for a.e. } x \in \Omega;$$

(H1) There exist $\mu > p$, $1 \le \sigma < p$, $a \in L^1(\Omega)$ *and a constant* $k \ge 0$ *such that*

$$\mu j(x, \xi) - j^0(x, \xi; \xi) \ge -a(x) - k|\xi|^\sigma, \quad \forall \xi \in \mathbb{R} \text{ and for a.e. } x \in \Omega;$$

(H2) Assume that

$$\liminf_{\substack{t \to +\infty \\ \eta \to \overline{\theta}}} \frac{1}{t^{p-1}} \int_\Omega -j^0\left(x, t\eta(x); -\overline{\theta}(x)\right) dx > 0,$$

$$\forall \overline{\theta} \in V_0 \ \text{ with } \ \|\overline{\theta}\|_{W_0^{1,p}(\Omega)} = 1.$$

Moreover, suppose that for a sequence $\{u_n\} \subset W_0^{1,p}(\Omega)$ *there exists* $\varepsilon_n \searrow 0$ *such that the conditions below are fulfilled:*

$$\int_\Omega |Du_n(x)|^{p-2} \left\langle Du_n(x), Dv(x) - Du_n(x) \right\rangle_{\mathbb{R}^N} dx$$
$$-\lambda_1 \int_\Omega |u_n(x)|^{p-2}\, u_n(x)\big(v(x) - u_n(x)\big)\, dx$$
$$+ \int_\Omega j^0\big(x, u_n(x); v(x) - u_n(x)\big)\, dx \ge -\varepsilon_n \|v - u_n\|_{W_0^{1,p}(\Omega)},$$
$$\forall v \in \mathrm{Lin}(\{u_n, \theta\}), \tag{4.5}$$

and

$$\frac{1}{p}\int_\Omega |Du_n(x)|^p\, dx - \frac{\lambda_1}{p}\int_\Omega |u_n(x)|^p\, dx + \int_\Omega j\big(x, u_n(x)\big)\, dx \le C, \quad C > 0. \tag{4.6}$$

where $\mathrm{Lin}(\{u_n, \theta\})$ *is the linear subspace of* $W_0^{1,p}(\Omega)$ *spanned by* $\{\theta, u_n\}$. *Then the sequence* $\{u_n\}$ *is bounded in* $W_0^{1,p}(\Omega)$, *i.e. there exists* $M > 0$ *such that*

$$\|u_n\|_{W_0^{1,p}(\Omega)} \le M. \tag{4.7}$$

Proof. Suppose on the contrary that the claim is not true, i.e. there exists a sequence $\{u_n\}_{n=1}^\infty \subset W_0^{1,p}(\Omega)$ with $\|u_n\|_{W_0^{1,p}(\Omega)} \to \infty$ for which (4.5) and (4.6) hold. Combining (4.6) and (4.5) with $v = 2u_n$ yields

$$C + \varepsilon_n \|u_n\|_{W_0^{1,p}(\Omega)} \ge \tfrac{\mu - p}{p}\Big(\|Du_n\|^p_{L^p(\Omega;\mathbb{R}^N)} - \lambda_1 \|u_n\|^p_{L^p(\Omega)}\Big)$$
$$+ \int_\Omega \Big(\mu j(u_n) - j^0(u_n; u_n)\Big)\, dx. \tag{4.8}$$

By the generalization of the Poincaré inequality (4.3) the decomposition results: $u_n = e_n\theta_n + \widehat{u}_n$, where $\widehat{u}_n \in \widehat{V}$, $e_n \ge 0$, $\theta_n \in \{\pm\theta\}$, $\|\theta\|_{W_0^{1,p}(\Omega)} = 1$, such that

$$\|Du_n\|^p_{L^p(\Omega;\mathbb{R}^N)} - \lambda_1 \|u_n\|^p_{L^p(\Omega)}$$
$$\ge c\left(e_n^{p-2}\int_\Omega |D\theta_n|^{p-2}\, |D\widehat{u}_n|^2\, dx + \|D\widehat{u}_n\|^p_{L^p(\Omega;\mathbb{R}^N)}\right). \tag{4.9}$$

Thus by (H1) we have

$$C + \varepsilon_n \|u_n\|_{W_0^{1,p}(\Omega)} \ge c\tfrac{\mu - p}{p} e_n^{p-2} \int_\Omega |D\theta|^{p-2}\, |D\widehat{u}_n|^2\, dx$$
$$+ c\tfrac{\mu - p}{p}\|D(\widehat{u}_n)\|^p_{L^p(\Omega;\mathbb{R}^N)} - c_1 \|u_n\|^\sigma_{L^p(\Omega)}. \tag{4.10}$$

Hence

$$C + \varepsilon_n(\|\widehat{u}_n\|_{W_0^{1,p}(\Omega)} + e_n) \ge c\tfrac{\mu - p}{p}\|D\widehat{u}_n\|^p_{L^p(\Omega;\mathbb{R}^N)} - c_1\|\widehat{u}_n\|^\sigma_{L^p(\Omega)} - c_2 e_n^\sigma. \tag{4.11}$$

Thus it follows that $e_n \to \infty$ because, otherwise, we would have the boundedness of $\{u_n\}$ in $W_0^{1,p}(\Omega)$. Consequently we arrive at the estimate

$$\frac{C}{e_n} + \varepsilon_n(\|\tfrac{\widehat{u}_n}{e_n}\|_{W_0^{1,p}(\Omega)} + 1) \geq e_n^{p-1} c^{\frac{\mu-p}{p}} \|D(\tfrac{\widehat{u}_n}{e_n})\|^p_{L^p(\Omega;\mathbb{R}^N)} \\ - e_n^{\sigma-1} c_1 \|\tfrac{\widehat{u}_n}{e_n}\|^\sigma_{L^p(\Omega)} - e_n^{\sigma-1} c_2. \tag{4.12}$$

which in view of $e_n \to \infty$ leads to the conclusion that

$$\|\tfrac{\widehat{u}_n}{e_n}\|_{W_0^{1,p}(\Omega)} \to 0. \tag{4.13}$$

Now let us turn back to (4.5). By passing to a subsequence one can suppose also that $\theta_n = \theta$ (or $\theta_n = -\theta$). Thus, substituting $v = \widehat{u}_n$ into (4.5) yields

$$e_n^p \int_\Omega \left|D(\tfrac{\widehat{u}_n}{e_n}) + D\theta\right|^{p-2} \left\langle D(\tfrac{\widehat{u}_n}{e_n}) + D\theta, -D\theta\right\rangle_{\mathbb{R}^N} dx$$

$$-e_n^p \lambda_1 \int_\Omega \left|\tfrac{\widehat{u}_n}{e_n} + \theta\right|^{p-2} (\tfrac{\widehat{u}_n}{e_n} + \theta)(-\theta)\, dx + e_n \int_\Omega j^0\big(e_n(\tfrac{\widehat{u}_n}{e_n} + \theta); -\theta\big)\, dx \geq -\varepsilon_n e_n.$$

Hence

$$\varepsilon_n \geq \frac{1}{e_n^{p-1}} \int_\Omega -j^0\big(e_n(\tfrac{\widehat{u}_n}{e_n} + \theta); -\theta\big)\, dx + \\ + \int_\Omega \left|D(\tfrac{\widehat{u}_n}{e_n}) + D\theta\right|^{p-2} \left\langle D(\tfrac{\widehat{u}_n}{e_n}) + D\theta, D\theta\right\rangle_{\mathbb{R}^N} dx \\ - \lambda_1 \int_\Omega \left|\tfrac{\widehat{u}_n}{e_n} + \theta\right|^{p-2} (\tfrac{\widehat{u}_n}{e_n} + \theta)\theta\, dx. \tag{4.14}$$

Now we are ready to pass to the limit with $n \to \infty$. For this purpose notice that in view of (4.13) it results

$$\lim_{n\to\infty} \Big\{ \int_\Omega \left|D(\tfrac{\widehat{u}_n}{e_n}) + D\theta\right|^{p-2} \left\langle D(\tfrac{\widehat{u}_n}{e_n}) + D\theta, D\theta\right\rangle_{\mathbb{R}^N} dx \\ -\lambda_1 \int_\Omega \left|\tfrac{\widehat{u}_n}{e_n} + \theta\right|^{p-2} (\tfrac{\widehat{u}_n}{e_n} + \theta)\theta\, dx \Big\} = \|D\theta\|^p_{L^p(\Omega;\mathbb{R}^N)} - \lambda_1 \|\theta\|^p_{L^p(\Omega)} = 0,$$

and by (iii) we have

$$\liminf_{n\to\infty} \frac{1}{e_n^{p-1}} \int_\Omega -j^0\big(e_n(\tfrac{\widehat{u}_n}{e_n} + \theta); -\theta\big)\, dx > 0.$$

Thus from (4.14) we arrive at the inequality $0 > 0$ which is a contradiction. Thus the proof of Lemma 4.1 is complete. □

Lemma 4.2. *Assume that (H0) and the hypotheses below hold:*

*(H3) The unilateral growth condition (Naniewicz 1994*a*): there exist $p < q < p^* = \frac{Np}{N-p}$, and a constant $\kappa \geq 0$ such that*

$$j^0(x, \xi; -\xi) \leq \kappa(1 + |\xi|^q), \quad \forall \xi \in \mathbb{R} \text{ and for a.e. } x \in \Omega;$$

(H4) Uniformly for a.e. $x \in \Omega$,

$$\liminf_{\xi \to 0} \frac{pj(x,\xi)}{|\xi|^p} \geq \phi(x) \geq 0,$$

with $\phi(x) \in L^\infty(\Omega)$ and $\phi(x) > 0$ on a set of positive measure.

Then there exists $\rho > 0$ such that

$$\mathcal{R}(u) := \tfrac{1}{p}\|Du\|^p_{L^p(\Omega;\mathbb{R}^N)} - \tfrac{\lambda_1}{p}\|u\|^p_{L^p(\Omega)} + \int_\Omega j(u)\,dx \geq \eta, \quad \eta = \text{const} > 0, \tag{4.15}$$

is valid for any $u \in W_0^{1,p}(\Omega)$ with $\|u\|_{W_0^{1,p}(\Omega)} = \rho$.

Proof. Suppose the assertion is not true. Thus there exist sequences $\{u_n\} \subset W_0^{1,p}(\Omega)$ and $\rho_n \searrow 0$ such that $\|u_n\|_{W_0^{1,p}(\Omega)} = \rho_n$ and $\mathcal{R}(u_n) \leq \rho_n^{p+1}$. So we have

$$\|Du_n\|^p_{L^p(\Omega;\mathbb{R}^N)} - \lambda_1\|u_n\|^p_{L^p(\Omega)} + \int_\Omega pj(u_n)\,dx \leq p\rho_n^{p+1}. \tag{4.16}$$

Further, from (H4) it follows that for any $\varepsilon > 0$, uniformly for all $x \in \Omega$ one can find $\delta > 0$ such that

$$pj(x,\xi) \geq \phi(x)\,|\xi|^p - \varepsilon\,|\xi|^p\,, \quad |\xi| \leq \delta.$$

Moreover, (H3) allows to conclude that (see Lemma 2.1, pp. 119-120, (Naniewicz 1997)):

$$j(x,\xi) \geq -\kappa_0(1 + |\xi|^q), \quad \forall \xi \in \mathbb{R}, \quad \kappa_0 = \text{const} > 0. \tag{4.17}$$

Thus it is easy to see that

$$pj(x,\xi) > (\phi(x) - \varepsilon)\,|\xi|^p - \gamma\,|\xi|^q\,, \quad \forall \xi \in \mathbb{R}, \tag{4.18}$$

for some positive $\gamma = \gamma(\delta) > 0$. Then by (4.16) it follows

$$\|Du_n\|^p_{L^p(\Omega;\mathbb{R}^N)} - \lambda_1\|u_n\|^p_{L^p(\Omega)} + \int_\Omega (\phi(x) - \varepsilon)\,|u_n(x)|^p\,dx \leq p\rho_n^{p+1} + \gamma \int_\Omega |u_n(x)|^q\,, dx. \tag{4.19}$$

Let us set $y_n = \frac{1}{\rho_n} u_n$. Dividing inequality (4.19) by ρ_n^p yields

$$\|Dy_n\|^p_{L^p(\Omega;\mathbb{R}^N)} - \lambda_1\|y_n\|^p_{L^p(\Omega)} + \int_\Omega (\phi(x) - \varepsilon)\,|y_n(x)|^p\,dx \leq p\rho_n + \gamma\rho_n^{q-p} \int_\Omega |y_n(x)|^q\,, dx. \tag{4.20}$$

Since $W_0^{1,p}(\Omega)$ is continuously embedded into $L^q(\Omega)$ we have

$$\|Dy_n\|^p_{L^p(\Omega;\mathbb{R}^N)} - \lambda_1\|y_n\|^p_{L^p(\Omega)} + \int_\Omega (\phi(x) - \varepsilon)\,|y_n(x)|^p\,dx \leq p\rho_n + \gamma_1\rho_n^{q-p}, \quad \gamma_1 = \text{const} > 0. \tag{4.21}$$

Taking into account that $\|y_n\|_{W_0^{1,p}(\Omega)} = 1$ we can suppose that for a subsequence (again denoted by the same symbol) $y_n \to y$ weakly in $W_0^{1,p}(\Omega)$ and $y_n \to y$ strongly in $L^p(\Omega)$ (the Rellich theorem) for some $y \in W_0^{1,p}(\Omega)$. Passing to the limit and the weak lower semicontinuity of the norm allows the conclusion

$$\|Dy\|^p_{L^p(\Omega;\mathbb{R}^N)} - \lambda_1 \|y\|^p_{L^p(\Omega)} + \int_\Omega (\phi(x) - \varepsilon)\,|y(x)|^p\, dx \le 0, \tag{4.22}$$

which is valid for an arbitrary $\varepsilon > 0$. Therefore we get

$$\|Dy\|^p_{L^p(\Omega;\mathbb{R}^N)} - \lambda_1 \|y\|^p_{L^p(\Omega)} + \int_\Omega \phi(x)\,|y(x)|^p\, dx \le 0. \tag{4.23}$$

Using the Rayleigh quotient characterization of λ_1 and (H4) leads to the equalities

$$\|Dy\|^p_{L^p(\Omega;\mathbb{R}^N)} = \lambda_1 \|y\|^p_{L^p(\Omega)}, \tag{4.24}$$

$$\int_\Omega \phi(x)\,|y(x)|^p\, dx = 0. \tag{4.25}$$

Now we show that $y \neq 0$. Indeed, from the results obtained it follows that

$$\|Dy_n\|^p_{L^p(\Omega;\mathbb{R}^N)} - \lambda_1 \|y_n\|^p_{L^p(\Omega)} \to 0$$

and by the compactness of the imbedding $W_0^{1,p}(\Omega) \subset L^p(\Omega)$ we get

$$\|y_n\|_{L^p(\Omega)} \to \|y\|_{L^p(\Omega)}.$$

Since $\|Dy_n\|_{L^p(\Omega;\mathbb{R}^N)} \ge c\|y_n\|_{W_0^{1,p}(\Omega)} = c$, $c > 0$ (the equivalence of the norms), we arrive at $\lambda_1\|y\|^p_{L^p(\Omega)} \ge c^p$ which establishes the assertion. Therefore, taking into account (4.24) we conclude that $y \neq 0$ is an λ_1-eigenfunction. Since $\phi(x) > 0$ on a set of positive measure (by (H4)), and, as it is well known (c.f. (Lindqvist 1990)), $|y(x)| > 0$ for a.e. $x \in \Omega$, we are led to the contradiction with (4.25). The proof of Lemma 4.2 is complete. □

Lemma 4.3. *Assume that (H0)-(H1) hold and that*

(H5) $\int_\Omega j(x,0)\, d\Omega \le 0$ *and either for some* $\overline{\theta} \in V_0$, $\overline{\theta} \neq 0$,

$$\liminf_{s\to+\infty} \int_\Omega j\big(x, s\overline{\theta}(x)\big)\, dx < 0, \tag{4.26}$$

or there exists $v_0 \in W_0^{1,p}(\Omega) \cap L^\infty(\Omega)$ *such that*

$$\liminf_{s\to+\infty} s^{-\sigma} \int_\Omega j\big(x, sv_0(x)\big)\, dx < \frac{k}{\sigma - \mu} \|v_0\|^\sigma_{L^\sigma(\Omega)}, \tag{4.27}$$

with the positive constants k, μ, σ *entering (H1).*

Then there exists $e \in W_0^{1,p}(\Omega) \cap L^\infty(\Omega)$, $e \neq 0$, *such that*

$$\mathcal{R}(se) \le 0, \quad \forall\, s \ge 1.$$

Proof. If (4.26) is fulfilled then the assertion holds for $e = s_0\overline{\theta}$ with sufficiently large $s_0 > 0$.

For the case (4.27) we follow the lines of (Motreanu & Panagiotopoulos 1999). For all $\tau \neq 0$, $x \in \Omega$ and $\xi \in \mathbb{R}$, the formula below of generalized gradient (with respect to τ) holds

$$\partial_\tau(\tau^{-\mu} j(x,\tau\xi)) = \tau^{-\mu-1}[-\mu j(x,\tau\xi) + \partial_\xi j(x,\tau\xi)(\tau\xi)],$$

for the constant $\mu > p$ fulfilling (H1). Since the function $\tau \mapsto \tau^{-\mu} j(x,\tau\xi)$ is differentiable a.e. on $\mathbb{R}$, the equality above and a classical property of Clarke's generalized directional derivative imply that

$$t^{-\mu} j(x,t\xi) - j(x,\xi) = \int_1^t \frac{d}{d\tau}(\tau^{-\mu} j(x,\tau\xi))d\tau$$

$$\leq \int_1^t \tau^{-\mu-1}[-\mu j(x,\tau\xi) + j^0(x,\tau\xi;\tau\xi)]d\tau, \quad \forall t > 1, \text{ a.e. } x \in \Omega, \ \xi \in \mathbb{R}.$$

In view of assumption (H1) we infer that

$$\begin{aligned} t^{-\mu} j(x,t\xi) - j(x,\xi) &\leq \int_1^t \tau^{-\mu-1}[a(x) + k\tau^\sigma |\xi|^\sigma]\, d\tau \\ &= \left[a(x)(-\frac{1}{\mu}t^{-\mu} + \frac{1}{\mu}) + k|\xi|^\sigma (\frac{1}{\sigma-\mu}t^{\sigma-\mu} - \frac{1}{\sigma-\mu})\right] \\ &\leq \mu^{-1}a(x) + (\mu-\sigma)^{-1}k|\xi|^\sigma, \quad \forall t > 1, \text{ a.e. } x \in \Omega, \ \xi \in \mathbb{R}. \end{aligned} \tag{4.28}$$

Set $\xi = sv_0(x)$ with $x \in \Omega$ and $s > 0$. We find from (4.28) the estimate

$$\begin{aligned} j(x,tsv_0(x)) &\leq t^\mu[j(x,sv_0(x)) + \mu^{-1}a(x) \\ &+(\mu-\sigma)^{-1}ks^\sigma |v_0(x)|^\sigma], \quad \forall t > 1, \ s > 0, \text{ a.e. } x \in \Omega. \end{aligned} \tag{4.29}$$

Combining (4.29) with (4.27) yields

$$\begin{aligned} \mathcal{R}(tsv_0) &\leq \tfrac{1}{p}t^p s^p \left(\|Dv_0\|^p_{L^p(\Omega;\mathbb{R}^N)} - \lambda_1\|v_0\|^p_{L^p(\Omega)}\right) \\ &+t^\mu s^\sigma \left[s^{-\sigma}\int_\Omega j(x,sv_0(x))dx + k(\mu-\sigma)^{-1}\|v_0\|^\sigma_{L^\sigma(\Omega)} + s^{-\sigma}\mu^{-1}\|a\|_{L^1(\Omega)}\right], \\ &\forall t > 1, \ s > 0. \end{aligned} \tag{4.30}$$

Assumption (4.27) allows to fix some number $s_0 > 0$ such that

$$s_0^{-\sigma}\int_\Omega j(x,s_0v_0(x))dx + k(\mu-\sigma)^{-1}\|v_0\|^\sigma_{L^\sigma(\Omega)} + s_0^{-\sigma}\mu^{-1}\|a\|_{L^1(\Omega)} < 0. \tag{4.31}$$

With such an $s_0 > 0$ we can pass to the limit as $t \to +\infty$ in (4.30) and obtain (in view of $\mu > p$) that $\mathcal{R}(ts_0v_0) \to -\infty$ as $t \to +\infty$. Consequently, setting $e = t_0s_0v_0$ with sufficiently large $t_0 > 0$ we establish the assertion. This completes the proof of Lemma 4.3. □

4.2 Finite dimensional approximation

Let us denote by Λ the family of all finite dimensional subspaces F of $W_0^{1,p}(\Omega)\cap L^\infty(\Omega)$ satisfying the conditions:

$$\begin{aligned} F\in\Lambda \Leftrightarrow\ F = V_0+\widehat{F} \\ \text{for some finite dimensional subspace } \widehat{F}\subset\widehat{V}\cap L^\infty(\Omega) \\ \text{and } e\in F, \end{aligned} \tag{4.32}$$

with $e\in W_0^{1,p}(\Omega)\cap L^\infty(\Omega)$ as explained in Lemma 4.3.

For every subspace $F\in\Lambda$ we introduce the functional $\mathcal{R}_F : F\to\mathbb{R}$ which is the restriction of $\mathcal{R}$ to F, i.e.

$$\mathcal{R}_F(v) = \tfrac{1}{p}\|v\|^p_{L^p(\Omega;\mathbb{R}^N)} - \tfrac{\lambda_1}{p}\|v\|^p_{L^p(\Omega)} + \int_\Omega j\big(x, v(x)\big)\,dx, \quad \forall\, v\in F. \tag{4.33}$$

It is obvious that the functional $\mathcal{R}_F$ is locally Lipschitz and its generalized gradient is expressed by

$$\partial\mathcal{R}_F(v)\subset i_F^\star A i_F v + \bar{i}_F^\star\partial J(v), \qquad \forall\, v\in F, \tag{4.34}$$

where $i_F : F\to W_0^{1,p}(\Omega)$ and $\bar{i}_F : F\to L^\infty(\Omega)$ are the inclusion maps with their dual projections $i_F^\star : W^{-1,p'}(\Omega)\to F^\star$ and $\bar{i}_F^\star : L^1(\Omega)\to F^\star$, respectively, while $A : W_0^{1,p}(\Omega)\to W^{-1,p'}(\Omega)$ is defined by

$$\begin{aligned}\langle Au, v\rangle_{W_0^{1,p}(\Omega)} = \int_\Omega |Du|^{p-2}\langle Du, Dv\rangle_{\mathbb{R}^N}\,d\Omega - \lambda_1\int_\Omega |u|^{p-2}uv\,d\Omega, \\ u, v\in W_0^{1,p}(\Omega). \end{aligned}\tag{4.35}$$

By $\partial J(\cdot)$ the generalized Clarke gradient of $J : L^\infty(\Omega)\to\mathbb{R}$ given by

$$J(v) = \int_\Omega j(x, v(x))\,dx, \quad \forall\, v\in L^\infty(\Omega)$$

has been denoted. Notice that in view of (H0), J is locally Lipschitz on $L^\infty(\Omega)$, so the generalized gradient $\partial J(\cdot)$ is well defined. The pairing over $F^\star\times F$ will be denoted by $\langle\cdot,\cdot\rangle_F$.

Proposition 4.1. *Assume the hypotheses (H0)-(H5). Then for each $F\in\Lambda$ Problem (P_F): Find $u_F\in F$ such as to satisfy the hemivariational inequality:*

$$\begin{aligned}\int_\Omega |Du_F|^{p-2}\langle Du_F, Dv - Du_F\rangle_{\mathbb{R}^N}\,d\Omega - \lambda_1\int_\Omega |u_F|^{p-2}u_F(v-u_F)\,d\Omega \\ + \int_\Omega j^0(u_F; v-u_F)\,d\Omega \geq 0, \quad \forall\, v\in F, \end{aligned}\tag{4.36}$$

has at least one solution $u_F\neq 0$. Moreover, there exist constants $M>0$, $\gamma_1>0$ and $\gamma_2>0$ not depending on $F\in\Lambda$ such that

$$\|u_F\|_{W_0^{1,p}(\Omega)}\leq M, \quad \forall\, F\in\Lambda \tag{4.37}$$

$$\gamma_1\leq\mathcal{R}(u_F)\leq\gamma_2, \quad \forall\, F\in\Lambda. \tag{4.38}$$

Proof. First we show that the functional $\mathcal{R}_F : F \to \mathbb{R}$ satisfies the Palais-Smale condition in the sense of (Chang 1981). Let $\{u_n\} \subset F$ and $\{w_n\} \subset F^\star$ be sequences such that $|\mathcal{R}_F(u_n)| \leq c$, for all $n \geq 1$, with a constant $c > 0$, and $w_n \in \partial\mathcal{R}_F(u_n)$, $\|w_n\|_{F^\star} = \varepsilon_n \to 0$ as $n \to \infty$. Since F is finite dimensional, it remains to show that $\{u_n\}$ is bounded in F. According to (4.34) we see that w_n can be expressed as follows

$$w_n = i_F^\star A u_n + \bar{i}_F^\star \chi_n, \qquad \text{with } \chi_n \in \partial J(u_n). \tag{4.39}$$

Let us notice that the hypothesis of Theorem 2.7.3 in Clarke (Clarke 1983), p. 80, is verified. Therefore we obtain

$$\partial J(v) \subset \int_\Omega \partial j(x, v(x))\, dx, \quad \forall\, v \in L^\infty(\Omega). \tag{4.40}$$

Thus

$$\begin{aligned}\langle A u_n, v - u_n\rangle_{W_0^{1,p}(\Omega)} + \int_\Omega j^0(u_n; v - u_n)\, d\Omega \geq \langle w_n, v - u_n\rangle_F \geq -\varepsilon_n \|v - u_n\|_F \\ \geq -c\varepsilon_n \|v - u_n\|_{W_0^{1,p}(\Omega)}, \quad \forall\, v \in F, \quad c = \text{const} > 0,\end{aligned}$$

because the norms $\|\cdot\|_F$ and $\|\cdot\|_{W_0^{1,p}(\Omega)}$ are equivalent in F (F is finite dimensional). Since $\mathrm{Lin}(\theta, u_n) \subset F$, the hypotheses of Lemma 4.1 are verified. Consequently $\{u_n\}$ is bounded in $W_0^{1,p}(\Omega)$ which means that

$$\|u_F\|_{W_0^{1,p}(\Omega)} \leq M_F \tag{4.41}$$

for some $M_F > 0$.

Following the lines of the proof of Lemma 4.2 (with $W_0^{1,p}(\Omega)$ replaced by F) we conclude the existence of positive constants $\rho_F > 0$ and $\eta_F > 0$ such that

$$\mathcal{R}_F(v) \geq \eta_F, \quad \forall\, v \in \{w \in F \colon \|w\|_F = \rho_F\}. \tag{4.42}$$

By Lemma 4.3 we know that $\mathcal{R}(te) \leq 0$ for any $t \geq 1$, therefore $\rho_F < \|e\|_F$. Thus taking into account that $\mathcal{R}_F(0) \leq 0$ and $\mathcal{R}_F(e) \leq 0$ we are allowed to apply the mountain pass theorem and deduce the existence of a critical point $u_F \in F$ of $\mathcal{R}_F$. This leads to the finite dimensional hemivariational inequality (4.36) (c.f. (Motreanu & Panagiotopoulos 1999)).

Let us recall that the critical value $\mathcal{R}_F(u_F)$ is characterized by (c.f. (Motreanu & Panagiotopoulos 1999))

$$\mathcal{R}_F(u_F) = \inf_{\gamma \in C_F} \max_{t \in [0,1]} \mathcal{R}_F(\gamma(t)), \tag{4.43}$$

where

$$C_F = \{\gamma \in C([0,1], F) : \ \gamma(0) = 0,\ \gamma(1) = e\},$$

is the the family of all continuous curves in F joining points 0 and e in F i.e. $\gamma(0) = 0$ and $\gamma(1) = e$, $\gamma(t) \subset F$. Further, from Lemma 4.2 it follows that for a certain positive $\rho > 0$ one can find $\eta > 0$ with

$$\mathcal{R}(v) \geq \eta, \quad \forall\, v \in S_\rho := \{v \in W_0^{1,p}(\Omega) \colon \|v\|_{W_0^{1,p}(\Omega)} = \rho\}, \tag{4.44}$$

while Lemma 4.3 ensures the existence of $e \in W_0^{1,p}(\Omega)$, $e \neq 0$, such that

$$\mathcal{R}(te) \leq 0, \quad \forall t \geq 1. \tag{4.45}$$

Therefore, for any $F \in \Lambda$, if $\gamma \in C_F([0,1]; F)$ then γ meets points of S_ρ which means that

$$\max_{t\in[0,1]} \mathcal{R}_F\big(\gamma(t)\big) \geq \eta. \tag{4.46}$$

Hence

$$\eta \leq \mathcal{R}(u_F) = \inf_{\gamma\in C_F} \max_{t\in[0,1]} \mathcal{R}_F(\gamma(t)) \leq \max_{t\in[0,1]} \mathcal{R}(te), \quad \forall F \in \Lambda \tag{4.47}$$

and (4.38) results.

Now we are ready to show that $M_F > 0$ in (4.41) is independent of $F \in \Lambda$. For this purpose suppose that a sequence $\{u_{F_n}\}_{F_n\in\Lambda}$ of solutions of (P_{F_n}) has the property that $\|u_{F_n}\|_{W_0^{1,p}(\Omega)} \to \infty$. Taking into account (4.36) and (4.47) it is easy to check that the hypotheses (4.6) and (4.5) of Lemma 4.1 hold (with F replaced by F_n and $\varepsilon_n = 0$). Following the lines of the proof of Lemma 4.1 we arrive at the contradiction which establishes the assertion. The proof of Proposition 4.1 is complete. □

For the restriction of J to F, $J_F := J|_F : F \to \mathbb{R}$, we have $\partial J_F(u_F) \subset \vec{i}_F^{\star} \partial J(u_F)$. Therefore Proposition 4.1 can be reformulated as follows.

Corollary 4.1. *Assume the hypotheses (H0)-(H5). Then for each $F \in \Lambda$ there exist $u_F \in F$ and $\chi_F \in L^1(\Omega)$ such that*

$$\int_\Omega |Du_F|^{p-2} \langle Du_F, Dv - Du_F\rangle_{\mathbb{R}^N}\, d\Omega - \lambda_1 \int_\Omega |u_F|^{p-2} u_F(v - u_F)\, d\Omega$$
$$+ \int_\Omega \chi_F(v - u_F)\, d\Omega = 0, \; \forall v \in F, \quad \text{and} \quad \chi_F \in \partial j(u_F) \;\; \text{a.e. in } \Omega. \tag{4.48}$$

According to the results obtained we know that to any $F \in \Lambda$ a pair $(u_F, \chi_F) \in F \times L^1(\Omega)$ can be assigned for which (4.48) holds. Moreover, the family $\{u_F\}_{F\in\Lambda}$ is uniformly bounded in $W_0^{1,p}(\Omega)$ ((4.37) holds). The question arises concerning the behavior of $\{\chi_F\}_{F\in\Lambda}$.

Proposition 4.2. *Assume that $(u_F, \chi_F) \in F \times L^1(\Omega)$ satisfies (4.48). Then the set $\{\chi_F\}_{F\in\Lambda}$ is weakly precompact in $L^1(\Omega)$.*

Proof. Since Ω is bounded, according to the Dunford-Pettis theorem (see, e.g., (Ekeland & Temam 1976), p. 239) it suffices to show that for each $\varepsilon > 0$ a number $\delta > 0$ can be determined such that for any $\omega \subset \Omega$ with $|\omega| < \delta$,

$$\int_\omega |\chi_F|\, dx < \varepsilon, \quad \forall F \in \Lambda. \tag{4.49}$$

Choose $\overline{q} \in (q, p^\star)$. Then the injection $W_0^{1,p}(\Omega) \subset L^{\overline{q}}(\Omega)$ is compact. Further, from (H3) it follows that there exists a function $\alpha : \mathbb{R}_+ \to \mathbb{R}_+$ such that (c.f. Remark 5.6, p. 156, (Naniewicz & Panagiotopoulos 1995) and Lemma 1, p. 95, (Motreanu & Naniewicz 2002))

$$j^0(x, \xi; \eta - \xi) \leq \alpha(r)(1 + |\xi|^q), \quad \forall \xi, \eta \in \mathbb{R}, |\eta| \leq r, r \geq 0. \tag{4.50}$$

Fix $r > 0$ and let $\eta \in \mathbb{R}$ be such that $|\eta| \le r$. Then, by (4.48), $\chi_F\,(\eta - u_F) \le j^0(x, u_F; \eta - u_F)$, from which we get

$$\chi_F\, \eta \le \chi_F\, u_F + \alpha(r)(1 + |u_F|^q) \quad \text{for a.e. } x \in \Omega. \tag{4.51}$$

Let us set $\eta \equiv r \operatorname{sgn} \chi_F(x)$ where $\operatorname{sgn} y = 1$ if $y > 0$, $\operatorname{sgn} y = 0$ if $y = 0$, $\operatorname{sgn} y = -1$ if $y < 0$. One obtains that $|\eta| \le r$ and $\chi_F(x)\eta = r\,|\chi_F(x)|$ for almost all $x \in \Omega$. Therefore from (4.51) it results

$$r\,|\chi_F| \le \chi_F\, u_F + \alpha(r)(1 + |u_F|^q).$$

Integrating this inequality over $\omega \subset \Omega$ yields

$$\int_\omega |\chi_F|\; dx \le \frac{1}{r}\int_\omega \chi_F\, u_F\, dx + \frac{1}{r}\alpha(r)|\omega| + \frac{1}{r}\alpha(r)|\omega|^{\frac{\overline{q}-q}{\overline{q}}}\, \|u_F\|^q_{L^{\overline{q}}(\Omega)}. \tag{4.52}$$

Consequently, from (4.37) and (4.52) it follows that

$$\int_\omega |\chi_F|\; dx \le \frac{1}{r}\int_\omega \chi_F\, u_F\, dx + \frac{1}{r}\alpha(r)|\omega| + \frac{1}{r}\alpha(r)|\omega|^{\frac{\overline{q}-q}{\overline{q}}}\, \gamma^q M^q, \tag{4.53}$$

where $\gamma > 0$ is a constant satisfying $\|\cdot\|_{L^{\overline{q}}(\Omega)} \le \gamma \|\cdot\|_{H^1_0(\Omega)}$ (which holds since $\hat{q} < p^\star$).

We claim

$$\int_\omega \chi_F\, u_F\, dx \le C \tag{4.54}$$

for some positive constant C not depending on $\omega \subset \Omega$ and $F \in \Lambda$. Indeed, from (4.50) we derive that

$$\chi_F\, u_F + \alpha(0)\big(|u_F|^q + 1\big) \ge 0 \quad \text{for a.e. in } \Omega.$$

Thus it follows

$$\int_\omega \chi_F\, u_F\, dx \le \int_\omega \big(\chi_F\, u_F + \alpha(0)\big(|u_F|^q + 1\big)\big)\, dx$$
$$\le \int_\Omega \big(\chi_F\, u_F + \alpha(0)\big(|u_F|^q + 1\big)\big)\, dx \le \int_\Omega \chi_F\, u_F\, dx + \overline{k}_1\big(\|u_F\|^q_{H^1_0(\Omega)} + |\Omega|\big),$$

where $\overline{k}_1 > 0$ is a constant. By (4.37) and (4.48) (with $v = 0$) it turns out that

$$\int_\Omega \chi_F\, u_F\; dx = -\int_\Omega |Du_F|^p\; dx + \lambda_1 \int_\Omega |u_F|^p\; dx \le 0,$$

The estimates above imply (4.54).

Further, (4.53) and (4.54) entail

$$\int_\omega |\chi_F|\; dx \le \frac{1}{r}C + \frac{1}{r}\alpha(r)|\omega| + \frac{1}{r}\alpha(r)|\omega|^{\frac{\overline{q}-q}{\overline{q}}}\, \gamma^q M^q, \quad \forall\, r > 0. \tag{4.55}$$

Corresponding to $\varepsilon > 0$, fix $r > 0$ with

$$\frac{1}{r}C < \frac{\varepsilon}{2} \tag{4.56}$$

and then take $\delta > 0$ small enough to have

$$\frac{1}{r}\alpha(r)|\omega| + \frac{1}{r}\alpha(r)|\omega|^{\frac{\overline{q}-q}{\overline{q}}}\gamma^q M^q < \frac{\varepsilon}{2} \tag{4.57}$$

provided that $|\omega| < \delta$. Using this together with (4.55) and (4.56) it follows that (4.49) is justified whenever $|\omega| < \delta$. This completes the proof. □

4.3 Main result

To formulate the main result we shall need the following hypothesis:

(H6) For any sequence $\{v_k\} \subset L^\infty(\Omega)$, $v_k \to 0$ strongly in $L^p(\Omega)$, if

$$\int_\Omega \min\{\psi(x)v_k(x) \colon \psi(x) \in \partial j(x, v_k(x))\}\, d\Omega \leq 0,$$

then

$$\limsup_{k\to\infty} \int_\Omega j(x, v_k(x))\, d\Omega \leq 0.$$

Theorem 4.1. *Assume the hypotheses (H0)-(H6). Then there exists $u \in W_0^{1,p}(\Omega)$ with $u \neq 0$ and $j(u) \in L^1(\Omega)$, such as to satisfy the hemivariational inequality*

$$\begin{aligned}\int_\Omega |Du|^{p-2}\langle Du, Dv - Du\rangle_{\mathbb{R}^N}\, d\Omega - \lambda_1 \int_\Omega |u|^{p-2} u(v-u)\, d\Omega \\ + \int_\Omega j^0(u; v-u)\, d\Omega \geq 0, \quad \forall\, v \in W_0^{1,p}(\Omega).\end{aligned} \tag{4.58}$$

Moreover, there exists $\chi \in L^1(\Omega)$ with the property that

$$\begin{aligned}\int_\Omega |Du|^{p-2}\langle Du, Dv - Du\rangle_{\mathbb{R}^N}\, d\Omega - \lambda_1 \int_\Omega |u|^{p-2} u(v-u)\, d\Omega \\ + \int_\Omega \chi(v-u)\, d\Omega = 0,\ \forall\, v \in W_0^{1,p}(\Omega) \cap L^\infty(\Omega),\end{aligned} \tag{4.59}$$

$$\chi u \in L^1(\Omega) \quad \text{and} \quad \chi \in \partial j(u) \quad \text{a.e. in } \Omega. \tag{4.60}$$

Proof. The proof is carried out in a sequence of steps.

Step 1. For every $F \in \Lambda$ we introduce

$$U_F = \{u_F \in W_0^{1,p}(\Omega) \colon \text{for some } \chi_F \in L^1(\Omega; \mathbb{R}^N),\ (u_F, \chi_F) \text{ is a solution of } (P_F)\}$$

and

$$W_F = \bigcup_{\substack{F' \in \Lambda \\ F' \supset F}} U_{F'}.$$

By Proposition 4.1, W_F is nonempty (even U_F is nonempty) and contained in the ball $B_M = \{v \in W_0^{1,p}(\Omega) \colon \|v\|_{W_0^{1,p}(\Omega)} \leq M\}$. We denote by weakcl$(W_F)$ the closure of W_F in the weak topology of $W_0^{1,p}(\Omega)$. Proposition 4.1 ensures that weakcl(W_F) is weakly compact in $W_0^{1,p}(\Omega)$. We claim that the family $\{\text{weakcl}(W_F)\}_{F\in\Lambda}$ has the finite intersection property. Indeed, if $F_1,\ldots,F_k \in \Lambda$ then $W_{F_1}\cap\ldots\cap W_{F_k} \supset W_F$, with $F = F_1+\ldots+F_k$ and the assertion follows. Thus we are allowed to conclude that there exists an element $u \in W_0^{1,p}(\Omega)$ with

$$u \in \bigcap_{F\in\Lambda} \text{weakcl}(W_F).$$

Let us choose $G \in \Lambda$ arbitrarily. Since $W_0^{1,p}(\Omega)$ is reflexive, one can extract an increasing sequence of subspaces $\{G_n\}$, each containing G, and for each n an element $u_n \in U_{G_n}$ such that $u_n \to u$ weakly in $W_0^{1,p}(\Omega)$ as $n \to \infty$ (Proposition 11, p.274, (Browder & Hess 1972)). Let us denote by $\{\chi_n\} \subset L^1(\Omega)$ the corresponding sequence with the property that for each n a pair (u_n,χ_n) is a solution of (P_{G_n}). By Proposition 4.2 we can suppose without loss of generality that $\chi_n \to \chi^G$ weakly in $L^1(\Omega)$ for some $\chi^G \in L^1(\Omega)$. Thus we have asserted that

$$u_n \to u \quad \text{weakly in } W_0^{1,p}(\Omega) \tag{4.61}$$

$$\chi_n \to \chi^G \quad \text{weakly in } L^1(\Omega) \tag{4.62}$$

and that (4.48) with F replaced by G_n reads

$$\langle Au_n, v-u_n\rangle_{W_0^{1,p}(\Omega)} + \int_\Omega \chi_n(v-u_n)\,d\Omega = 0, \quad \forall v \in G_n, \tag{4.63}$$

where $A : W_0^{1,p}(\Omega) \to W^{-1,p'}(\Omega)$ is defined by (4.35).

Step 2. Now we prove that $\chi^G \in \partial j(u)$ a.e in Ω. Since $W_0^{1,p}(\Omega)$ is compactly imbedded into $L^p(\Omega)$, due to (4.37) one may suppose that

$$u_n \to u \quad \text{strongly in } L^p(\Omega). \tag{4.64}$$

This implies that for a subsequence of $\{u_n\}$ (again denoted by the same symbol) one gets $u_n \to u$ a.e. in Ω. Thus Egoroff's theorem can be applied from which it follows that for any $\varepsilon > 0$ a subset $\omega \subset \Omega$ with $|\omega| < \varepsilon$ can be determined such that $u_n \to u$ uniformly in $\Omega\setminus\omega$ with $u \in L^\infty(\Omega\setminus\omega)$. Let $v \in L^\infty(\Omega\setminus\omega)$ be an arbitrary function. From the estimate

$$\int_{\Omega\setminus\omega} \chi_n v\,d\Omega \le \int_{\Omega\setminus\omega} j^0(u_n;v)\,d\Omega$$

combined with the weak convergence in $L^1(\Omega)$ of χ_n to χ^G, (4 64) and with the upper semicontinuity of

$$L^\infty(\Omega\setminus\omega) \ni u_n \longmapsto \int_{\Omega\setminus\omega} j^0(u_n;v)\,d\Omega$$

it follows

$$\int_{\Omega\setminus\omega} \chi^G v\,d\Omega \le \int_{\Omega\setminus\omega} j^0(u;v)\,d\Omega, \quad \forall v \in L^\infty(\Omega\setminus\omega).$$

But the last inequality amounts to saying that $\chi^G \in \partial j(u)$ a.e. in $\Omega \setminus \omega$. Since $|\omega| < \varepsilon$ and ε was chosen arbitrarily,

$$\chi^G \in \partial j(u) \quad \text{a.e. in } \Omega, \tag{4.65}$$

as claimed.

Step 3. Now it will be shown that

$$\limsup_{n\to\infty} \int_\Omega j^0(u_n; v - u_n)\, d\Omega \le \int_\Omega j^0(u; v-u)\, d\Omega \tag{4.66}$$

holds for any $v \in W_0^{1,p}(\Omega) \cap L^\infty(\Omega)$. It can be supposed that $u_n \to u$ a.e. in Ω, since $u_n \to u$ in $L^q(\Omega)$. Fix $v \in L^\infty(\Omega)$ arbitrarily. In view of $\chi_n \in \partial j(u_n)$ and (4.50) we get

$$j^0(u_n; v - u_n) \le \alpha(\|v\|_{L^\infty(\Omega)})(1 + |u_n|^q). \tag{4.67}$$

From Egoroff's theorem it follows that for any $\varepsilon > 0$ a subset $\omega \subset \Omega$ with $|\omega| < \varepsilon$ can be determined such that $u_n \to u$ uniformly in $\Omega \setminus \omega$. One can also suppose that ω is small enough to fulfill $\int_\omega \alpha(\|v\|_{L^\infty(\Omega)})(1 + |u_n|^q)\, d\Omega \le \varepsilon$, $n = 1, 2, \ldots$, and $\int_\omega \alpha(\|v\|_{L^\infty(\Omega)})(1 + |u|^q)\, d\Omega \le \varepsilon$. Hence

$$\int_\Omega j^0(u_n; v - u_n)\, d\Omega \le \int_{\Omega\setminus\omega} j^0(u_n; v - u_n)\, d\Omega + \varepsilon$$

which by Fatou's lemma and upper semicontinuity of $j^0(\cdot\,;\cdot)$ yields

$$\limsup_{n\to\infty} \int_\Omega j^0(u_n; v - u_n)\, d\Omega \le \int_\Omega j^0(u; v - u)\, d\Omega + 2\varepsilon.$$

By arbitrariness of $\varepsilon > 0$ one obtains (4.66), as required.

Step 4. Now we show that

$$\chi^G u \in L^1(\Omega). \tag{4.68}$$

$$\liminf_{n\to\infty} \int_\Omega \chi_n u_n\, d\Omega \ge \int_\Omega \chi^G u\, d\Omega \tag{4.69}$$

For this purpose let $\{\epsilon_k\} \subset L^\infty(\Omega)$ be such that (Hedberg 1978):

$$\begin{aligned} &\{(1-\epsilon_k)u\} \subset W_0^{1,p}(\Omega) \cap L^\infty(\Omega), \quad 0 \le \epsilon_k \le 1 \\ &\widetilde{u}_k := (1-\epsilon_k)u \to u \text{ strongly in } W_0^{1,p}(\Omega) \text{ as } k \to \infty. \end{aligned} \tag{4.70}$$

Without loss of generality it can be assumed that $\widetilde{u}_k \to u$ a.e. in Ω. Since it is already known that $\chi^G \in \partial j(u)$, one can apply (H3) to obtain $\chi^G(-u) \le j^0(u; -u) \le \kappa(1 + |u|^q)$. Hence

$$\chi^G \widetilde{u}_k = (1-\epsilon_k)\chi^G u \ge -\kappa(1 + |u|^q). \tag{4.71}$$

This implies that the sequence $\{\chi^G \widetilde{u}_k\}$ is bounded from below by integrable function and $\chi^G \widetilde{u}_k \to \chi^G u$ a.e. in Ω. On the other hand, one gets

$$\int_\Omega \chi_n(\widetilde{u}_k - u_n)\, d\Omega \le \int_\Omega j^0(u_n; \widetilde{u}_k - u_n)\, d\Omega.$$

Thus passing to the limit with $n \to \infty$ yields

$$\int_\Omega \chi^G \widetilde{u}_k \, d\Omega - \liminf_{n\to\infty} \int_\Omega \chi_n u_n \, d\Omega \le \limsup_{n\to\infty} \int_\Omega j^0(u_n; \widetilde{u}_k - u_n) \, d\Omega,$$

and due to (4.66) we are led to the estimate

$$\begin{aligned} \int_\Omega \chi^G \widetilde{u}_k \, d\Omega &\le \liminf_{n\to\infty} \int_\Omega \chi_n u_n \, d\Omega + \int_\Omega j^0(u; \widetilde{u}_k - u) \, d\Omega \\ &\le \liminf_{n\to\infty} \int_\Omega \chi_n u_n \, d\Omega + \int_\Omega j^0(u; -\epsilon_k u) \, d\Omega \\ &\le \liminf_{n\to\infty} \int_\Omega \chi_n u_n \, d\Omega + \int_\Omega \epsilon_k \kappa (1 + |u|^q) \, d\Omega \le C, \quad C = \text{const}. \end{aligned}$$

Thus by Fatou's lemma we are allowed to conclude that $\chi^G u \in L^1(\Omega)$, i.e. (4.68) holds. Taking into account that $\epsilon_k \to 0$ a.e. in Ω as $k \to \infty$ (passing to a subsequence if necessary) we establish (4.69), as required.

Step 5. It will be shown that

$$\begin{aligned} &\langle Au, v - u \rangle_{W_0^{1,p}(\Omega)} + \int_\Omega \chi^G (v - u) \, d\Omega = 0, \quad \forall v \in \bigcup_{n=1}^\infty G_n \supset G \\ &\chi^G \in \partial j(u). \end{aligned} \tag{Q^G}$$

Since A is bounded and $\{u_F\}_{F\in\Lambda} \subset \{v \in W_0^{1,p}(\Omega) \colon \|v\|_{W_0^{1,p}(\Omega)} \le M\}$, there exists $K > 0$ such that $\{Au_F\}_{F\in\Lambda} \subset \{l \in W^{-1,p'}(\Omega) \colon \|l\|_{W^{-1,p'}(\Omega)} \le K\}$. From (4.63) it follows that for any fixed $G \in \Lambda$ we get

$$\left| \int_\Omega \chi^G v \, d\Omega \right| \le K \|v\|_V, \quad \forall v \in \bigcup_{n=1}^\infty G_n, \; \chi^G \in \partial j(u), \tag{4.72}$$

because $\{G_n\}$ is an increasing sequence. Further, by making use of (4.68) and (4.69) we have $\chi^G u \in L^1(\Omega)$ and

$$\limsup_{n\to\infty} \langle Au_n, u_n - u \rangle_{W_0^{1,p}(\Omega)} \le \int_\Omega \chi^G (v - u) \, d\Omega, \quad \forall v \in \bigcup_{n=1}^\infty G_n. \tag{4.73}$$

Since $u_n \in G_n$ and $u_n \to u$ weakly in $W_0^{1,p}(\Omega)$, the closure of $\bigcup_{n=1}^\infty G_n$ in the strong topology of $W_0^{1,p}(\Omega)$, $\overline{\bigcup_{n=1}^\infty G_n}$, must contain u. Thus there exists a sequence $\{w_i\} \subset \bigcup_{n=1}^\infty G_n$ converging strongly to u in $W_0^{1,p}(\Omega)$ as $i \to \infty$. We claim that for such a sequence,

$$\int_\Omega \chi^G w_i \, d\Omega \to \int_\Omega \chi^G u \, d\Omega \quad \text{as } i \to \infty. \tag{4.74}$$

Indeed, let $\{\widetilde{u}_k\}_{k=1}^\infty$ be given by (4.70). From (4.71) it follows

$$-\kappa(1 + |u|^q) \le \chi^G \widetilde{u}_k \le |\chi^G u|, \quad k = 1, 2 \ldots, \tag{4.75}$$

with the bounds $-\kappa(1+|u|^q)$ and $|\chi^G u|$ being integrable in Ω. Thus there exists a constant $C>0$ such that

$$\left|\int_\Omega \chi^G \widetilde{u}_k \, d\Omega\right| \le C\|\widetilde{u}_k\|_{W_0^{1,p}(\Omega)}, \quad k=1,2,\ldots. \tag{4.76}$$

Denote by $\mathcal{A}$ a linear subspace spaned by $\{\widetilde{u}_k\}_{k=1}^\infty$ and define a linear functional $\widehat{l}_{\chi^G} : \bigcup_{n=1}^\infty G_n + \mathcal{A} \to \mathbb{R}$ by the formula

$$\widehat{l}_{\chi^G}(v) := \int_\Omega \chi^G v \, d\Omega, \quad v \in \bigcup_{n=1}^\infty G_n + \mathcal{A}.$$

Taking into account (4.72) and (4.76), from the Hahn-Banach theorem it follows that $\widehat{l}_{\chi^G}$ admits its linear continuous extension onto $W_0^{1,p}(\Omega)$, $l_{\chi^G} \in W^{-1,p'}(\Omega)$. By the dominated convergence,

$$\int_\Omega \chi^G \widetilde{u}_k \, d\Omega \to \int_\Omega \chi^G u \, d\Omega, \quad \text{as } k \to \infty,$$

so we get $l_{\chi^G}(u) = \int_\Omega \chi^G u \, d\Omega$ which, in particular, implies (4.74), as claimed.

Taking into account (4.73) and (4.74) we conclude

$$\limsup_{n\to\infty} \langle Au_n, u_n - u\rangle_{W_0^{1,p}(\Omega)} \le 0, \tag{4.77}$$

which by the pseudomonotonicity of A implies

$$Au_n \to Au \ \text{ weakly in } W_0^{1,p}(\Omega) \tag{4.78}$$

$$\langle Au_n, u_n\rangle_{W_0^{1,p}(\Omega)} \to \langle Au, u\rangle_{W_0^{1,p}(\Omega)}. \tag{4.79}$$

Hence from (4.63) we are led to (Q^G), as desired. Notice that (4.78) and (4.79) imply the strong convergence $u_n \to u$ in $W_0^{1,p}(\Omega)$.

Step 6. It remains to show that there exists $\chi \in \partial j(u)$ with the associated linear functional defined by

$$\widehat{l}_\chi(v) := \int_\Omega \chi v \, d\Omega, \quad \forall v \in W_0^{1,p}(\Omega) \cap L^\infty(\Omega),$$

admitting a continuous extension $l_\chi \in W^{-1,p'}(\Omega)$ such that

$$Au + l_\chi = 0, \quad \langle l_\chi, u\rangle_{W_0^{1,p}(\Omega)} = \int_\Omega \chi u \, d\Omega. \tag{4.80}$$

For every $G \in \Lambda$ let us introduce

$$V^{(G)} = \{\chi^G \in L^\infty(\Omega) \colon (Q^G) \text{ holds}\}$$

and

$$Z^{(G)} = \bigcup_{\substack{G' \in \Lambda \\ G' \supset G}} V^{(G')}.$$

As in the proof of Proposition 4.2 we show that the family $\{\chi^G\}_{G\in\Lambda}$ is weakly precompact in $L^1(\Omega)$. Denoting by weakcl$(Z^{(G)})$ the closure of $Z^{(G)}$ in the weak topology of $L^1(\Omega)$ we prove analogously that the family $\{\text{weakcl}(Z^{(G)})\}_{G\in\Lambda}$ has the finite intersection property. Thus there exists an element $\chi \in \partial j(u)$ such that for any $G \in \Lambda$ it holds

$$\langle Au, v\rangle_{W_0^{1,p}(\Omega)} + \int_\Omega \chi\, v\, d\Omega = 0, \quad \forall v \in G.$$

Since $G \in \Lambda$ has been chosen arbitrarily and Λ is dense in $W_0^{1,p}(\Omega)$, (4.80) results, as desired.

Step 7. It remains to show (4.58). From (4.59) we obtain easily its validity for any $v \in W_0^{1,p}(\Omega) \cap L^\infty(\Omega)$.

Let us consider the case $j^0(u; v-u) \in L^1(\Omega)$ with $v \in W_0^{1,p}(\Omega)$. There exists a sequence $\widetilde{v}_k = (1-\epsilon_k)v$ such that $\{\widetilde{v}_k\} \subset W_0^{1,p}(\Omega) \cap L^\infty(\Omega)$, $\widetilde{v}_k \to v$ strongly in $W_0^{1,p}(\Omega)$. Since, as already has been established,

$$\langle Au, \widetilde{v}_k - u\rangle_{W_0^{1,p}(\Omega)} + \int_\Omega j^0(u; \widetilde{v}_k - u)\, d\Omega \geq 0,$$

so in order to show (4.58) it remains to deduce that

$$\limsup_{k\to\infty} \int_\Omega j^0(u; \widetilde{v}_k - u)\, d\Omega \leq \int_\Omega j^0(u; v-u)\, d\Omega.$$

For this purpose let us observe that $\widetilde{v}_k - u = (1-\epsilon_k)(v-u) + \epsilon_k(-u)$ which combined with the convexity of $j^0(u;\cdot)$ yields the estimate

$$j^0(u; \widetilde{v}_k - u) \leq (1-\epsilon_k) j^0(u; v-u) + \epsilon_k j^0(u; -u) \leq \left|j^0(u; v-u)\right| + \kappa(1+|u|^q).$$

Thus Fatou's lemma implies the assertion.

Consider the case $j^0(u; v-u) \notin L^1(\Omega)$. Recall that if $j^0(u; v-u) \notin L^1(\Omega)$ then according to the convention that $+\infty - \infty = +\infty$ we have

$$\int_\Omega j^0(u; v-u)\, d\Omega =$$
$$= \begin{cases} +\infty & \text{if } \int_\Omega [j^0(u; v-u)]^+\, d\Omega = +\infty \\ -\infty & \text{if } \int_\Omega [j^0(u; v-u)]^+\, d\Omega < +\infty \text{ and } \int_\Omega [j^0(u; v-u)]^-\, d\Omega = +\infty, \end{cases}$$

where the notation has been used: $r^+ := \max\{r, 0\}$ and $r^- := \max\{-r, 0\}$ for any $r \in \mathbb{R}$. Thus, if $\int_\Omega j^0(u; v-u)\, d\Omega = +\infty$ then (4.58) holds immediately.

Now we show that the case $\int_\Omega j^0(u; v-u)\, d\Omega = -\infty$ is not allowed for any $v \in W_0^{1,p}(\Omega)$. Indeed, if we suppose that for some $v \in W_0^{1,p}(\Omega)$, $\int_\Omega j^0(u; v-u)\, d\Omega = -\infty$, then one can find a sequence $\widetilde{v}_k = (1-\epsilon_k)v$ such that $\{\widetilde{v}_k\} \subset W_0^{1,p}(\Omega) \cap L^\infty(\Omega)$, $\widetilde{v}_k \to v$ strongly in $W_0^{1,p}(\Omega)$. Since, as already has been established,

$$\langle Au, \widetilde{v}_k - u\rangle_{W_0^{1,p}(\Omega)} + \int_\Omega j^0(u; \widetilde{v}_k - u)\, d\Omega \geq 0,$$

we get

$$\int_\Omega j^0(u;\widetilde{v}_k-u)\,d\Omega \geq \langle Au,-\widetilde{v}_k+u\rangle_{W_0^{1,p}(\Omega)} \geq -C, \quad C=\text{const},$$

and consequently

$$\int_\Omega [j^0(u;\widetilde{v}_k-u)]^+\,d\Omega \geq \int_\Omega [j^0(u;\widetilde{v}_k-u)]^-\,d\Omega - C. \tag{4.81}$$

By the hypothesis we have $\int_\Omega [j^0(u;v-u)]^-\,d\Omega = +\infty$ and $\int_\Omega [j^0(u;v-u)]^+\,d\Omega < +\infty$. Since

$$j^0(u;\widetilde{v}_k-u) \leq (1-\epsilon_k)j^0(u;v-u) + \epsilon_k j^0(u;-u) \leq (1-\epsilon_k)j^0(u;v-u) \\ +\kappa(1+|u|^q),$$

so we obtain

$$\int_\Omega [j^0(u;v_k-u)]^+\,d\Omega \leq \int_\Omega [j^0(u;v-u)]^+\,d\Omega + \int_\Omega \kappa(1+|u|^q)\,d\Omega \leq D, \quad D=\text{const},$$

which combined with (4.81) yields

$$\int_\Omega [j^0(u;\widetilde{v}_k-u)]^-\,d\Omega \leq C+D.$$

The application of Fatou's lemma concludes

$$\int_\Omega [j^0(u;v-u)]^-\,d\Omega \leq C+D,$$

which is a contradiction with the assumption that $\int_\Omega j^0(u;v-u)\,d\Omega = -\infty$. This contradiction completes the proof of (4.58).

Step 8. In order to show that $j(u) \in L^1(\Omega)$ it is enough to use (4.17) and (4.38) to get

$$\int_\Omega j(u_n)\,d\Omega \leq \gamma_2 - \tfrac{1}{p}\|Du_n\|^p_{L^p(\Omega;\mathbb{R}^N)} + \tfrac{\lambda_1}{p}\|u_n\|^p_{L^p(\Omega)} \leq \gamma_2$$

and

$$j(u_n) \geq -\kappa_0(1+|u_n|^q).$$

Since $j(u_n) \to j(u)$ a.e. in Ω as $n\to\infty$, we are allowed to apply Fatou's lemma which yields the assertion.

Step 9. The existence of a nontrivial solution $u \neq 0$ follows from (H6). Indeed, if we suppose that $u=0$ then we have $\{u_n\} \subset W_0^{1,p}(\Omega)\cap L^\infty(\Omega)$ and $u_n \to 0$ strongly in $W_0^{1,p}(\Omega)$. By making use of (4.63) with $v=2u_n$ and the Rayleigh quotient characterization of λ_1, it follows

$$\int_\Omega \min\{\psi u_n : \psi \in \partial j(u_n)\}\,d\Omega \leq \int_\Omega \chi_n u_n\,d\Omega \leq 0.$$

Hence, by (H6),

$$\limsup_{n\to\infty} \int_\Omega j(u_n)\, d\Omega \le 0$$

and consequently,

$$\limsup_{n\to\infty} \mathcal{R}(u_n) \le 0,$$

which contradicts to (4.38). This contradiction yields the assertion. The proof of Theorem 4.1 is complete. □

From (4.59) and (4.60) we obtain the result.

Corollary 4.2. *Assume the hypotheses (H0)-(H6). Then the problem: find $u \in W_0^{1,p}(\Omega)$ and $\chi \in L^1(\Omega)$ such that*

$$(P) \qquad \begin{cases} \Delta_p u + \lambda_1 |u|^{p-2} u &= \chi & \text{in the distributional sense} \\ \chi &\in \partial j(u) & \text{a.e. in } \Omega \\ \chi u &\in L^1(\Omega) & \\ j(u) &\in L^1(\Omega) & \\ u &= 0 & \text{on } \partial\Omega \text{ (in the sense of traces).} \end{cases}$$

has at least one nontrivial solution ($u \neq 0$).

Remark 4.1. The energy functional $\mathcal{R}$ is finite at a solution u of (P), i.e. $\mathcal{R}(u) = \|Du\|^p_{L^p(\Omega;\mathbb{R}^N)} - \lambda_1 \|u\|^p_{L^p(\Omega)} + \int_\Omega j(u)\, d\Omega \in \mathbb{R}$.

Remark 4.2. In the case of the unilateral growth condition as formulated in (H3), the function $J(v) = \int_\Omega j(v)\, d\Omega$, $v \in W_0^{1,p}(\Omega)$, is not upper semicontinuous. Thus the problem concerning the existence of a nontrivial solution of (P) arises because we are not allowed to conclude by making use of the estimate (4.38) that $\mathcal{R}(u) \ge \eta_1 > 0$. To overcome this difficulty the hypothesis (H6) has been introduced.

Note that when the classical growth condition $|\partial j(\xi)| \le c(1 + |\xi|^{q-1})$ $\forall \xi \in \mathbb{R}$, holds then the upper semicontinuity of J is ensured.

5 Signorini problem

Let us consider a linear elastic body which in its undeformed state ocupies an open, bounded, connected subset Ω of $\mathbb{R}^3$ with Lipschitz boundary Γ. Ω is referred to a fixed Cartesian coordinate system $0x_1x_2x_3$. Denote by $S = \{S_i\}$ the stress vector on Γ. We recall that $S_i = \sigma_{ij} n_j$, where $\sigma = \{\sigma_{ij}\}$ is the stress tensor and $n = \{n_i\}$ is the outward unit normal vector on Γ. The vector S may be decomposed into a normal component S_N and a tangential component S_T with respect to Γ, i.e.,

$$S_N = \sigma_{ij} n_j n_i \quad \text{and} \quad S_{T_i} = \sigma_{ij} n_j - (\sigma_{ij} n_i n_j) n_i,$$

where the summation convention holds.
Analogously to S_N and S_T, u_N and u_T denote the normal and the tangential components of the displacement vector u with respect to Γ, i.e.

$$u_N = u_i n_i \quad \text{and} \quad u_{T_i} = u_i - (u_j n_j) n_i.$$

S_N and u_N are considered as positive if they are parallel to n.
Γ is decomposed into three nonoverlaping parts Γ_F, Γ_U and Γ_S, i.e. $\overline{\Gamma} = \overline{\Gamma_F} \cup \overline{\Gamma_U} \cup \overline{\Gamma_S}$. On Γ_U the Signorini boundary conditions are prescribed, which hold when an elastic body is in contact with a rigid support. They read

$$\begin{aligned} &\text{if } u_N < 0, \quad \text{then } S_N = 0; \\ &\text{if } u_N = 0, \quad \text{then } S_N \leq 0, \quad \text{on } \Gamma_U \end{aligned}$$

or equivalently

$$S_N \leq 0, \quad u_N \leq 0, \quad \text{and} \quad S_N u_N = 0, \quad \text{on } \Gamma_U.$$

The surface forces $F = (F_i)$ are given on Γ_F, $F_i \in L^2(\Omega)$, i.e.

$$S_i = \sigma_{ij} n_j = F_i, \quad \text{on } \Gamma_F.$$

Let us assume that on Γ_S the general nonmonotone possibly multivalued boundary condition holds:

$$-S \in \partial j(u), \quad \text{on } \Gamma_S,$$

where j is a given locally Lipschitz function.

Assuming small deformations, the equilibrium equations take the form

$$\sigma_{ij,j} + b_i = 0, \quad \text{on } \Omega,$$

where $b_i \in L^2(\Omega)$ is the volume force vector, while the strain-displacement relationship can be written as

$$\varepsilon_{ij} = \frac{1}{2}(u_{i,j} + u_{j,i}).$$

Let us assume further that the body is linear elastic, i.e. that

$$\sigma_{ij} = C_{ijhk} \varepsilon_{hk},$$

where $C = \{C_{ijhk}\}$, $i, j, h, k = 1, 2, 3$, $C_{ijhk} \in L^\infty(\Omega)$, is the elasticity tensor which has the well-known symmetry and ellipticity properties

$$C_{ijhk} = C_{jihk} = C_{khij}$$

$$C_{ijhk} \varepsilon_{ij} \varepsilon_{hk} \geq c \varepsilon_{ij} \varepsilon_{hk} \quad \forall \varepsilon = \{\varepsilon_{ij}\} \in \mathbb{R}^6, \quad c = \text{const} > 0.$$

In order to give a variational formulation of the foregoing problem, we introduce the following notations:

$$V = H^1(\Omega; \mathbb{R}^3);$$

$$A : V \to V^\star, \quad \langle Au, v \rangle := \int_\Omega C_{ijhk} \varepsilon_{ij}(u) \varepsilon_{hk}(v) d\Omega, \quad u, v \in V;$$

$$f \in V^\star, \quad \langle f, v \rangle := \int_\Omega b_i v_i d\Omega + \int_{\Gamma_F} F_i v_i d\Gamma, \quad v \in V;$$

$$K \subset V, \quad K := \{v \in V : v_N \leq 0 \text{ on } \Gamma_U\}.$$

The variational formulation of the aforementioned problem takes the form:
Find $u \in K$ such that

$$\int_{\Omega} C_{ijhk}\varepsilon_{ij}(u)\varepsilon_{hk}(v-u)d\Omega$$

$$-\int_{\Omega} b_i(v_i-u_i)d\Omega - \int_{\Gamma_F} F_i(v_i-u_i)d\Gamma + \int_{\Gamma_S} j^0(u;v-u)d\Gamma \geq 0,$$

$$\forall v \in K,$$

or equivalently,

$$\langle Au - f, v-u\rangle + \Phi(v) - \Phi(u) + \int_{\Gamma_S} j^0(u;v-u)d\Gamma \geq 0, \quad \forall v \in V, \tag{5.1}$$

where $\Phi : V \to \mathbb{R} \cup \{+\infty\}$ is the indicator function of K, i.e.

$$\Phi(v) = \begin{cases} 0 & \text{if } v \in K. \\ +\infty & \text{if } v \notin K. \end{cases}$$

Thus we have put our considerations into the framework of Section 2.

Let us denote by $\mathcal{R} \subset V$ the space of all rigit-body displacements, i.e. if $r \in \mathcal{R}$, then $r(x) = a + b \wedge x$, $a, b \in \mathbb{R}^3$, $x \in \Omega$. From the second Korn inequality and the ellipticity of the linear-ellasticity tensor it follows that each elememt $u \in V$ can be represented as $u = \widehat{u} + r$, $r \in \mathcal{R}$, for which the inequality holds:

$$\int_{\Omega} C_{ijhk}\varepsilon_{ij}(u)\varepsilon_{hk}(u)d\Omega \geq c\|\widehat{u}\|_V^2, \quad \forall u = \widehat{u} + r,\ r \in \mathcal{R}.$$

This means that A fulfills the semicorecivity condition (2.1). Further, let us assume that j satisfies the unilateral growth conditions:

$$j^0(\xi;\eta-\xi) \leq \alpha(\rho)(1+|\xi|^s),\ \ \forall\, \xi,\eta \in \mathbb{R}^3,\ |\eta| \leq \rho,\ \rho \geq 0, \tag{5.2}$$

$$j^0(\xi;-\xi) \leq \widehat{k}\,|\xi|^{\sigma}, \quad \forall\, \xi \in \mathbb{R}^3, \quad 1 \leq \sigma < 2. \tag{5.3}$$

for some $s < 6$ and nondecreasing function $\alpha : \mathbb{R}_+ \to \mathbb{R}_+$. Let $j^{\infty} : \mathbb{R}^3 \to \mathbb{R} \cup \{+\infty\}$ be the recession function of j, given by

$$j^{\infty}(\xi) = \liminf_{\substack{\eta\to\xi \\ t\to+\infty}} [-j^0(t\eta;-\eta)],\ \ \xi \in \mathbb{R}^3.$$

Suppose that the body forces b_i and surface tractions F_i satisfy the compatibility condition of the form

$$\int_{\Omega} b_i r_i d\Omega + \int_{\Gamma_F} F_i r_i d\Gamma < \int_{\Omega} j^{\infty}(r)d\Omega, \quad \forall\, r \in (\mathcal{R} \cap K) \setminus \{0\}. \tag{5.4}$$

Since K is a convex, closed cone, the recession function corresponding to the indicator function of K, coincides with the same function, i.e. $\Phi^\infty = \Phi$, where $\Phi^\infty : V \to \mathbb{R}\cup\{+\infty\}$ is defined by

$$\Phi^\infty(u) = \lim_{\lambda\to+\infty} \frac{\Phi(u_0+\lambda u) - \Phi(u_0)}{\lambda}, \quad u_0 \in \mathrm{Dom}\,\Phi.$$

Thus (5.4) implies the compatibility condition (2.12).

Finally, let us check that the hypothesis (H) of Theorem 2.1 is fulfilled. Indeed, if $u \in K$ then $u_N \le 0$ on Γ_S and $\Phi(u) = 0$. By the truncation result for vector valued Sobolev spaces, (Naniewicz 1997), there exists a sequence of functions $\{\varepsilon_n\} \subset L^\infty(\Omega)$ with $0 \le \varepsilon_n \le 1$ such that

$$\{(1-\varepsilon_n)u\} \subset V \cap L^\infty(\Omega;\mathbb{R}^3)$$
$$(1-\varepsilon_n)u \to u \quad \text{strongly in } V.$$

Since $u_N \le 0$ on Γ_S we have $((1-\varepsilon_n)u)_N = (1-\varepsilon_n)u_N \le 0$ on Γ_S, which means that $(1-\varepsilon_n)u \in K$. Hence $\Phi(u) = \Phi((1-\varepsilon_n)u) = 0$ and the hypothesis (H) holds.

Now on the basis of Theorem 2.1 and Theorem 2.70 we are ready to formulate the following existence result.

Theorem 5.1. *Assume that* (5.2)-(5.4) *hold. Then the variational-hemivariational inequality: Find* $u \in K$ *such that*

$$\int_\Omega C_{ijhk}\varepsilon_{ij}(u)\varepsilon_{hk}(v-u)d\Omega$$
$$-\int_\Omega b_i(v_i-u_i)d\Omega - \int_{\Gamma_F} F_i(v_i-u_i)d\Gamma + \int_{\Gamma_S} j^0(u;v-u)d\Gamma \ge 0,$$
$$\forall v \in K,$$

admits solutions. Moreover, there exists $\chi \in L^1(\Gamma_S;\mathbb{R}^3)$, $\chi \in \partial j(u)$ *on* Γ_S, $\chi_i u_i \in L^1(\Gamma_S)$, *such that for* l_χ *defined by*

$$\langle l_\chi, v\rangle := \int_{\Gamma_S} \chi_i v_i d\Gamma, \quad v \in V \cap L^\infty(\Gamma_S;\mathbb{R}^3),$$

it follows that

$$f - Au - l_\chi \in \widehat{\partial}\Phi(u),$$

where $\widehat{\partial}\Phi$ *is a* $\mathcal{L}(\mathcal{V})$*-subgradient of* Φ *at* u *in the sense of Pallaschke-Rolewicz, (Pallaschke & Rolewicz 1997).*

References

Adly, S. & Goeleven, D. (2000), 'A discretization theory for a class of semi-coercive unilateral problems', *Numerische Mathematik* **87**, 1–34.

Ambrosetti, A. & Rabinowitz, P. H. (1973), 'Dual variational methods in critical point theory and applications', *J. Funct. Anal.* **14**, 349–381.

Anane, A. (1988), Etude des valeurs propres et de la résonance pour l'opérateur p-Laplacien, PhD thesis, Université Libre de Bruxelles, Brussels.

Baiocchi, C., Buttazzo, G., Gastaldi, F. & Tomarelli, F. (1988), 'General existence theorems for unilateral problems in continuum mechanics', *Arch. Rational Mech. Anal.* **100**, 149–188.

Brézis, H. (1968), 'Équations et inéquations non-linéaires dans les espaces véctoriels en dualité', *Comm. Pure Appl. Math.* **18**, 115–176.

Brézis, H. & Nirenberg, L. (1978), 'Characterizations of the ranges of some nonlinear operators and applications to boundary value problems', *Ann. Scuola Normale Superiore Pisa, Classe di Scienze* **V**, 225–326.

Browder, F. E. & Hess, P. (1972), 'Nonlinear mappings of monotone type in Banach spaces', *J. Funct. Anal.* **11**, 251–294.

Chang, K. C. (1981), 'Variational methods for non-differentiable functionals and their applications to partial differential equations', *J. Math. Anal. Appl.* **80**, 102–129.

Clarke, F. H. (1983), *Optimization and Nonsmooth Analysis*, John Wiley & Sons.

Dem'yanov, V. F., Stavroulakis, G. E., Polyakova, L. N. & Panagiotopoulos, P. D. (1996), *Quasidifferentiability and Nonsmooth Modelling in Mechanics, Engineering and Economics*, Kluwer Academic Publishers.

Duvaut, G. & Lions, J. L. (1972), *Les Inéquations en Mécanique et en Physique*, Dunod.

Ekeland, I. & Temam, R. (1976), *Convex Analysis and Variational Problems*, North-Holland.

Fleckinger-Pellé, J. & Tkáč, P. (2002), 'An improved Poincaré inequality and the p-Laplacian at resonance', *Advances in Differential Equations* **7**, 951–971.

Gasiński, L. & Papageorgiou, N. S. (2000), 'Nonlinear hemivariational inequalities at resonance', *J. Math. Anal. Appl.* **244**, 200–213.

Goeleven, D. (1996), *Noncoercive Variational Problems and Related Topics*, Vol. 357 of *Pitman Research Notes in Mathematics Series*, Longman.

Goeleven, D., Motreanu, D. & Panagiotopoulos, P. D. (1997), 'Multiple solutions for a class of eigenvalue problems in hemivariational inequalities', *Nonlinear Anal.* **29**, 9–26.

Goeleven, D. & Théra, M. (1995), 'Semicoercive variational hemivariational inequalities', *J. Global Optim.* **6**, 367–381.

Haslinger, J., Miettinen, M. & Panagiotopoulos, P. D. (1999), *Finite Element Method for Hemivariational Inequalities. Theory, Methods and Applications*, Kluwer Academic Publishers.

Hedberg, L. I. (1978), 'Two approximation problems in function spaces', *Ark. Mat.* **16**, 51–81.

Lindqvist, P. (1990), 'On the equation $\text{div}(|Dx|^{p-2}Dx) + \lambda|x|^{p-2}x = 0$', *Proceedings of the American Math. Society* **109**(1).

Motreanu, D. & Naniewicz, Z. (1996), 'Discontinuous semilinear problems in vector-valued function spaces', *Differential and Integral Equations* **9**, 581–598.

Motreanu, D. & Naniewicz, Z. (2001), 'A topological approach to hemivariational inequalities with unilateral growth condition', *J. Appl. Anal.* **7**, 23–41.

Motreanu, D. & Naniewicz, Z. (2002), Semilinear hemivariational inequalities with Dirichlet boundary condition, *in* Y. Gao, D & R. W. Ogden, eds, 'Advances in Mechanics and Mathematics: AMMA 2002', Advances in Mechanics and Mathematics, Kluwer Academic Publishers, pp. 89–110.

Motreanu, D. & Panagiotopoulos, P. D. (1995), 'Nonconvex energy functions, Related eigenvalue hemivariational inequalities on the sphere and applications', *J. Global Optimiz.* **6**, 163–177.

Motreanu, D. & Panagiotopoulos, P. D. (1996), 'On the eigenvalue problem for hemivariational inequalities: existence and multiplicity of solutions', *J. Math. Anal. Appl.* **197**, 75–89.

Motreanu, D. & Panagiotopoulos, P. D. (1999), *Minimax Theorems and Qualitative Properties of the Solutions of Hemivariational Inequalities*, Kluwer Academic Publishers.

Naniewicz, Z. (1994*a*), 'Hemivariational inequalities with functions fulfilling directional growth condition', *Applicable Analysis* **55**, 259–285.

Naniewicz, Z. (1994*b*), 'Hemivariational inequality approach to constrained problems for star-shaped admissible sets', *J. Optim. Theory Appl.* **83**, 97–112.

Naniewicz, Z. (1995*a*), 'Hemivariational inequalities with functionals which are not locally Lipschitz', *Nonlinear Anal.* **25**, 1307–1320.

Naniewicz, Z. (1995*b*), 'On variational aspects of some nonconvex nonsmooth global optimization problem', *J. Global Optim.* **6**, 383–400.

Naniewicz, Z. (1997), 'Hemivariational inequalities as necessary conditions for optimality for a class of nonsmooth nonconvex functionals', *Nonlinear World* **4**, 117–133.

Naniewicz, Z. (2003), 'Pseudomonotone semicoercive variational-hemivariational inequalities with unilateral growth condition', *Control & Cybernetics* **32**, 223–244.

Naniewicz, Z. & Panagiotopoulos, P. D. (1995), *Mathematical Theory of Hemivariational Inequalities and Applications*, Marcel Dekker.

Pallaschke, D. & Rolewicz, S. (1997), *Foundation of Mathematical Optimization*, Kluwer Academic Publishers.

Panagiotopoulos, P. D. (1981), 'Non-convex superpotentials in the sense of F.H. Clarke and applications', *Mech. Res. Comm.* **8**, 335–340.

Panagiotopoulos, P. D. (1983), 'Noncoercive energy function, hemivariational inequalities and substationarity principles', *Acta Mech.* **48**, 160–183.

Panagiotopoulos, P. D. (1985), *Inequality Problems in Mechanics and Applications. Convex and Nonconvex Energy Functions*, Birkhäuser Verlag.

Panagiotopoulos, P. D. (1993), *Hemivariational Inequalities. Applications in Mechanics and Engineering*, Springer-Verlag.

Pop, G., Panagiotopoulos, P. D. & Naniewicz, Z. (1997), 'Variational-hemivariational inequalities for multidimensional superpotential laws', *Numer. Funct. Anal. Optim.* **18**, 827–856.

Rabinowitz, H. (1986), *Minimax Methods in Critical Point Theory with Applications to Differential Equations*, CBMS Regional Conf. Sec. in Math. No. 65, Amer. Math Soc., Providence, R. I.

Mathematical Programs with Equilibrium Constraints: Theory and Numerical Methods.

Jiří V. Outrata

Institute of Information Theory and Automation
Czech Academy of Science, Prague, Czech Republic

Abstract. The lecture notes deal with optimization problems, where a generalized equation (modeling an equilibrium) arises among the constraints. The main attention is paid to necessary optimality conditions and methods to the numerical solution of such problems. The applications come from continuum mechanics.

1 Introduction

These lecture notes are devoted to a relatively new class of optimization problems which play an important role in mechanics. This class is now commonly entitled *mathematical programs with equilibrium problems* (MPECs) and our knowledge about it has grown remarkably fast especially in the last two decades. Nevertheless, this area still offers a considerable number of open questions motivated by both theoretical considerations as well as real-life problems. An MPEC can be written in an abstract form

$$\begin{array}{ll} \text{minimize} & f(x,y) \\ \text{subject to} & \\ & y \in S(x) \\ & (x,y) \in \kappa, \end{array} \tag{1}$$

where x is the *control* or *design* variable, y is the *state* variable, f is the objective, κ is a closed set of feasible control-state pairs and S is a closed-valued multifunction. The relation

$$y \in S(x) \tag{2}$$

says that y is a solution of an equilibrium problem, parameterized by x. Problems of this kind can be found already in (von Stackelberg 1934), where the equilibrium is another (lower-level) optimization problem (in variable y). One speaks then about a *bilevel* optimization problem (Dempe 2002) and we frequently meet such MPECs eg in shape optimization. Relation (2) can model, however, completely different equilibria governed by variational inequalities, fixed point problems etc., and it is exactly this relation (called also *equilibrium constraint*) which makes MPECs so specific.

The literature devoted to MPECs consists currently of three monographs (Luo, Pang & Ralph 1996),(Outrata, Kočvara & Zowe 1998),(Dempe 2002) and hundreds of papers dealing with a broad spectrum of questions arising in this context. These notes follow approximately the setup of (Outrata et al. 1998), but they employ another tools from nonsmooth analysis and handle a substantially broader spectrum of equilibria. Similarly as in (Outrata et al. 1998), however, the considered equilibria are modeled by *generalized equations* (GEs)

$$0 \in F(x,y) + Q(y), \tag{3}$$

where F is a single-valued and Q is a set-valued map. All considered MPECs are finite-dimensional and so the obtained results are applicable in continuum mechanics only after a suitable discretization. In case of proposed numerical methods this is not a drawback. Concerning optimality conditions, the available infinite-dimensional results are substantially less sharp than their finite-dimensional counterparts and so we decided to keep all problems of these notes finite-dimensional. We omit completely the existence questions associated with (1), because the known results (based on various simple modifications of the Weierstrass Theorem) are not very deep and can easily be found in (Luo et al. 1996). So our main attention is concentrated on 1st-order necessary optimality conditions and methods for the numerical solution of (1) with S given by various equilibria arising in mechanics.

The outline is as follows: In Section 1 we collect the fundamentals from nonsmooth and set-valued analysis needed throughout the whole sequel. Section 2 deals with the modeling of basic types of equilibria and some simple existence and uniqueness questions. With the constraint system in (1) we can associate multifunctions playing a crucial role in optimality conditions and stability analysis. Their Lipschitzian properties are investigated in Section 3. The results of Section 1 and 3 enable then to derive easily rather sharp necessary optimality conditions, which is done in Section 4. Section 5 and the first part of Section 6 are then focused on numerical methods. More precisely, Section 5 concerns the *implicit programming approach* (ImP) that, in connection with bundle methods of nonsmooth optimization, leads to an effective numerical procedure. In Section 6.1 we then describe briefly the *nonlinear programming* (NLP) approach which leads to several effective methods capable to cope with properly multi-valued maps S. Section 6.2 concentrates on MPECs, where the underlying equilibrium has an evolutionary nature; such problems arise eg in connection with delamination.

We employ the following notation which is rather standard in MPEC literature and nonsmooth analysis. Concretely, $\mathbb{R}^n_+, \mathbb{R}^n_-$ is the nonnegative and the nonpositive orthant of $\mathbb{R}^n$, respectively, and $\overline{\mathbb{R}}$ is the extended real line. For an $[m \times n]$ matrix A and index sets $I \subset \{1, 2, \ldots, m\}, J \subset \{1, 2, \ldots, n\}, A_{I,J}$ denotes the submatrix of A with rows and columns specified by I, J, respectively. A_I is the submatrix of A with rows specified by I. Similarly, for a vector $d \in \mathbb{R}^n, d_I$ is the subvector composed from the components $d^i, i \in I$. Furthermore, epi$f$ is the epigraph of a real-valued function f, for a multifunction Φ, domΦ and GphΦ denote its domain and its graph, and for a set Ω, intΩ, clΩ, convΩ and $\delta_\Omega(\cdot)$ mean its interior, closure, convex hull and indicatory function, respectively. $\mathbb{B}$ is the unit ball, E is the unit matrix and for a finite set J, $|J|$ denotes its cardinality. D^0 is the negative polar to a cone D and, for a continuously differentiable map $F[\mathbb{R}^n \to \mathbb{R}^m], \nabla F(z)$ denotes the Jacobian of F at $z \in \mathbb{R}^n$. If $m = 1, \nabla F(z)$ is the gradient (and thus a column vector). If F is only directionally differentiable at $z, F'(z; d)$ is its directional derivative at z in the direction d. Having two vectors u, v of the same dimension, $\max\{u, v\}$ and $\min\{u, v\}$ denote their componentwise maximum and minimum. Finally, $o[\mathbb{R}_+ \to \mathbb{R}]$ is a function with the property

$$\lim_{t \downarrow 0} \frac{o(t)}{t} = 0$$

and "$\bullet$" denotes the Hadamard product.

The reference list by far does not contain all works, relevant with respect to this topic. An interested reader can find further useful references in the above mentioned monographs and in the commented bibliography (Dempe 2003). The following acronyms will be used throughout the whole text.

MPEC	mathematical program with equilibrium constraints
MPCC	mathematical program with complementarity constraints
ImP	implicit programming approach
NLP	nonlinear programming
KKT	Karush-Kuhn-Tucker
CQ	constraint qualification
MFCQ	Mangasarian-Fromowitz CQ
GMFCQ	generalized MFCQ
SCQ	Slater CQ
ESCQ	extended SCQ
LICQ	linear independence CQ
ELICQ	extended LICQ
MPEC - LICQ	LICQ for MPEC
VI	variational inequality
QVI	quasi-variational inequality
HVI	hemivariational inequality
CP	complementarity problem
NCP	nonlinear complementarity problem
ICP	implicit complementarity problem
GE	generalized equation
NE	nonsmooth equation
SRC	strong regularity condition
SSOSC	strong second order sufficient condition
BT	bundle-trust
QP	quadratic programming
SQP	sequential QP
PC^1	piecewise continuously differentiable
lsc	lower semicontinuous
usc	upper semicontinuous

2 Basic tools from nonsmooth analysis

Nonsmooth analysis represents a powerful tool enabling us to work with single- and set-valued mappings without classical differentiability assumptions. In some problems already the data exhibit a lack of differentiability, in another problems the nondifferentiability is inherent even if all data are sufficiently smooth (eg when dealing with the so-called value functions). In a fully convex framework the main part of this theory has been completed already in the sixties. Without convexity assumptions, however, the situation becomes substantially more difficult and some approaches can be applied only to certain special classes of problems. As a result, we dispose currently with several approaches with various degrees of generality and various advantages and drawbacks.

Nevertheless, as shown in the excellent monograph by (Rockafellar & Wets 1998), most of them can be derived on basis of a general framework which has been gradually built up in the last three decades of the past century. Following this framework, we describe the local behaviour of sets, real-valued functions and set-valued mappings (multifunctions) by means of notions defined either in the original space (at multifunctions it is the space, where their graphs are living), or in its topological dual. In this way one describes the local behaviour of sets via tangent and normal cones, the local behaviour of real-valued functions by means of directional derivatives and subdifferentials and the local behaviour of multifunctions through graphical derivatives and coderivatives. One speaks about six pillars of nonsmooth analysis. Not all existing approaches provide us with all of them, but within the same approach all available "pillars" are based on the same principle and closely related. In this series of lectures we will make use of two approaches that grew up from the classical notion of strict derivative; it is the approach of F.H. Clarke ((Clarke 1983)) and B.S. Mordukhovich ((Mordukhovich 1988)). In the latter one disposes merely with the dual notions (normal cones, subdifferentials and coderivatives) but they enjoy a number of excellent properties. We will introduce, however, only those notions which are indispensable for the understanding of the main part of these lectures. Thereafter we will turn our attention to Lipschitzian properties of multifunctions, playing a crucial role both in optimality conditions as well as in the numerical solution of optimization problems. The first lecture will be closed by stating necessary optimality conditions for several types of frequently arising optimization problems.

2.1 Approach of B. Mordukhovich

Consider a closed set $\Pi \in \mathbb{R}^l$ and a point $\bar{z} \in \Pi$. The cone

$$\widehat{N}_\Pi(\bar{z}) := \{\xi \in \mathbb{R}^l | \langle \xi, z - \bar{z} \rangle \leq o(\|z - \bar{z}\|) \forall z \in \Pi\}$$

is called the *Fréchet normal cone* to Π at $\bar{z}$. This cone is convex and closed and it is the negative polar to the *contingent (Bouligand)* cone

$$T_\Pi(\bar{z}) := \limsup_{\lambda \downarrow 0} \frac{\Pi - \bar{z}}{\lambda}. \tag{4}$$

If Π is convex, $\widehat{N}_\Pi(\bar{z})$ coincides with the classical normal cone in the sense of convex analysis.

Definition 1.1 The cone

$$N_\Pi(\bar{z}) := \limsup_{z \xrightarrow{\Pi} \bar{z}} \widehat{N}_\Pi(z)$$

is called the *limiting normal cone* to Π at $\bar{z}$.

The " lim sup" in the above definitions is the upper limit of multifunctions in the sense of Painléve-Kuratowski, cf. (Aubin & Frankowska 1990). $N_\Pi(\bar{z})$ is also closed, but not necessarily convex. By definition, one always has the inclusion

$$\widehat{N}_\Pi(\bar{z}) \subset N_\Pi(\bar{z}).$$

If this inclusion becomes equality, we say that Π is *normally regular* at $\bar{z}$. Each convex set is normally regular at all its points. Frequently, Π amounts to the intersection $g^{-1}(\Lambda) \cap \Xi$, where Ξ, Λ are closed sets in $\mathbb{R}^l, \mathbb{R}^p$, respectively, and $g[\mathbb{R}^l \to \mathbb{R}^p]$ is a continuously differentiable operator. In this case the following statement can be useful.

Theorem 1.1 (Rockafellar & Wets 1998, Thm.6.14). Let $\bar{z} \in \Pi = g^{-1}(\Lambda) \cap \Xi$. Then one has

$$\widehat{N}_\Pi(\bar{z}) \supset (\nabla g(\bar{z}))^T \widehat{N}_\Lambda(g(\bar{z})) + \widehat{N}_\Xi(\bar{z}). \tag{5}$$

Under the qualification condition

$$\left.\begin{array}{c} (\nabla g(\bar{z}))^T u^* \in -N_\Xi(\bar{z}) \\ u^* \in N_\Lambda(g(\bar{z})) \end{array}\right\} \Rightarrow u^* = 0 \tag{6}$$

it holds further that

$$N_\Pi(\bar{z}) \subset (\nabla g(\bar{z}))^T N_\Lambda(g(\bar{z})) + N_\Xi(\bar{z}). \tag{7}$$

By combining of (5) and (7) it follows that

$$N_\Pi(\bar{z}) = \widehat{N}_\Pi(\bar{z}) = (\nabla g(\bar{z}))^T N_\Lambda(g(\bar{z})) + N_\Xi(\bar{z}) \tag{8}$$

whenever condition (6) is fulfilled, Ξ is normally regular at $\bar{z}$ and Λ is normally regular at $g(\bar{z})$. We also immediately observe that condition (6) holds true provided $\Xi = \mathbb{R}^l$ and $\nabla g(\bar{z})$ is surjective. Equation (8) is valid without any assumptions, provided Ξ, Λ are convex polyhedra and g is affine.

In some cases $\Pi = \Pi_1 \times \Pi_2$ with $\Pi_1 \in \mathbb{R}_1^{l_1}, \Pi_2 \in \mathbb{R}^{l_2}, l_1 + l_2 = l$. Accordingly $z = (z_1, z_2)$ with $z_1 \in \Pi_1$ and $z_2 \in \Pi_2$. Then one has (Mordukhovich 1988, Prop.1.6)

$$N_\Pi(z) = N_{\Pi_1}(z_1) \times N_{\Pi_2}(z_2). \tag{9}$$

We turn now our attention to real-valued functions.

Definition 1.2 Consider a lsc function $f[\mathbb{R}^l \to \bar{\mathbb{R}}]$ and assume that $z \in \mathrm{dom} f$. The sets

$$\hat{\partial} f(z) = \{\xi \in \mathbb{R}^n | (\xi, -1) \in \widehat{N}_{\mathrm{epi} f}(z, f(z))\},$$
$$\partial f(z) = \{\xi \in \mathbb{R}^n | (\xi, -1) \in N_{\mathrm{epi} f}(z, f(z))\}$$

are called the *Fréchet subdifferential* and the *(lower) limiting subdifferential* of f at z, respectively.

$\partial f(z)$ is a closed, possibly nonconvex set which is bounded iff f is Lipschitz on a neighborhood of z. Putting $f = \delta_\Pi$, we observe that $\mathrm{epi}\delta_\Pi = \Pi \times \mathbb{R}_+$ and consequently, by virtue of (9),

$$\begin{aligned} \partial \delta_\Pi(z) &= \{\xi \in \mathbb{R}^n | (\xi, -1) \in N_{\mathrm{epi}\delta_\Pi}(z, 0)\} \\ &= \{\xi \in \mathbb{R}^n | (\xi, -1) \in N_\Pi(z) \times N_{\mathbb{R}_+}(0)\} = N_\Pi(z) \end{aligned}$$

whenever $z \in \Pi$. So, we have established a two-way relation between the limiting normal cone and the limiting subdifferential by means of the epigraph of the considered function and the indicatory function of the considered set. The set $\partial f(z)$ is a singleton iff f is strictly differentiable at z. In such a case $\partial f(z)$ amounts to the strict derivative of f at z.

Consider now a multifunction $\Phi[\mathbb{R}^l \rightsquigarrow \mathbb{R}^p]$ having a closed graph and assume that $b \in \Phi(a)$, i.e., $(a,b) \in \text{Gph}\Phi$. The local behaviour of Φ around (a,b) can be described via another multifunction, denoted $D^*\Phi(a,b)$, defined by

$$D^*\Phi(a,b)(\eta) := \{\xi \in \mathbb{R}^l | (\xi, -\eta) \in N_{\text{Gph}\Phi}(a,b)\},\ \eta \in \mathbb{R}^p. \tag{10}$$

This new multifunction is entitled the *coderivative* of Φ at (a,b). If the limiting normal cone in (10) is replaced by the Fréchet normal cone, we write $\widehat{D}^*\Phi$ and speak about the *regular coderivative.* If Φ is single-valued, we write simply $D^*\Phi(a)(\cdot)$, because $b = \Phi(a)$. If Φ is even continuously differeentiable at a, then one has $D^*\Phi(a)(\eta) = (\nabla\Phi(a))^T\eta\ \forall\, \eta \in \mathbb{R}^p$. In the single-valued case we can consider the composition

$$\varphi(z) := f(\Phi(z)), \tag{11}$$

where the outer function f maps $\mathbb{R}^p$ into $\overline{\mathbb{R}}$. Consider $\bar{z} \in \mathbb{R}^l$ such that $\Phi(\bar{z}) \in \text{dom} f$. Then one has, cf. (Rockafellar & Wets 1998, Thm.10.49),

$$\partial\varphi(\bar{z}) \subset D^*\Phi(\bar{z})(\partial f(\Phi(\bar{z}))), \tag{12}$$

provided Φ is Lipschitz around $\bar{z}$ and f is Lipschitz around $\Phi(\bar{z})$. If f is even continuously differentiable around $\Phi(\bar{z})$, it holds

$$\partial\varphi(\bar{z}) \supset \widehat{D}^*\Phi(\bar{z})(\nabla f(\Phi(\bar{z}))). \tag{13}$$

2.2 Approach of F. Clarke

In this section we confine ourselves only to two notions: The Clarke's subdifferential and the generalized Jacobian. Moreover, we will consider only Lipschitz mappings which play a crucial role in applications, where these notions will be employed.

Definition 1.3. Let $F[\mathbb{R}^l \to \mathbb{R}^p]$ be Lipschitz near $\bar{z} \in \mathbb{R}^l$ and let ω_F denote the set, where F fails to be differentiable. Let

$$\partial_B F(\bar{z}) := \left\{ \lim_{i\to\infty} \nabla F(z_i) | z_i \to \bar{z},\ z_i \notin \omega_F \right\}.$$

Then the set $\bar{\partial}F(\bar{z}) := \text{conv}\partial_B F(\bar{z})$ is called the *Clarke's subdifferential* of F at $\bar{z}$, provided $p = 1$, and the *generalized Jacobian* of F at $\bar{z}$ otherwise.

The Clarke's subdifferential can also be defined via the directional derivative

$$F^0(\bar{z}; h) := \limsup_{z\to\bar{z}, \lambda\downarrow 0} \frac{F(z + \lambda h) - F(z)}{\lambda}.$$

Indeed, it holds that

$$\bar{\partial}F(\bar{z}) = \{\xi \in \mathbb{R}^l | \langle \xi, h \rangle \leq F^0(\bar{z}; h), \ \forall h \in \mathbb{R}^l\}.$$

Observe that this set is nonempty, closed and convex. Further, one has

$$\bar{\partial}F(\bar{z}) = \operatorname{conv} \partial F(\bar{z}). \tag{14}$$

If $p > 1$, then for all $y^* \in \mathbb{R}^p$

$$(\bar{\partial}F(\bar{z}))^T y^* = \operatorname{conv} D^*F(\bar{z})(y^*). \tag{15}$$

So, in the Lipschitz case the Clarke's notions are closely related to their limiting counterparts. Using generalized Jacobians, one can enrich the statements concerning the composition (11). According to (Clarke 1983, Thm.2.6.6) one has namely that

$$\bar{\partial}\varphi(\bar{z}) = (\bar{\partial}\Phi(\bar{z}))^T \nabla f(\Phi(\bar{z})), \tag{16}$$

whenever f is continuously differentiable around $\Phi(\bar{z})$. The Clarke's notions can very well be applied to the so-called piecewise $C^1(PC^1)$-maps.

Definition 1.4 Let U be an open subset of $\mathbb{R}^l$. A function $F[U \to \mathbb{R}^p]$ is called a PC^1 – *function* on U if it is continuous and if for every $z_0 \in U$ there exists an open neighborhood $\mathcal{O} \subset U$ and a finite number of continuously differentiable functions $F_i[\mathcal{O} \to \mathbb{R}^p]$, $i = 1, 2, \ldots, k$, such that for every $z \in \mathcal{O}$ one has $F(z) \in \{F_1(z), \ldots, F_k(z)\}$. The single functions F_i are called *selections* or *pieces* for F at z_0, and the set

$$I_F(z_0) := \{i \in \{1, 2, \ldots, k\} \mid F(z_0) = F_i(z_0)\}$$

is the *active index set* at z_0. The selections F_i, $i \in I_F(z_0)$, are called *active selections (pieces)* for F at z_0. A selection F_i is *essentially active* at z_0, provided

$$z_0 \in \operatorname{cl}(\operatorname{int}\{z \in \mathcal{O} \mid F(z) = F_i(z)\}).$$

The respective index set of essentially active selections at z_0 is denoted by $I^e_F(z_0)$ and plays an important role in connection with the generalized Jacobian of F at z_0. Indeed, in (Scholtes 1994) it was proved that every PC^1-function is locally Lipschitz and

$$\partial F(z_0) = \operatorname{conv}\{\nabla F_i(z_0) \mid i \in I^e_F(z_0)\}. \tag{17}$$

When deriving necessary optimality conditions for nonsmooth optimization problems, it is, by virtue of (14),(15), definitely better to work with the limiting notions than with the Clarke's subdifferentials and generalized Jacobians. The resulting conditions are then generally sharper, i.e., more selective. On the other hand, when solving nondifferentiable problems numerically, then the convexity of the Clarke's notions plays an important role. In fact, numerical methods based on the Clarke's subdifferential are dominating in nonsmooth minimization problems, whereas the generalized Jacobian is widely used in the solution of nonsmooth equations.

2.3 Lipschitz properties of multifunctions

When dealing with MPECs, one has to work with various multifunctions exhibiting various Lipschitz-like properties. It is one of the tasks of nonsmooth analysis to detect these properties, because they are of a great importance both in optimality conditions as well as in numerical procedures. Moreover, it is also an interesting issue to learn, how stable is a computed solution with respect to small perturbations of the problem data. For our purposes, the three notions defined below are the most relevant.

Definition 1.5 Consider a multifunction $\Gamma[\mathbb{R}^l \rightsquigarrow \mathbb{R}^p]$.

(i) Γ is called *(locally) upper Lipschitz* at $\bar{z} \in \mathrm{dom}\Gamma$ (with modulus L), if there is a neighborhood $\mathcal{U}$ of $\bar{z}$ and a real $L \geq 0$ such that

$$\Gamma(z) \subset \Gamma(\bar{z}) + L\|z - \bar{z}\|\mathbb{B} \text{ for all } z \in \mathcal{U}.$$

(ii) We say that Γ has the *Aubin property* around $(a, b) \in \mathrm{Gph}\Gamma$ (with modulus L), if there are neighborhoods $\mathcal{U}$ of a, $\mathcal{V}$ of b and a real $L \geq 0$ such that

$$\Gamma(a') \cap \mathcal{V} \subset \Gamma(a'') + L\|a' - a''\|\mathbb{B} \text{ for all } a', a'' \in \mathcal{U}.$$

(iii) We say that Γ has a *Lipschitz single-valued localization* at $(a, b) \in \mathrm{Gph}\Gamma$, provided there are neighborhoods $\mathcal{U}$ of a, $\mathcal{V}$ of b and a Lipschitz single-valued map $\sigma[\mathcal{U} \rightarrow \mathbb{R}^p]$ such that $b = \sigma(a)$ and $\Gamma(a') \cap \mathcal{V} = \{\sigma(a')\}$ for all $a' \in \mathcal{U}$.

As stated in the next theorem, property (i) is possessed by a large class of the so-called polyhedral multifunctions. A multifunctions $P[\mathbb{R}^l \rightsquigarrow \mathbb{R}^p]$ is called *polyhedral*, provided its graph is the union of finitely many convex polyhedra, termed *components*.

Theorem 1.2 (Robinson 1981) Let $P[\mathbb{R}^l \rightsquigarrow \mathbb{R}^p]$ be polyhedral. Then there is a constant λ such that P is (locally) upper Lipschitz with modulus λ at each $z \in \mathrm{dom}P$. In particular, if $\Omega \subset \mathbb{R}^l$ is a convex polyhedron, then the map

$$P : z \mapsto \begin{cases} N_\Omega(z) \text{ if } z \in \Omega \\ \emptyset \text{ otherwise,} \end{cases}$$

being polyhedral, is (locally) upper Lipschitz at each $z \in \Omega$.

From Theorem 1.2 it follows that a single-valued polyhedral map P is necessarily Lipschitz in the standard sense on $\mathrm{dom}P$. For property (ii) we dispose with a characterization in terms of coderivatives. Indeed, Γ has the Aubin property around $(a, b) \in \mathrm{Gph}\Gamma$ iff $D^*\Gamma(a, b)(0) = \{0\}$, cf. (Mordukhovich 1994). This characterization is known as Mordukhovich criterion. If Γ has a concrete structure, then this characterization, together with rules of the coderivative calculus enables to obtain a sufficient condition ensuring the Aubin property. Consider eg the multifunction $\Sigma[\mathbb{R}^p \rightsquigarrow \mathbb{R}^l]$ defined by

$$\Sigma(u) := \{z \in \Xi \,|\, g(z) + u \in \Lambda\}, \tag{18}$$

where g, Ξ, Λ fulfill the assumptions posed before Theorem 1.1.

Theorem 1.3. Let $\bar{z} \in \Sigma(\bar{u})$ and assume that the qualification condition (6) holds true. Then Σ has the Aubin property around $(\bar{u}, \bar{z})$.

Property (iii) is closely related to the implicit programming approach investigated in Section 5. Also for this property nonsmooth analysis provides us with a suitable characterization, cf. (Klatte & Kummer 2003). Instead of it, however, we will apply later two other criteria which are tailored to the concrete structure of Γ being investigated.

2.4 Necessary optimality conditions

In this section we will present a general framework suitable for deriving optimality conditions in MPECs. We start with an abstract optimization problem

$$\begin{array}{ll} \text{minimize} & f(z) \\ \text{subject to} & \\ & z \in \Pi, \end{array} \tag{19}$$

where $z \in \mathbb{R}^l$, $f[\mathbb{R}^l \to \mathbb{R}]$ is locally Lipschitz and $\Pi \subset \mathbb{R}^l$ is closed. If $\widehat{z}$ is a local minimizer in (19), then, following (Mordukhovich 1988), one has

$$0 \in \partial f(\widehat{z}) + N_\Pi(\widehat{z}), \tag{20}$$

and the main problem consists in the computation of the limiting normal cone $N_\Pi(\widehat{z})$ (or an upper approximation of it). In MPECs the set Π has frequently the structure investigated in Theorem 1.1, i.e.,

$$\Pi = \{z \in \Xi \,|\, g(z) \in \Lambda\}. \tag{21}$$

Theorem 1.4. Let $\widehat{z}$ be a local minimizer in (19), where Π is given by (21) and g, Ξ, Λ fulfill the assumptions posed before Theorem 1.1. Further assume that the qualification condition (6) holds true (with $\bar{z}$ replaced by $\widehat{z}$). Then there exists a Karush-Kuhn-Tucker (KKT) vector $u^* \in N_\Lambda(g(\widehat{z}))$ such that

$$0 \in \partial f(\widehat{z}) + (\nabla g(\widehat{z}))^T u^* + N_\Xi(\widehat{z}). \tag{22}$$

If g is affine and Ξ, Λ are convex polyhedral, then relation (22) holds true without any qualification conditions.

Proof. Under the posed assumptions one has that

$$N_\Pi(\widehat{z}) \subset (\nabla g(\widehat{z}))^T N_\Lambda(g(\widehat{z})) + N_\Xi(\widehat{z}).$$

The existence of a KKT vector $u^* \in N_\Lambda(g(\widehat{z}))$ thus follows from (20). □

If $\Lambda = \mathbb{R}^p_+$ condition (6) can be written down in the form

$$\left.\begin{array}{l} 0 \in \sum_{i=1}^{p} u^{*i} \nabla g^i(\bar{z}) + N_\Xi(\bar{z}) \\ u^{*i} \in \mathbb{R}^p_+,\ u^{*i} = 0 \text{ if } g^i(\bar{z}) < 0 \end{array}\right\} \Rightarrow u^* = 0,$$

where we recognize the dual form of the Mangasarian-Fromowitz constraint qualification (MFCQ), known from mathematical programming. In equilibrium problems, Λ is usually a complicated nonconvex set and therefore we will use the acronym GMFCQ. If f is continuously differentiable, then, following (Hiriart-Urruty 1979), it is possible to replace condition (20) by a sharper one, namely

$$0 \in \nabla f(\widehat{z}) + \widehat{N}_{\Pi}(\widehat{z}). \tag{23}$$

The hurdle consists now in the computation of $\widehat{N}_{\Pi}(\widehat{z})$. In this regard, however, additional structural assumptions on Π can be helpful.

Theorem 1.5. (Flegel, Kanzow & Outrata 2004) Let $\widehat{z}$ be a local minimizer in (19), where f is continuously differentiable. Let $\Pi = g^{-1}(\Lambda)$ with g like in Theorem 1.3 and assume that Λ is the union of polyhedral convex sets $\Lambda_i, i = 1, 2, \ldots, q$. Finally suppose that the constraint qualification

$$T_{g^{-1}(\Lambda)}(\widehat{z}) = \{h \in \mathbb{R}^l \mid \nabla g(\widehat{z})h \in T_{\Lambda}(g(\widehat{z}))\} \tag{24}$$

holds true and with $I(\widehat{z}) := \{i \in \{1, 2, \ldots, q\} \mid g(\widehat{z}) \in \Lambda_i\}$ the condition

$$\bigcap_{i \in I(\widehat{z})} (\nabla g(\widehat{z}))^T \widehat{N}_{\Lambda_i}(g(\widehat{z})) = (\nabla g(\widehat{z}))^T \bigcap_{i \in I(\widehat{z})} \widehat{N}_{\Lambda_i}(g(\widehat{z})) \tag{25}$$

is fulfilled. Then there exists a KKT vector $u^* \in \widehat{N}_{\Lambda}(g(\widehat{z}))$ such that

$$0 \in \nabla f(\widehat{z}) + (\nabla g(\widehat{z}))^T u^*. \tag{26}$$

Proof. By our assumptions $\Lambda = \bigcup_{i=1}^{q} \Lambda_i$ with all sets Λ_i being convex polyhedra. Thus, by (24), one has

$$\begin{aligned} T_{g^{-1}(\Lambda)}(\widehat{z}) &= \{h \in \mathbb{R}^l \mid \nabla g(\widehat{z})h \in T_{\bigcup_{i=1}^{q} \Lambda_i}(g(\widehat{z}))\} \\ &= \bigcup_{i \in I(\widehat{z})} \{h \in \mathbb{R}^l \mid \nabla g(\widehat{z})h \in T_{\Lambda_i}(g(\widehat{z}))\}. \end{aligned}$$

Indeed, the contingent cone to the union $\bigcup_{i=1}^{q} \Lambda_i$ at $g(\widehat{z})$ amounts to the union of the contingent cones to the single sets Λ_i for $i \in I(\widehat{z})$ (at $g(\widehat{z})$). Since the polar to a union is the intersection of the polars, we infer that

$$\widehat{N}_{g^{-1}(\Lambda)}(\widehat{z}) = \bigcap_{i \in I(\widehat{z})} \left[(\nabla g(\widehat{z}))^{-1}(T_{\Lambda_i}(g(\widehat{z})))\right]^0.$$

The polars on the right-hand side of the preceding equation can be computed due to the polyhedrality of Λ_i; one obtains that

$$\left[(\nabla g(\widehat{z}))^{-1}(T_{\Lambda_i}(g(\widehat{z})))\right]^0 = (\nabla g(\widehat{z}))^T \widehat{N}_{\Lambda_i}(g(\widehat{z})).$$

It remains to observe that

$$\widehat{N}_\Lambda(g(\widehat{z})) = \bigcap_{i\in I(\widehat{z})} \widehat{N}_{\Lambda_i}(g(\widehat{z})),$$

employ condition (25), and we are done. □

It is easy to see that condition (25) is fulfilled whenever $\nabla g(\widehat{z})$ is surjective or $|I(\widehat{z})| = 1$. Surjectivity of $\nabla g(\widehat{z})$ guarantees simultaneously the satisfaction of condition (24). In classical mathematical programming Λ amounts to the carthesian product of zeros (for equality constraints) and a nonpositive orthant (for inequality constraints). In this case condition (24) is called the *Abadie* constraint qualification and is known to be one of the weakest constraint qualifications at all. We will keep this terminology also in the considered general situation. It has been proved eg in (Henrion & Outrata to appear) that (24) is fulfilled whenever the map

$$u \mapsto \{z \in \mathbb{R}^l \mid g(z) + u \in \Lambda\}$$

possesses the Aubin property around $(0, \widehat{z})$. Since (GMFCQ) implies the Aubin property by Theorem 1.3, we have the implication

$$\text{(GMFCQ)} \Rightarrow \text{(Abadie)}$$

exactly like in the classical situation. It follows that under (GMFCQ) the optimality conditions (22) can be sharpened to the form (26) whenever f is continuously differentiable, $\Xi = \mathbb{R}^l$, Λ has the structure specified above and condition (25) is fulfilled.

Some MPECs can be converted to the form

$$\begin{array}{ll} \text{minimize} & f(x,y) \\ \text{subject to} & \\ & y = S(x) \\ & x \in \omega, \end{array} \tag{27}$$

where $x \in \mathbb{R}^n, y \in \mathbb{R}^m, f[\mathbb{R}^n \times \mathbb{R}^m \to \mathbb{R}]$ and $S[\mathbb{R}^n \to \mathbb{R}^m]$ are locally Lipschitz and $\omega \subset \mathbb{R}^n$ is closed. In such a case one can apply the following statement.

Theorem 1.6. Let $(\widehat{x}, \widehat{y})$ be a locally optimal pair in (27). Then there exist vectors $(\xi, \eta) \in \partial f(x, y)$ such that

$$0 \in \xi + D^*S(\widehat{x})(\eta) + N_\omega(\widehat{x}). \tag{28}$$

Proof. Problem (27) can be converted to the form

$$\begin{array}{ll} \text{minimize} & h(x) \\ \text{subject to} & \\ & x \in \omega, \end{array}$$

where $h := f \circ \Phi$ with $\Phi(x) := \begin{bmatrix} x \\ S(x) \end{bmatrix}$. Therefore, for an arbitrary pair $(a, b) \in \mathbb{R}^n \times \mathbb{R}^m$ one has that

$$D^*\Phi(x)(a, b) = a + D^*S(x)(b).$$

The statement follows now from inclusion (12) and optimality condition (20). □

In some MPECs considered in Chapters 4 and 5, f is continuously differentiable, but we are able to compute only an upper approximation of the coderivative $D^*S(\widehat{x})(\nabla_y f(\widehat{x}, \widehat{y}))$. This can be the set $(\bar{\partial} S(\widehat{x}))^T \nabla_y f(\widehat{x}, \widehat{y})$ or even its upper approximation. In this way we arrive naturally at weaker (less selective) optimality conditions.

3 Mathematical modeling of discretized equilibrium problems

Variational inequalities (VIs) and complementarity problems (CPs) provide a convenient and elegant tool for modeling of manifold equilibria. In numerous situations, however, it is advantageous to work with another models, above all with the so-called generalized and nonsmooth equations. Following Robinson, under *generalized equation* (GE) we understand the relation

$$0 \in F(y) + Q(y),$$

where F is a single- and Q is a set-valued mapping. In our models F is usually continuously differentiable. If F is merely locally Lipschitz and Q disappears, we speak about a *nonsmooth equation* (NE). In this chapter we will

(i) reformulate several fundamental types of VIs to the form of GEs and NEs;

(ii) state some basic existence and uniqueness results, and

(iii) describe briefly an efficient numerical method, associated with the NE model.

3.1 Basic reformulations

This section is devoted to the modeling of equilibria that are standardly described via variational inequalities of the first and second kind, hemivariational inequalities (HVIs) and quasi-variational inequalities (QVIs). As a particular case of VIs of the first kind we will introduce complementarity problems and mixed complementarity problems. A special attention is paid to (discretized) contact problems with Coulomb friction which can be modeled in various ways. Let $y \in \mathbb{R}^m, F[\mathbb{R}^m \to \mathbb{R}^m]$ be continuous and $\Omega \subset \mathbb{R}^m$ be convex and closed. The problem

$$\left.\begin{array}{ll}\text{find} & \bar{y} \in \Omega \text{ such that} \\ & \langle F(\bar{y}), y - \bar{y} \rangle \geq 0 \ \forall\, y \in \Omega\end{array}\right\} \tag{29}$$

is called the *variational inequality of the first kind.* Ω is termed the *constraint set.* If $\Omega = \mathbb{R}^m_+$, (29) amounts to a *complementarity problem* and if Ω is a box in $\mathbb{R}^m$, we speak about a *mixed complementarity problem.* In dependence on the nature of F these CPs can be linear (if F is affine) or nonlinear (otherwise). If F is monotone, continuously differentiable and the Jacobian ∇F is symmetric, then (29) is equivalent to the convex program

$$\begin{array}{ll} \text{minimize} & f(y) \\ \text{subject to} & \\ & y \in \Omega, \end{array} \tag{30}$$

where $F = \nabla f$. Equilibria of this type arise frequently in connection with (discretized) contact problems in linear elasticity. One readily deduces that (29) amounts to the GE

$$0 \in F(y) + N_\Omega(y) \tag{31}$$

and to the NE

$$\mathrm{Proj}_\Omega(y - F(y)) - y = 0. \tag{32}$$

Since $\mathrm{Proj}_{\mathbb{R}^m_+}(v) = \max\{v, 0\}$, equation (32) attains in case of a CP a particularly simple form

$$\min\{F(y), y\} = 0. \tag{33}$$

Consider now instead of (29) the problem

$$\left.\begin{array}{l} \text{find } \bar{y} \in \mathbb{R}^m \text{ such that} \\ \langle F(\bar{y}), y - \bar{y}\rangle + \varphi(y) - \varphi(\bar{y}) \geq 0 \;\forall\, y \in \mathbb{R}^m, \end{array}\right\} \tag{34}$$

where $\varphi[\mathbb{R}^m \to \overline{\mathbb{R}}]$ is proper convex and lsc. This problem is entitled the *variational inequality of the 2nd kind* and generalizes the model (29). Indeed, it suffices to put $\varphi = \delta_\Omega$ to obtain (29). With other choices of φ, however, we can model rather complicated equilibria arising eg in (discretized) contact problems with given friction. By virtue of the convexity of φ, (34) can also be written down in an equivalent form:

$$\left.\begin{array}{l} \text{Find } y \in \mathbb{R}^m \text{ such that} \\ \langle F(\bar{y}), y - \bar{y}\rangle + \varphi'(\bar{y}, y - \bar{y}) \geq 0 \;\forall\, y \in \mathbb{R}^m. \end{array}\right\} \tag{35}$$

From (35) we can now immediately derive the corresponding GE

$$0 \in F(y) + \partial\varphi(y). \tag{36}$$

If φ in (34) is not convex, but locally Lipschitz, it is possible to replace the classical directional derivative in (35) by the directional derivative of Clarke. In this way we arrive at a *hemivariational inequality.* In fact, in the literature (eg (Panagiotopoulos 1993)) HVI has been introduced in a more general form. In the sequel, however, we will come out with this simplified formulation. As the counterpart to (35) we obtain now the GE

$$0 \in F(y) + \bar{\partial}\varphi(y). \tag{37}$$

By *constrained hemivariational inequality* we will mean the problem:

$$\left.\begin{array}{l} \text{Find } \bar{y} \in \Omega \text{ such that} \\ \langle F(\bar{y}), y - \bar{y}\rangle + \varphi^0(\bar{y}; y - \bar{y}) \geq 0 \;\forall\, y \in \Omega \end{array}\right\}, \tag{38}$$

where $\varphi[\mathbb{R}^m \to \mathbb{R}]$ is locally Lipschitz and the constraint set Ω is closed and convex. In this case the equivalent GE attains the form

$$0 \in F(y) + \bar{\partial}\varphi(y) + N_\Omega(y). \tag{39}$$

In this reasoning the convexity of Ω plays an important role. HVIs are frequently used to the modeling of the behaviour of laminates or sandwiches. If F is continuously differentiable and ∇F is symmetric, problem (38) is a first-order necessary optimality condition for the nonsmooth program

$$\begin{array}{ll} \text{minimize} & (f+\varphi)(y) \\ \text{subject to} & \\ & y \in \Omega, \end{array} \tag{40}$$

where again $F = \nabla f$. To be precise, the solutions of (38) are then *C(larke)- stationary* points of the program (40). Since in some cases not all these points are acceptable equilibria from the physical point of view, it is then reasonable to replace (39) by the GE

$$0 \in F(y) + \partial\varphi(y) + N_\Omega(y), \tag{41}$$

where the Clarke's subdifferential is replaced by the (smaller) limiting one. The GE (41), however, does not have a corresponding HVI due to the nonconvexity of the limiting subdifferential.

Let $\Gamma[\mathbb{R}^m \rightsquigarrow \mathbb{R}^m]$ be a closed- and convex-valued multifunction, F be like in (29) and consider the problem:

$$\left.\begin{array}{l} \text{Find } \bar{y} \in \Gamma(\bar{y}) \text{ such that} \\ \langle F(\bar{y}), y - \bar{y}\rangle \geq 0 \,\forall y \in \Gamma(\bar{y}) \end{array}\right\}. \tag{42}$$

Since the constraint set (counterpart to Ω in (29)) depends now on y, we speak about a *quasi-variational inequality.* If $\Gamma(y) = g(y) + \mathbb{R}^m_+$ with all components $g^i[\mathbb{R}^m \to \mathbb{R}], i = 1, 2, \ldots, m$, continuous, we arrive at the *implicit complementarity problem* (ICP):

$$\left.\begin{array}{l} \text{Find } \bar{y} \in \mathbb{R}^m \text{ such that} \\ F(\bar{y}) \geq 0, \bar{y} \geq g(\bar{y}), \langle F(\bar{y}), \bar{y} - g(\bar{y})\rangle = 0. \end{array}\right\} \tag{43}$$

Problems of the type (43) arise eg in (discretized) contact problems with compliant obstacles or in filtration through pourous media. (42) amounts evidently to the GE

$$0 \in F(y) + N_{\Gamma(y)}(y) \tag{44}$$

(with a rather difficult set-valued part) or to the NE

$$\text{Proj}_{\Gamma(y)}(y - F(y)) - y = 0. \tag{45}$$

In ICPs one has

$$\text{Proj}_{\Gamma(y)}(v) = \max\{v, g(y)\}.$$

Eq. (45) attains then the form

$$\max\{y - F(y), g(y)\} - y = 0$$

which is equivalent to

$$\min\{F(y), y - g(y)\} = 0. \tag{46}$$

From eq. (46) one immediately deduces the original formulation (43). Due to its simplicity, eq. (46) is frequently used in various contexts. Naturally, for $g \equiv 0$ we obtain the standard CP. As we have seen, most equilibria can be modeled as variational inequalities and generalized equations. By using of duality theory and artificial variables the modeling possibilities can be further enriched. This holds typically for contact problems with Coulomb friction. In (Haslinger & Panagiotopoulos 1984) one finds a model in form of a QVI, whereas in (A.M.Al-Fahed, Stavroulakis & Panagiotopoulos 1991) the same problem is modeled as a linear CP. In the context of MPECs it is convenient to start with the GE (36) corresponding to the underlying (discretized) contact problem with given friction. In this problem $y = (y_1, y_2, y_3)^T, F = (F_1, F_2, F_3)^T$, φ depends on y_1 and a nonnegative parameter ν (coefficient of the given friction) and the constraint set Ω concerns only y_3. Correspondingly, the respective GE (36) attains the form

$$\begin{array}{rcl} 0 & \in & F_1(y) + \partial\varphi_\nu(y_1) \\ 0 & = & F_2(y) \\ 0 & \in & F_3(y) + N_\Omega(y_3), \end{array} \tag{47}$$

where $\varphi_\nu(y_1) = \nu \bullet \widetilde{\varphi}(y_1)$ with a convex continuous function $\widetilde{\varphi}$. If we now put in (47) ν equal to y_3 (multiplier responsible for the nonpenetrability condition), we obtain a GE describing a (discretized) contact problem with Coulomb friction. This procedure demonstrates the usefulness of the GE modeling framework.

3.2 Some elementary existence and uniqueness results

In this section we confine ourselves to VIs of the first kind and to HVIs. The presented results can be found in standard literature so that our aim is rather to illustrate the advantages of various reformulations presented in the previous section.
Consider first the VI (29).

Theorem 2.1. Assume that F is continuous and either

(i) Ω is compact;

(ii) F is *coercive* on Ω, i.e., $\exists y_0 \in \mathbb{R}^m$ such that

$$\frac{\langle F(y) - F(y_0), y - y_0\rangle}{\|y - y_0\|} \to \infty \text{ for } \|y\| \to \infty,\ y \in \Omega.$$

Then the VI (29) possesses a solution.

Sketch of the proof. Let (i) hold true. Then we can consider the NE (32) which can be written down as the equation

$$\Phi(y) = y$$

with $\Phi(y) := \mathrm{Proj}_\Omega(y - F(y))$. Since the projection onto a convex set is continuous, Φ is a continuous map from Ω into Ω and the statement follows from the Brouwer's Fixed

Point Theorem. To prove the statement under assumption (ii), we show first that the existence of a solution to the VI (29) is equivalent to the existence of a positive real r and a solution y_r to the GE

$$0 \in F(y) + N_{\Omega \cap r\mathbb{B}}(y) \tag{48}$$

such that $\|y_r\| < r$. The existence of r and y_r can then be ensured eg via the coercivity assumption (ii), cf. (Ekeland & Temam 1976). □

Of course, one could construct also a combination of conditions (i), (ii) which would generalize the above statement. Assumption (ii) is fulfilled provided F is *strongly monotone* on Ω, i.e., $\exists \alpha > 0$ such that

$$\langle F(v) - F(w), v - w \rangle \geq \alpha \|v - w\|^2 \ \forall \ v, w \in \Omega.$$

Concerning the existence of solutions to the HVI (38), we can make use of the GE (39) and employ another strong tool from nonlinear analysis, namely the Ky Fan's Inequality.

Theorem 2.2. Assume that F is continuous and Ω is compact. Then the HVI (38) possesses a solution.

For the proof it is important to note that the multifunction

$$y \mapsto F(y) + \bar{\partial}\varphi(y)$$

is usc and nonempty-, convex- and compact-valued due to the properties of the Clarke's subdifferential. This enables to apply (Aubin 1993, Thm.9.9) to the GE (39). If F is continuously differentiable and ∇F is symmetric, then one can employ the reformulation (40). In this optimization problem the objective is continuous and the constraint set is compact, whence the existence by virtue of the Weierstrass Theorem. This argumentation remains valid also for the GE (41) for which the preceding idea of the proof does not work.

Concerning the question of uniqueness of the solutions to equilibrium problems analyzed above, we restrict ourselves merely to the VIs of the first kind, because there is no chance to ensure uniqueness at HVIs by means of realistic assumptions. On the basis of the original formulation (29) it is easy to prove that this VI has at most one solution, provided F is *strictly monotone* on Ω, i.e.,

$$\langle F(v) - F(w), v - w \rangle > 0 \ \forall v, w \in \Omega, v \neq w.$$

Since the strong monotonicity apparently implies the strict monotonicity, the strong monotonicity of F on Ω ensures both the existence as well as the uniqueness of solutions to the VI (29). It is thus an important issue to find verifiable criteria for this property under, possibly, some additional assumptions. If F happens to be continuously differentiable, then one can enforce its strong monotonicity on Ω via a suitable condition imposed on the Jacobian of F. One usually requires that ∇F is *uniformly positive definite* on Ω, i.e., $\exists \alpha > 0$ such that

$$\langle d, \nabla F(y)d \rangle \geq \alpha \|d\|^2 \ \forall \ y \in \Omega \ \forall d \in \mathbb{R}^m.$$

For affine F, however, this condition leads to positive definiteness of the defining matrix which is unduly strong in some cases.

Theorem 2.3. Let $F(y) = Ay + b$ with an $m \times m$ matrix A and $b \in \mathbb{R}^m$. Assume that Ω is a cone and A is *strictly copositive* with respect to $\Omega - \Omega$, i.e.,

$$\langle d, Ad \rangle > 0 \ \forall \ d \in \Omega - \Omega, d \neq 0.$$

Then the respective VI (29) has a unique solution.

If, additionally, $\Omega = \mathbb{R}^m_+$, we dispose with a full characterization of the case, when the respective linear CP possesses a unique solution for each $b \in \mathbb{R}^m$: This happens iff A is a P-matrix ((Murty 1988)). In many cases

$$\Omega = \{y \in \mathbb{R}^m | H(y) = 0, \ G(y) \leq 0\}, \tag{49}$$

where $H[\mathbb{R}^m \to \mathbb{R}^l]$ is an affine map and all components G^i of $G[\mathbb{R}^m \to \mathbb{R}^s]$ are convex and continuously differentiable. Let the classic Slater constraint qualification

(SCQ) : $\exists \bar{y} \in \mathbb{R}^m$ such that $H(\bar{y}) = 0$ and $G(\bar{y}) \in \mathrm{int}\mathbb{R}^s_+$;

be valid. Then we can compute the normal cone $N_\Omega(y)$ at $y \in \Omega$ and obtain that

$$N_\Omega(y) = \left\{(\nabla H(y))^T \mu + (\nabla G(y))^T \lambda | \mu \in \mathbb{R}^l, \lambda \in \mathbb{R}^s_+, \langle \lambda, G(y) \rangle = 0\right\}.$$

Consequently, the GE (31) attains then the form

$$\left.\begin{array}{ll} 0 & = \mathcal{L}(y, \mu, \lambda) \\ 0 & = H(y) \\ 0 & \in -G(y) + N_{\mathbb{R}^s_+}(\lambda), \end{array}\right\} \tag{50}$$

where

$$\mathcal{L}(y, \mu, \lambda) := F(y) + (\nabla H(y))^T \mu + (\nabla G(y))^T \lambda$$

is the *Lagrangian* associated with the analyzed VI. In (50) we have a substantially simpler constraint set than Ω, but we have to do with additional variables μ, λ. The GE (50) is called the KKT form of a VI and plays an important role in the sequel. Concerning the existence and uniqueness of its solutions, we have the following simple statement, where

$$I(y) := \{i \in \{1, 2, \ldots, s\} | G^i(y) = 0\}$$

denotes the *set of active indices* at y.

Theorem 2.4. Let F be strongly monotone on Ω given by (49) and $\bar{y}$ be the unique solution of the corresponding GE (31). Assume that the linear independence constraint qualification (LICQ) holds true at $\bar{y}$, i.e., the matrix

$$\begin{bmatrix} \nabla H(\bar{y}) \\ (\nabla G(\bar{y}))_{I(\bar{y})} \end{bmatrix}$$

has the full row rank. Then the GE (50) possesses a unique solution $(\widehat{y}, \widehat{\mu}, \widehat{\lambda})$ with $\widehat{y} = \bar{y}$.

3.3 Nonsmooth Newton's method

In section 2.1 we have seen that various equilibrium problems can be written down as the NE

$$C(z)=0, \tag{51}$$

where the function $C[\mathbb{R}^m \to \mathbb{R}^m]$ arises at the left-hand side of (32) or (45). If C enjoys some properties specified below, we can compute the respective equilibria by a "nonsmooth" variant of the Newton's method. Naturally such Newton's methods are based on first-order approximations of C (directional derivatives, generalized Jacobians, limit sets of Thibault); see eg (Kummer 1992),(Qi 1993). We shortly sketch the variant due to (Qi & Sun 1993) which has been used in numerous mechanical applications.

We assume throughout this section that C is *semismooth* on $\mathbb{R}^m$, i.e., C is locally Lipschitz and for each $z \in \mathbb{R}^m$ the limit

$$\lim_{\substack{V\in\bar{\partial}C(z+th')\\ h'\to h,t\downarrow 0}} \{Vh'\} \tag{52}$$

exists for all $h \in \mathbb{R}^m$. This implies in particular that C is directionally differentiable on $\mathbb{R}^m$. Moreover, by virtue of the Lipschitz continuity, if $z \in \mathbb{R}^m$ is fixed and $h \in \mathbb{R}^m$ is variable, then one has

$$C(z+h)-C(z)-C'(z;h)=o(\|h\|), \tag{53}$$

i.e., the positively homogeneous map $C'(z;\cdot\,)$ has the same approximation property as the Fréchet derivative; cf. (Shapiro 1990). We speak of *Bouligand differentiability* and make use of (53) in the main convergence statement (Theorem 2.6. below). From semismoothness it further follows that for any $V \in \bar{\partial}C(z+h)$

$$Vh-C'(z;h)=o(\|h\|). \tag{54}$$

Besides semismoothness, we will require from C still another property.

Definition 2.1. We say that C is *strongly BD-regular* at z if all matrices $V \in \partial_B C(z)$ are nonsingular.

Lemma 2.5. Assume that C is strongly BD-regular at z. Then there exist a neighbourhood $\mathcal{U}$ of z and a constant $c \geq 0$ such that all matrices $V \in \partial_B C(v)$ with $v \in \mathcal{U}$ are nonsingular and

$$\|V^{-1}\| \leq c. \tag{55}$$

Proof. First we show the existence of a neighbourhood $\mathcal{U}$ and a constant c such that the Jacobians $\nabla C(v)$ are nonsingular for all $v \in \mathcal{U}\backslash\omega_C$ (the set where C is not differentiable) and

$$\|\nabla C(v)\|^{-1} \leq c. \tag{56}$$

If this claim is not true then there is a sequence $z_k \to z$, $z_k \notin \omega_C$ such that either all Jacobians $\nabla C(z_k)$ are singular or $\|(\nabla C(z_k))^{-1}\| \to \infty$. Since C is Lipschitz near z, the

Jacobians $\nabla C(z_k)$ are bounded in a neighbourhood of z. Hence, there is a subsequence of $\{z_k\}$, say $\{z_{k'}\}$, such that $\{\nabla C(z_{k'})\}$ converges to a matrix V. This V must be singular. By the definition, $V \in \partial_B C(z)$ and this contradicts the strong BD-regularity of C at z. Hence, inequality (55) holds for all $v \in \mathcal{U} \setminus \omega_C$. At the points $v \in \mathcal{U} \cap \omega_C$ it suffices to observe that each $V \in \partial_B C(v)$ is the limit of a sequence $\nabla C(z_i)$ with $z_i \to v$ and $z_i \in \mathcal{U} \setminus \omega_C$ for all i. □

The Newton iteration based on generalized Jacobians is defined by

$$z_{k+1} = z_k - V_k^{-1} C(z_k), \quad k = 0, 1, \ldots \tag{57}$$

where $V_k \in \partial C(z_k)$. In (Qi & Sun 1993) a variant is studied, where the V_k in (57) is taken from $\partial_B C(z_k)$. This allows to weaken the assumptions ensuring local superlinear convergence. Recall that a sequence $\{x_k\}$ is said to converge *superlinearly* to x^*, if

$$\|x_{k+1} - x^*\| = o\left(\|x_k - x^*\|\right).$$

The resulting algorithm can be written in the form:

Nonsmooth Newton's Method: *Choose a starting point* (58)
$z_0 \in I\!R^m, \varepsilon > 0$ *and put* $k := 0$.

(1) *If* $\|C(z_k)\| \leq \varepsilon$, *then STOP.*

(2) *Compute an arbitrary matrix* $V_k \in \partial_B C(z_k)$ *and put*

$$z_{k+1} = z_k - V_k^{-1} C(z_k).$$

(3) *Put* $k := k + 1$ *and go back to (1).*

The next theorem gives a convergence result for the above method in the case $\varepsilon = 0$.

Theorem 2.6. Suppose that z^* is a solution of (51) and C is strongly BD-regular at z^*. Then the Newton's method (58) is well-defined and converges locally superlinearly to z^*.

Proof. Let $\mathcal{U}$ be a neighborhood of z^* specified by Lemma 2.5. (with $z = z^*$) and assume that $z_k \in \mathcal{U}$. Then, by Lemma 2.5, the kth Newton iteration in step (2) is well defined, and

$$\begin{aligned}
\|z_{k+1} - z^*\| &= \|z_k - z^* - V_k^{-1} C(z_k)\| \\
&= \left\|V_k^{-1}\left[V_k(z_k - z^*) - C(z_k) + C(z^*) + C'(z^*; z_k - z^*) - C'(z^*; z_k - z^*)\right]\right\| \\
&\leq \left\|V_k^{-1}\left[C(z_k) - C(z^*) - C'(z^*; z_k - z^*)\right]\right\| \\
&\qquad + \left\|V_k^{-1}\left[V_k(z_k - z^*) - C'(z^*; z_k - z^*)\right]\right\| \\
&\leq \|V_k^{-1}\| \left[\|C(z_k) - C(z^*) - C'(z^*; z_k - z^*)\|\right. \\
&\qquad \left. + \|V_k(z_k - z^*) - C'(z^*; z_k - z^*)\|\right].
\end{aligned}$$

By (53)

$$C(z_k) - C(z^*) - C'(z^*; z_k - z^*) = o(\|z_k - z^*\|)$$

and, due to (54),

$$V_k(z_k - z^*) - C'(z^*; z_k - z^*) = o(\|z_k - z^*\|).$$

Consequently, there is a function $\varphi(t) = o(t)$ (not dependent on k) such that

$$\|z_{k+1} - z^*\| \leq \varphi(\|z_k - z^*\|).$$

This implies the superlinear convergence of $\{z_k\}$ to z^*, provided z_0 is sufficiently close to z^* and all iterates z_k stay in $\mathcal{U}$. To this purpose, however, it suffices to start in such a ball $\mathcal{B} := z^* + \rho\mathbb{B}$, which satisfies the following conditions:

(i) $\mathcal{B} \subset \mathcal{U}$;

(ii) for all $\nu \in [0, \rho]$ one has $\varphi(\nu) \leq k\nu$ with a $k \in (0, 1)$.

The statement has been proved. □

Under additional assumptions on C, even local quadratic convergence can be proved for (58). Although we had no difficulties with the convergence of this method in our mechanical problems, the method may not converge at all, similarly as the classic Newton's method in the smooth case. To achieve global convergence, various modifications of the presented method (and also other nonsmooth Newton variants) are available. These questions go, however, beyond the scope of these lecture notes.

4 Lipschitzian properties of solution maps

In this type of analysis it is very convenient to work with the GE model, introduced in the previous section. The model has to be enriched, however, by a parameter $x \in \mathbb{R}^n$ which may generally enter both the single-valued as well as the set-valued part of the respective GE. Later this parameter will take over the role of a control variable. We confine ourselves here only to equilibria, where this parameter arises merely in the single-valued part. So, we will investigate the *solution map* $S[\mathbb{R}^n \rightsquigarrow \mathbb{R}^m]$ defined by

$$S(x) := \{y \in \vartheta | 0 \in F(x, y) + Q(y)\}, \tag{59}$$

where the closed set $\vartheta \subset \mathbb{R}^m$ specifies the "nonequilibrial" constraints imposed on y. Throughout this chapter $F[\mathbb{R}^n \times \mathbb{R}^m \to \mathbb{R}^m]$ is continuously differentiable and $Q[\mathbb{R}^m \rightsquigarrow \mathbb{R}^m]$ has a closed graph. Let $(\bar{x}, \bar{y}) \in \mathrm{Gph} S$ be a *reference point.* In Section 3.1 we will derive conditions ensuring the Aubin property of S around $(\bar{x}, \bar{y})$, whereas Section 3.2 is devoted to the existence of a Lipschitz single-valued localization of S at $(\bar{x}, \bar{y})$. The former property is important above all in stability considerations. Indeed, if S possesses the Aubin property around $(\bar{x}, \bar{y})$ and x is close to $\bar{x}$, then there is an equilibrium $y \in S(x)$ and $\|y - \bar{y}\| \leq L\|x - \bar{x}\|$ for some $L \geq 0$. The Lipschitz single-valued localizations play an important role above all in ImP.

4.1 Aubin property

Clearly, $S(x) = \{y \in \vartheta | g(x,y) \in \Lambda\}$, where

$$g(x,y) = \begin{bmatrix} y \\ -F(x,y) \end{bmatrix} \quad \text{and} \quad \Lambda = \text{Gph}Q. \tag{60}$$

Therefore we can proceed in a similar way as in the case of the map (18), but this time our analysis will be finer. We say that a multifunction $\Gamma[\mathbb{R}^l \rightsquigarrow \mathbb{R}^p]$ is *calm* at $(a,b) \in \text{Gph}\Gamma$ (with modulus L), provided there exist neighborhoods $\mathcal{U}$ of a, $\mathcal{V}$ of b and a real $L \geq 0$ such that

$$\Gamma(a') \cap \mathcal{V} \subset \Gamma(a) + L\|a' - a\|\mathbb{B} \;\forall\; a' \in \mathcal{U}.$$

By comparing this property with Definition 1.5 we immediately conclude that calmness is a weaker property than both the (local) upper Lipschitz continuity at a as well as the Aubin property around (a,b). To be able to apply the Mordukhovich criterion to S, we have first to compute an upper approximation of $D^*S(\bar{x},\bar{y})(0)$. To this purpose we introduce an auxiliary map $P[\mathbb{R}^m \rightsquigarrow \mathbb{R}^n \times \mathbb{R}^m]$ defined by

$$P(\xi) := \{(x,y) \in \mathbb{R}^n \times \vartheta | \xi + \begin{bmatrix} y \\ -F(x,y) \end{bmatrix} \in \text{Gph}Q\}. \tag{61}$$

Clearly, $\text{Gph}S = P(0)$.

Lemma 3.1. Let P be calm at $(0,\bar{x},\bar{y})$. Then one has for all $y^* \in \mathbb{R}^m$ the inclusion

$$D^*S(\bar{x},\bar{y})(y^*) \subset \left\{ \begin{array}{l} x^* \in \mathbb{R}^n \text{ with} \\ \left| \begin{bmatrix} x^* \\ y^* \end{bmatrix} \in \begin{bmatrix} 0 & -(\nabla_x F(\bar{x},\bar{y}))^T \\ E & -(\nabla_y F(\bar{x},\bar{y}))^T \end{bmatrix} N_{\text{Gph}Q}(\bar{y}, -F(\bar{x},\bar{y})) + \begin{bmatrix} 0 \\ N_\vartheta(\bar{y}) \end{bmatrix} \right. \end{array} \right\}.$$

The Mordukhovich criterion leads now directly to the following statement.

Theorem 3.2. Let P be calm at $(0,\bar{x},\bar{y})$ and the qualification condition

$$\begin{bmatrix} x^* \\ 0 \end{bmatrix} \in \begin{bmatrix} 0 & -(\nabla_x F(\bar{x},\bar{y}))^T \\ E & -(\nabla_y F(\bar{x},\bar{y}))^T \end{bmatrix} N_{\text{Gph}Q}(\bar{y}, -F(\bar{x},\bar{y})) + \begin{bmatrix} 0 \\ N_\vartheta(\bar{y}) \end{bmatrix} \Rightarrow x^* = 0 \tag{62}$$

is fulfilled. Then S has the Aubin property around $(\bar{x},\bar{y})$.

Unfortunately, it is not easy to ensure the calmness of P at $(0,\bar{x},\bar{y})$ except of the case when F is affine, ϑ is a convex polyhedron and Q is polyhedral. Then, namely, one can employ Theorem 1.2 which implies that P is even (locally) upper Lipschitz at 0. A standard way consists in the application of the Mordukhovich criterion also to P along the lines of Theorem 1.3. In this way we arrive at the following statement.

Theorem 3.3. Assume that the qualification condition

$$\left.\begin{array}{l} 0 \in u - (\nabla_y F(\bar{x},\bar{y}))^T v + N_\vartheta(\bar{y}) \\ (u,v) \in N_{\mathrm{Gph}Q}(\bar{y}, -F(\bar{x},\bar{y})) \end{array}\right\} \Rightarrow u = 0, v = 0 \tag{63}$$

is fulfilled. Then S has the Aubin property around $(\bar{x},\bar{y})$.

If $\vartheta = \mathbb{R}^m$, condition (63) reduces to the form

$$\left.\begin{array}{l} 0 = u - (\nabla_y F(\bar{x},\bar{y}))^T v \\ (u,v) \in N_{\mathrm{Gph}Q}(\bar{y}, -F(\bar{x},\bar{y})) \end{array}\right\} \Rightarrow v = 0.$$

Proof. P has the same structure as the multifunction Σ in (18). Consequently, by virtue of Theorem 1.3, P has the Aubin property around $(0,\bar{x},\bar{y})$, whenever the implication

$$\left.\begin{array}{l} 0 \in \begin{bmatrix} 0 & -(\nabla_x F(\bar{x},\bar{y}))^T \\ E & -(\nabla_y F(\bar{x},\bar{y}))^T \end{bmatrix} \begin{bmatrix} u \\ v \end{bmatrix} + \begin{bmatrix} 0 \\ N_\vartheta(\bar{y}) \end{bmatrix} \\ (u,v) \in N_{\mathrm{Gph}Q}(\bar{y}, -F(\bar{x},\bar{y})) \end{array}\right\} \Rightarrow u = 0, v = 0 \tag{64}$$

holds true. Thus, condition (63) ensures both condition (64) (and hence the calmness of P at $(0,\bar{x},\bar{y})$) as well as condition (62). The result follows from Theorem 3.2. The simplification of (63) in the case $\vartheta = \mathbb{R}^m$ is evident. □

The condition (63) is, however, sometimes too restrictive and excludes many situations, where S does have the Aubin property. Therefore it is important to develop specific calmness criteria, weaker than condition (64), cf. (Henrion, Jourani & Outrata 2002), (Henrion & Outrata to appear). Conditions (62),(63) are helpful only in the case, when we are able to express the limiting normal cone $N_{\mathrm{Gph}Q}(\bar{y}, -F(\bar{x},\bar{y}))$ in terms of the original data, associated with the considered equilibrium. This can be a difficult task as shown in (Dontchev & Rockafellar 1996) and (Mordukhovich & Outrata 2001). To illustrate the procedure, consider the case with $Q(y) = N_{\mathbb{R}^m_+}(y)$ corresponding to a complementarity problem. Then we can exploit a simple result from (Outrata 1999) (Lemma 3.4 below). In its formulation and also in numerous other results, presented in the sequel, we will associate with each pair $(y,z) \in \mathrm{Gph}N_{\mathbb{R}^m_+}$ the index sets

$$\begin{array}{ll} L(y,z) := & \{i \in \{1,2,\ldots,m\} | y^i > 0\} \\ I_+(y,z) := & \{i \in \{1,2,\ldots,m\} | y^i = 0, z^i < 0\} \\ I_0(y,z) := & \{i \in \{1,2,\ldots,m\} | y^i = 0, z^i = 0\}. \end{array}$$

They are related to the constraint $y \geq 0$, and therefore we speak about the index set of *inactive, strongly* and *weakly active* inequalities, respectively. The set $I_0(y,z)$ is also sometimes called the *degenerate* index set. The arguments at L, I_+, I_0 are omitted whenever these sets arise as subscripts or whenever it cannot cause any confusion.

Lemma 3.4. Consider a pair $(a, b) \in \text{Gph} N_{\mathbb{R}^m_+}$ Then one has

$$N_{\text{Gph} N_{\mathbb{R}^m_+}}(a, b) = \mathop{\text{X}}_{i=1}^{m} N_{\text{Gph} N_{\mathbb{R}_+}}(a^i, b^i),$$

where

$$N_{\text{Gph} N_{\mathbb{R}_+}}(a^i, b^i) = \begin{cases} 0 \times \mathbb{R} \text{ for } i \in L(a, b) \\ \mathbb{R} \times 0 \text{ for } i \in I_+(a, b) \\ (0 \times \mathbb{R}) \cup (\mathbb{R} \times 0) \cup (\mathbb{R}_- \times \mathbb{R}_+) \text{ for } i \in I_0(a, b). \end{cases}$$

If $\vartheta = \mathbb{R}^m$, condition (63) attains in this situation the form

$$\left.\begin{array}{l} 0 = u - ((\nabla_y F(\bar{x}, \bar{y}))_{L \cup I_0})^T v_{L \cup I_0} \\ u_L = 0 \\ \text{for } i \in I_0 \text{ either } u^i v^i = 0 \\ \qquad\qquad \text{or } u^i < 0 \text{ and } v^i > 0 \end{array}\right\} \Rightarrow v_{L \cup I_0} = 0. \tag{65}$$

In the above relations all index sets L, I_+, I_0 have the argument $(\bar{y}, -F(\bar{x}, \bar{y}))$. Of course, also the verification of condition (65) is not easy; one has to employ the structural properties of the partial Jacobian $\nabla_y F(\bar{x}, \bar{y})$.

4.2 Lipschitz single-valued localization

We immediately infer that S possesses a Lipschitz single-valued localization at $(\bar{x}, \bar{y})$, whenever it has the Aubin property around $(\bar{x}, \bar{y})$ and for each x close to $\bar{x}$ the GE

$$0 \in F(x, y) + Q(y) \tag{66}$$

possesses a unique solution in $\vartheta \cap \mathcal{O}$, where $\mathcal{O}$ is a neighborhood of $\bar{y}$. Another way is to employ a powerful result from the landmark paper (Robinson 1980). In this context, however, we have to assume that $\vartheta = \mathbb{R}^m$.

Definition 3.1. We say that the GE (66) satisfies the *Strong Regularity Condition* (SRC) at $(\bar{x}, \bar{y})$ provided the multifunction $\sum[\mathbb{R}^m \rightsquigarrow \mathbb{R}^m]$, defined by

$$\sum(\xi) = \{y \in \mathbb{R}^m | \xi \in F(\bar{x}, \bar{y}) + \nabla_y F(\bar{x}, \bar{y})(y - \bar{y}) + Q(y)\}$$

possesses a Lipschitz single-valued localization at $(0, \bar{y})$.

Theorem 3.5. Assume that the GE (66) satisfies (SRC) at $(\bar{x}, \bar{y})$. Then the corresponding solution map S possesses a Lipschitz single-valued localization at $(\bar{x}, \bar{y})$.

In this way we have converted the analysis of S to the analysis of a simpler map $\sum$, where the single-valued part of the underlying GE is affine. Thus, $\sum$ is (locally) upper Lipschitz at 0 whenever Q is polyhedral and then it suffices to show that the GE

$$\xi \in F(\bar{x}, \bar{y}) + \nabla_y F(\bar{x}, \bar{y})(y - \bar{y}) + Q(y) \tag{67}$$

possesses a unique solution y for all ξ from a neighborhood of 0. If $Q(y) = N_\Omega(y)$ with Ω being a convex polyhedron, the GE (67) can be significantly simplified. We recall that the cone

$$K(y, z) := T_\Omega(y) \cap z^\perp$$

is termed the *critical* cone of Ω with respect to y and z.

Theorem 3.6. Let Ω be a convex polyhedron. Then the GE

$$\xi \in F(\bar{x}, \bar{y}) + \nabla_y F(\bar{x}, \bar{y})(y - \bar{y}) + N_\Omega(y) \tag{68}$$

possesses a unique solution y for all ξ from a neighborhood of 0 iff the GE

$$\xi \in \nabla_y F(\bar{x}, \bar{y})h + N_{K(\bar{y}, F(\bar{x}, \bar{y}))}(h) \tag{69}$$

possesses a unique solution h for all $\xi \in \mathbb{R}^m$.

Proof. (68) can clearly be rewritten to the form

$$y = \text{Proj}_\Omega(\xi - F(\bar{x}, \bar{y}) - \nabla_y F(\bar{x}, \bar{y})(y - \bar{y}) + y) \tag{70}$$

If we choose y sufficiently close to $\bar{y}$ and ξ sufficiently close to 0, the norm of the vector

$$\eta := \xi - \nabla_y F(\bar{x}, \bar{y})(y - \bar{y}) + y - \bar{y}$$

can be made arbitrarily small. Since

$$\bar{y} = \text{Proj}_\Omega(\bar{y} - F(\bar{x}, \bar{y})),$$

we infer that

$$\begin{aligned} \text{Proj}_\Omega(\xi - F(\bar{x}, \bar{y}) - \nabla_y F(\bar{x}, \bar{y})(y - \bar{y}) + y) &= \text{Proj}_\Omega(\bar{y} - F(\bar{x}, \bar{y}) + \eta) \\ &= \bar{y} + \text{Proj}_{K(\bar{y}, F(\bar{x}, \bar{y}))}(\eta), \end{aligned}$$

where the last equality follows from the directional differentiability of the projection onto convex polyhedra ((Robinson 1991)). We conclude that with $h := y - \bar{y}$ equation (70) amounts to

$$h = \text{Proj}_{K(\bar{y}, F(\bar{x}, \bar{y}))}(\xi - \nabla_y F(\bar{x}, \bar{y})h + h)$$

which is exactly the GE (69). It remains to observe that, since (69) is positively homogeneous in h, the restriction of ξ to a neighborhood of 0 has no importance. □

In the considered case the problem of the Lipschitz single-valued localization has thus been reduced to the question of existence and uniqueness of solutions to (69). The application of the results from Section 2 leads now a number of useful results. We confine ourselves, however, only to the following two situations playing an important role in the sequel.

Theorem 3.7. Consider the NCP governed by the GE

$$0 \in F(x, y) + N_{\mathbb{R}^m_+}(y). \tag{71}$$

This GE fulfills the (SRC) at the reference point $(\bar{x}, \bar{y})$ iff the matrix $(\nabla_y F(\bar{x}, \bar{y}))_{L,L}$ is nonsingular and its Schur complement in the matrix

$$\begin{bmatrix} (\nabla_y F(\bar{x}, \bar{y}))_{L,L} & (\nabla_y F(\bar{x}, \bar{y}))_{L,I_0} \\ (\nabla_y F(\bar{x}, \bar{y}))_{I_0,L} & (\nabla_y F(\bar{x}, \bar{y}))_{I_0,I_0} \end{bmatrix}$$

is a P-matrix. (Again, the index sets L, I_0 have the argument $(\bar{y}, -F(\bar{x}, \bar{y}))$).

Proof. We have to show that the above condition is equivalent to the unique solvability of the GE

$$\xi \in \nabla_y F(\bar{x}, \bar{y})h + N_K(h)$$

for each $\xi \in \mathbb{R}^m$; thereby

$$\begin{aligned} K = T_{\mathbb{R}^m_+}(\bar{y}) \cap (F(\bar{x}, \bar{y}))^\perp &= \{h \in \mathbb{R}^m | h_{I_+} \geq 0,\ h_{I_0} \geq 0\} \cap \{h \in \mathbb{R}^m | h_{I_+} = 0\} \\ &= \{h \in \mathbb{R}^m | h_{I_+} = 0,\ h_{I_0} \geq 0\}. \end{aligned}$$

Consequently,

$$N_K(h) = \{\eta \in \mathbb{R}^m | \eta_L = 0,\ \ \eta_{I_0} \leq 0,\ \langle \eta_{I_0}, h_{I_0} \rangle = 0\},$$

and the respective GE (69) reduces to the form

$$\begin{aligned} \xi_L &= (\nabla_y F(\bar{x}, \bar{y}))_{L,L} h_L + (\nabla_y F(\bar{x}, \bar{y}))_{L,I_0} h_{I_0} \\ \xi_{I_0} &= (\nabla_y F(\bar{x}, \bar{y}))_{I_0,L} h_L + (\nabla_y F(\bar{x}, \bar{y}))_{I_0,I_0} h_{I_0} + N_{\mathbb{R}^a_+}(h_{I_0}), \end{aligned}$$

where $a = |I_0|$. The result follows now from the characterization of the unique solvability of a linear CP, mentioned in the previous section, cf. (Robinson 1980). □

Next we turn our attention to the case, where Ω is given by (49) and apply the technique leading to the GE (50). This technique enables even to generalize the original model and allows that x enters also the functions H, G defining Ω. So,we start with the GE

$$0 \in F(x, y) + N_{\Omega(x)}(y) \tag{72}$$

where the multifunction $\Omega[\mathbb{R}^n \rightsquigarrow \mathbb{R}^m]$ is defined by

$$\Omega(x) := \{y \in \mathbb{R}^m | H(x, y) = 0, G(x, y) \leq 0\},$$

and assume that $H[\mathbb{R}^n \times \mathbb{R}^m \to \mathbb{R}^l], G[\mathbb{R}^n \times \mathbb{R}^m \to \mathbb{R}^s]$ are continuously differentiable and for all $x \in \mathbb{R}^n, H(x, \cdot)$ is affine and all components $G^i(x, \cdot)$ are convex. Moreover, for the reference parameter value $\bar{x}$ we modify the (SCQ) as follows:

(ESCQ): $\exists\ \bar{y}_{\bar{x}} \in \mathbb{R}^m$ such that $H(\bar{x}, \bar{y}_x) = 0$ and $G(\bar{x}, \bar{y}_x) \in \operatorname{int} \mathbb{R}^s_+$.

In this way, for x close to $\bar{x}$, the GE (72) admits the KKT form

$$\left.\begin{array}{rcl} 0 & = & \mathcal{L}(x,y,\mu,\lambda) \\ 0 & = & H(x,y) \\ 0 & \in & -G(x,y) + N_{\mathbb{R}^s_+}(\lambda), \end{array}\right\} \tag{73}$$

where $\mathcal{L}(x,y,\mu,\lambda) := F(x,y) + (\nabla_y H(x,y))^T\mu + (\nabla_y G(x,y))^T\lambda$. Note that in (73) the parameter x does not enter the multi-valued part but, in contrast to (72), we have to do with additional variables μ, λ. As in case of NCPs we will need the index sets

$$\begin{array}{rl} L(x,y,\lambda) := & \{i \in \{1,2,\ldots,s\} | \lambda^i > 0\} \\ I_+(x,y,\lambda) := & \{i \in \{1,2,\ldots,s\} | \lambda^i = 0, G^i(x,y) < 0\} \\ I_0(x,y,\lambda) := & \{i \in \{1,2,\ldots,s\} | \lambda^i = 0, G^i(x,y) = 0\}, \end{array}$$

related to the constraint $\lambda \geq 0$. Again, their arguments are omitted whenever these sets arise as subscripts or whenever it cannot cause any confusion. To be able to present the next result, we also have to associate with the reference point $(\bar{x},\bar{y},\bar{\mu},\bar{\lambda})$ the matrix

$$D(\bar{x},\bar{y},\bar{\mu},\bar{\lambda}) := \begin{bmatrix} \nabla_y \mathcal{L}(\bar{x},\bar{y},\bar{\mu},\bar{\lambda}) & (\nabla_y H(\bar{x},\bar{y}))^T & (\nabla_y G(\bar{x},\bar{y}))^T \\ \nabla_y H(\bar{x},\bar{y}) & 0 & 0 \\ -\nabla_y G(\bar{x},\bar{y}) & 0 & 0 \end{bmatrix},$$

and, for an arbitrary subset N of $I_0(\bar{x},\bar{y},\bar{\lambda})$, the matrix

$$D_{(N)}(\bar{x},\bar{y},\bar{\mu},\bar{\lambda}) := \begin{bmatrix} \nabla_y \mathcal{L}(\bar{x},\bar{y},\bar{\mu},\bar{\lambda}) & (\nabla_y H(\bar{x},\bar{y}))^T & ((\nabla_y G(\bar{x},\bar{y}))_N)^T \\ \nabla_y H(\bar{x},\bar{y}) & 0 & 0 \\ (\nabla_y G(\bar{x},\bar{y}))_N & 0 & 0 \end{bmatrix}.$$

Theorem 3.8. The GE (73) fulfills the (SRC) at $(\bar{x},\bar{y},\bar{\mu},\bar{\lambda})$ iff the matrix $D_{(L)}(\bar{x},\bar{y},\bar{\mu},\bar{\lambda})$ is nonsingular and its complement in $D_{(L\cup I_0)}(\bar{x},\bar{y},\bar{\mu},\bar{\lambda})$ is a P-matrix.

This statement can be proved along the same lines as Theorem 3.7., cf. eg (Outrata et al. 1998, Theorem 5.7). It remains to ensure the above requirements by means of reasonable assumptions imposed on the problem data.

Theorem 3.9. Suppose that the constraint qualification

(ELICQ): The gradients $\nabla_y H^i(\bar{x},\bar{y}), i = 1,2,\ldots,l$, and

$\nabla_y G^i(\bar{x},\bar{y}), i \in L(\bar{x},\bar{y}) \cup I_0(\bar{x},\bar{y},\bar{\lambda})$, are linearly independent;

is fulfilled. Further assume that $\nabla_y \mathcal{L}(\bar{x},\bar{y},\bar{\mu},\bar{\lambda})$ is strictly copositive on

$$\ker \nabla_y H(\bar{x},\bar{y}) \cap \ker((\nabla_y G(\bar{x},\bar{y}))_L).$$

Then the requirement of Theorem 3.8 is satisfied and, consequently, the GE (73) fulfills the (SRC) at $(\bar{x},\bar{y},\bar{\mu},\bar{\lambda})$.

The proof follows directly from Theorem 2.3. If the GE (73) represents first-order optimality conditions of an optimization problem, then the second condition of Theorem 3.9 is just the *strong second-order sufficient condition*, well-known from mathematical programming. Therefore it is sometimes also referred to as the (SSOSC) at $(\bar{x}, \bar{y}, \bar{\mu}, \bar{\lambda})$.

If σ is a Lipschitz single-valued localization of S at $(\bar{x}, \bar{y})$, one could describe its local behaviour by $\bar{\partial}\sigma(\bar{x})$ or an upper approximation of this set. The computation of an upper approximation of $\bar{\partial}\sigma(\bar{x})$ is particularly simple, provided σ is a PC^1-function. We explain this technique again by means of an NCP and put $\vartheta = \mathbb{R}^m$ for the sake of simplicity. From a continuity argument it is clear that in a neighborhood of the reference point one definitely has

$$\begin{array}{lll} F^i(x,y) \geq 0 & y^i = 0 & \text{for } i \in I_+ \cup M_1 \\ F^i(x,y) = 0 & y^i \geq 0 & \text{for } i \in L \cup M_2 \end{array} \tag{74}$$

for some partition(s) M_1, M_2 of $I_0(\bar{y}, -F(\bar{x}, \bar{y}))$. Ignoring the inequalities in (74), we arrive at a system of nonlinear equations

$$\begin{array}{rcll} F^i(x,y) & = & 0 & \text{for } i \in L \cup M_2 \\ y^i & = & 0 & \text{for } i \in I_+ \cup M_1 \end{array} \tag{75}$$

to which the classical implicit function theorem can be applied whenever the matrix (after a suitable reorganization)

$$\begin{bmatrix} (\nabla_y F(\bar{x}, \bar{y}))_{L \cup M_2} & \\ 0 & E_{I_+ \cup M_1} \end{bmatrix} \tag{76}$$

is nonsingular. If matrices (76) are nonsingular for all possible partitions M_1, M_2, we immediately infer that σ is a PC^1-function. Indeed, let us introduce the index set $\mathbb{K}(\bar{x}, \bar{y})$ of all possible partitions of I_0 so that to each pair M_1, M_2 we assign an index $j \in \mathbb{K}(\bar{x}, \bar{y})$. Further, denote by $\sigma_j, j \in \mathbb{K}(\bar{x}, \bar{y})$, the implicit function specified by the corresponding system (75). Clearly, $\bar{y} = \sigma_j(\bar{x})$ for all $j \in \mathbb{K}(\bar{x}, \bar{y})$, the single functions σ_j are continuously differentiable and there exists a neighborhoods $\mathcal{U}$ of $\bar{x}$ such that $\sigma(x) = \sigma_j(x)$ for some $j \in \mathbb{K}(\bar{x}, \bar{y})$ and all $x \in \mathcal{U}$. Hence, σ is a PC^1-function on $\mathcal{U}$. On the basis of (17) we are now able to compute an upper approximation of $\bar{\partial}\sigma(\bar{x})$ in the form

$$\bar{\partial}\sigma(\bar{x}) \subset \operatorname{conv}\{\nabla\sigma_j(\bar{x}) | j \in \mathbb{K}(\bar{x}, \bar{y})\}. \tag{77}$$

This so-called PC^1 approach to the computing of $\bar{\partial}\sigma(\bar{x})$ is exploited above all in cases, where the existence of a Lipschitz single-valued localization is ensured by means of the strong regularity condition. Then we benefit from the fact that (SRC) implies the regularity for all systems (75). We return to this procedure in Section 5, devoted to ImP.

5 Optimality conditions

Consider the MPEC

$$\begin{array}{ll} \text{minimize} & f(x,y) \\ \text{subject to} & \\ & 0 \in F(x,y) + Q(y) \\ & (x,y) \in \kappa, \end{array} \tag{78}$$

where $f[\mathbb{R}^n \times \mathbb{R}^m \rightarrow \mathbb{R}]$ is an objective function, the equilibrium constraint is modeled by a GE studied in the previous section and $\kappa \subset \mathbb{R}^n \times \mathbb{R}^m$ is a closed set which comprises all other constraints imposed on x and y. In particular, κ can be the Carthesian product $\omega \times \vartheta$, where $\omega \subset \mathbb{R}^n$ is the set of *feasible controls* and $\vartheta \subset \mathbb{R}^m$ is the set of *feasible state variables.* To derive optimality conditions, we can put $z = (x, y)$ and rewrite the constraints in (78) to the form

$$\{z \in \kappa | g(z) \in \Lambda\}$$

where g and Λ are defined in (60). This enables a direct application of Theorems 1.4, 1.5 which will be done in Section 4.1. If the GE in (78) admits a Lipschitz single-valued localization at a (locally) optimal pair $(\widehat{x}, \widehat{y})$ and the nonequilibrium constraint κ admits the product form $\omega \times \vartheta$, we can make use of Theorem 1.6 and arrive at optimality conditions of a similar type. This technique, however, combined with the PC^1 approach from the previous section, leads to different optimality conditions, closely related with ImP. This is the topic of Section 4.2. All optimality conditions will be illustrated by means of MPCC, i.e. the problem (78), where the GE models an NCP.

5.1 Optimality conditions via mathematical programming

Throughout the sequel $(\widehat{x}, \widehat{y})$ is a (locally) optimal pair in (78). Moreover, we will assume that the objective f is Lipschitz near $(\widehat{x}, \widehat{y})$.

Theorem 4.1. Let the qualification condition

$$\left.\begin{array}{c} \begin{bmatrix} 0 & -(\nabla_x F(\widehat{x}, \widehat{y}))^T \\ E & -(\nabla_y F(\widehat{x}, \widehat{y}))^T \end{bmatrix} \begin{bmatrix} u \\ v \end{bmatrix} \in -N_\kappa(\widehat{x}, \widehat{y}) \\ (u, v) \in N_{\mathrm{Gph}Q}(\widehat{y}, -F(\widehat{x}, \widehat{y})) \end{array}\right\} \Rightarrow \left\{ \begin{array}{lcl} u & = & 0 \\ v & = & 0 \end{array}\right. \tag{79}$$

be fulfilled. Then there exists a pair $(\widehat{\xi}, \widehat{\eta}) \in \partial f(\widehat{x}, \widehat{y})$, a KKT vector $\widehat{v} \in \mathbb{R}^m$ and a pair $(\widehat{w}, \widehat{z}) \in N_\kappa(\widehat{x}, \widehat{y})$ such that

$$\begin{array}{rcccccc} 0 = \widehat{\xi} & + & (\nabla_x F(\widehat{x}, \widehat{y}))^T \widehat{v} & + & \widehat{w} & & \\ 0 = \widehat{\eta} & + & (\nabla_y F(\widehat{x}, \widehat{y}))^T \widehat{v} & + & D^*Q(\widehat{y}, -F(\widehat{x}, \widehat{y}))(\widehat{v}) & + & \widehat{z}. \end{array} \tag{80}$$

Proof. (79) is exactly condition (6) for the constraint structure being considered. It enables to express an upper approximation of the limiting normal cone to the whole constraint system in (78) in the form

$$\begin{bmatrix} 0 & -(\nabla_x F(\widehat{x}, \widehat{y}))^T \\ E & -(\nabla_y F(\widehat{x}, \widehat{y}))^T \end{bmatrix} N_{\mathrm{Gph}Q}(\widehat{y}, -F(\widehat{x}, \widehat{y})) + N_\kappa(\widehat{x}, \widehat{y}) =$$
$$\left\{ \begin{bmatrix} (\nabla_x F(\widehat{x}, \widehat{y}))^T v + w \\ (\nabla_y F(\widehat{x}, \widehat{y}))^T v + D^*Q(\widehat{y}, -F(\widehat{x}, \widehat{y}))(v) + z \end{bmatrix} \text{ with } \right.$$
$$\left. v \in \mathbb{R}^m, (w, z) \in N_\kappa(\widehat{x}, \widehat{y}) \right\}.$$

It remains thus to apply relation (20). □

Instead of the qualification condition (79) it suffices to require that the map $\widetilde{P}[\mathbb{R}^m \rightsquigarrow \mathbb{R}^n \times \mathbb{R}^m]$, defined by

$$\widetilde{P}(\xi) := \left\{ (x,y) \in \kappa \left| \xi + \begin{bmatrix} y \\ -F(x,y) \end{bmatrix} \in \mathrm{Gph} Q \right. \right\},$$

is calm at $(0, \widehat{x}, \widehat{y})$, cf. (Henrion et al. 2002). From this it follows that the qualification condition (79) can be omitted, whenever F is affine (in both variables), κ is a convex polyhedron and Q is polyhedral. It is also important to point out that the qualification condition (79) is fulfilled whenever $\kappa = \omega \times \mathbb{R}^m$ and the GE in (78) satisfies (SRC) at $(\widehat{x}, \widehat{y})$, cf. (Outrata 2000). Points $(\widehat{x}, \widehat{y})$ satisfying (80) are called *M(ordukhovich)-stationary.* Similarly as in the statement of Theorem 3.3, relations (79),(80) have to be expressed in terms of the problem data. We show next the form of these relations provided $Q(y) = N_{\mathbb{R}^m_+}(y)$, i.e., the considered equilibrium is an NCP.

Theorem 4.2. Suppose that the qualification condition

$$\left.\begin{array}{l} \begin{bmatrix} 0 & -((\nabla_x F(\widehat{x}, \widehat{y}))_{L \cup I_0})^T \\ E & -((\nabla_y F(\widehat{x}, \widehat{y}))_{L \cup I_0})^T \end{bmatrix} \begin{bmatrix} u \\ v_{L \cup I_0} \end{bmatrix} \in -N_\kappa(\widehat{x}, \widehat{y}) \\ u_L = 0 \text{ and for } i \in I_0 \text{ either } u^i v^i = 0 \\ \text{or } u^i < 0 \text{ and } v^i > 0 \end{array} \right\} \Rightarrow \left\{ \begin{array}{lcl} u & = & 0 \\ v_{L \cup I_0} & = & 0 \end{array} \right. \tag{81}$$

holds true. Then there exists a pair $(\widehat{\xi}, \widehat{\eta}) \in \partial f(\widehat{x}, \widehat{y})$, a pair of KKT vectors $(\widehat{u}, \widehat{v}) \in \mathbb{R}^m \times \mathbb{R}^m$ and a pair $(\widehat{w}, \widehat{z}) \in N_\kappa(\widehat{x}, \widehat{y})$ such that $\widehat{u}_L = 0, \widehat{v}_{I_+} = 0$,

$$\begin{array}{lcl} 0 & = & \widehat{\xi} - ((\nabla_x F(\widehat{x}, \widehat{y}))_{L \cup I_0})^T \widehat{v}_{L \cup I_0} + \widehat{w} \\ 0 & = & \widehat{\eta}_{L \cup I_0} + \widehat{u}_{L \cup I_0} - ((\nabla_y F(\widehat{x}, \widehat{y}))_{L \cup I_0, L \cup I_0})^T \widehat{v}_{L \cup I_0} + \widehat{z}_{L \cup I_0}, \end{array} \tag{82}$$

and for $i \in I_0$ one has either $\widehat{u}^i \widehat{v}^i = 0$ or $\widehat{u}^i < 0$ and $\widehat{v}^i > 0$.

The index sets L, I_+, I_0 in the above statement are related to the argument $(\widehat{y}, -F(\widehat{x}, \widehat{y}))$. Its proof follows immediately from Theorem 4.1 and Lemma 3.4. Condition (81) is generally not easy to verify. In (Outrata 1999) this condition has been suitably reformulated and the resulting condition seems to be more easily tractable, at least in low dimensional problems. Furthermore, there is a number of easily verifiable conditions which imply (81), cf. (Ye 2000). One of them is the MPEC-LICQ (linear independence constraint qualification) which will be presented in the next part of this section. If f is continuously differentiable, we can think about the strengthening of the above statements (Theorems 4.1, 4.2) along the lines of Theorem 1.5. To this purpose we will assume that

$$\kappa := \{(x,y) \in \mathbb{R}^n \times \mathbb{R}^m | q(x,y) \leq 0\}, \tag{83}$$

where the map $q[\mathbb{R}^n \times \mathbb{R}^m \to \mathbb{R}^t]$ is continuously differentiable. This enables to write down the overall constraint system from (78) in the compact form

$$\left\{ z \in \mathbb{R}^n \times \mathbb{R}^m \left| \begin{bmatrix} g(z) \\ q(z) \end{bmatrix} \in GphQ \times \mathbb{R}^t_- \right. \right\}. \tag{84}$$

Let $J(\widehat{x}, \widehat{y})$ be the index set of inequalities from (83) that are active at $(\widehat{x}, \widehat{y})$.

Theorem 4.3. Assume that Q is polyhedral and the matrix

$$\begin{bmatrix} 0 & -(\nabla_x F(\hat{x},\hat{y}))^T & ((\nabla_x q(\hat{x},\hat{y}))_J)^T \\ E & -(\nabla_y F(\hat{x},\hat{y}))^T & ((\nabla_y q(\hat{x},\hat{y}))_J)^T \end{bmatrix} \tag{85}$$

has the maximum column rank. Then there are multipliers $(\hat{u},\hat{v},\hat{\mu}) \in \mathbb{R}^m \times \mathbb{R}^m \times \mathbb{R}^a_+$, $a = |J(\hat{x},\hat{y})|$ such that

$$\begin{array}{l} 0 = \nabla_x f(\hat{x},\hat{y}) - (\nabla_x F(\hat{x},\hat{y}))^T \hat{v} + ((\nabla_x q(\hat{x},\hat{y}))_J)^T \hat{\mu} \\ 0 = \nabla_y f(\hat{x},\hat{y}) + \hat{u} - (\nabla_y F(\hat{x},\hat{y}))^T \hat{v} + ((\nabla_y q(\hat{x},\hat{y}))_J)^T \hat{\mu} \\ (\hat{u},\hat{v}) \in \hat{N}_{\mathrm{Gph}Q}(\hat{y}, -F(\hat{x},\hat{y})). \end{array}$$

Proof. Since Q is polyhedral, the constraints have the structure investigated in Theorem 1.5. The rank condition, imposed on the matrix (85) ensures condition (25). Moreover, it ensures also condition (24), cf. (Rockafellar & Wets 1998, Exercise 6.7). Therefore, it suffices to observe that (with $\hat{z} = (\hat{x},\hat{y})$)

$$\hat{N}_{(\mathrm{Gph}Q \times \mathbb{R}^t_-)}(g(\hat{z}), q(\hat{z})) = \hat{N}_{\mathrm{Gph}Q}(g(\hat{z})) \times \hat{N}_{\mathbb{R}^t_-}(q(\hat{z})),$$

and

$$\hat{N}_{\mathbb{R}^t_-}(q(\hat{z})) = \{\mu \in \mathbb{R}^t_+ | \mu^i = 0 \text{ for } i \notin J(\hat{z})\}.$$

Thus the Fréchet normal cone to the set (84) amounts to

$$\left\{ \begin{array}{l} \begin{bmatrix} 0 & -(\nabla_x F(\hat{z}))^T & ((\nabla_x q(\hat{z}))_J)^T \\ E & -(\nabla_y F(\hat{z}))^T & ((\nabla_y q(\hat{z}))_J)^T \end{bmatrix} \begin{bmatrix} u \\ v \\ \mu \end{bmatrix} \\ \text{with } (u,v) \in \hat{N}_{\mathrm{Gph}Q}(\hat{y}, -F(\hat{x},\hat{y})), \mu \geq 0 \end{array} \right\},$$

and we are done. □

Condition imposed on the matrix (85) is the MPEC-LICQ mentioned above. Let us illustrate the difference between the limiting and the Fréchet normal cones by means of the MPCC, investigated in Theorem 4.2.

Theorem 4.4. Suppose that the matrix

$$\begin{bmatrix} 0 & -((\nabla_x F(\hat{x},\hat{y}))_{L\cup I_0})^T & ((\nabla_x q(\hat{x},\hat{y}))_J)^T \\ (E_{I_+ \cup I_0})^T & -((\nabla_y F(\hat{x},\hat{y}))_{L\cup I_0})^T & ((\nabla_y q(\hat{x},\hat{y}))_J)^T \end{bmatrix} \tag{86}$$

has the maximum column rank. Then there are multipliers $(\hat{u},\hat{v},\hat{\mu}) \in \mathbb{R}^m \times \mathbb{R}^m \times \mathbb{R}^a_+$ such that $\hat{u}_L = 0, \hat{v}_{I_+} = 0$,

$$\begin{array}{rcl} 0 & = & \nabla_x f(\hat{x},\hat{y}) - ((\nabla_x F(\hat{x},\hat{y}))_{L\cup I_0})^T \hat{v}_{L\cup I_0} + (\nabla_x q(\hat{x},\hat{y}))_J)^T \hat{\mu} \\ 0 & = & (\nabla_y f(\hat{x},\hat{y}))_{L\cup I_0} + \hat{u}_{L\cup I_0} - ((\nabla_y F(\hat{x},\hat{y}))_{L\cup I_0, L\cup I_0,})^T \hat{v}_{L\cup I_0} \\ & & + (\nabla_y q(\hat{x},\hat{y}))_{J, L\cup I_0})^T \hat{\mu} \end{array}$$

and for $i \in I_0$ one has $\widehat{u}^i \leq 0$, $\widehat{v}^i \geq 0$.

Proof. Q is polyhedral and so, again, Theorem 1.5 can be applied. The condition, imposed on the matrix (86) implies in particular the linear independence of the vectors $\nabla q^i(\widehat{x}, \widehat{y}), i \in J$, so that

$$N_\kappa(\widehat{x}, \widehat{y}) = \{((\nabla q(\widehat{x}, \widehat{y}))_J)^T \mu | \mu \geq 0\}.$$

Therefore the condition, imposed on the matrix (86) implies the satisfaction of condition (81). The latter is exactly the (GMFCQ) for the considered MPEC; therefore the respective Abadie CQ (24) holds true as well. Next we observe that for a pair $(a, b) \in \mathrm{Gph}\widehat{N}_{\mathbb{R}^m_+}$ one has

$$\widehat{N}_{\mathrm{Gph}N_{\mathbb{R}^m_+}}(a, b) = \bigtimes_{i=1}^{m} \widehat{N}_{\mathrm{Gph}N_{\mathbb{R}_+}}(a^i, b^i), \tag{87}$$

$$\widehat{N}_{\mathrm{Gph}N_{\mathbb{R}_+}}(a^i, b^i) = \begin{cases} 0 \times \mathbb{R} & \text{for} \quad i \in L(a, b) \\ \mathbb{R} \times 0 & \text{for} \quad i \in I_+(a, b) \\ \mathbb{R}_- \times \mathbb{R}_+ & \text{for} \quad i \in I_0(a, b). \end{cases}$$

Thus $u_L = 0, v_{I_+} = 0$ for all $(u, v) \in \widehat{N}_{\mathrm{Gph}N_{\mathbb{R}^m_+}}(\widehat{y}, -F(\widehat{x}, \widehat{y}))$ and we claim that the condition, imposed on that matrix (86), implies also condition (25). Indeed, the equation

$$\begin{bmatrix} 0 & -((\nabla_x F(\widehat{x}, \widehat{y}))_{L \cup I_0})^T & ((\nabla_x q(\widehat{x}, \widehat{y}))_J)^T \\ (E_{I_+ \cup I_0})^T & -((\nabla_y F(\widehat{x}, \widehat{y}))_{L \cup I_0})^T & ((\nabla_y q(\widehat{x}, \widehat{y}))_J)^T \end{bmatrix} \begin{bmatrix} u_{I_+ \cup I_0} \\ v_{L \cup I_0} \\ \mu \end{bmatrix} = 0$$

implies immediately that $u_{I_+ \cup I_0}, v_{L \cup I_0}, \mu = 0$ so that altogether the linear mapping

$$\begin{bmatrix} 0 & -(\nabla_x F(\widehat{x}, \widehat{y}))^T & ((\nabla_x q(\widehat{x}, \widehat{y}))_J)^T \\ E & -(\nabla_y F(\widehat{x}, \widehat{y}))^T & ((\nabla_y q(\widehat{x}, \widehat{y}))_J)^T \end{bmatrix} \begin{bmatrix} \cdot \\ \cdot \\ \cdot \end{bmatrix}$$

is injective on the Fréchet normal cone to each (convex polyhedral) component of $\mathrm{Gph}N_{\mathbb{R}^m_+} \times \mathbb{R}^t_-$. This is sufficient for condition (25) to hold. It remains to modify formulas (82) in such a way that $\widehat{\xi} = \nabla_x f(\widehat{x}, \widehat{y}), \widehat{\eta} = \nabla_y f(\widehat{x}, \widehat{y})$,

$$\begin{bmatrix} \widehat{w} \\ \widehat{z} \end{bmatrix} = \begin{bmatrix} ((\nabla_x q(\widehat{x}, \widehat{y}))_J)^T \mu \\ ((\nabla_y q(\widehat{x}, \widehat{y}))_J)^T \mu \end{bmatrix},$$

and $(\widehat{u}, \widehat{v}) \in \widehat{N}_{\mathrm{Gph}N_{\mathbb{R}^m_+}}(\widehat{y}, -F(\widehat{x}, \widehat{y}))$. □

When comparing Theorems 4.2 and 4.4, the most important difference concerns the multipliers $\widehat{u}^i, \widehat{v}^i, i \in I_0$. In Theorem 4.4 they have necessarily certain signs whereas in Theorem 4.2 they merely satisfy a weaker condition of combinatorial nature. The points satisfying the optimality conditions from Theorem 4.4 are called *strongly stationary*, cf. (Scheel & Scholtes 2000). It is also important to note that under strong stationarity there cannot be a direction h in the contingent cone to the set (84) at $\widehat{z}$, for which

$$\langle \nabla f(\widehat{z}), h \rangle < 0.$$

This is not true in case of the M-stationarity.

5.2 Optimality conditions via ImP

ImP is applicable under the following two principal assumptions:

(i) $\kappa = \omega \times \vartheta$;

(ii) the map S defined in (59) possesses a Lipschitz single-valued localization σ at $(\widehat{x}, \widehat{y})$.

Typically, $\vartheta = \mathbb{R}^m$ and condition (ii) is implied by (SRC) at $(\widehat{x}, \widehat{y})$, cf. (Outrata et al. 1998). Another possibility is to impose condition (63) and the local single-valuedness of S around $\widehat{x}$. Under (i),(ii) it is possible to rewrite (78) locally to the form

$$\begin{array}{ll} \text{minimize} & \Theta(x) \\ \text{subject to} & \\ & x \in \omega, \end{array} \tag{88}$$

where $\Theta(x) := f(x, \sigma(x))$. Reformulation (88) is the essential step in ImP. Concerning optimality conditions, the only hurdle consists now in the computation of $\partial\Theta(\widehat{x})$.

Theorem 4.5. Let assumptions (i),(ii) be fulfilled and suppose that the multifunction P (given by (61)) is calm at $(0, \widehat{x}, \widehat{y})$. Then there exists a pair $(\widehat{\xi}, \widehat{\eta}) \in \partial f(\widehat{x}, \widehat{y})$, a KKT vector $\widehat{v} \in \mathbb{R}^m$ and vectors $\widehat{w} \in N_\omega(\widehat{x}), \widehat{z} \in N_\vartheta(\widehat{y})$ such that equations (80) hold true.

Proof. By definition, $D^*\sigma(\widehat{x}) = D^*S(\widehat{x})$. Due to incl. (12) one has under assumption (ii) that

$$\partial\Theta(\widehat{x}) \subset \{\xi + D^*S(\widehat{x})(\eta) | (\xi, \eta) \in \partial f(\widehat{x}, \widehat{y})\}.$$

By virtue of Lemma 3.1 under assumption (i) and the imposed calmness condition

$$D^*S(\widehat{x})(\eta) \subset \left\{ \begin{array}{l} (\nabla_x F(\widehat{x}, \widehat{y}))^T v \text{ with} \\ 0 \in \eta + (\nabla_y F(\widehat{x}, \widehat{y}))^T v + D^*Q(\widehat{x}, -F(\widehat{x}, \widehat{y}))(v) + N_\vartheta(\widehat{y}) \end{array} \right\}.$$

The (local) optimality of $(\widehat{x}, \widehat{y})$ yields

$$0 \in \partial\Theta(\widehat{x}) + N_\omega(\widehat{x}),$$

and it suffices to combine all three relations together. □

Observe that the calmness of P is a different condition than the calmness of $\widetilde{P}$ required (essentially) in Theorem 4.1. If $\vartheta = \mathbb{R}^m$ and assumption (ii) is ensured via (SRC) at $(\widehat{x}, \widehat{y})$, then the respective map P is calm at $(0, \widehat{x}, \widehat{y})$ ((Outrata 2000, Proposition 3.2)). The relation

$$0 \in \eta + (\nabla_y F(\widehat{x}, \widehat{y}))^T v + D^*Q(\widehat{x}, -F(\widehat{x}, \widehat{y}))(v) + N_\vartheta(\widehat{y}) \tag{89}$$

is called the *adjoint generalized equation.* ImP can also be suitably combined with the PC^1 approach, described at the end of Section 3. This combination leads to a number of optimality conditions for problem (78) with manifold equilibria, cf. (Outrata et al. 1998). We illustrate the nature of these conditions by means of the MPCC, discussed already in Theorems 4.2, 4.4.

Theorem 4.6. Consider the problem (78), where f is continuously differentiable, $\kappa = \omega \times \mathbb{R}^m$ and $Q(y) = N_{\mathbb{R}^m_+}(y)$. Suppose that (SRC) is fulfilled at $(\widehat{x}, \widehat{y})$ and for all $j \in \mathbb{K}(\widehat{x}, \widehat{y})$ the vectors $\widehat{p}_j$ are the (unique) solutions of the adjoint equations

$$\begin{bmatrix} (\nabla_y F(\widehat{x}, \widehat{y}))_{L \cup M_2} & \\ 0 & E_{I_+ \cup M_1} \end{bmatrix}^T p_j + \nabla_y f(\widehat{x}, \widehat{y}) = 0, \tag{90}$$

where M_1, M_2 is the partition of $I_0(\widehat{y}, -F(\widehat{x}, \widehat{y}))$, corresponding to j. Then one has

$$0 \in \nabla_x f(\widehat{x}, \widehat{y}) + \operatorname{conv} \left\{ \begin{bmatrix} (\nabla_x F(\widehat{x}, \widehat{y}))_{L \cup M_2} \\ 0 \end{bmatrix}^T \widehat{p}_j \,\middle|\, j \in \mathbb{K}(\widehat{x}, \widehat{y}) \right\} + N_\omega(\widehat{x}). \tag{91}$$

Proof. By virtue of the generalized Jacobian Chain Rule ((Clarke 1983)) and incl. (77) it suffices to show that

$$\begin{bmatrix} (\nabla_x F(\widehat{x}, \widehat{y}))_{L \cup M_2} \\ 0 \end{bmatrix}^T \widehat{p}_j = (\nabla \sigma_j(\widehat{x}))^T \nabla_y f(\widehat{x}, \widehat{y}) \text{ for all } j \in \mathbb{K}(\widehat{x}, \widehat{y}),$$

where σ_j is the implicit function specified by the equation system

$$\begin{aligned} F^i(x, y) &= 0 \quad \text{for} \quad i \in L \cup M_2 \\ y^i &= 0 \quad \text{for} \quad i \in I_+ \cup M_1. \end{aligned}$$

Due to the classical Implicit Function Theorem one has

$$\nabla \sigma_j(\widehat{x}) = - \begin{bmatrix} (\nabla_y F(\widehat{x}, \widehat{y}))_{L \cup M_2} & \\ 0 & E_{I_+ \cup M_1} \end{bmatrix}^{-1} \begin{bmatrix} (\nabla_x F(\widehat{x}, \widehat{y}))_{L \cup M_2} \\ 0 \end{bmatrix},$$

where the existence of the inverse is ensured by (SRC) at $(\widehat{x}, \widehat{y})$. Thus,

$$\begin{aligned} &(\nabla \sigma_j(\widehat{x}))^T \nabla_y f(\widehat{x}, \widehat{y}) = \\ &= \begin{bmatrix} (\nabla_x F(\widehat{x}, \hat{y}))_{L \cup M_2} \\ 0 \end{bmatrix}^T \left(\begin{bmatrix} (\nabla_y F(\widehat{x}, \widehat{y}))_{L \cup M_2} & \\ 0 & E_{I_+ \cup M_1} \end{bmatrix}^{-1} \right)^T \nabla_y f(\widehat{x}, \widehat{y}) \\ &= \begin{bmatrix} (\nabla_x F(\widehat{x}, \widehat{y}))_{L \cup M_2} \\ 0 \end{bmatrix} \widehat{p}_j, \end{aligned}$$

which was to be proved. □

The terminology "adjoint equations", used for eqs.(90), comes from optimal control. Indeed, due to assumption (ii), problem (78) can be viewed as a standard optimal control problem. By the technique, illustrated in Theorem 4.6, it is possible to derive optimality conditions for MPECs with substantially more difficult equilibria than NCP, cf. (Beremlijski, Haslinger, Kočvara & Outrata 2002). Due to the relation between coderivatives and generalized Jacobians, however, these conditions are less sharp than the conditions involving coderivatives. We speak about the *C(larke)-stationarity*, which is weaker than the M-stationarity. On the other hand, conditions of the type (90),(91) are intimately connected with a nonsmooth approach to the numerical solution of problems (78), described in the next section.

6 Numerical solution of MPECs via ImP

There is a number of approaches for the numerical solution of MPECs with various degrees of generality. In these lecture notes we will pay our attention only to two rather general methods which seem to be perspective in problems coming from mechanics. The first one relies on the connection of ImP with bundle methods of nonsmooth optimization. In fact, to the numerical solution of (88) various optimization techniques can be used, cf. eg (Luo et al. 1996). The connection with bundle methods, however, proved to be especially efficient in a large variety of shape optimization problems, cf. (Outrata et al. 1998),(Beremlijski et al. 2002). In Section 6.1 we will discuss the essential question of the computation of subgradients required by the bundle methods. Section 6.2 is then devoted to a brief description of the classical bundle idea and Section 6.3 illustrates the proposed technique by a difficult MPEC: A shape optimization of contact problems with Coulomb friction.

6.1 Computation of subgradients

Consider problem (88) and assume that ω (the set of feasible controls) is of a simple nature (eg a convex polyhedron or a box). Then a standard bundle method of nonsmooth optimization can be applied to its numerical solution, provided we are able to compute to each $x \in \omega$ the corresponding value $\Theta(x)$ and one arbitrary vector from $\bar{\partial}\Theta(x)$. To facilitate the explanation, we will assume that either our problem does not have any state constraints ($\vartheta = \mathbb{R}^m$), or it is possible to augment them to the objective by means of a locally Lipschitz penalty. Moreover, we will suppose that the solution map

$$S(x) = \{y \in \mathbb{R}^m | 0 \in F(x,y) + Q(y)\}$$

is single-valued on ω. We start with the situation, where the Lipschitz continuity of S over ω is ensured by (SRC), valid at each point $\bar{x} \in \omega$.

Theorem 5.1. Let $\bar{y} = S(\bar{x}), (\bar{\xi}, \bar{\eta}) \in \widehat{\partial} f(\bar{x}, \bar{y})$ and $\bar{p}$ is a solution of the "regular" adjoint GE

$$0 \in \bar{\eta} + (\nabla_y F(\bar{x}, \bar{y}))^T p + \widehat{D}^* Q(\bar{y}, -F(\bar{x}, \bar{y}))(p). \tag{92}$$

Then one has

$$\bar{\xi} + (\nabla_x F(\bar{x}, \bar{y}))^T \bar{p} \in \partial\Theta(\bar{x}) \subset \bar{\partial}\Theta(\bar{x}). \tag{93}$$

If f is regular at $(\bar{x}, \bar{y})$ and Gph$Q$ is normally regular at $(\bar{y}, -F(\bar{x}, \bar{y}))$, then it holds even

$$\begin{aligned} \partial\Theta(\bar{x}) \quad = \quad & \{\xi + (\nabla_x F(\bar{x}, \bar{y}))^T p | 0 \in \eta + (\nabla_y F(\bar{x}, \bar{y}))^T p + D^* Q(\bar{y}, -F(\bar{x}, \bar{y}))(p), \\ & (\xi, \eta) \in \partial f(\bar{x}, \bar{y})\}. \end{aligned}$$

Proof. As in the proof of Theorem 1.6 we can invoke formulas (12),(13) which yield

$$\begin{aligned} \{\xi + \widehat{D}^* S(\bar{x})(\eta) | (\xi, \eta) \in \bar{\partial} f(\bar{x}, \bar{y})\} \quad & \subset \quad \widehat{\partial}\Theta(\bar{x}) \subset \partial\Theta(\bar{x}) \subset \\ & \subset \quad \{\xi + D^* S(\bar{x})(\eta) | (\xi, \eta) \in \partial f(\bar{x}, \bar{y})\}. \end{aligned}$$

According to (Rockafellar & Wets 1998, Thm.10.6) it holds

$$\begin{aligned}\widehat{D}^*S(\bar{x})(\eta) \supset \{(\nabla_x F(\bar{x},\bar{y}))^T p \; | 0 \in \eta + (\nabla_y F(\bar{x},\bar{y}))^T p + \widehat{D}^* Q(\bar{y}, -F(\bar{x},\bar{y}))(p),\\ (\xi,\eta) \in \widehat{\partial} f(\bar{x},\bar{y}) \text{ with some } \xi \in \mathbb{R}^n\}\end{aligned} \tag{94}$$

and, under a suitable qualification condition, one has

$$\begin{aligned}D^*S(\bar{x})(\eta) \subset \{(\nabla_x F(\bar{x},\bar{y}))^T p \; | 0 \in \eta + (\nabla_y F(\bar{x},\bar{y}))^T p + D^* Q(\bar{y}, -F(\bar{x},\bar{y}))(p),\\ (\xi,\eta) \in \partial f(\bar{x},\bar{y}) \text{ with some } \xi \in \mathbb{R}^n\},\end{aligned} \tag{95}$$

cf. also Lemma 3.1. The first statement follows now from (94), whereas the second one results from the equality of the right-hand sides at (94),(95) under the assumed strong regularity and additional conditions. □

We conclude that one either has to solve the "regular" adjoint GE (92) or require that f is regular at $(\bar{x},\bar{y})$ and Gph$Q$ is normally regular at $(\bar{y}, -F(\bar{x},\bar{y}))$. In the latter case (92) amounts to the usual adjoint GE (89) (at $(\bar{x},\bar{y})$). Since (89) is mostly easier to solve than (92), it is important to realize that incl. (95) becomes equality whenever $\nabla_x F(\bar{x},\bar{y})$ is surjective. Indeed, one can then invoke a result from (Rockafellar & Wets 1998), according to which

$$N_{\text{Gph}S}(\bar{x},\bar{y}) = \begin{bmatrix} 0 & -(\nabla_x F(\bar{x},\bar{y}))^T \\ E & -(\nabla_y F(\bar{x},\bar{y}))^T \end{bmatrix} N_{\text{Gph}Q}(\bar{y}, -F(\bar{x},\bar{y})),$$

whenever the matrix

$$\begin{bmatrix} 0 & E \\ -\nabla_x F(\bar{x},\bar{y}) & -\nabla_y F(\bar{x},\bar{y}) \end{bmatrix}$$

has the full row rank. This helps in the case when f is continuously differentiable. On the basis of (16) one has then

$$\nabla_x f(\bar{x},\bar{y}) + (\nabla_x F(\bar{x},\bar{y}))^T \bar{p} \in \bar{\partial}\Theta(\bar{x})$$

with $\bar{p}$ being a solution of (89) (at $(\bar{x},\bar{y})$).

The above idea can sometimes be applied also in the case, when the considered equilibrium is governed by the GE

$$0 \in F(x,y) + Q(y), \tag{96}$$

with $F(x,y) = \begin{bmatrix} F_1(x,y) \\ F_2(x,y) \end{bmatrix}$, $y = (y_1, y_2) \in \mathbb{R}^{m_1} \times \mathbb{R}^{m_2}$, $Q(y) = Q_1(y_1) \times Q_2(y_2)$; thereby F_i maps $\mathbb{R}^n \times \mathbb{R}^m$ into $\mathbb{R}^{m_i}$ and Q_i maps $\mathbb{R}^{m_i}$ into (subsets of) $\mathbb{R}^{m_i}, i = 1,2$. Then, namely, Gph$Q_1$ is often rather simple and the relation

$$0 \in F_1(x,y) + Q_1(y_1)$$

can be used to the elimination of y_1 from the other one, i.e.

$$0 \in F_2(x,y) + Q_2(y_2).$$

The surjectivity argument needs then to be applied only to a "reduced" GE. We illustrate now the above theory again by means of an NCP, i.e., we put $Q(y) = N_{\mathbb{R}^m_+}(y)$.

Theorem 5.2. Consider a reference point $(\bar{x}, \bar{y}) \in \mathrm{Gph} S$ and suppose that $(\bar{\xi}, \bar{\eta}) \in \widehat{\partial} f(\bar{x}, \bar{y})$.

(i) Let the adjoint variable $\bar{p} \in \mathbb{R}^m$ be a feasible point of the system

$$\left.\begin{array}{l} p_{I_+} = 0, p_{I_0} \geq 0 \\ 0 = \bar{\eta}_L + (((\nabla_y F(\bar{x}, \bar{y}))_{L \cup I_0})^T p_{L \cup I_0} \\ \bar{\eta}_{I_0} + (((\nabla_y F(\bar{x}, \bar{y}))_{L \cup I_0})^T p_{L \cup I_0} \leq 0. \end{array}\right\} \tag{97}$$

Then one has

$$\xi + (\nabla_x F(\bar{x}, \bar{y}))^T \bar{p} \in \bar{\partial} \Theta(\bar{x}). \tag{98}$$

(ii) Let f be continuously differentiable, the matrix

$$\begin{bmatrix} 0 & E_{I_+ \cup I_0} \\ -\nabla_x F(\bar{x}, \bar{y})_{L \cup I_0} & -\nabla_y F(\bar{x}, \bar{y})_{L \cup I_0} \end{bmatrix}$$

have the maximum row rank and M_1, M_2 be an arbitrary partition of I_0. Then the relation (98) is fulfilled whenever $\bar{p}$ is a solution of the linear system

$$\left.\begin{array}{l} 0 = p_{I_+ \cup M_1} \\ 0 = \bar{\eta}_{L \cup M_2} + ((\nabla_y F(\bar{x}, \bar{y}))_{L \cup M_2, L \cup M_2})^T p_{L \cup M_2} \end{array}\right\}. \tag{99}$$

Proof. Statement (i) follows from the first assertion of Theorem 5.1, because (97) amounts exactly to (92) by virtue of (87). To prove statement (ii), note that the adjoint equation (89) attains here the form

$$\left.\begin{array}{l} 0 = p_{I_+}, 0 = u_L \\ 0 = \bar{\eta}_{L \cup I_0} + ((\nabla_y F(\bar{x}, \bar{y}))_{L \cup I_0, L \cup I_0})^T p_{L \cup I_0} + u_{L \cup I_0} \\ \text{for } i \in I_0 \text{ either } p^i u^i = 0 \text{ or } p^i > 0 \text{ and } u^i < 0. \end{array}\right\} \tag{100}$$

We can thus partition I_0 into M_1, M_2 and put $p^i = 0$ for $i \in M_1, u^i = 0$ for $i \in M_2$. Therefore u^i is free for $i \in M_1$ and we obtain exactly the equation system (99). □

As in Section 3, the index sets L, I_+, I_0 are related to the argument $(\bar{y}, -F(\bar{x}, \bar{y}))$. One immediately sees that (99) is substantially simpler than the system in statement (i). For each choice of M_1, M_2, the solvability of (99) (for each $\bar{\eta}_{L \cup I_0}$) is ensured by (SRC) at $(\bar{x}, \bar{y})$. The same procedure as in (ii) can be derived by means of the PC^1 approach.

It remains to comment on the situation, when we are not able to fulfill the assumptions needed for the computation of an element from $\bar{\partial}\Theta(\bar{x})$ via the standard adjoint GE (in (95)). In such a case it is possible that we supply the used bundle method a false subgradient information. Theoretically, the convergence of a general bundle method in this case has been investigated in (Mikhalevich, Gupal & Norkin 1987) and especially in (Dempe 2002). It turned out that the method provides then a sequence converging to a point which stationarity is weaker than the C-stationarity. This is, in a sense, a positive result, because it justifies the combination of ImP with a bundle method without too restrictive assumptions.

To ensure the convergence of bundle method, one also needs the co-called weak semismoothness of the minimized function which excludes some pathological functions of the type $x^2 \sin \frac{1}{x}$. We say that a function $f[\mathbb{R}^p \to \mathbb{R}]$ is *weakly semismooth* at $\bar{x} \in \mathbb{R}^p$, provided it is Lipschitz near $\bar{x}$ and the limit

$$\lim_{\substack{\xi_i \in \bar{\partial} f(\bar{x}+\theta_i h) \\ \theta_i \downarrow 0}} \langle \xi_i, h \rangle \tag{101}$$

exists for all $h \in \mathbb{R}^p$. This assumption is automatically fulfilled, provided S is a PC^1 -map and the objective is smooth. Otherwise, this assumption should be tested but its violation is very unlikely.

6.2 Bundle idea in nonsmooth optimization

Bundle methods have been originally designed for unconstrained minimization of convex continuous functions. Substantial further developments have lead, however, to variants which are applicable also to the minimization of (nonconvex) locally Lipschitz functions subject to various constraints (including, eg, nonsmooth inequalities). In the following we explain briefly the essential idea in the convex framework and thereafter point out some substantial changes indispensable for the minimization of nonconvex functions. Finally, we mention some developments which essentially contributed to the considerable performance of modern implementations.

Assume first that $f[\mathbb{R}^n \to \mathbb{R}]$ is convex continuous and consider a sequence of points $x_1, x_2, \ldots, x_k$ and the associated bundle of subgradients $g_1, g_2, \ldots, g_k$ with $g_i \in \partial g(x_i), i = 1, 2, \ldots, k$. Put

$$\alpha_{k,i} := f(x_k) - f(x_i) - \langle g_i, x_k - x_i \rangle \tag{102}$$

so that $\alpha_{ki} \geq 0$ due to the convexity of f. This enables to introduce a new function $\widehat{f}[\mathbb{R}^n \to \mathbb{R}]$ defined by

$$\begin{aligned} \widehat{f}(x) &:= \max_{i=1\ldots k} \{f(x_i) + \langle g_i, x - x_i \rangle\} \\ &= f(x_k) + \max_{i=1\ldots k} \{-f(x_k) + f(x_i) + \langle g_i, x - x_k \rangle + \langle g_i, x_k - x_i \rangle\} \\ &= f(x_k) + \max_{i=1\ldots k} \{-\alpha_{k,i} + \langle g_i, x - x_k \rangle\}. \end{aligned} \tag{103}$$

$\widehat{f}$ is piecewise affine and serves as a lower approximation (model) of f on a neighborhood of x_k. Note that if $\alpha_{k,i}$ is large, then the respective affine component $-\alpha_{k,i} + \langle g_i, x - x_k \rangle$ can hardly be active at x close to x_k. We set now

$$d := x - x_k$$

and employ the model function $\widehat{f}$ in the *direction finding subproblem*, where one attempts to compute the "best" descent direction $\widehat{d}$. Since $\widehat{f}$ can be quite different from f at points, far from x_k, it is reasonable not to permit too large directions d. Therefore we compute $\widehat{d}$ as the second component of the (unique) solution $(\widehat{v}, \widehat{d})$ of the quadratic program

$$\begin{array}{ll} \text{minimize} & v + \frac{1}{2t}\|d\|^2 \\ \text{subject to} & \\ & \langle g_i, d\rangle - \alpha_{k,i} \leq v, \;\; i = 1, 2. \ldots, k, \end{array} \tag{104}$$

where the parameter $t > 0$ specifies the neighborhood of x_k on which we consider $\widehat{f}$ to be an acceptable approximation of f. The first component $\widehat{v}$ estimates the difference $f(x_k + \widehat{d}) - f(x_k)$. It is an easy task to verify that the optimization problem (in variable $\lambda \in \mathbb{R}^k$)

$$\begin{array}{ll} \text{minimize} & \frac{1}{2}\|\sum_{i=1}^{k} \lambda^i g_i\|^2 + \frac{1}{t}\sum_{i=1}^{k} \lambda^i \alpha_{k,i} \\ \text{subject to} & \\ & \sum_{i=1}^{k} \lambda^i = 1, \;\; \lambda^i \geq 0 \end{array} \tag{105}$$

is the standard Lagrangian dual to (104). To its numerical solution various efficient techniques can be used and one has

$$\begin{aligned} \widehat{d} &= -t \sum_{i=1}^{k} \widehat{\lambda}^i g_i \\ \widehat{v} &= -\tfrac{1}{t}\|\widehat{d}\|^2 - \sum_{i=1}^{k} \widehat{\lambda}^i \alpha_{k,i}, \end{aligned} \tag{106}$$

where $\widehat{\lambda}$ is a solution of (105). When k (the dimension of the bundle) is not too high, it is reasonable to solve (104) via its dual (105) and compute its solution by (106). As the next step in classical bundle methods one performs a special line search ((Lemaréchal 1981)) along the direction $\widehat{d}$ which answers a new point $y_{k+1} = x_k + \theta\widehat{d}, \theta \in [0, 1]$, and a new subgradient $g^+ \in \partial f(y_{k+1})$ such that

$$\langle g^+, \widehat{d}\rangle \geq m_1 \widehat{v} \tag{107}$$

and either

$$f(y_{k+1}) \leq f(x_k) + m_2 \theta \widehat{v} \tag{108}$$

or

$$\alpha_{k,k+1} \leq m_3 \sum_{i=1}^{k} \widehat{\lambda}^i \alpha_{k,i}. \tag{109}$$

In (107)-(109) the given parameters m_1, m_2, m_3 are positive and fulfill the inequalities $m_2 < m_1, m_1 + m_3 < 1$. Inequality (108) ensures a "sufficient" decrease of the objective value when moving from x_k to y_{k+1}; so, if it holds, we set $x_{k+1} := y_{k+1}$ and include g^+ to the bundle. This is called a *serious step*. Otherwise we perform a *null step* which means that we do not leave x_k but improve our model $\widehat{f}$ by including a new affine component generated by y_{k+1} and g^+. The above conditions ensure that in both cases the new direction finding subproblem will produce a descent direction, different from $\widehat{d}$. Indeed, in case of the serious step

$$\langle g^+, \widehat{d}\rangle - \alpha_{k+1,k+1} = \langle g^+, \widehat{d}\rangle \geq m_1\widehat{v} > \widehat{v}$$

and in case of the null step

$$\langle g^+, \widehat{d}\rangle - \alpha_{k,k+1} \geq m_1\widehat{v} - m_3\sum_{i=1}^{k}\widehat{\lambda}^i\alpha_{k,i} = m_1(-\frac{1}{t}\|\widehat{d}\|^2)$$
$$+(m_1+m_3)(-\sum_{i=1}^{k}\widehat{\lambda}^i\alpha_{k,i}) > \widehat{v}.$$

Consequently, the pair $(\widehat{v}, \widehat{d})$ is infeasible with respect to the respective direction finding subproblems. The line search from (Lemaréchal 1981) is consistent in the sense that it produces a serious step or a null step after a finite number of computations of function values and subgradients. The whole method is now organized in such a way that the stabilizing parameter t is changed in the course of iterations. The main convergence statement claims that after a finite number of null steps we either end up with a serious step or verify an approximate optimality of the current iterate (x_k) in the sense that

$$\left\|\sum_{i=1}^{k}\widehat{\lambda}^i g_i\right\| \leq \mathcal{E} \qquad \text{and} \qquad \sum_{i=1}^{k}\widehat{\lambda}^i\alpha_{k,i} \leq \mathcal{E}, \tag{110}$$

where $\mathcal{E}$ is an a priori specified accuracy. Note that inequalities (110) imply that

$$f(x_k) \leq f(x) + \mathcal{E}\|x - x_k\| + \mathcal{E} \quad \text{for all} \quad x \in \mathbb{R}^n. \tag{111}$$

Indeed, on the basis of (103),

$$f(x) \geq f(x_k) + \langle g_i, x - x_k\rangle - \alpha_{k,i} \quad \text{for all} \quad i.$$

For an arbitrary set of coefficients $\lambda^i \geq 0$, $\sum_{i=1}^{k}\lambda_i = 1$ one has

$$\sum_{i=1}^{k}\langle\lambda_i g_i, x - x_k\rangle \leq f(x) - f(x_k) + \sum_{i=1}^{k}\lambda_i\alpha_{k,i},$$

which immediately implies (111).

For an appropriate treatment of t (or other quantities with the same role) various strategies have been proposed, cf. (Lemaréchal, Strodiot & Bihain 1981),(Lemaréchal 1989).

To prevent a permanent increase of the number of subgradients in the bundle (and thus the dimension of problem (105)) one resets the bundle, mostly after each serious or null step. A successful resetting strategy consists in selecting a fixed number of subgradients with the smallest values of $\alpha_{k,i}$. From the rest one constructs the so-called aggregated subgradient by using the solution of (105). In this way it is possible to work with a bundle of a reasonable size without destroying the convergence of the method.

If we do not require f to be convex, the reals $\alpha_{k,i}$ need not be nonnegative and $\widehat{f}$ need not be a lower approximation of f around x_k. This hurdle can be overcome by the following trick: We replace $\alpha_{k,i}$ with different quantities $\beta_{k,i}$, defined by

$$\beta_{k,i} := \max\{\alpha_{k,i}, c_0\|x_k - x_i\|\},$$

where c_0 is a suitable positive number. Clearly, the reals $\beta_{k,i}$ are nonnegative by definition and one can introduce a modified model function

$$\widetilde{f}(x) := f(x_k) + \max\{-\beta_{k,i} + \langle g_i, x - x_k\rangle\}. \tag{112}$$

Note that whenever $\|x_k - x_i\|$ is too large, the corresponding affine component $-\beta_{k,i} + \langle g_i, x - x_k\rangle$ is very unlikely to be active at x close to x_k.

To ensure a proper behaviour of the line search in this case, one needs to assume that f is weakly semismooth, cf. (101). With this additional assumption all statements can be proved in a similar manner like in the convex case. We have to confess, however, that the model (112) is much less satisfactory than its convex counterpart. Also the inequalities (110) do not imply the approximate optimality (111). All we get from (110) in the nonconvex case is that 0 "lies up to $\mathcal{E}$" in the convex hull of certain subgradients computed at the points not too far away from x_k. This corresponds to "almost" stationarity in smooth optimization.

The numerical method described above proved to be rather efficient in numerous nonsmooth optimization problems. At the end of the eighties the bundle idea has been successfully coped with the trust region philosophy. This lead to an improved treatment of t during the iteration process and enabled to eliminate completely the line search in the convex case. The resulting Bundle-Trust (BT) algorithm ((Schramm & Zowe 1992)) is able to minimize convex piecewise affine functions in a finite number of iterations. In the nonconvex case, however, the line search could not be completely eliminated: It has to be applied if the modified conditions (108), (109) do not permit to perform the serious step or the null step and also the trust region (specified by t) cannot be decreased. Nevertheless, this branch of the algorithm is only an emergency exit. The BT algorithm has been successfully applied to the numerical solution of all MPECs in (Outrata et al. 1998). As further important developments let us mention a class of methods relying on the Yosida approximation of the minimized function ((Lemaréchal & Sagastizábal 1997)) and the method of (Lukšan & Vlček 1998) which replaces the model (103) by a piecewise quadratic model. This method proved its efficiency eg in a class of equilibria, governed by HVIs, where the piecewise affine model (103) is quite inadequate.

Concerning possible constraints we have to distinguish between affine and nonaffine equations and inequalities. The affine case can be treated by a suitable modification of the direction finding subproblems. The nonaffine one, however, requires a more sophisticated approach. In this respect, eg the *filter idea*, proposed in (Fletcher & Leyffer 2002) seems to be rather efficient.

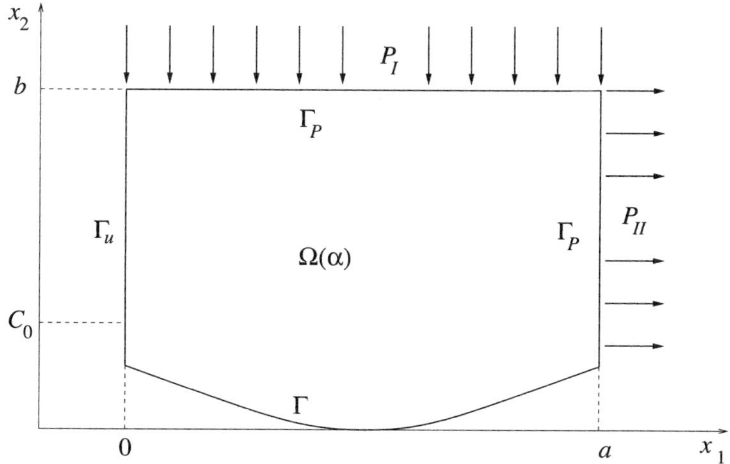

Figure 1: The elastic body and applied loads.

6.3 Shape optimization of contact problems with Coulomb friction

To demonstrate the performance of ImP combined with a bundle method, we have chosen a rather difficult MPEC with an equilibrium mentioned already in Section 2.1. This MPEC is thoroughly analyzed in (Beremlijski et al. 2002); here we given only a brief description of the numerical realization of the method and provide the results of some test examples. The equilibrium is governed by the GE

$$\left.\begin{array}{rcl} 0 & \in & A_{\tau\tau}(x)u_\tau + A_{\tau\nu}(x)u_\nu - l_\tau(x) + \widetilde{Q}(u_\tau, \lambda_\nu) \\ 0 & = & A_{\nu\tau}(x)u_\tau + A_{\nu\nu}(x)u_\nu - l_\nu(x) \\ 0 & \in & u_\nu + x + N_{\mathbb{R}^p_+}(\lambda_\nu) \end{array}\right\} \tag{113}$$

with the control variable $x \in \mathbb{R}^n$ specifying the shape of the contact boundary Γ and the state variable

$$y := (u_\tau, u_\nu, \lambda_\nu) \in \mathbb{R}^p \times \mathbb{R}^p \times \mathbb{R}^p$$

consisting from the normal and the tangential component of the displacements on the contact boundary Γ and the normal component of the contact stress vector on Γ, see Fig. 1. This figure also shows the distribution of external loads P_I, P_{II} on the boundary Γ_P. Further, Γ_u is the part of the boundary with prescribed Dirichlet condition. $A_{\tau,\tau}, A_{\tau,\nu}, A_{\nu\tau}$ and $A_{\nu\nu}$ are blocks of the appropriate restriction of the *stiffness matrix* to Γ. These matrices depend on x in a continuously differentiable way. This holds true also for the vectors l_τ, l_ν reflecting the action of external forces on Γ. The multivalued part in the first line of (113) is given by (cf. 47)

$$\widetilde{Q}(u_\tau, \lambda_\nu) = \lambda_\nu \bullet \partial j(u_\tau), \;\; j(u_\tau) = \mathcal{F} \sum_{i=1}^{p} |u_\tau^i|,$$

where $\mathcal{F}$ is the *friction coefficient.*

The shape optimization problem is defined as follows:

$$\begin{array}{ll} \text{minimize} & f(x,y) \\ \text{subject to} & \\ & y \text{ solves the GE (113)} \\ & x \in \omega \end{array}$$

with

$$\omega = \{x \in \mathbb{R}^n \quad | \; 0 \le x^i \le C_0, i = 0,1,\ldots,n; \\ |x^{i+1} - x^i| \le \tfrac{C_1}{n}, i = 0,1,\ldots,n-1; \sum_{i=0}^{n} x^i = C_2(n+1)\},$$

where C_0, C_1, C_2 are given positive constants. The equality constraint in the definition of ω has a physical meaning of preserving the body volume.

As it has been proved in (Beremlijski et al. 2002), the respective solution map S is single-valued and locally Lipschitz on ω. Consequently, we are entitled to apply ImP. The resulting program (88) was solved by the BT algorithm from (Schramm & Zowe 1992). In every BT iteration we have to solve the equilibrium problem which we formulate to this purpose as a fixed-point problem. For that, we use the splitting variant of the fixed-point method introduced in (Haslinger, Dostál & Kučera 2002). This is basically the method of successive approximations where, at each step, we solve the contact problem with a given friction. The iterative process then updates the coefficient of the given friction. The problem with the given friction is solved using the so-called reciprocal variational formulation that leads to a quadratic programming problem with simple box constraints. For its solution we used a special splitting technique, based on a version of the Gauss-Seidel algorithm.

To compute a subgradient information required by BT one can use the upper approximation of D^*S on the right-hand side of (95). Suitable elements from D^*Q are selected on the basis of index sets describing the position of the considered point on the graph of Q. Concerning $\widetilde{Q}$, we have to distinguish for each node in contact with the obstacle (superscript i) among the *slip*

$$u^i_\tau \neq 0,$$

the *strong stick*

$$u^i_\tau = 0, \; |(l_\tau(x) - A_{\tau\tau}(x)u_\tau - A_{\tau\nu}(x)u_\nu)^i| < \mathcal{F}\lambda^i$$

and the *weak stick*

$$u^i_\tau = 0, \; |(l_\tau(x) - A_{\tau\tau}(x)u_\tau - A_{\tau\nu}(x)u_\nu)^i| = \mathcal{F}\lambda^i.$$

The adjoint GE becomes then a linear equation systems whose solvability follows from the properties of the stiffness matrix. To verify that a vector computed in this way is indeed a subgradient of the respective composite objective Θ, an additional testing is needed; however, in the test runs we did not observe any difficulties of BT caused by a false subgradient information.

Next, we will present results of a numerical example. The shape of the unloaded elastic body $\mathcal{O}(x), x \in \omega$, is defined by

$$\mathcal{O}(x) = \{(\xi_1, \xi_2) \in \mathbb{R}^2 \,|\, \xi_1 \in (0, a), \mathcal{B}_x(\xi_1) < \xi_2 < b\},$$

where $\mathcal{B}_x$ is a suitable approximation of the shape of Γ, dependent on x, see Figure 2.

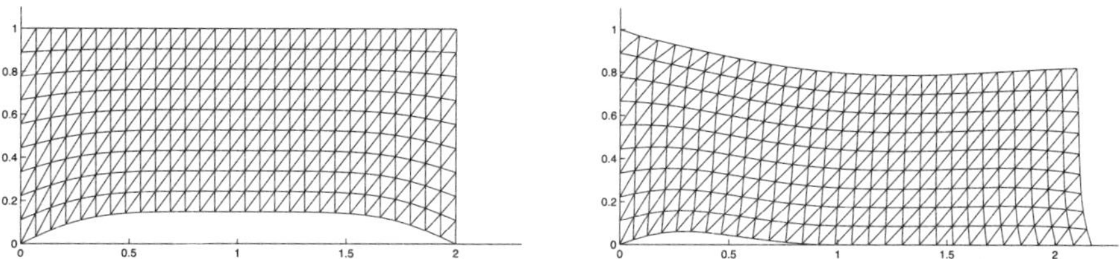

Figure 2: Initial design, unloaded and loaded case.

We try to identify the contact normal stress distribution with a given function $\bar{\lambda}_\nu$. The shape optimization problem can be written as

$$\begin{array}{ll} \text{minimize} & \|\bar{\lambda}_\nu - \lambda_\nu\|^2 \\ \text{subject to} & \\ & x \in \omega. \end{array}$$

We discretized $\mathcal{O}(x)$ by a regular 29×9 mesh, i.e., we had 261 nodes and 522 unknowns in the state problem. The results are shown in Figure 3.

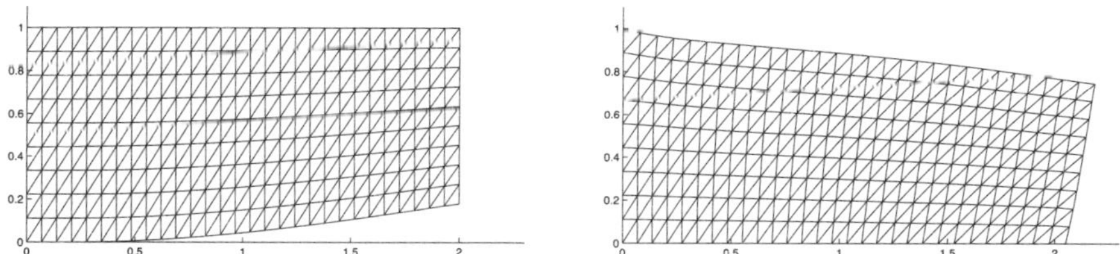

Figure 3: Optimal design, unloaded and loaded case.

7 Nonlinear programming approach and evolutionary equilibrium

Numerous MPECs do not fulfill the principal assumptions needed for the application of ImP, cf. Section 4.2. To their numerical solution, various alternative approaches have been proposed and tested. Among them we find methods based on the penalization of the lower-level constraint, on the penalization of the whole equilibrium constraint and on the PC^1 approach. Recently, however, it turned out that, at least for MPCCs, one can modify some standard NLP methods to enable them to cope with the original problem formulation

$$\begin{array}{ll} \text{minimize} & f(x,y) \\ \text{subject to} & \\ & F(x,y) \geq 0, \ y \geq 0, \ \langle F(x,y), y \rangle = 0 \\ & (x,y) \in \kappa. \end{array} \tag{114}$$

In fact, in this so called NLP approach it is convenient to generalize the complementarity constraints from (114) to the form

$$F(x,y) \geq 0, \ H(x,y) \geq 0, \ \langle F(x,y), H(x,y) \rangle = 0, \tag{115}$$

where $H[\mathbb{R}^n \times \mathbb{R}^m \to \mathbb{R}^m]$ is continuously differentiable. In Section 6.1 we will briefly discuss the main issues associated with this recent development.

In some mechanical applications it is important to take into account the evolution nature of the considered process. One of the reasons is the necessity to respect the irreversible changes occuring during the process. Such a situation arises eg in delamination, where the modeling via a stationary HVI does not always lead to a physically acceptable solution. It seems to be better to replace a stationary HVI by a more sophisticated model (Kočvara, Mielke & Roubíček n.d.) which leads, after a suitable discretization, to a finite sequence of coupled GEs associated with the single time levels. MPECs with such evolutionary equilibria have been considered in (Kocvara & Outrata 2004), where necessary optimality conditions are derived. Concerning the numerical solution of these problems, the NLP approach seems to be useful, but it is important to take the special structure, induced by the time discretization, into account. This work has not been done yet, but one can employ the ideas used in the adaptation of SQP methods to optimal control problems. In Section 6.2 we will present the evolution model from (Kocvara & Outrata 2004) and discuss the optimality conditions.

7.1 NLP approach

Consider an MPCC with the equilibrium constraint given by (115). Put again $z := (x,y)$ and assume that κ is given by (83). Then we face a standard NLP with inequalities and one difficult equality

$$\langle F(z), H(z) \rangle = 0. \tag{116}$$

It is exactly this equality which is responsible for all hurdles associated with this type of MPCC. In particular, the KKT vectors can be unbounded, which destroys our estimates of the rate of convergence. In what follows we will employ SQP methods as a solver. Therefore we strengthen the smoothness requirements and assume that the functions f, F, H and q are twice continuously differentiable. The NLP approach has been initiated by the paper (Scholtes 2001), where the author suggests to replace (116) by the harmless inequality

$$\langle F(z), H(z) \rangle \leq \tau$$

with some $\tau > 0$ and to approach a solution of the considered MPCC by solutions of the respective nonlinear programs with $\tau \downarrow 0$. Under certain conditions this sequence converges to various types of stationary points. In the subsequent works, however, an SQP method has been applied directly to the considered MPCC, where (116) has been handled as the inequality

$$\langle F(z), H(z)\rangle \le 0. \tag{117}$$

Of course, this formal change has no influence on the constraint qualifications. Nevertheless, for modern SQP implementations this reformulation proved to be a useful step: both the failure quota as well as the convergence improved. When analysing this phenomenon, the mathematicians encountered a distinguished, so-called basic KKT vector to which SQP methods converge provided all QP subproblems remain consistent. This is, unfortunately, not always the case and so, various devices have been proposed to maintain their consistency or to return to a consistent QP after an inconsistency has been realized. In (Anitescu 2000) the usage of the so-called *elastic mode* is recommended. This consists in relaxing of (117) and adding of $\langle F(z), H(z)\rangle$ to the objective with a suitable penalty parameter. This parameter is updated during the iteration process together with the relaxation parameter. Another way how to cope with the QP inconsistency is to relax only the linearization of (117). This way has been adopted in the successful code Filter MPEC together with a restoration mode which enforces consistency of the next QP subproblem, cf. (Fletcher, Leyffer, Ralph & Scholtes 2002), (Fletcher & Leyffer 2004).

Among the assumptions, needed to ensure a local superlinear convergence of SQP methods in the above setting we find also a respective variant of MPEC-LICQ which amounts to the requirement that the matrix

$$\left[((\nabla H(\hat{z}))_{I_+\cup I_0})^T - ((\nabla F(\hat{z}))_{L\cup I_0})^T \;\; ((\nabla q(\hat{z}))_J)^T\right]$$

has the maximum column rank. Thereby, $\hat{z}$ is an accumulation point of the sequence generated by the used SQP algorithm, J is the active index set for the inequality $q(z) \le 0$ (at $\hat{z}$) and the index sets L, I_+, I_0 are this time associated with the point $(H(\hat{z}), -F(\hat{z}))$. It follows from Theorem 4.4 that, under MPEC-LICQ, $\hat{z}$ is strongly stationary whenever it is a (local) minimum of the solved MPEC. In (Scholtes 2001) it has been proved that MPEC-LICQ is generically fulfilled so that it cannot be considered as a too restrictive requirement.

Let us list the main advantages and drawbacks of the NLP approach.

+ The NLP approach can directly handle state constraints (unlike the ImP technique).

+ There are several well-developed, robust and sophisticated NLP codes, both academic and commercial. Also, some MPECs allow for formulations using tools like GAMS or AMPL. In this way, one can easily generate new problems and solve them using standard codes.

On the other hand,

- The fact that the NLP approach works on Cartesian product of the variables x and y prevents from using special solvers for the equilibrium problems (finite elements solvers, multigrid, etc.). Both variables, control and state, are treated in the same way and any structure in the problem is ignored (in the current methods).

- The NLP approach is limited to MPECs with equilibria governed by (generalized) NCPs.

It can be expected that the performance of modern NLP methods (not only SQP) with respect to MPECs will increase. This will happen in particular due to modifications respecting the concrete structure of the considered equilibrium. An example of a specially structured equilibrium can be found in the next section.

7.2 Evolutionary equilibria

A thorough study of the delamination process, modeled usually by an elliptic HVI, has lead to a more sophisticated model ((Kočvara et al. n.d.)) which is able to grasp better the irreversibility of changes occurring during the process (breakage of the adhesive). After a complete discretization this model can be embeded into a class of *evolutionary equilibria* defined below.

Let T be a natural number specifying the terminal time, $\bar{y}_0 \in \mathbb{R}^m$ be an initial condition and $x \in \mathbb{R}^n$ be a control vector. Further, we are given a family of continuously differentiable functions $F_i[\mathbb{R}^n \times \mathbb{R}^m \times \mathbb{R}^m \to \mathbb{R}^m]$ and closed-graph multifunctions $Q_i[\mathbb{R}^n \times \mathbb{R}^m \times \mathbb{R}^m \rightsquigarrow \mathbb{R}^m], i = 1, 2, \ldots, T$. With these data we associate a control-dependent *evolutionary equilibrium* via the solution map $S[\mathbb{R}^n \rightsquigarrow \mathbb{R}^m]$ defined by

$$S(x) := \{y_T \in \mathbb{R}^n \mid 0 \in F_i(x, y_{i-1}, y_i) + Q_i(x, y_{i-1}, y_i), i = 1, 2, \ldots, T, y_0 = \bar{y}_0\}. \quad (118)$$

So, the solution y_i of the GE

$$0 \in F_i(x, y_{i-1}, y_i) + Q_i(x, y_{i-1}, y_i) \quad (119)$$

depends both on the control x (which is the same for all time levels) as well as on the solution y_{i-1} of the preceding GE (or on $\bar{y}_0$). In fact, to each x we assign a set of *trajectories* $(y_1, y_2, \ldots, y_T)$ corresponding to the *evolution* of the considered equilibrium, but only the terminal vectors y_T are considered as the image $S(x)$.

The GE (119) could be the KKT system for a convex program, where both the objective as well as the functions in the constraints depend on x and on a KKT pair from the previous GE. If all objectives are strictly convex (with respect to the unknown variable) for all values of the parameters, then S becomes single-valued. In the delamination process we have indeed to do with convex programs, but they admit nonunique solutions and so ImP cannot be applied to an MPEC with such an evolutionary equilibrium. Observe that this equilibrium does not amount to a discrete-time optimal control problem, because we do not perform an optimization over the time interval $[0, T]$; instead of it we optimize at each time level separately.

Consider now the MPEC

$$\begin{array}{ll} \text{minimize} & \varphi(x, y_T) \\ \text{subject to} & \\ & 0 \in F_i(x, y_{i-1}, y_i) + Q_i(x, y_{i-1}, y_i), i = 1, 2, \ldots, T \\ & y_0 = \bar{y}_0 \\ & x \in \omega, \end{array} \quad (120)$$

where the objective $\varphi[\mathbb{R}^n \times \mathbb{R}^m \rightarrow \mathbb{R}]$ is locally Lipschitz and the set of admissible controls ω is nonempty and closed. In contrast to the previous notation y means now the trajectory $(y_1, y_2, \ldots, y_T)$. Further, we put $\bar{m} = Tm$ and denote by $\nabla_j F_i, j = 1, 2, 3$, the partial Jacobian of F_i with respect to the jth variable. Finally, $\widetilde{Q}_1(x, y_1) := Q_1(x, \bar{y}_0, y_1)$.

Theorem 6.1 Let $(\widehat{x}, \widehat{y}_T)$ be a (local) solution of the MPEC (120) and $\widehat{y}$ be the corresponding trajectory. Put $\widehat{y}_0 = \bar{y}_0$, denote $\widehat{c}_i = -F_i(\widehat{x}, \widehat{y}_{i-1}, \widehat{y}_i), i = 1, 2, \ldots, T$, and define the multifunction $P[\mathbb{R}^{\bar{m}} \times \mathbb{R}^{nT} \times \mathbb{R}^{(m-1)T} \times \mathbb{R}^{\bar{m}} \rightsquigarrow \mathbb{R}^n \times \mathbb{R}^{\bar{m}}]$ by

$$\begin{aligned} &P(\xi_1, \xi_2, \ldots, \xi_T, \alpha_1, \alpha_2, \ldots, \alpha_T, \beta_2, \beta_3, \ldots, \beta_T, \gamma_1, \gamma_2, \ldots, \gamma_T) \\ &:= \{(x, y) \in \omega \times \mathbb{R}^{\bar{m}} \,|\xi_1 \in F_1(x, \widehat{y}_0, y_1) + \widetilde{Q}_1(x - \alpha_1, y_1 - \gamma_1),\ \xi_i \in F_i(x, y_{i-1}, y_i) \\ &+ Q_i(x - \alpha_1, y_{i-1} - \beta_i, y_i - \gamma_i), i = 2, 3, \ldots, T\}. \end{aligned}$$

Assume that P is calm at $(0, \widehat{x}, \widehat{y})$. Then there exist four sequences of adjoint vectors $p_1, p_2, \ldots, p_T, q_2, q_3, \ldots, q_T, v_1, v_2, \ldots, v_T, w_1, w_2, \ldots, w_T$ and subgradients $(\kappa, \eta) \in \partial\varphi(\widehat{x}, \widehat{y}_T)$ such that

$$\begin{aligned} &(p_1, v_1) \in D^*\widetilde{Q}_1(\widehat{x}, \widehat{y}_1, \widehat{c}_1)(w_1) \\ &(p_i, q_i, v_i) \in D^*Q_i(\widehat{x}, \widehat{y}_{i-1}, \widehat{y}_i, \widehat{c}_i)(w_i), i = 2, 3, \ldots, T \end{aligned} \tag{121}$$

and the adjoint equation system

$$\left.\begin{aligned} &0 = \eta + (\nabla_3 F_T(\widehat{x}, \widehat{y}_{T-1}, \widehat{y}_T))^T w_T + v_T \\ &0 = (\nabla_3 F_{T-1}(\widehat{x}, \widehat{y}_{T-2}, \widehat{y}_{T-1}))^T w_{T-1} + v_{T-1} \\ &\quad + (\nabla_2 F_T(\widehat{x}, \widehat{y}_{T-1}, \widehat{y}_T))^T w_T + q_T \\ &\ldots\ldots \\ &0 = (\nabla_3 F_1(\widehat{x}, \widehat{y}_0, \widehat{y}_1))^T w_1 + v_1 + (\nabla_2 F_2(\widehat{x}, \widehat{y}_1, \widehat{y}_2))^T w_2 + q_2 \end{aligned}\right\} \tag{122}$$

is fulfilled. Moreover, one has

$$0 \in \kappa + \sum_{i=1}^{T} \left[\nabla_1 F_i(\widehat{x}, \widehat{y}_{i-1}, \widehat{y}_i))^T w_i + p_i\right] + N_\omega(\widehat{x}). \tag{123}$$

Proof. The constraints in (120) can be written down in the form

$$\Phi(x, y) \in \Lambda,\ x \in \omega,$$

where

$$\Phi(x, y) = \begin{bmatrix} x \\ y_1 \\ -F_1(x, \bar{y}_0, y_1) \\ x \\ y_1 \\ y_2 \\ -F_2(x, y_1, y_2) \\ \ldots\ldots \\ x \\ y_{T-1} \\ y_T \\ -F_T(x, y_{T-1}, y_T) \end{bmatrix} \quad \text{and} \quad \Lambda = Gph\widetilde{Q}_1 \times \mathop{\mathsf{X}}_{i=2}^{T} GphQ_i.$$

Now one can invoke Theorem 1.4 with the CQ (6) replaced by the assumed calmness of P. This yields the existence of a KKT vector

$$b = (p_1, v_1, -w_1, p_2, q_2, v_2, -w_2, \ldots, p_T, q_T, v_T, -w_T,) \in N_\Lambda(\Phi(\widehat{x}, \widehat{y}))$$

and subgradients $(\kappa, \eta) \in \partial\varphi(\widehat{x}, \widehat{y})$ such that

$$0 \in \begin{bmatrix} \kappa \\ 0 \\ \vdots \\ 0 \\ \eta \end{bmatrix} + (\nabla\Phi(\widehat{x}, \widehat{y}))^T b + \begin{bmatrix} N_\omega(\widehat{x}) \\ 0 \\ \vdots \\ 0 \end{bmatrix}. \tag{124}$$

Since $N_\Lambda(\Phi(\widehat{x}, \widehat{y})) = N_{Gph\widetilde{Q}_1}(\widehat{x}, \widehat{y}_1, \widehat{c}_1) \times X_{i=2}^T N_{GphQ_i}(\widehat{x}, \widehat{y}_{i-1}, \widehat{y}_i, \widehat{c}_i)$, see (Mordukhovich 1988, Prop.1.6), it follows that $(p_1, v_1, -w_1) \in N_{Gph\widetilde{Q}_1}(\widehat{x}, \widehat{y}_1, \widehat{c}_1)$ and $(p_i, q_i, v_i, -w_i) \in N_{GphQ_i}(\widehat{x}, \widehat{y}_{i-1}, \widehat{y}_i, \widehat{c}_i)$, $i = 2, 3, \ldots, T$. In this way relations (121) have been established. The first line of (124) leads now directly to relation (123), whereas the remaining T lines generate the adjoint system (122). □

As mentioned in the text below Theorem 3.2, the calmness assumption is automatically fulfilled provided ω is convex polyhedral, all functions F_i are affine and all maps Q_i are polyhedral. An other possibility is to impose the respective (GMFCQ) which attains the following form:

(GMFCQ): The system

$$\begin{aligned} 0 &= (\nabla_3 F_T(\widehat{x}, \widehat{y}_{T-1}, \widehat{y}_T))^T w_T + v_T \\ 0 &= (\nabla_3 F_{T-1}(\widehat{x}, \widehat{y}_{T-2}, \widehat{y}_{T-1}))^T w_{T-1} + v_{T-1} \\ &\quad + (\nabla_2 F_T(\widehat{x}, \widehat{y}_{T-1}, \widehat{y}_T))^T w_T + q_T \\ &\cdots\cdots \\ 0 &= (\nabla_3 F_1(\widehat{x}, \widehat{y}_0, \widehat{y}_1))^T w_1 + v_1 + (\nabla_2 F_2(\widehat{x}, \widehat{y}_1, \widehat{y}_2))^T w_2 + q_2 \\ 0 &\in \sum_{i=1}^T \left[(\nabla_1 F_i(\widehat{x}, \widehat{y}_{i-1}, \widehat{y}_i))^T w_i + p_i\right] + N_\omega(\widehat{x}) \end{aligned}$$

with $(p_1, v_1) \in D^*\widetilde{Q}_1(\widehat{x}, \widehat{y}_1, \widehat{c}_1)(w_1)$ and $(p_i, q_i, v_i) \in D^*Q_i(\widehat{x}, \widehat{y}_{i-1}, \widehat{y}_i, \widehat{c}_i)(w_i)$, $i = 2, 3, \ldots, T$, possesses only the trivial solution $p_1 = p_2 = \ldots = p_T = 0, q_2 = q_3 = \ldots = q_T = 0, v_1 = v_2 = \ldots = v_T = 0$ and $w_1 = w_2 = \ldots = w_T = 0$.

In this way we ensure even the Aubin property of P around $(0, \widehat{x}, \widehat{y})$. The optimality conditions of Theorem 6.1 attain a simpler form provided the maps Q_i do not depend on all three variables x, y_{i-1}, y_i.

Corollary 6.2. Let all assumption of Theorem 6.1 be fulfilled and assume that the maps $Q_i, i = 1.2. \dots, T$, do not depend on x. Then there exist three sequences of adjoint vectors $q_2, q_3, \dots, q_T, v_1, v_2, \dots, v_T, w_1, w_2, \dots, w_T$ and subgradients $(\kappa, \eta) \in \partial\varphi(\widehat{x}, \widehat{y}_T)$ such that

$$\begin{array}{l} v_1 \in D^*\widetilde{Q}_1(\widehat{y}_1, \widehat{c}_1)(w_1) \\ (q_i, v_i) \in D^*Q_i(\widehat{y}_{i-1}, \widehat{y}_i, \widehat{c}_i)(w_i),\ i = 2, 3, \dots, T, \end{array}$$

the adjoint equation system (122) is satisfied, and

$$0 \in \kappa + \sum_{i=1}^{T} (\nabla_1 F_i(\widehat{x}, \widehat{y}_{i-1}, \widehat{y}_i))^T w_i + N_\omega(\widehat{x}). \tag{125}$$

Corollary 6.3. Let all assumptions of Theorem 6.1 be fulfilled and assume that for $i = 1, 2, \dots, T$, the maps Q_i depend exclusively on variables y_i, respectively. Then there exist two sequences of adjoint vectors $v_1, v_2, \dots, v_T$, $w_1, w_2, \dots, w_T$ and subgradients $(\kappa, \eta) \in \partial\varphi(\widehat{x}, \widehat{y}_T)$ such that

$$v_i \in D^*Q_i(\widehat{y}_i, \widehat{c}_i)(w_i),\ i = 1, 2, \dots, T,$$

and the adjoint equation system

$$\left.\begin{array}{l} 0 = \eta + (\nabla_3 F_T(\widehat{x}, \widehat{y}_{T-1}, \widehat{y}_T))^T w_T + v_T \\ 0 = (\nabla_3 F_{T-1}(\widehat{x}, \widehat{y}_{T-2}, \widehat{y}_{T-1}))^T w_{T-1} + v_{T-1} + (\nabla_2 F_T(\widehat{x}, \widehat{y}_{T-1}, \widehat{y}_T))^T w_T \\ \dots\dots \\ 0 = (\nabla_3 F_1(\widehat{x}, \widehat{y}_0, \widehat{y}_1))^T w_1 + v_1 + (\nabla_2 F_2(\widehat{x}, \widehat{y}_1, \widehat{y}_2))^T w_2 \end{array}\right\} \tag{126}$$

is fulfilled. Moreover, relation (125) holds true.

The next example illustrates the nature of evolutionary equilibria by means of a simple academic problem, close to a delamination process.

Example 6.1. Consider a four-string mechanical system depicted on Figure 4. In reality, the strings are at the same horizontal position, here they are plotted with some gap for presentation reasons. The elasticity modulae of the strings are $e_1, \dots, e_4$. Assume that $e_3 = e_4$ and denote it by e. The end-nodes of the strings are denoted n_0, n_1, n_2. The vertical displacements at the nodes are u^0, u^1, u^2. Node n_0 is fixed ($u^0 = 0$), node n_2 is subjected to nonzero Dirichlet boundary condition (prescribed displacements) $u^2 = \bar{u}$.

The left-hand strings are elastic and "never" break. The right-hand strings are also elastic but can break when the relative displacement reaches a certain value.

The Dirichlet condition $\bar{u}$ depends on time. The equilibrium state at the ith time

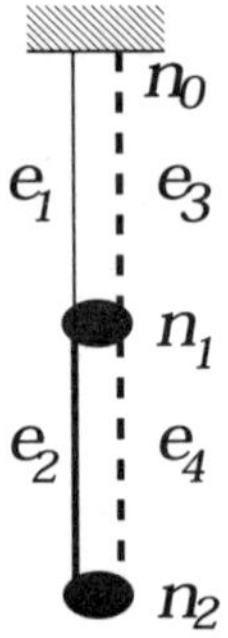

Figure 4: Four-string system.

level is obtained by solving the optimization problem (in variables u_i, ζ_i)

$$\begin{array}{ll} \text{minimize} & g(e_1, e_2, \zeta_{i-1}, u_i, \zeta_i) := \\ & \sum_{j=1}^{2} \left[e_j(u_i^j - u_i^{j-1})^2 + \zeta_i^j e(u_i^j - u_i^{j-1})^2 + (\zeta_{i-1}^j - \zeta_i^j)^2 ed \right] \\ \text{subject to} & \\ & u_i^2 \geq \bar{u}_i \\ & u_i^j - u_i^{j-1} \geq 0, \ j = 1, 2 \\ & \zeta_{i-1}^i \geq \zeta_i^j \geq 0, \ j = 1, 2, \end{array} \tag{127}$$

where $\zeta_i \in [0, 1] \times [0, 1]$ is the *damage parameter* indicating the state of the breakable strings: 1 means that the respective string is intact, whereas 0 indicates that the string is broken. The values from $(0, 1)$ do not have a physical meaning in this example, but can be interpreted as a partial damage of the adhesive in delamination. The first term in the objective is the strain energy of the elastic strings 1 and 2. The second term is the strain energy of the breakable strings 3 and 4 and the third term is the dissipation energy. Number d is the energy dissipated by breaking one string (this is given in our case). The first from the last two inequalities among the constraints reflects the fact that a breakage is irreversible.

We put now $x := (e_1, e_2)$ and introduce the notation

$$G(\zeta_{i-1}, u_i, \zeta_i) := \begin{bmatrix} u_i^1 - u_i^2 \\ \zeta_i^1 - \zeta_{i-1}^1 \\ \zeta_i^2 - \zeta_{i-1}^2 \end{bmatrix},$$

$$\Omega_i = \left\{ a \in \mathbb{R}^4 \mid a^1 \geq 0, a^2 \geq \bar{u}_i, \ a^3 \geq 0, \ a^4 \geq 0 \right\}.$$

This enables to write down the KKT system for (127) in the following form, where the partial derivatives are denoted by integers, indicating the positions of the respective variables.

Theorem 6.4. Let $x = \widetilde{x}$ and $\zeta_{i-1} = \widetilde{\zeta}_{i-1}$ be given and $(\widetilde{u}_i, \widetilde{\zeta}_i)$ be a solution of the respective problem (127). Then there exists a KKT vector $\widetilde{\lambda}_i \in \mathbb{R}^3_+$ such that

$$\begin{array}{l} 0 \in \nabla_{3,4} g(\widetilde{x}, \widetilde{\zeta}_{i-1}, \widetilde{u}_i, \widetilde{\zeta}_i) + (\nabla_{2,3} G(\widetilde{\zeta}_{i-1}, \widetilde{u}_i, \widetilde{\zeta}_i))^T \widetilde{\lambda}_i + N_{\Omega_i}(\widetilde{u}_i, \widetilde{\zeta}_i) \\ 0 \in -G(\widetilde{\zeta}_{i-1}, \widetilde{u}_i, \zeta_i) + N_{\mathbb{R}^3_+}(\widetilde{\lambda}_i). \end{array} \qquad (128)$$

Note that in the above optimality conditions we do not need any constraint qualification, because all functions arising in the constraints are affine. The GE (128) is already in the required form (119); it suffices to put

$$\begin{array}{rcl} y_i : & = & (u_i, \zeta_i, \lambda_i), \\ F(x, y_{i-1}, y_i) : & = & \left[\begin{array}{l} \nabla_{3,4} g(x, \zeta_{i-1}, u_i, \zeta_i) + (\nabla_{2,3} G(\zeta_{i-1}, u_i, \zeta_i))^T \lambda_i \\ -G(\zeta_{i-1}, u_i, \zeta_i) \end{array} \right] \end{array}$$

and

$$Q_i(y_i) := \left[\begin{array}{l} N_{\Omega_i}(u_i, \zeta_i) \\ N_{\mathbb{R}^3_+}(\lambda_i) \end{array} \right].$$

Following (Kocvara & Outrata 2004), we can now formulate an academic MPEC with the above equilibrium, where one aims to find the elasticity modulae e_1, e_2 in such a way that the dissipation energy at the terminal time T will be maximized. Since the objectives in (127) are not strictly convex, ImP cannot be applied to the numerical solution of this MPEC. On the other hand, due to the time discretization, the dimension $\bar{m} = 7T$ can be quite large even in this very simple example. So we see that the requirements, posed on an NLP method used to the numerical solution of MPECs with evolutionary equilibria, are rather demanding.

References

A.M.Al-Fahed, Stavroulakis, G. & Panagiotopoulos, P. (1991), 'Hard and soft fingered robot grippers. The linear complementarity approach', *ZAMM-Zeit. Angew. Math. Mech.* **71**, 257–265.

Anitescu, M. (2000), On solving mathematical programs with complementarity constraints as nonlinear programs, Technical Report ANL/MCS-P864-1200, Argonne National Laboratory.

Aubin, J. & Frankowska, H. (1990), *Set-valued Analysis*, Birkhäuser.

Aubin, J.-P. (1993), *Optima and Equilibria*, Springer-Verlag.

Beremlijski, R., Haslinger, J., Kočvara, M. & Outrata, J. (2002), 'Shape optimization in contact problems with Coulomb friction', *SIAM J. Optimization* **13**, 561–587.

Clarke, F. (1983), *Optimization and Nonsmooth Analysis*, J. Wiley & Sons.

Dempe, S. (2002), *Foundation of Bilevel Programming*, Kluwer Acad. Publ.

Dempe, S. (2003), 'Annotated bibliography on bilevel programming and mathematical programs with equilibrium constraints', *Optimization* **52**, 333–359.

Dontchev, A. & Rockafellar, R. (1996), 'Characterizations of strong regularity for variational inequalities over polyhedral convex sets', *SIAM J. Optimization* **7**, 1087–1105.

Ekeland, I. & Temam, R. (1976), *Convex Analysis and Variational Problems*, North-Holland Publ. Comp.

Flegel, M., Kanzow, C. & Outrata, J. (2004), Optimality conditions for disjunctive programs with application to mathematical programs with equilibrium constraints, Technical Report 256, Math. Inst. of Julius-Maximilian University Wuerzburg.

Fletcher, R. & Leyffer, S. (2002), 'Nonlinear programming without a penalty function', *Mathematical Programming* **91**, 239–269.

Fletcher, R. & Leyffer, S. (2004), 'Solving mathematical programs with complementarity constraints as nonlinear programs', *Optimization Methods and Software* **19**, 15–40.

Fletcher, R., Leyffer, S., Ralph, D. & Scholtes, S. (2002), Local convergence of SQP methods for mathematical programs with equilibrium constraints, Technical Report NA/209, University of Dundee.

Haslinger, J., Dostál, Z. & Kučera, R. (2002), 'On a splitting type algorithm for the numerical realization of contact problems with Coulomb friction', *Comput. Methods Appl. Mech. Engrg.* **191**, 2261–2281.

Haslinger, J. & Panagiotopoulos, P. (1984), 'The reciprocal variational approach to the Signorini problem with friction. Approximation results', *Proc. of the Royal Society of Edinburgh, Sect.A* **98**, 365–383.

Henrion, R., Jourani, A. & Outrata, J. (2002), 'On the calmness of a class of multifunctions', *SIAM J. Optimization* **13**, 603–618.

Henrion, R. & Outrata, J. (to appear), 'Calmness of constraint systems with applications', *Mathematical Programming* .

Hiriart-Urruty, J.-B. (1979), 'Refinements of necessary optimality conditions in nondifferentiable programming I', *Appl. Math. Optim.* **5**, 63–82.

Klatte, D. & Kummer, B. (2003), Strong Lipschitz stability of stationary solutions for nonlinear programs and non-polyhedral variational inequalities under MFCQ, Technical Report 2003-9, Humboldt-Universitaet zu Berlin.

Kočvara, M., Mielke, A. & Roubíček, T. (n.d.), 'A rate-independent approach to the delamination problem', *Mathematics and Mechanics of Solids (to appear)* .

Kocvara, M. & Outrata, J. (2004), On the modeling and control of delamination processes, *in* J. Cagnol & J.-P. Zolésio, eds, 'Control and Boundary Analysis', Marcel Dekker.

Kummer, B. (1992), Newton's method based on generalized derivatives for nonsmooth functions: Convergence analysis, *in* W. Oettli & D. Pallaschke, eds, 'Advances in Optimization', Springer-Verlag.

Lemaréchal, C. (1981), A view of line searches, *in* W. Oettli & J. Stoer, eds, 'Optimization and Optimal Control', Springer-Verlag.

Lemaréchal, C. (1989), Nondifferentiable optimization, *in* G. Nemhauser, A. Kan & M. Todd, eds, 'Handbooks in Operations Research and Management Science, Volume I', North-Holland Publ. Comp.

Lemaréchal, C. & Sagastizábal, C. (1997), 'Variable metric bundle method. From conceptual to implementable forms', *Mathematical Programming* **76**, 393–410.

Lemaréchal, C., Strodiot, J.-J. & Bihain, A. (1981), On a bundle algorithm for nonsmooth optimization, *in* O. Mangasarian, R. Meyer & S. Robinson, eds, 'Nonlinear Programming 4', Academic Press.

Lukšan, L. & Vlček, J. (1998), 'A bundle Newton method for nonsmooth unconstrained minimization', *Mathematical Programming* **83**, 373–391.

Luo, Z., Pang, J.-S. & Ralph, D. (1996), *Mathematical Programs with Equilibrium Constraints*, Cambridge Univ. Press.

Mikhalevich, V., Gupal, A. & Norkin, V. (1987), *Methods of Nonconvex Optimization (in Russian)*, Nauka.

Mordukhovich, B. (1988), *Approximation Methods in Problems of Optimization and Control (In Russian)*, Nauka.

Mordukhovich, B. (1994), 'Lipschitzian stability of constraint systems and generalized equations', *Nonl. Anal.: Theory, Methods & Appl.* **22**, 173–206.

Mordukhovich, B. & Outrata, J. (2001), 'On second-order subdifferentials and their applications', *SIAM J. Optimization* **12**, 139–169.

Murty, K. (1988), *Linear Complementarity, Linear and Nonlinear Programming*, Heldermann.

Outrata, J. (1999), 'Optimality conditions for a class of mathematical programs with equilibrium constraints', *Mathematics of Operations Research* **24**, 627–644.

Outrata, J. (2000), 'A generalized mathematical program with equilibrium constraints', *SIAM J. Control Optimization* **38**, 1623–1638.

Outrata, J., Kočvara, M. & Zowe, J. (1998), *Nonsmooth Approach to Optimization Problems with Equilibrium Constraints*, Kluwer Acad. Publ.

Panagiotopoulos, P. (1993), *Hemivariational Inequalities. Applications in Mechanics and Engineering*, Springer Verlag.

Qi, L. (1993), 'Convergence analysis of some algorithms for solving nonsmooth equations', *Mathematics of Operations Research* **17**, 227–244.

Qi, L. & Sun, J. (1993), 'A nonsmooth version of Newton's method', *Mathematical Programming* **58**, 353–368.

Robinson, S. (1980), 'Strongly regular generalized equations', *Mathematics of Operations Research* **5**, 43–62.

Robinson, S. (1981), 'Some continuity properties of polyhedral multifunctions', *Mathematical Programming Study* **14**, 206–214.

Robinson, S. (1991), 'An implicit-function theorem for a class of nonsmooth functions', *Math. Oper. Res.* **16**, 292–309.

Rockafellar, R. & Wets, R. (1998), *Variational Analysis*, Springer-Verlag.

Scheel, H. & Scholtes, S. (2000), 'Mathematical programs with complementarity constraints: Stationarity, optimality, and sensitivity', *Mathematics of Operations Research* **25**, 1–22.

Scholtes, S. (1994), Introduction to piecewise differentiable equations, Technical Report (Habilitation Thesis), Universitaet Karlsruhe.

Scholtes, S. (2001), 'Convergence properties of a regularization scheme for mathematical programs with complementarity constraints', *SIAM J. Optimization* **11**, 918–936.

Schramm, H. & Zowe, J. (1992), 'A version of the bundle idea for minimizing a nonsmooth function: conceptual idea, convergence analysis, numerical results', *SIAM J. Optimization* **2**, 121–152.

Shapiro, A. (1990), 'On concepts of directional differentiability', *J. Optim. Theory Appl.* **66**, 477–487.

von Stackelberg, H. (1934), *Marktform und Gleichgewicht*, Springer-Verlag.

Ye, J. (2000), 'Constraint qualifications and necessary optimality conditions for optimization problems with variational inequality constraints', *SIAM J. Optimization* **10**, 943–962.

Applied Nonsmooth Mechanics of Deformable Bodies.

Georgios E. Stavroulakis

Department of Production Engineering and Management, Technical University of Crete, Chania, Greece
and
Department of Civil Engineering, Technical University of Braunschweig, Germany

Abstract. In this article the interconnections between nonsmooth optimization, nonsmooth analysis and complementarity problems, from the one side, and modern computational mechanics, from the other side, are outlined. The arising problems are, in general, variational and hemivariational inequalities. A short discussion of suitable numerical algorithms for the approximation of their solution, in connection with finite element or boundary element techniques, follows. Related topics of existence, stability and path-following of the solution are briebly discussed.

1 Introduction

Unilateral contact is certainly the most well-known source of nonsmooth effects in mechanics. In fact, it is an introductory example of a whole class of highly nonlinear effects. They have a variable, exhitation-dependent mechanical behaviour, which can be described by an appropriate selection of classical models. The transition points between the various models make the problem nonsmooth. Therefore we deals with a problem of *nonsmooth mechanics*.

Unilateral contact analysis has been the subject of several monographs and review articles in mechanics during the last decades. In most of them the finite element method is used. The contact problem leads to a complementarity problem or a variational inequality. Accordingly, the solution of the problem is based on some iterative (mostly trial and error) method. The reader may consult the monographs of Panagiotopoulos (1985), Johnson (1987), Kikuchi and Oden (1988), Antes and Panagiotopoulos (1992), Zhong (1993) and Pfeiffer and Glocker (1996) and the recent works of Brogliato (1999), Fischer (2000), Laursen (2002) and Wriggers (2002). Among the recent review articles in this area one may mention those of Wriggers (1995), Klarbring (1999), Barber and Ciavarella (2000) and Pfeiffer and Glocker (2000). A mathematical presentation, which is near to the one followed in this text, can be found in Curnier (1999) and especially for the unilateral contact case Klarbring (1999). Unilateral contact problems treated with the boundary element method have been presented in Antes and Panagiotopoulos (1992) and Stavroulakis and Antes (2003).

The links between nonsmooth analysis, optimization and mechanics, which provide a more general theoretical framework for the formulation and study of nonsmooth nonlinearities, including the case of unilateral contact and the stick-slip motion in friction, are discussed in this text. Further material can be found in the previous chapters of this volume, in Panagiotopoulos (1993), Demyanov et al. (1996), Mistakidis and Stavroulakis (1998), Brogliato (1999), Frémond (2002) and the references given there.

The adoption of refined models, which are based on nonsmooth mechanics, allows for the development of powerful solution algorithms. For instance, in the case of unilateral contact mechanics, one is able to formulate inequality constrained optimization problems or complementarity problems. Analogous results exists for elastoplasticity and other nonlinear computational mechanics models. From the technical papers in this area, which can be used as starting point of further study, one may mention the following works: in unilateral contact elastostatics, Bathe and Bouzinov (1997), Kim and Kwak (1996) and Klarbring (1986), in unilateral contact with plasticity, for example, Tin-Loi and Xia (2001), in unilateral contact elastodynamics, Vola et al. (1998), Moreau (1999), Song et al. (2001) and Czekanski and Meguid (2001).

Some representative industrial applications are described in the following publications: multibody dynamical problems in machines and machine components, including unilateral contacts, have been studied in Pfeiffer and Hajek (1992), Pfeiffer and Glocker (1996), Pfeiffer (1998), Klisch (1999), Wolfsteiner and Pfeiffer (2000) and Kim and Kwak (2000), the modelling of complex assemply processes, for instance in Blanze et al. (1996), of robotics, Pfeiffer (1998), of rolling contact, Gonzalez and Abascal (2000), of dampers in turbine blades, Pfeiffer and Hajek (1992), and in metal forming processes Chabrand et al. (1998), unilateral or frictional effects in fracture mechanics, see among other, Barber and Ciavarella (2000), Leblond (2000), Zozulya and Gonzalez-Chi (2000), large-scale of granular materials in Jean (1999).

2 Theoretical Framework: NSO, NSA, VI, HI and CP

The world is certainly nonsmooth, therefore classical assumptions about smoothness can not be accepted for the study of certain applications. An answer to this need is reflected into the developments of NSO (nonsmooth optimization) and NSA (nonsmooth analysis). In mechanics and optimization, variational methods have both a tradition and a usefulness. For nonsmooth models the classical variational equations must be replaced by VI (variational inequalities), HI (hemivariational inequalities) and similar constructions, Panagiotopoulos (1993). Roughly speaking, inequalities reflect the fact that we are dealing with systems with direction-dependent (unilateral) behaviour. In some cases, one may simplify the mathematical description and develop models leading to linear or nonlinear CP (complementarity problems). These latter problems are well-known in several branches of science and engineering, from economics to transportation modelling, and have been studied from both theoretical and algorithmic point of view.

2.1 Elements of nonsmooth (convex and nonconvex) analysis

Smooth and Nonsmooth Functions Let f be a real-valued finite function defined on the real line R. The most powerful and widely used tool to study the properties of f is the notion of derivative. Function f is called differentiable at $x \in R$ if there exists its derivative $f'(x)$ at x, which is defined by

$$f'(x) = \lim_{\alpha \to 0} \frac{1}{\alpha}\big[f(x+\alpha) - f(x)\big]. \tag{1}$$

If this limit exists for every point of some open set $S \in R$ the function f is called differentiable on S.

Among the variety of applications of the derivative one recalls here the first order approximation (linearization) of f in the neighborhood of a point x:

$$f(x+\Delta) = f(x) + f'(x)\Delta + o_x(\Delta) \tag{2}$$

with

$$\frac{o_x(\Delta)}{\Delta} \longrightarrow 0 \text{ as } \Delta \to 0 \tag{3}$$

Moreover, x^* is a minimum of the function f if

$$f'(x^*) = 0. \tag{4}$$

Relation (4) defines a *stationary point* of f, since it also holds true for a maximum and for a saddle point of f. In fact, higher order derivatives are checked in order to specify the nature of the stationary point.

One-sided differentials Assume now that the limit (1) does not exist, but at the same time the following directional derivatives exist: the right-hand side derivative $f'_+(x)$ and the left-hand side derivative $f'_-(x)$ of f at x. They are defined by:

$$f'_+(x) = \lim_{\alpha\downarrow 0} \frac{1}{\alpha}\big[f(x+\alpha) - f(x)\big],\ f'_-(x) = \lim_{\alpha\uparrow 0} \frac{1}{\alpha}\big[f(x+\alpha) - f(x)\big]. \tag{5}$$

Here $\alpha \downarrow 0$ means that $\alpha \to 0$, by taking positive values $\alpha > 0$ and $\alpha \uparrow 0$ means that $\alpha \to 0$, with negative values $\alpha < 0$.

For a function f to be differentiable at x it is necessary and sufficient that $f'_+(x) = f'_-(x)$.

The *directional derivative* of a function f at point x and in the direction $x \in R$ is defined by:

$$f'(x, g) = \lim_{\alpha\downarrow 0} \frac{1}{\alpha}\big[f(x+\alpha g) - f(x)\big], \tag{6}$$

if this limit exists. This is a proper extension of the notion of the derivative. For example, it can be used to linearize a given function (cf. (2)) along a direction g. In this case relation (2) holds along a given direction, a different value holds for the opposite direction, etc, so that it provides the basis for a quasi-linearization of the function f.

From the definition one may easily see that a necessary condition for a directionally differentiable function f to attain a minimum at point x^* is that:

$$f'(x^*, g) \geq 0 \qquad \forall g \in R. \tag{7}$$

If strict inequality holds in (7) for every direction g not equal to zero, the condition becomes also sufficient for x^* to be a strict local minimum of f. On the other hand, a necessary condition for a directionally differentiable function f to attain a maximum at point x^* is that:

$$f'(x^{**}, g) \leq 0 \qquad \forall g \in R, \tag{8}$$

with analogous implications for a strict local maximum.

A point x^* which satisfies relation (7) is called an inf-stationary point of f , while a point x^{**} satisfying (8) is called a sup-stationary point. It is interesting to observe that for a nonsmooth function first order optimality conditions may, in some cases, become sufficient for a minimum or a maximum.

Convex Analysis Let us assume a convex nondifferentiable function $\pi(x)$. The set-valued *sub-differential* of this function, denoted by ∂ is written as

$$w \in \partial\pi(x). \tag{9}$$

Relation (9) is by definition equivalent to the variational inequality

$$w(x)y \leq \pi(x+y) - \pi(x), \forall x, y \in R^n, \tag{10}$$

which defines for all elements of the subdifferential set (9) at a given point x, the subgradients $w(x)$, a lower linear approximation of the considered function, which holds for all small variations y around pont x. In fact the name *sub*-gradient characterizes their main property. The directional derivative of function $\pi(x)$ at point x in the direction y reads

$$\pi^{'}(x,y) = \max_{w\in\partial\pi(x)} \langle w, y\rangle\,. \tag{11}$$

In this case, $\langle .,. \rangle$ denotes the inner product of two vectors. For a piecewice smooth function which is defined as the maximum of two smooth functions,

$$\pi(x) = \max\left\{\pi_1(x), \pi_2(x)\right\}, \tag{12}$$

the convex analysis subdifferential at the point of nondifferentiability (x_0 such that $\pi_1(x_0) = \pi_2(x_0)$) is constructed by means of the convex hull of the classical gradients of the two smooth constituents, i.e., $\partial\pi(x_0) = co\left\{\nabla\pi_1(x_0), \nabla\pi_2(x_0)\right\}.$

A subdifferential relation which can be described by the convex analysis tools is the one describing the mechanical behaviour of a locking spring. For the mathematical modeling of the following nonlinear, monotone constitutive law

$$s = \begin{cases} s_2 + c_1 e, & \text{for } e \geq 0, \\ \left[-s_1, s_2\right], & \text{for } e = 0, \\ -s_1 + c_2 e, & \text{for } e \leq 0, \end{cases} \tag{13}$$

between scalar stress s and strain e, components of the corresponding tensors, one needs the introduction of a nondifferentiable, convex potential (the so-called superpotential)

$$\pi(e) = \max\left\{\pi_1(e), \pi_2(e)\right\} = \max\left\{s_2 + \frac{1}{2}c_1 e^2, -s_1 + \frac{1}{2}c_2 e^2\right\}. \tag{14}$$

At the point of nondifferentiability, $e = 0$, the vertical branch of relation (13), i.e. the interval $[-s_1, s_2]$, is produced by the convex analysis subdifferential ∂ of the nonsmooth function $\pi(e)$, which is constructed by means of the convex hull of the classical gradients of the two smooth constituents, i.e., $\partial\pi(e=0) = co\left\{\nabla\pi_1(e), \nabla\pi_2(e)\right\}.$

As we will see later on, for the description of unilateral contact mechanisms one replaces e with the normal u_{cn} and s with the corresponding normal traction t_{cn} in the previous relations (cf. (55)). The corresponding superpotential has a vertical branch going up to $+\inf$ (cf. (57)).

Nonconvex Analysis Among the many available generalizations of convex analysis, let us describe the *Clarke's generalized derivative.* Let a function f be locally Lipschitz at $x \in X$ and let y be a vector in X. The directional differential in the sense of F.H. Clarke of f at x in the direction y, denoted by $f^0(x,y)$, is defined by the relation:

$$f^0(x,y) = \limsup_{\substack{\mu \to 0_+ \\ h \to 0}} \frac{f(x+h+\mu y) - f(x+h)}{\mu}.$$

$f^0(x,y)$ is also called generalized directional differential.

By means of the directional differential $f^0(x,y)$ one can now define the generalized gradient $\bar{\partial} f(x) : X \to X^\star$:

$$\bar{\partial} f(x) = \{x^\star : x^\star \in X^\star, f^0(x, x_1 - x) \geq (x^\star, x_1 - x) \quad \forall x_1 \in X\}, \tag{15}$$

or the equivalent definition:

$$\bar{\partial} f(x) = \{x^\star : x^\star \in X^\star, (x^\star, -1) \in N_{epi\, f}(x, f(x))\}, \tag{16}$$

where $N_C(x)$ denotes the normal cone to a set C at point x and $epif$ is the epigraph of the function f.

Note that $\bar{\partial} f(\cdot)$ is a multivalued mapping, it is a nonempty convex, closed and bounded subset of $X^\star$ and the following relation holds true:

$$f^0(x,y) = \max\{\langle y, x^\star \rangle : x^\star \in \bar{\partial} f(x)\}. \tag{17}$$

The notation $\bar{\partial}$ is used here (and in the most of the work of P.D. Panagiotopoulos), while in other sources the notation ∂_{CL}, in honour of F.H. Clarke who proposed it, is used. When misunderstanding is not expected, the notation ∂, which is usually reserved for the convex analysis subdifferential, is also used. It should also be noted here that $\bar{\partial}$ should not be confused with the superdifferential used within the theory of quasidifferentiability in the sense of V.F. Demyanov.

Relation (15) can be used to define the generalized gradient $\bar{\partial} f(x)$ for any type of function $f : X \to \bar{R}$ which is finite at the point x. Note that $\bar{\partial} f(x)$ may be empty. The above definition of $\bar{\partial} f(x)$ for any function $f : X \to \bar{R}$ makes sense, because the normal cone $N_C(x)$ can be defined with respect to any set epi f. The generalized directional differential $f^\uparrow(x;y)$ at x in the direction y is defined by the relation:

$$f^\uparrow(x,y) = \sup\{\langle y, x^\star \rangle : x^\star \in \bar{\partial} f(x)\}. \tag{18}$$

Thus one may write the relation:

$$\bar{\partial} f(x) = \{x^\star : x^\star \in X^\star, f^\uparrow(x, x_1 - x) \geq \langle x^\star, x_1 - x \rangle \quad \forall x_1 \in X\}. \tag{19}$$

The directional differential $f^\uparrow(x;y)$ is also called directional differential in the sense of Rockafellar (1981). Note that $\bar{\partial} f(x) = \emptyset$ if $f^\uparrow(x,0) = -\infty$, and if $f^\uparrow(x,y)$ is finite for every y then $\bar{\partial} f(x) \neq \emptyset$.

It should be noted that for a convex function f one has:

$$f^\uparrow(x,y) = \liminf_{\tilde{y} \to y} f'(x, \tilde{y}) \quad \forall y \in X, \tag{20}$$

where $f'(\cdot,\cdot)$ denotes the one-sided directional Gâteaux differential. Moreover, for a locally Lipschitz function f at point x one has:

$$f^{\uparrow}(x,y) = f^{0}(x,y) \quad \forall y \in X \tag{21}$$

and for a continuously differentiable f:

$$\bar{\partial} f(x) = \{\text{grad} f(x)\}. \tag{22}$$

Examples For a convex function f one gets the classical subdifferential of convex analysis

$$\bar{\partial} f(x) = \partial f(x) \tag{23}$$

and for a concave and bounded below on a neighbourhood of x function f:

$$\bar{\partial} f(x) = -\partial(-f)(x) \tag{24}$$

at every point x where f is finite. The indicator function I_C of a (generally nonconvex) set C, is defined by $I_C(x) = \{0 \text{ if } x \in C, \infty \text{ otherwise}\}$. It can be proved that

$$\bar{\partial} I_C(x) = N_C(x) \tag{25}$$

and

$$I_C^{\uparrow}(x,y) = I_{T_{C(x)}}(y). \tag{26}$$

In the finite dimensional case $X \equiv R^n$, for a locally Lipschitz function f at a point x, $\bar{\partial} f(x)$ is the convex hull of all points $y \in R^n$ of the form

$$y = \lim_{i \to \infty} \text{grad} f(x_i), \tag{27}$$

where x_i converges as $i \to \infty$ to x, avoiding the nondifferentiability points and any other points of a set of measure zero (in the sense of Lebesgue) and such that $\text{grad} f(x_i)$ converges.

For a maximum-type function f, i.e., when the function is defined by means of continuously differentiable functions $\varphi_i = \varphi_i(x)$, $i = 1,\ldots,m$, $x \in R^n$ by the relation: $f = \max\{\varphi_i,\ldots,\varphi_m\}$, one has:

$$\bar{\partial} f(x) = \text{co}\{\text{grad}\, \varphi_i(x) : \; i \in I(x)\}, \tag{28}$$

where $I(x) = \{i : \varphi_i(x) = f(x)\}$ is the active index set.

The normal cone to a set defined by: $C = \{x \in R^n | f(x) \leq 0\}$ at a point x_0 with $f(x_0) = 0$ is described by the relation

$$N_C(x_0) \subset \{\lambda x^{\star} : x^{\star} \in R^n, \lambda \geq 0, x^{\star} \in \bar{\partial} f(x_0)\},$$

whenever f is Lipschitzian on a neighbourhood of x_0 and $0 \notin \bar{\partial} f(x_0)$. The notion of $\bar{\partial}$-regularity assures that directional derivative information can be regained from the F.H. Clarke's notion. For a locally Lipschitz function one requires that $f^{0}(x,y) = f'(x,y) \quad \forall y \in X$, holds at x. This definition is equivalent to the statement that epi f is regular at $\big(x, f(x)\big)$. For instance, a convex function and a maximum type function are $\bar{\partial}$-regular at a point x where they are finite. For example, for the max-type function $f = \max\{\varphi_1, \ldots, \varphi_m\}$ one has

$$N_C(x_0) = \bar{\partial} I_C(x_0) = \{z : z = \sum_{i=1}^{m} \lambda_i \mathrm{grad}\, \varphi_i(x_0), \lambda_i \geq 0,\ \varphi_i(x_0) \leq 0,\ \lambda_i \varphi_i(x_0) = 0\}, \quad (29)$$

if $0 \notin \bar{\partial} f(x_0)$. The above relation permits the extension of the Lagrange multiplier rule for optimization problems subjected to the nonconvex inequality constraints $\varphi_i(x) \leq 0,\ i = 1, \ldots, m$. This becomes obvious, e.g. if one considers the search for a local minimum problem of a continuously differentiable function $g : R^n \to R$ over $C = \{x \in R^n : \varphi_i(x) \leq 0,\ i = 1, \ldots, m\}$. A necessary condition is $0 \in \bar{\partial}(g + I_C)(x)$ which implies that

$$-\mathrm{grad}\, g(x) \in \bar{\partial} I_C(x), \quad (30)$$

which together with (29) leads to the Lagrange multiplier rule.

D.C. functions Let us consider nonconvex and nonsmooth functions which can be expressed as a difference of convex functions (d.c. functions):

$$\pi(x) = \pi_1(x) - \pi_2(x). \quad (31)$$

By using elements of the subdifferentials of the convex constituents, i.e.,

$$w = w_1 - w_2, w_1 \in \partial\pi_1(x), w_2 \in \partial\pi_2(x), \quad (32)$$

the directional derivative of the d.c. function $\pi(x)$ can be expressed by

$$\begin{aligned} \pi'(x, y) &= \max_{w_1 \in \partial\pi_1(x)} \langle w_1, y\rangle - \max_{w_2 \in \partial\pi_2(x)} \langle w_2, y\rangle \\ &= \max_{w_1 \in \partial\pi_1(x)} \langle w_1, y\rangle + \min_{w_2 \in -\partial\pi_2(x)} \langle w_2, y\rangle . \end{aligned} \quad (33)$$

Exploiting the latter relation one may write a system of variational inequalities, or an implicit variational inequality, that characterizes the critical points or the local minima of a d.c. nonconvex function. In a more general setting, one may use the theory of the quasidifferentials and express the directional derivatives of an arbitrary function by means of an ordered pair of convex sets $D\pi(x) = [\underline{\partial}\pi(x), \overline{\partial}\pi(x)]$, the subdifferential $\underline{\partial}$ and the superdifferential $\overline{\partial}$ as follows:

$$\pi'(x, y) = \max_{w_1 \in \underline{\partial}\pi_1(x)} \langle w_1, y\rangle + \min_{w_2 \in \overline{\partial}\pi_2(x)} \langle w_2, y\rangle \quad (34)$$

As one easily recognizes, the d.c. case is a special case of quasidifferentiable functions.

Two examples of nonmonotone relations with complete vertical branches, which require the use of nonconvex superpotentials for their description, are given now.

Let us consider the following nonmonotone, possibly multivalued one-dimensional law, i.e. a law with a graph having complete falling branches:

$$s = \begin{cases} 0 & \text{for } e < -e_1, \\ [-s_1, 0] & \text{for } e = -e_1, \\ \frac{s_1}{e_1} e & \text{for } -e_1 < e < e_1, \\ [0, s_1] & \text{for } e = e_1, \\ 0 & \text{for } e > e_1. \end{cases}$$

This law can be derived by differentiating the following nonsmooth and nonconvex potential (which is also a d.c. function):

$$\begin{aligned}\pi(e) &= \min\left\{\frac{1}{2}c_1e^2, \frac{1}{2}c_1e_1^2\right\}\\ &= \frac{1}{2}c_1e^2 - \left\{\begin{array}{ll} 0, & \text{for } e < e_1, \\ s_1e + \frac{1}{2}c_1(e-e_1)^2, & \text{for } e > e_1 \end{array}\right\}\\ &= \pi_1(e) - \pi_2(e), \end{aligned} \tag{35}$$

with $c_1 = \frac{s_1}{e_1}$.

QD functions The more general case of structured, nonconvex functions is the class of QD (quasidifferentiable) functions in the sense of Demyanov and Rubinov (see, among others, Demyanov et al. (1996), in Demyanov and Rubinov (2000)). A function $f : R \to R$ is called quasidifferentiable (q.d) at a point x if it is directionally differentiable at x and there exists a pair of closed intervals $\underline{\partial} f(x) = [v_1, v_2]$ and $\overline{\partial} f(x) = [w_1, w_2]$ such that

$$f'(x, g) = \max_{v \in \underline{\partial} f(x)} vg + \min_{w \in \overline{\partial} f(x)} wg \quad \forall g \in R. \tag{36}$$

The pair of the intervals $Df(x) = [\underline{\partial} f(x), \overline{\partial} f(x)]$ is called a quasidifferential of f at x. The set $\underline{\partial} f(x)$ is called the subdifferential and the set $\overline{\partial} f(x)$ the superdifferential of f at x.

It is clear that a quasidifferential is not uniquely defined. In fact, if a function f is quasidifferentiable at x and $Df(x)$ is its quasidifferential at this point, i.e., $Df(x) = [\underline{\partial} f(x), \overline{\partial} f(x)]$, then every pair of the form $[\underline{\partial} f(x) + C, \overline{\partial} f(x) - C]$, where C is an interval $C = [c_1, c_2] \in R$, with $c_1 \leq c_2$ is also a quasidifferential of f at x. In fact, the quasidifferential is a class of equivalent ordered pairs of convex sets.

Necessary and sufficient optimality conditions For a quasidifferentiable function the necessary and sufficient optimality conditions (see (7)–(8)) can be written as follows.

Let $Df(x) = [\underline{\partial} f(x), \overline{\partial} f(x)]$ be a quasidifferential of f at x. A necessary condition for function f to attain a minimum at point x^* is that:

$$-\overline{\partial} f(x^*) \subset \underline{\partial} f(x^*). \tag{37}$$

The condition

$$-\overline{\partial} f(x^*) \subset \text{int}\, \underline{\partial} f(x^*) \tag{38}$$

is sufficient for x^* to be a strict local minimum of f. Analogously, a necessary condition for a maximum of f at x^{**} is that:

$$-\underline{\partial} f(x^{**}) \subset \overline{\partial} f(x^{**}), \tag{39}$$

with an analogous result for a sufficient condition for a strict local maximum:

$$-\underline{\partial} f(x^{**}) \subset \text{int}\, \overline{\partial} f(x^{**}). \tag{40}$$

Quasidifferentiable functions A directionally differentiable function f defined on an open set $X \subset R^n$ is called quasidifferentiable at a point $x \in X$, if there exists an ordered pair of convex compact sets $[U, V]$ in $R^n \times R^n$ which produces the directional derivative of the function by:

$$f'(x, g) = f'_x(g) = \max_{h \in U}(h, g) + \min_{h \in V}(h, g), \forall g \in R^n. \tag{41}$$

Clearly, the first term in the r.h.s. of (41) is a sublinear function while the second term is a superlinear function. Thus, the directional derivative of a quasidifferentiable function belongs to the space L of functions which can be written as the sum of a sublinear function and a superlinear function. Moreover with an element $[U, V]$ of the space of compact sets it is associated the class of equivalent ordered pairs of compact convex sets.

Thus, the class of equivalent ordered pairs of convex compact sets $[U, V]$ of (41) (the quqsidifferential $Df(x)$ of f at x) fully describes the first order derivative of the directionally differentiable function f and gives rise to the quasi-linearization (41) and, subsequently, to a qualitative and quantitative first order approximation of f in the sense of (2).

Critical points. Substationarity. The notion of substationarity plays an important role in the theory of hemivariational inequalities because it permits the formulation of the propositions of substationary potential and complementary energy which generalize the corresponding classical minimum energy propositions in mechanics, Panagiotopoulos (1985), Panagiotopoulos (1993). Point x_0 is a substationarity point of a functional $f : X \to \bar{R}$ if

$$0 \in \bar{\partial} f(x_0). \tag{42}$$

Equivalently one has:

$$f^{\uparrow}(x_0, y) \geq 0 \quad \forall y \in X. \tag{43}$$

Substationarity points are all the classical stationarity points, all the local minima, a large class of local maxima, as well as all the saddle points. Point x is said to be a substationarity point of f with respect to a set C, if $f + I_C$ is substationary at x.

Quasidifferential and Clarke subdifferential Under appropriate assumptions on f, and for appropriate choice of the elements of the subdifferential and the subdifferential of f at x, one may calculate estimates of the Clarke subdifferential, by using the information provided for a quasidifferentiable function. The results are, in general, of the form:

$$A \subset \partial_{CL} f(x) \subset B, \tag{44}$$

with suitable sets A and B. Details can be found in Demyanov and Rubinov (1995), pp.143-159. It should be noted here that the margins of the above estimates may be large in some cases, and accordingly the results may become useless for certain applications. Further research is needed at this point, which is related in some sense with the areas of interval analysis and fuzzy logic.

2.2 Variational and hemivariational inequalities

Classical, weak formulation of boundary value problems lead to variational equalities. The underlying principle is the one of smooth approximation of a given relation at this point. The most known case is the linearization. In fact, in the neighborhood of a given point, one linearizes all given nonlinear relations and by this way produces a set of linear equations which characterize, within some accuracy, the studied mechanical problem. For problems of nonsmooth mechanics one needs quasilinearization. The approximation depends on the direction in which one looks for a solution. The techniques of convex analysis, if applicable, or of more general nonsmooth analysis, are required. The corresponding approximations have the form of inequalities. In the area of convexity, cf. relation (10), one gets variational inequalities. For nonconvex problems one has, for example, hemivariational inequalities or implicit system of variational inequalities, cf. relation (33). Many other cases exist, like the implicitly dependent variational inequalitie or the quasivariational inequalities, which have been mentioned previously in the context of frictional contact modelling. More details on these notions and their use in nonsmooth mechanics are given, among others, in Panagiotopoulos (1985), Antes and Panagiotopoulos (1992), Panagiotopoulos (1993), Demyanov et al. (1996) and Mistakidis and Stavroulakis (1998). It should be emphasized that convex analysis is a more or less well-known field of nonsmooth modelling and optimization. Thus all models based on convexity can be considered as being classical nonsmooth mechanics' material. On the other hand, a lot of notions for nonsmooth analysis in the nonconvex case have been proposed. It seems that each one of them is, at least, suitable for some class of applications. Thus, no complete theory exists, for the time being, for the general nonconvex case. The here outlined methods reflect the author's research work. Since this field is very active, one should expect that other new notions of nonsmooth analysis will be used in the near future as well.

For applications, it seems that in most cases a resolution into piecewise differentiable (or even linear) models is appropriate. The result is the formulation of linear and nonlinear complementarity problems. Therefore, for practical applications, the theoretical and numerical study of these problems is of interest. Some information along these lines is provided later.

Convex energy functions and variational inequalities This part appears in other places of these lecture notes and will not be discussed further.

Nonconvex energy functions. The nonconvex superpotentials resulting by integrating discontinuous functions $\beta \in L^{\infty}_{loc}(R)$ play an important role in the formulation of hemivariational inequalities for several types of mechanical problems. After some technical details about filling in gaps in a multifunction, nonmonotone laws in mechanics which admit a nonconvex energy function will be introduced. First, we need the following *filling-in* of the gaps arising in a multifunction. Suppose that $\beta : R \to R$ is a function such that $\beta \in L^{\infty}_{loc}(R)$, i.e. a function which is essentially bounded on any bounded interval of R. For any $\rho > 0$ and $\xi \in R$ let us define

$$\bar{\beta}_\rho(\xi) = \underset{|\xi_1-\xi|\leq\rho}{ess, inf}\, \beta(\xi_1), \quad \bar{\bar{\beta}}_\rho(\xi) = \underset{|\xi_1-\xi|\leq\rho}{ess, sup}\, \beta(\xi_1). \tag{45}$$

The monotonicity properties of $\rho \to \bar{\beta}_\rho(\xi)$ and $\rho \to \bar{\bar{\beta}}_\rho(\xi)$ imply that the limits as $\rho \to 0_+$ exist. Therefore one may write that: $\bar{\beta}(\xi) = \lim_{\rho\to 0_+} \bar{\beta}_\rho(\xi)$ and $\bar{\bar{\beta}}(\xi) = \lim_{\rho\to 0_+} \bar{\bar{\beta}}_\rho(\xi)$. Furthermore one defines the multivalued function:

$$\tilde{\beta}(\xi) = [\bar{\beta}(\xi), \bar{\bar{\beta}}(\xi)], \tag{46}$$

where $[\cdot,\cdot]$ denotes an interval between the two given arguments. Then, a locally Lipschitz function j can be determined, up to an additive constant, by the relation

$$j(\xi) = \int_0^{\xi} \beta(\xi_1) d\xi_1, \tag{47}$$

such that $\bar{\partial} j(\xi) \subset \tilde{\beta}(\xi)$. If moreover $\beta(\xi_{\pm})$ exist for each $\xi \in R$, then one has $\bar{\partial} j(\xi) = \tilde{\beta}(\xi)$. Then we can define the *Superpotential nonmonotone laws*. Let us assume that one has an one-dimensional mechanical law which is described by the graph of a discontinuous function. For instance, a force-displacement law ($S - u$) is considered, which may correspond to an one-dimensional nonlinear spring law, to a nonlinear boundary condition, etc. The law is considered to be of the form $\beta : u \rightarrow -S$ where $u \in R$ and $S \in R$. The procedure of (45)-(47) is used in order to define a locally Lipschitz nonconvex superpotential energy function $j(u)$. The mechanical law is produced, in turn, by using the previously introduced generalized subdifferential operator $\partial_{CL} = \overline{\partial}$ and the nonconvex superpotential j as follows:

$$-f \in \partial_{CL} j(u) \tag{48}$$

By definition (48) is equivalent to the inequality

$$j^{\uparrow}(u, u^{\star} - u) \geq \langle -f, u^{\star} - u \rangle, \quad \forall u^{\star} \in U \tag{49}$$

for $u \in U$, which has been called by P.D. Panagiotopoulos hemivariational inequality, and to the inclusion

$$(-f, -1) \in N_{\text{epi}\, j}(u, j(u)). \tag{50}$$

For j Lipschitzian, $j^{\uparrow}$ in (48) is replaced by j^0 Obviously, if j is a convex superpotential, one has superpotential laws which can be described by monotone graphs with complete vertical branches. The procedure would be identical in that case as well, where ∂_{CL} is replaced by the subdifferential of convex analysis. The result would be a variational inequality. Note also that extensions to multi-dimensional laws (e.g., material constitutive relations) can be considered within this formulation.

Now we are able to introduce the *Hemivariational inequality* problem. An abstract coercive hemivariational inequality is written first. Let V be a real Hilbert space with the property that $V \subset [L^2(\Omega)]^n \subset V^{\star}$, where $V^{\star}$ denotes the dual space of V, and the injections are continuous and dense. Let moreover a boundary value problem be defined in an open, bounded subset Ω of R^n. Here $(\cdot, \cdot)$ denotes the $[L^2(\Omega)]^n$ inner product and the duality pairing, $|| \cdot ||$ the norm of V and $| \cdot |_2$ is the $[L^2(\Omega)]^n$-norm. One should recall that the form $(\cdot, \cdot)$ extends uniquely from $V \times L^2[(\Omega)]^n$ to $V \times V^{\star}$. Moreover let $L : V \rightarrow L^2(\Omega)$, $Lu = \hat{u}$, $\hat{u}(x) \in R$ be a linear continuous mapping. Further, assume that $l \in V^{\star}$, that $L : V \rightarrow L^2(\Omega)$ and that $\tilde{V} = \{v \in V : \hat{v} \in L^{\infty}(\Omega)\}$ is dense in V for the V-norm, and has a Galerkin base in V. It is also assumed that $a(\cdot, \cdot)$: $V \times V \rightarrow R$ is a bilinear symmetric continuous form which is coercive, i.e. there exists $c > 0$ constant such that

$$a(v, v) \geq c||v||^2 \qquad \forall v \in V. \tag{51}$$

A coercive hemivariational inequality problem reads: Find $u \in V$ such that

$$a(u, v - u) + \int_{\Omega} j^0(\hat{u}, \hat{v} - \hat{u}) d\Omega \geq (l, v - u), \ \forall v \in V. \tag{52}$$

For example, a linear elastostatic structural analysis problem with additional nonlinear elements of nonmonotone type which admit a nonconvex superpotential $j(u)$ can be written in the hemivariational inequality form (52). In this context u are the displacements at the various points of the structure, which occupies Ω in its undeformed configuration. For a plate problem one has $\Omega \subset R^2$, for a three-dimensional continuum $\Omega \subset R^3$, etc. Moreover the functional space V is dictated from the kind of the assumed application and from the (natural, support) boundary conditions of the structure. The operator L and the energy form $a(u, u)$ depend on the mechanical theory used for the elastic part of the structure, while l denotes the external loading. Finally, coercivity usually means that a well-defined mechanical theory has been assumed and sufficient boundary conditions are assigned so that, for example, no rigid body motions of the structure are allowed. Finally, one should note that the familiar form of the principle of virtual work (i.e., a variational equality) can be obtained back from the hemivariational inequality (52) if the effect of the nonlinear terms is neglected, i.e., if the second term in the l.h.s. of (52) is absent and if an equality is assumed in the place of the inequality.

2.3 On linear and nonlinear complementarity problems

A nonlinear complementarity problem has the following general form: find $x \in R^n$ such as to satisfy the following set of relations:

$$x \geq 0, F(x) \geq 0, x^T F(x) = 0, \tag{53}$$

where $F(x) : R^n \to R^n$ is a continuously differentiable function. In the case of a linear function $F(x)$, one gets the linear complementarity problem in the standard form. This is the case, for example, of boundary nonlinearities of the unilateral contact type. Note that, in general, a nonlinear complementarity problem (53) is equivalent to the variational inequality: find $x \in \mathrm{R}^n_+$ such as to satisfy

$$\langle F(x), x^* - x\rangle \geq 0, \ \forall\, x^* \in \mathbb{R}^n_+\,. \tag{54}$$

More general relations where the set R^n_+ is replaced by a convex closed subset C of $\mathbb{R}^n$ can also be considered.

Recall that if the function $F(x)$ is the gradient of a continuously differentiable convex function $\Phi(x)$, i.e., $F(x) = \partial\Phi(x)$, then the above problems (53) and (54), can be identified to be the solvability relations of an inequality constrained optimization problem with respect to the function $\Phi(x)$. For a linear function $F(x) = Bx + c$, one gets a linear complementarity problem.

Complementarity problems arise in several models of engineering. Details on these models as well as on solution methods can be found, among others, in Harker and Pang (1990), Billups and Murty (2000). Among the various applications of variational inequalities one mentions here applications in fluid mechanics Oden and Kikuchi (1980), in traffic equlibria Dafermos (1980) and in economics Nagurney (1994) and Nagurney and Siokos (1999).

2.4 Applications in Engineering Modeling

Unilateral Contact From the general modeling of a structure, either from a finite element method or by using a boundary element technique, linear or nonlinear relations between boundary displacements $\mathbf{u}$ and tractions $\mathbf{t}$ can be written. The classical case is that one has given boundary displacements on a part of the boundary and given loading on the complementary part of it, i.e., some elements of vector $\mathbf{u}$ and the complementary elements of vector $\mathbf{t}$ are known. One then solves the resulting system of equations for the unknown variables. The complete description of unilateral contact, which involve no-penetration inequalities, no-tensile contact stresses inequalities and complementarity relations which exclude, pointwise, the one of the previous two cases, reads

$$\mathbf{y}_n = \mathbf{d} - \mathbf{u}_{cn} \geq \mathbf{0},\ \mathbf{t}_{cn} \geq \mathbf{0},\ \mathbf{y}_n^T \mathbf{t}_{cn} = 0. \tag{55}$$

Here, $\mathbf{u}_{cn}$ is the subvector of $\mathbf{u}$ with the displacements normal to the boundary and $\mathbf{t}_{cn}$ the corresponding subvector of tractions $\mathbf{t}$, for the potential contact boundary. Normal boundary quantities are expressed in an appropriate local coordinate system such that the outward direction is positive. Moreover, $\mathbf{d}$ is the vector of initial gaps (e.g., initial distance from a rigid support or an initial crack opening).

The set of initial (linear or nonlinear) relations between boundary displacements and tractions, and (55) constitutes a linear or nonlinear complementarity problem (LCP resp. NCP). The solution of this problem gives automatically the contact-separation areas and the corresponding contact forces.

From another point of view the unilateral contact mechanism can be seen as a multivalued, monotone contact law:

$$-t_n = \begin{cases} 0, & \text{for } u_{cn} \leq d, \\ [0, +\infty], & \text{for } u_{cn} = d. \end{cases} \tag{56}$$

The latter relation can be produced by subdifferentiating the following potential function, which has the form of an indicator function $I_{U_{ad}}$ for the set of kinematically admissible displacements: $U_{ad} = \left\{ u_{cn} \in R^1, u_{cn} - d \leq 0 \right\}$,

$$\phi_n(u_{cn}) = I_{U_{ad}}(u_{cn}) = \begin{cases} 0, & \text{for } u_{cn} \leq d, \\ +\infty, & \text{for } u_{cn} = d. \end{cases} \tag{57}$$

Thus, one has the subdifferential unilateral law

$$-t_{cn} \in \partial I_{U_{ad}}(u_{cn}). \tag{58}$$

One should keep in mind that in mechanics, the directional derivative mentioned there plays the role of the virtual work (for a small virtual displacement or deformation δu). In this framework, the unilateral contact relation (58) is connected to the following variational inequality

$$-t_n \delta u \leq I_{U_{ad}}(u_{cn} + \delta u) I_{U_{ad}}(u_{cn}), \forall \delta u \in R^1, \tag{59}$$

or, by taking into account the definition of the indicator function, equivalently, to

$$-t_n \delta u \leq 0, \forall \delta u \in U_{ad}. \tag{60}$$

The above variational inequalities describe local nonlinear effects which are expressed by means of monotone, possibly multivalued, subdifferential laws (like the ones used here for unilateral contact problems and in the later for more general multivalued boundary or constitutive relations). They can be integrated along the whole nonlinear boundary, introduced into the principle of virtual work of mechanics and they lead to variational formulations of structural analysis problems which have the form of variational inequalities.

The formulation of the unilateral contact interactions in the previous set of complementarity relations, or in the form of a set- valued subdifferential relation is best suited for use within the framework of energy methods in mechanics. This is the case where a structural analysis problem is seen as a minimization of the potential energy of the structure. For small displacement, small deformation linear elastostatics, the potential energy function is quadratic. The noncompatibility relations of unilateral contact introduce linear inequality restrictions in this optimization problem. The linear complementarity problem, which has been constructed here directly, arises then naturally from the Karush-Kuhn-Tucker optimality conditions of the inequality constrained quadratic programming problem. This point is discussed further at the end of the present section. Recent review articles which provide further information, and appropriate solution algorithms borrowed from the area of numerical optimization include Mistakidis and Stavroulakis (1998), Klarbring (1999), Mijar and Arora (2000a) and Mijar and Arora (2000b).

Frictional Stick-Slip Effects A 'static Coulomb friction' model, which is analogous to the holonomic Hencky elastoplasticity, is used here in order to demonstrate the nonsmooth modeling of the frictional stick-slip effect. This permits us to follow the discussion of the previous section, cf. relations (58), and, in particular, to continue using the convex analysis setting introduced there. On the other hand, this simplification is in accordance with the used practice which is based on the solution of a quasistatic problem within each time step.

The assumed friction law has the form (for an one- dimensional friction element within a two-dimensional mechanical model)

$$\begin{aligned} u_{ct} &= 0 \text{ iff } t_{ct} \leq \mu \mid t_{cn} \mid, \\ u_{ct} &\geq 0 \text{ for } -t_{ct} = \mu t_{cn}, \\ u_{ct} &\leq 0 \text{ for } -t_{ct} = -\mu t_{cn}, \end{aligned} \tag{61}$$

For a given contact limit traction t_{cn}, one writes the subdifferential law

$$-t_{ct} \in \partial\phi_t(u_{ct}; t_{cn}), \tag{62}$$

where the following convex and nonsmooth, max-type subdifferential is considered (as a function of the first argument, parametrized by the second argument)

$$\phi(u_{ct}; t_{cn}) = \max\{\mu t_{cn}, -\mu t_{cn}\}\,. \tag{63}$$

The connection with variational inequalities, cf. (59), is straightforward. In the general case of coupled contact and friction, the contact traction t_{cn} is not known apriori. Therefore, due to the implicit dependence on t_{cn}, one gets implicit variational inequalities which are connected with quasivariational inequalities (see, for example, Telega (1988) or Baiocchi and Capelo (1984) for more details in this area).

Let us add here that for coupled unilateral contact and frictional problems there does not exist a potential function from which the formulation of the problem can be derived. This is well-known from analogous, nonassociated plasticity theories. To resolve this problem a class of iterative algorithms has been proposed, where separate contact and friction subproblems are formulated and solved within each iteration step. Clearly these problems are of the potential form studied here. Updating between various steps is done in various ways (see, for example, initial publications outlined in Panagiotopoulos (1985) and subsequent investigations in Mistakidis and

Stavroulakis (1998), Bisegna et al. (2001)). Another affiliated interesting proposal is the so-called bipotential problem approach (e.g. Saxce and Feng (1998)).

Let us finally mention that by using appropriate slack variables one can write the law of (61) in the primitive complementarity form used for the contact relation (55):

$$\begin{aligned} u_{ct} &= \lambda_1 - \lambda_2, \\ -t_{ct} - \tau_2 &= -\mu t_{cn}, \\ -t_{ct} + \tau_1 &= \mu t_{cn}, \\ \lambda_1 \geq 0,\ \tau_1 \geq 0,\ \lambda_1\tau_1 &= 0, \\ \lambda_2 \geq 0,\ \tau_2 \geq 0,\ \lambda_2\tau_2 &= 0. \end{aligned} \tag{64}$$

More complicated, monotone laws are easily formulated within the previous framework. Multi-dimensional laws can be formulated by following the example of the plasticity or elasto-plasticity with hardening. In particular, frictional laws for three-dimensional problems which involves the Coulomb friction cone lead to nonlinear complementarity problems, as it is discussed in the literature (see, e.g., Al-Fahed and Stavroulakis (2000)). Analogous considerations hold for nonmonotone laws, as it will be discussed in the next section for a model delamination problem.

Variational formulation of general subdifferential laws Let us assume a monotone possibly multivalued (i.e., with complete vertical branches) relation (a law) between the quantities u and $-f$. To be more precise, one may think about a nonlinear boundary law which connects boundary reactions $-f$ with boundary displacements u in mechanics. Let a convex l.s.c. and proper function Φ exists, the convex superpotential in the sense of J.J. Moreau, and that the previously mentioned law is written in the subdifferential form:

$$-f \in \partial\Phi(u). \tag{65}$$

Here ∂ denotes the subdifferential of convex analysis. Function $\Phi(u)$ can be considered as the potential energy corresponding to the mechanical law (65). By definition, (65) is equivalent to the following variational inequality:

$$\Phi(u^\star) - \Phi(u) \geq -\langle f, u^\star - u\rangle, \quad \forall u^\star \in R. \tag{66}$$

For example, if Φ is the indicator I_K of a convex closed interval K of R, then one has

$$-f \in \partial I_K(u). \tag{67}$$

This is a unilateral constraint as one easily recognizes by considering the equivalent variational inequality (for $u \in K$):

$$\langle f, u^\star - u\rangle \geq 0, \quad \forall u^\star \in K. \tag{68}$$

Indeed, if $u^\star - u$ is an admissible variation of u (in the sense that it satisfies (68), then the same does not hold for the variation $u - u^\star$. Of course in the one-dimensional case $< \cdot, \cdot >$ is a simple multiplication. For multidimensional problems it will be an inner product.

Elastoplasticity The relations of elastoplasticity are based on the assumption that (a) stresses are restricted within a (usually convex) set of elastic behaviour and (b) plastic yield appears if the boundary of the previous set (yield locus) is attained. Plastic strain (or strain rate) may be produced from a separate potential. The either-or question arises at this boundary is similar to the one arising in unilateral contact mechanics. In fact it leads to similar complementarity problems. The link between convex analysis and elastoplasticity is well known from eary works of Profs. J.J. Moreau and G. Maier. Technical details can be found in Duvaut and Lions (1972), Panagiotopoulos (1985), Sewell (1987), Maier and Novati (1990), Gao (2000), Mistakidis and Stavroulakis (1998), Christensen (2000), Harker and Pang (1990), Demyanov et al. (1996), Tin-Loi and Xia (2001).

Adhesive Contact Let us consider the following one-dimensional law which models delamination effects with complete destruction for both compression and tension (or, equivalently expressed, perfect damage for crushing and cracking). For notational simplicity the same traction limit t_1 and delamination yield displacement u_1 are assumed. The nonmonotone law with complete vertical (i.e., falling) branches reads

$$-t_{cn} = \begin{cases} 0 & \text{for } u_{cn} < -u_1, \\ [0, t_1] & \text{for } u_{cn} = -u_1, \\ c_1 = \frac{-t_1}{u_1} u_{cn} & \text{for } -u_1 < u_{cn} < u_1, \\ [-t_1, 0] & \text{for } u_{cn} = u_1, \\ 0 & \text{for } u_{cn} > u_1. \end{cases} \tag{69}$$

This law can be derived by differentiating the following nonsmooth and nonconvex potential, which can be expressed as a difference of convex functions (i.e., it is a so-called d.c. function), i.e.,

$$\begin{aligned} \overline{\phi}_n(u_{cn}) &= \min\left\{\frac{1}{2} c_1 {u_{cn}}^2, \frac{1}{2} c_1 u_1^2\right\} \\ &= \frac{1}{2} c_1 u_{cn}^2 - \left\{ \begin{matrix} 0, & \text{for } u_{cn} < u_1, \\ s_1 u_{cn} + \frac{1}{2} c_1 (u_{cn} - u_1)^2, & \text{for } u_{cn} > u_1 \end{matrix} \right\} \\ &= \phi_1(u_{cn}) - \phi_2(u_{cn}). \end{aligned} \tag{70}$$

As it has already been mentioned, the boundary law (69) can be expressed as a complementarity law by means of appropriate slack variables (see, Bolzon et al. (1997b), for recent work in this direction). On the other hand, in previous works of the first author (see, Panagiotopoulos and Stavroulakis (1992); Stavroulakis (1993); Mistakidis and Stavroulakis (1998)) the difference convex structure of (70) is exploited so that the law (69) can be written as

$$\begin{aligned} -t_{cn} &= w_1 - w_2, \\ w_1 \in \partial \phi_1(u_{cn}), &\; w_2 \in \partial \phi_2(u_{cn}). \end{aligned} \tag{71}$$

The main advantage of using structured (i.e., the d.c. structure) nonconvex approach becomes obvious if one considers the energy optimization approach to nonsmooth mechanics. In that case, elements of d.c. optimization theory and algorithms can be used for the effective solution of nonconvex problems. The links with nonconvex variational inequalities and with hemivariational inequalities, as well as with the quasidifferentiability concept of nonsmooth optimization are briefly addressed in in the next two sections. For more information the reader should consult the references given there.

It should be noted that the d.c. approach of this section can be used for the modeling of other nonmonotone laws in mechanics, like the slip-dependent frictional effects (the so- called Stribeck curve, see Pfeiffer and Glocker (1996), pages 5 and 236) or constitutive laws of layers made of shape memory alloys (see, e.g., Pagano and Alart (1999), or the paper Polyakova and Stavroulakis (2000) which are of importance for smart materials and structures.

Variational formulation of quasidifferential laws Let us assume now a nonmonotone possibly multivalued relation. By means of a real-valued, quasidifferentiable superpotential energy function Φ, one may expresses this relation in the form:

$$\begin{aligned} -f &= w_1 + w_2, \\ \text{with } \{w_1, w_2\} &\in D\Phi(u) = [\underline{\partial}\Phi(u), \overline{\partial}\Phi(u)] \end{aligned} \tag{72}$$

By definition, the previous relation is equivalent to:

$$\begin{aligned} \langle -f, u^* - u \rangle = &\max_{w_1^* \in \underline{\partial}\Phi(u)} \langle w_1^*, u^* - u \rangle + \\ &\min_{w_2^* \in \overline{\partial}\Phi(u)} \langle w_2^*, u^* - u \rangle, \ \forall u^* \in U, \end{aligned} \tag{73}$$

for $u \in U$, or with the system of variational inequalities

$$\langle -f, u^* - u \rangle - \langle w_2^*, u^* - u \rangle \leq \max_{w_1^* \subset \underline{\partial}\Phi(u)} \langle w_1^*, u^* - u \rangle,$$

$$\forall u^* \in U, \ \forall w_2^* \in \overline{\partial}\Phi(u), \tag{74}$$

and

$$\langle -f, u^* - u \rangle - \langle w_1^*, u^* - u \rangle \geq \min_{w_2^* \in \overline{\partial}\Phi(u)} \langle w_2^*, u^* - u \rangle,$$

$$\forall u^* \in U, \ \forall w_1^* \in \underline{\partial}\Phi(u). \tag{75}$$

Space U is in general a subspace of R^n and depends on the considered application.

Analogously, one treats multi-dimensional relations (for example, boundary adhesive layers) or constitutive laws (e.g., materials with softening effects). A number of concrete examples have been given in Chapter 3 of Demyanov et al. (1996).

Optimization and Contact Analysis A simple demonstration of the links between optimization (with inequality constraints), complementarity problems and the more general nonsmooth and possibly nonconvex optimization with mechanics. In mechanics, a frictionless unilateral contact problem for a linear elastic structure in the framework of small displacements and deformations which is discretized by the finite element method, is formulated as the minimum of the potential energy function. Additional inequality constraints which arise from the unilateral contact (kinematic nonpenetration) relations must be taken into account as well. Thus, the problem reads: find displacements $u \in U_{ad}$ such that

$$u = \arg \min_{v \in U_{ad}} \left\{ \Phi(v) = \frac{1}{2} v^T K v + p^T v \right\}, \tag{76}$$

where, for instance, the set of admissible displacements reads $U_{ad} = \{u \in \mathbb{R}^n : Au - b \leq 0\}$, K is the stiffness matrix of the discretized structure, u is the discrete nodal displacement vector of the finite element assemblage, p is the loading vector and $\Phi(u)$ is the potential energy of the structure. Moreover, n is the number of displacement degrees of freedom of the finite element model. By writing down the solvability relations of the linearly constrained quadratic programming minimization problem (76), the following mixed linear complementarity problem arises:

$$\begin{aligned} &Ku + A^T \lambda = p, \\ &\quad Au - b \leq 0, \ \lambda \geq 0, \ \lambda^T (Au - b) = 0. \end{aligned} \tag{77}$$

Here, the vector of discrete contact forces λ (i.e., the Lagrange multipliers) appears.

Further elaboration of the problem (77) (static condensation of the displacement variables u outside of the contact nodes) and the introduction of a set of nonnegative slack variables y, leads to the following standard linear complementarity problem (LCP):

$$\begin{aligned} &y = b - Au = -AK^{-1}A^T \lambda + b - AK^{-1} p, \\ &y \geq 0, \ \lambda \geq 0, \ y^T \lambda = 0. \end{aligned} \tag{78}$$

Here, the additional assumption that K is invertible, i.e., that no rigid body displacements or rotations arise, has been done for simplicity (cf. Stavroulakis et al. (1991), Goeleven et al. (1997) for more general cases). Further material on the links between mathematical programming and optimization with unilateral contact analysis can be found in Klarbring (1999), Curnier (1999)

3 BVPs in Nonsmooth Mechanics

3.1 Variational Inequalities

Let us first consider abstract variational formulations of a boundary value problem which is defined in a subset Ω of R^n, $n = 1, 2, \ldots, n$ with boundary Γ. Let V be a real Hilbert space and V' be its dual space. Let $a(\cdot, \cdot) \colon V \times V \to R$ be a symmetric, continuous and coercive bilinear form and $(l, \cdot)$ be a continuous linear form on V. An abstract variational problem reads: find $u \in V$ such that

$$a(u, v - u) = (l, v - u), \forall v \in V. \tag{79}$$

Let moreover K be a closed convex subset of V and assume that a solution of the boundary value problem within the set K is sought. It can be shown that this solution is characterized by the following abstract variational inequality (of the G. Fichera type, according to Panagiotopoulos (1985), p. 188): find $u \in K \subset V$ such that

$$a(u, v - u) \geq (l, v - u), \forall v \in K. \tag{80}$$

For a convex, l.s.c. proper functional Φ on V one may define the more general variational inequality (Brezis (1972)): find $u \in V$ such that

$$\alpha(u, v) + \Phi(v) - \Phi(u) \geq (l, v - u), \forall v \in V. \tag{81}$$

It is obvious that (80) is a special case of (81), with $\Phi = I_K$, where the indicator function of the set K is defined by $I_K(v) = \{0 \text{ if } v \in K, +\infty \text{ otherwise}\}$.

Hemivariational Inequalities Let moreover j: $R \to R$ denotes a locally Lipschitz function and let $j^0(u, v-u)$ denotes the generalized gradient of the nonconvex and nonsmooth function j. By using the previously given definition, we arrive at the hemivariational inequality problem: find $u \in V$ such as to satisfy the inequality

$$a(u, v - u) + \int_\Omega j^0(u, v - u) d\Omega \ \geq \ (l, v - u), \ \forall v \in V. \tag{82}$$

3.2 Implicit variational inequalities and quasivariational inequalities

If one assumes for instance that the linear form $(l, \cdot)$ or the set K in the previous relations depend on the solution u, one gets various types of implicit variational inequalities or quasi-variational inequalities.

Let the set K depend on the solution u. Then from (80) one gets the quasivariational inequality: find $u \in K(u) \subset V$ such that

$$\alpha(u, v) \geq (l, v - u), \ \forall v \in K(u).$$

Along the same lines one formulates from (81) the implicit variational inequality: find $u \in V$ such that

$$\alpha(u, v) + \Phi(u; v) - \Phi(u; u) \geq (l, v - u), \ \forall v \in V.$$

Here the first argument in $\Phi(\cdot, \cdot)$ is tackled as a parameter. An application in mechanics is given in the next section.

Finally, in analogy to the previous extensions, for a continuous mapping $h(u)$ the following quasi-hemivariational inequality (which may also be characterized as implicit hemivariational inequality) problem can be written (see Naniewicz and Panagiotopoulos (1995), p. 128): find $u \in V$ such as to satisfy the inequality

$$a(u, v - u) + h(u) j^0(u, v - u) d \ \geq \ l(v - u), \forall v \in V. \tag{83}$$

3.3 Mechanical example: coupled unilateral contact problem with friction

Let $\Omega \in R^3$ be an open bounded subset occupied by a deformable body in its undeformed state. On the assumption of small deformations one writes the virtual work relation (for $u \in V$)

$$\int_\Omega \sigma_{ij} \varepsilon_{ij}(v - u) d\Omega = \int_\Omega f_i(v_i - u_i) d\Omega + \int_\Gamma S_N(v_N - u_N) d\Gamma + \int_\Gamma S_{T_i}(v_{T_i} - u_{T_i}) d\Gamma \qquad \forall v \in V. \tag{84}$$

Here V denotes the function space of the displacements, which in general is an appropriate subset of $H^1(\Omega)$ and f_i, S_N, $S_{T_i} \in L_2(\Gamma)$. Recall here that the abstract bilinear form $\alpha(\cdot,\cdot)$ in this case of linear elasticity reads:

$$\alpha(u,v) = \int_\Omega C_{ijhk}\varepsilon_{ij}(u)\varepsilon_{hk}(v)d\Omega. \tag{85}$$

Here $C = \{C_{ijhk}\}$, $i,j,h,k = 1,2,3$, is the elasticity tensor which satisfies the well-known symmetry and ellipticity properties.

The underlying elastostatic equilibrium equation boundary value problem has the form:

$$\sigma_{ij,j} + f_i = 0, \tag{86}$$

where the f_i is the volume force vector. One recalls here the strain-displacement relation (small deformation theory):

$$\varepsilon_{ij} = \frac{1}{2}(u_{i,j} + u_{j,i}). \tag{87}$$

The material follows a linearly elastic constitutive law:

$$\sigma_{ij} = C_{ijhk}\varepsilon_{hk} \tag{88}$$

Recall here that on the assumption that classical support conditions hold on Γ (i.e., say $u_N = 0$ and $u_{T_i} = 0,\ i = 1,2,3$) one gets the following variational equality: find $u \in V_0 = \{v \big| v \in V$, with $v_N = 0$, and $v_{T_i} = 0$ on $\Gamma\}$ such that

$$\alpha(u,v) = \int_\Omega f_i v_i d\Omega \quad \forall v \in V_0. \tag{89}$$

Signorini-Coulomb unilateral frictional contact Let us assume the pointwise unilateral contact relations (known as Signorini condition, for the frictionless unilateral contact case):

$$-S_N \geq 0,\ u_N - g \leq 0,\ \ -S_N(u_N - g) = 0, \text{ in } \Gamma. \tag{90}$$

Here, the inequalities on the boundary tractions correspond to the mechanical restriction that no tensile tractions are permitted. Moreover, the normal boundary displacements are not permitted to be higher than a given initial distance g, which means that no penetration is allowed. Finally, the complementarity relation expresses the physical fact that either contact is realized or a separation takes place.

A simplified static version of the Coulomb's friction law connects the tangential (frictional) forces S_{T_i} with the normal (contact) forces S_N by the relation

$$\gamma = \mu|S_N| - |S_T|,\ \geq 0 \tag{91}$$

Here $| * |$ denotes the norm in R^3 and μ is the friction coefficient. The friction mechanism is considered to work in the following way: If $|S_T| < \mu|S_N|$ (i.e. $\gamma > 0$) the slipping value γ must be equal to zero and if $|S_T| = \mu|S_N|$ (i.e. $\gamma = 0$) then we have slipping in the opposite direction of S_T. Explicitly we have:

$$\begin{array}{ll} \text{If } \gamma > 0 \text{ then} & y_T = 0 \\ \text{if } \gamma = 0 \text{ then} & \text{there exists } \sigma > 0 \\ & \text{such that } y_{Ti} = -\sigma S_{Ti} \end{array} \tag{92}$$

where $i = 1, 2, 3$ indicate the components of vector S_T with respect to a reference Cartesian coordinate system.

The contact law (90) can be written in the superpotential form:

$$-S_N \in \partial \mathbf{I}_{U_{ad}}(u_N) = \partial \Phi_N(u) = \mathcal{N}_{U_{ad}}(u_N). \tag{93}$$

Here, the following set of admissible displacements is introduced:

$$U_{ad} = \{u \in V \mid u_N - g \leq 0\} \tag{94}$$

and the notions of the convex analysis subdifferential and of the normal cone to a set have been used. The corresponding variational inequality reads

$$-S_N(u_N)(v_N - u_N) \leq 0, \ \forall v_N \in U_{ad} \tag{95}$$

For the friction law one writes, analogously:

$$-S_T \in \partial_{u_N}\left(\mu\,|S_N|\,|u_N|\right) = \partial \Phi_T(S_N; u_T), \tag{96}$$

where the involved potential is nondifferentiable (due to the absolute value nonlinearity of $|u_N|$) and depends on the normal contact traction S_N, thus, implicitly, on the solution of the problem u, i.e. one considers the parametrized potential $\Phi_T(u; u_T) = \Phi_T(S_N; u_T) = \mu\,|S_N|\,|u_N|$. The corresponding variational inequality reads

$$-S_T(u_T)(v_T - u_T) \leq \Phi_T(u; v_T) - \Phi_T(u; u_T), \ \forall v_T \in U_{ad}. \tag{97}$$

By combining relations (84), (86) and (93) one gets the implicit variational inequality: find $u \in U_{ad}$ such that

$$\int_\Omega \sigma_{ij}\varepsilon_{ij}(v - u)d\Omega + \Phi_T(u; v_T) - \Phi_T(u; u_T) \geq \int_\Omega f_i(v_i - u_i)d\Omega \ \forall v \in U_{ad}. \tag{98}$$

A dual problem in terms of stresses provides us a corresponding quasi-variational inequality problem. For simplicity, a two-dimensional problem is considered further. Moreover, the following set of admissible boundary tractions for the Signorini-Coulomb unilateral contact problem is assumed:

$$S_{ad} = \left\{(S_N, S_T) \mid g_j(S_N, S_T) \leq 0, j = 1, 2\right\}, \tag{99}$$

where the constraint functions have the form:

$$\begin{aligned} g_1(S_N, S_T) &= \mu S_N - S_T, \\ g_1(S_N, S_T) &= \mu S_N + S_T. \end{aligned}$$

Moreover one needs the set of admissible stresses (which include the boundary tractions): $\Sigma(\sigma) = \{\sigma \text{ such that } \sigma_{ij,j} + f_i = 0\} \cap S_{ad}$. It may be shown that in this case the previous problem is expressed in the form of the quasi-variational inequality: find $\sigma \in \Sigma(\sigma)$ such that

$$\int_{\Omega} \varepsilon_{ij}(\tau_{ij} - \sigma_{ij})d\Omega \geq 0, \forall \tau \in \Sigma(\sigma)$$

3.4 Multivalued, monotone laws and variational inequalities

Let us assume now that on Γ the general monotone multivalued boundary condition

$$-S \in \partial j(u), \tag{100}$$

holds. Here $j(u)$ is assumed to be a convex superpotential and ∂ denotes the subdifferential of convex analysis. Moreover, all (normal and tangential) contributions of boundary displacements u and tractions S are included in (100), which holds as a multi-dimensional boundary condition at each point of the boundary Γ. Relation (100) is, by definition of the subdifferential, equivalent to:

$$j(v) - j(u) \geq -S_i(v_i - u_i), \ \forall v = \{v_i\} \in R^3. \tag{101}$$

By using (101) one gets the variational inequality: find $u \in V$ with $j(u) < \infty$, such that

$$\alpha(u, v-u) + \int_{\Gamma} (j(v) - j(u))d\Gamma \geq \int_{\Omega} f_i(v_i - u_i)d\Omega, \tag{102}$$

$\forall v \in V$ with $j(v) < \infty$.

3.5 Multivalued, nonmonotone laws and hemivariational inequalities

In this case the basic building element is the definition of boundary conditions and material laws based on Clarke subdifferential. For instance, let on Γ the nonmonotone, possibly multivalued boundary condition

$$-S \in \partial_{CL} j(u) \tag{103}$$

holds, where j is a locally Lipschitz superpotential functional. By using the inequality

$$j^0(u, v-u) \geq -S_i(v_i - u_i), \ \forall v = \{v_i\} \in R^3, \tag{104}$$

which defines on Γ the condition (103), one gets the following hemivariational inequality: Find $u \in V$ such as to satisfy the inequality

$$\alpha(u, v-u) + \int_{\Gamma} j^0(u, v-u)d\Gamma \geq \int_{\Omega} f_i(v_i - u_i)d\Omega \quad \forall v \in V. \tag{105}$$

If instead of (103) one assumes on Γ that:

$$-S_N \in \partial_{CL} j_N(u_N), \ -S_T \in \partial_{CL} j_T(u_T) \tag{106}$$

then one gets analogously the hemivariational inequality:
Find $u \in V$ such that

$$\alpha(u, v-u) + \int_\Gamma j_N^0(u_N, v_N - u_N)d\Gamma \tag{107}$$

$$+ \int_\Gamma j_T^0(u_T, v_T - u_T)d\Gamma \geq \int_\Omega f_i(v_i - u_i)d\Omega \quad \forall v \in V.$$

The last type of variational expressions involving $j^0(\cdot,\cdot)$ or $j_N^0(\cdot,\cdot)$ and $j_T^0(\cdot,\cdot)$ have been called hemivariational inequalities by the P.D. Panagiotopoulos, who introduced and studied them in mechanics Panagiotopoulos (1985), Panagiotopoulos (1993), Naniewicz and Panagiotopoulos (1995). Note that in the more general case in which j or j_N and j_T are not locally Lipschitz $j^0(\cdot,\cdot)$ in (105) and $j_N^0(\cdot,\cdot), j_T^0(\cdot,\cdot)$ in (108) are replaced by $j^\uparrow(\cdot,\cdot)$ and $j_N^\uparrow(\cdot,\cdot), j_T^\uparrow(\cdot,\cdot)$. Moreover a combination of monotone subdifferential laws and nonmonotone laws (cf. (103)) for different (nonoverlapping) parts of the boundary Γ is possible. One then gets variational-hemivariational inequality problems.

The solution of variational problems, like the variational equalities, or the hemivariational inequalities derived previously, satisfies the operator equations of the problem, e.g. the equation of equilibrium, and the boundary conditions of the problem in a weak sense. This means, roughly speaking, that these relations are satisfied in an integral form and not pointwise. Analogous considerations are familiar within the weak formulations used in the finite element method.

3.6 QD Laws and Systems of Variational Inequalities

Let us assume that on Γ the nonmonotone, possibly multivalued boundary condition (73) holds, where Φ is a quasidifferentiable functional. It has the form

$$-S = w_1 + w_2\ , \ \text{with}\{w_1, w_2\} \in Dj(u) = \{\underline{\partial}\Phi(u), \bar{\partial}\Phi(u)\}. \tag{108}$$

Then one has, by definition, the relation (74), where $\Phi'(u,v) = \langle -S, v\rangle$. Finally, by an analogous way, one has the variational problem: find $u \in V,\ \ w_1, w_2 \in W$ such as to satisfy the relation

$$\alpha(u, v-u) - \int_\Omega f_i(v_i - u_i)d\Omega + \max_{w_1^*(x)\in\underline{\partial}\Phi(u(x))\ a.e.\ on\ \Gamma} \langle w_1^*, v-u\rangle +$$

$$+ \min_{w_2^*(x)\in\bar{\partial}\Phi(u(x))\ a.e.\ on\ \Gamma} \langle w_2^*, v-u\rangle = 0, \forall\, v \in V. \tag{109}$$

The function spaces V and W depend on the studied application. For instance, for three dimensional elastostatics the following choice has been proposed in Demyanov et al. (1996): $V = [H^1(\Omega)]^3$, $W = [L^2(\Gamma)]^3$. A more general formulation, also proposed in the previously given original publication, would be to assume that $w_1, w_2 \in [H^{-1/2}(\Gamma)]^3$. Then in the left hand side of (109) one should replace $w_1(x) \in \bar{\partial}\Phi(u(x))$ a.e. on Γ by $w_1 \in \bar{\partial}F(u)$ and $w_2(x) \in \underline{\partial}\Phi(u(x))$ a.e. on Γ by $w_2 \in \underline{\partial}F(u)$, where one assumes that

$$F(u) = \begin{cases} \int_\Gamma \Phi(u(x))d\Gamma, & \text{if } \Phi(\cdot) \in L^2(\Gamma) \\ \infty, & \text{otherwise.} \end{cases} \tag{110}$$

Then instead of (109) one has the following problem: find $u \in [H^1(\Omega)]^3,\ w_1, w_2 \in [H^{-1/2}(\Gamma)]^3$ such that

$$\alpha(u, v-u) - \int_\Omega f_i(v_i - u_i)d\Omega + \max_{w_1^*}\{\langle w_1^* + w_2^*, v-u\rangle \mid w_1^* \in \underline{\partial}F(u)\} +$$

$$+ \min_{w_2^*}\{\langle w_1^* + w_2^*, v-u\rangle \mid w_1^* \in \underline{\partial}F(u)\} = 0, \forall\, v \in [H^1(\Omega)]^3. \tag{111}$$

One should mention that the related questions concerning the extension of QD-superpotentials to function spaces remain still open.

Moreover we can write the min–max form which reads: find $u \in [H^1(\Omega)]^3$ such as to satisfy the relation:

$$\alpha(u, v-u) - \int_\Omega f_i(v_i - u_i)d\Omega + \min_{w_2^* \in \underline{\partial}F(u)} \max_{w_1^* \in \overline{\partial}F(u)} \{\langle w_1^* + w_2^*, v-u\rangle = 0,$$

$$\forall\, v \in [H^1(\Omega)]^3.$$

If in particular the superpotential F can be expressed as the difference of two convex functions, i.e. if $F = \Phi_1 - \Phi_2$, with Φ_1 and Φ_2 convex, then one has

$$\underline{\partial}F = \partial\Phi_1, \quad \overline{\partial}F = -\partial\Phi_2,$$

where ∂ is the subdifferential of the convex analysis. In this case the following system of variational inequalities results, as it can easily be shown by using the definition of the subdifferential: find $u \in [H^1(\Omega)]^3$, such as to satisfy

$$\alpha(u, v-u) - \int_\Omega f_i(v_i - u_i)d\Omega - \langle w_2^\star, v-u\rangle + \Phi_1(v) - \Phi_1(u) \geq 0,$$

$$\forall\, v \in [H^1(\Omega)]^3 \tag{112}$$

for all $w_2^\star \in [H^{-1/2}(\Gamma)]^3$ such that

$$\langle w_2^\star, v-u\rangle \leq \Phi_2(v) - \Phi_2(u),\ \forall\, v \in [H^1(\Omega)]^3. \tag{113}$$

4 Discretized Problems

The solution of the previously outlined variational problems can be approximated numerically, after discretization, by using finite element or boundary element techniques. As an example, we present here the formulation of discretized hemivariational inequalities for nonlinear material laws in a matricial form, which is more familiar to the matrix structural analysis community. More details can be found in Panagiotopoulos (1993), Antes and Panagiotopoulos (1992), Mistakidis and Stavroulakis (1998), Haslinger et al. (1999).

As an example, let us consider the static analysis of an elastic structure with both classical, linearly elastic and degrading elements within the finite element / matrix structural analysis

framework.
The stress equilibrium equations read:

$$\overline{G}\overline{s} = \begin{bmatrix} G \; G_n \end{bmatrix} \begin{bmatrix} s \\ s_n \end{bmatrix} = p \tag{114}$$

where $\overline{G}$ is the equilibrium matrix of the discretized structure which takes into account the stress contribution of the linear s and nonlinear s_n elements and p is the loading vector.

The strain–displacement compatibility equations take the form:

$$\overline{e} = \begin{bmatrix} e \\ e_n \end{bmatrix} = \overline{G}^T u = \begin{bmatrix} G^T \\ G_n^T \end{bmatrix} u, \tag{115}$$

where e, u are the deformation and displacement vectors respectively.

The linear material constitutive law for the structure reads:

$$s = K_0(e - e_0). \tag{116}$$

where K_0 is the natural and stiffness flexibility matrix and e_0 is the initial deformation vector.

The nonlinear material law is considered in the form:

$$s_n \in \partial_{CL}\phi_n(e_n), \tag{117}$$

where $\phi_n(\cdot)$, is a general nonconvex superpotential and summation over all nonlinear elements gives the total strain energy contribution of them as:

$$\Phi_n(e_n) = \sum_{i=1}^{q} \phi_n^{(i)}(e_n). \tag{118}$$

Classical support boundary conditions complete the description of the problem.

The discretized form of the virtual work equation reads:

$$s^T(e^* - e) + s_n^T(e_n^* - e_n) = p^T(u^* - u), \; u^*, e_n^*. \tag{119}$$

Entering the elasticity law (116) into the virtual work equation (119), and using (115) we get:

$$\begin{aligned} u^T G K_0^T G^T(u^* - u) - (p + GK_0 e_0)^T(u^* - u) + \\ + s_n^T(e_n^* - e_n) = 0, \; \forall u^* \in V_{ad}, \end{aligned} \tag{120}$$

where $K = G^T K_0^T G$ denotes the stiffness matrix of the structure, $\overline{p} = p + GK_0 e_0$ denotes the nodal equivalent loading vector and V_{ad} includes all support boundary conditions of the structure.

Further one considers the nonlinear elements (117) in the inequality form:

$$s_n^T(e_n^* - e_n) \leq \Phi_n^o(e_n^* - e_n), \; \forall e_n^* \tag{121}$$

where $\Phi_n^o(e_n^* - e_n)$ is the directional derivative of the potential Φ_n. Thus the following discretized hemivariational inequality is obtained:
Find kinematically admissible displacements $u \in V_{ad}$ such that

$$u^T K(u^* - u) - \overline{p}^T(u^* - u) + \Phi_n^o(u_n^* - u_n) \geq 0, \; \forall u^* \in V_{ad}. \tag{122}$$

Equivalently a substationarity problem for the total potential energy can be written:
Find $u \in V_{ad}$ such that:

$$\Pi(u) = \underset{v \in V_{ad}}{\text{stat}} \left\{\Pi(v)\right\}. \tag{123}$$

Here the potential energy reads $\Pi(v) = \frac{1}{2}v^T K v - \overline{p}^T v + \Phi_n(v)$, where the first two terms (quadratic potential) are well-known in the structural analysis community. The corresponding energy optimization problems or, in genera, stationarity problems, are analogous to the already simple examples of contact analysis.

5 Algorithms

For the numerical solution of the highly nonlinear problems of nonsmooth mechanics, in principle, a general-purpose, iterative algorithm, let us say of a Newton-type, can be used. Besides that nonsmoothness has to be taken into account, convergence is not guaranteed and, for nonconvex problems, one is not sure which solution has been found. Nevertheless this attempt is followed by many authors, since it is not far away from standard nonlinear computational mechanics techniques (see, g.e., Bathe (1996)).

For certain classes of problems, specialized algorithms can be constructed. A typical case is the unilateral contact phenomenon which, within a small deformation and displacement setting, leads to linear complementarity problems. Details can be found in the literature on contact analysis and its generalizations, see for example Brogliato (1999), Chabrand et al. (1998), Glocker (2001).

For general nonsmooth and possibly nonconvex problems a small number of specialized algorithms have been proposed, with variable efficiency. Every algorithm of this class depends on nonsmooth and nonconvex optimization techniques, see for example, Demyanov and Rubinov (1995), Czekanski and Meguid (2001). The variety of interesting phenomena which may arise in nonsmooth and nonconvex problems have been demonstrated on small scale problems by means of detailed calculations in Kurutz (1993) and Kurutz (1994).

5.1 Condensation

If nonsmooth effects are restricted on some boundaries or interfaces, standard static or dynamic substructuring (condensation) techniques can be applied for the drastic reduction of the problem. This reduction of dimensions is critical for the nonlinear applications studied.

5.2 Transofmation to CPs by Using Slack Variables. Lemke's LCP Algorithm

The complementarity problem formulation of unilateral effects provides a unified, general formulation which includes the either- if decisions of the corresponding nonlinear mechanisms. Instead of trying to solve the problem by adjusting incomplete equations by trial-and-error methods, one can use dedicated algorithms for the solution of the mechanical problem. They are iterative, due to the nonlinearity of the problem. Nevertheless, the large experience on solving those problems in other scientific disciplines can be used in a beneficial way.

Linear complementarity problems can be solved by simplex-type methods. The most known (and robust) algorithm is the Lemke's complementary pivoting algorithm (see Murty (1988) and the mechanical interpretation of the algorithm given in Stavroulakis et al. (1991)). Other techniques of nonsmooth programming are also appropriate. In order to avoid using very specialized algorithms, which make the production of general purpose software complicated, the use of special (smooth) penalty functions which transform the complementarity problem into a set of nonlinear equations is also very promising. This is a special case of the smoothing technique, which has been tested in various finite element contact mechanics analyses and is used in almost all contact analysis parts of large finite element method packages. The idea is discussed in some details in another part of this chapter. According to our experience Stavroulakis and Antes (2000), its use with the boundary element method is not always numerically efficient. The reason is that the arising system of nonlinear equations can not be solved with high accuracy, especially in the case of frictional contact problems. If one considers that smoothing, in every case, hides the either-if information of the unilateral mechanism, one should be very careful with the use of this method before sufficient numerical experience has been accumulated. On the other hand, its use in frictionless contact problems for (at least) two-dimensional applications, can be recommended.

For linear problems with unilateral contact joints with or without friction the use of slack variables has been demonstrated in the following publications: Al-Fahed and Stavroulakis (1991), Al-Fahed et al. (1992), Al-Fahed and Stavroulakis (2000).

5.3 NCP- and LCP- functions. Smoothing approach

The nonclassical character of the nonlinearities which can be described by relation (53), and especially the combinatorial character introduced by the last complementarity relation, requires the use of nonstandard solution techniques for the numerical treatment of problems which involve nonlinearities of this kind. In order to integrate the nonlinear relation (53) into a general purpose nonlinear equations solver and, thus, to allow the treatment of inequality mechanics' applications by classical, smooth computational mechanics' tools, a nonlinear equation reformulation of (53) is required. Thus, a function $\phi(x, F(x)) : R^2 \rightarrow R^1$ is sought for which the following equivalence is true: $\phi(x, F(x)) = 0 \Longleftrightarrow x, F(x)$ satisfy the relations (53). A function with this property is called a NCP-function. By this way, an underdetermined system of linear or nonlinear equations between variables x and $F(x)$ of (53) is derived, based on the physics of the studied problem. Adding one nonlinear NCP-function for each couple of complementary variables enforces the relations of (53) and makes the whole system of equations solvable.

For example, the following Fischer-Burmeister function (Fisher (1992))

$$\phi_{FB}(a, b) = \sqrt{a^2 + b^2} - (a + b). \tag{124}$$

can be used. More details on this direction of studying and solving complementarity problems can be found in Fisher (1992), Fukushima (1992), Facchinei et al. (1996), Kanzow (1994) and Kanzow (1996).

It is worth-mentioning that analogous smoothing techniques have been proposed in contact mechanics (in connection with finite element formulations). Details can be found in Klarbring and Bjoerkman (1988), Park and Kwack (1994), Bathe (1996) (page 628), Bathe and Bouzinov

(1997), Zavarise et al. (1998), Leung et al. (1998), Christensen et al. (1998), Christensen and Klarbring (1999). Static unilateral contact problems in two-dimensional elastic structures with cracks have been solved with this method and boundary elements by the authors, see Stavroulakis and Antes (2000). A few dynamic analysis results have been presented in Stavroulakis and Antes (1999).

Let us outline the application of this method here, by using as a model a frictionless contact problems formulated with the BEM. A schematic system of equations, between boundary displacements u_N, u_T, and boundary tractions r_N, r_T, for the unilateral contact boundary, and the other boundary unknowns x reads:

$$\begin{bmatrix} \mathbf{H}_{ff} & \mathbf{H}_{fN} & \mathbf{H}_{fT} \\ \mathbf{H}_{cf} & \mathbf{H}_{cN} & \mathbf{H}_{cT} \end{bmatrix} \begin{bmatrix} \mathbf{x} \\ \mathbf{u}_N \\ \mathbf{u}_T \end{bmatrix} = \begin{bmatrix} \mathbf{f}_f \\ \mathbf{f}_c \end{bmatrix} + \begin{bmatrix} \mathbf{G}_{fN} & \mathbf{G}_{fT} \\ \mathbf{G}_{cN} & \mathbf{G}_{cT} \end{bmatrix} \begin{bmatrix} \mathbf{r}_N \\ \mathbf{r}_T \end{bmatrix}, \tag{125}$$

Here, as usual in the classical boundary element method, vector x includes all unknown elements of boundary displacements or tractions at the classical boundary, i.e. the boundary with given support or external loading. Unilateral contact and no-friction assumption are introduced by the following LCP-functions:

$$\begin{aligned} \tilde{\phi}(\mathbf{u}_N, \mathbf{r}_N) &= \mathbf{0}, \\ \mathbf{r}_T &= \mathbf{0}. \end{aligned} \tag{126}$$

Here one has a system of $2n$ equations from BEM, n_c nonlinear equations for the unilateral contact part and n_c trivial equations for the known frictional tractions (set to zero, for the frictionless case). This system is solved for the $2(n - n_c)$ unknowns of vector $\mathbf{x}$ and of the unknowns of the vectors $\mathbf{u}_N$, $\mathbf{r}_N$, $\mathbf{u}_T$, each one being of dimension equal to n_c.

For frictional contact problems a simple way is the formulation of a LCP based on the slack variables technique, Al-Fahed and Stavroulakis (2000)). Then one may propose the use of implicit LCP-functions, schematically written as:

$$\tilde{\phi}_T(\mathbf{u}_T, \mathbf{r}_T, \mathbf{u}_N, \mathbf{r}_N) = \mathbf{0}. \tag{127}$$

Details can be found in Stavroulakis and Antes (2000).

6 Related Topics and Further Research

6.1 Existence of solution

Noncoercive problems arise in mechanics, for example, when the existing classical boundary conditions are not sufficient to prevent (even infinitesimal) rigid body displacements and rotations of the structure. In classical, equality mechanics, one may write conditions that guarantee the existence of a solution to the considered boundary value problem. Nevertheless, uniqueness of some quantities may be lost. One may consider, as an example, a free elastic body subjected to self-equilibrated external forces. In this case stresses and deformations of the elastic body can be determined, but its displacements are only determined modulo some rigid body displacements and rotations. In the presence of inequality (e.g., unilateral contact) constraints analogous relations have also been provided by G. Fichera (see, e.g., Panagiotopoulos (1985), Chapter 4). It is interesting to observe that some inequality constraints may be activated and stabilize the body. Unilateral supports can be realized by fingers in a robotic grasping application: according to their placement, they may or may not stabilize a given object, depending on the applied loading. In another context, a stone wall or a vault constructed from stones without mortar is stable due to the interplay between self-weight and unilateral (no-tension) interfaces. Solvability may pose restrictions on the intensity and direction of any additional loading, comming for instance from a traffic loading on a stone bridge or from an earthquake. This strong relation between between geometrical form and loading was quite well-known, at least to the masons of the middle-age (see, Leftheris (2004)).

Theoretically, one asks if a finite solution of the engineering problem exists for a given case. It is interesting that the complementarity problem formulation of unilateral problems allows for a systematic investigation of this question. Solvability conditions for linear complementarity problems have been investigated in Murty (1988), Goeleven et al. (1997). Analogous results exist if one considers the equivalent variational inequality or (if applicable) the energy minimization problem. In this respect the reader is referred to the classical work of Fichera (1972) (see also chapter four in Panagiotopoulos (1985)). An application in the area of robotic hands has been given in Al-Fahed and Stavroulakis (1991), Al-Fahed et al. (1992), Al-Fahed and Stavroulakis (2000) and Stavroulakis et al. (1991). In that case the authors were able to study, in a systematic way, the (previously known) notions of form and force grasping within a robotic hand, using solvability conditions of the unilateral contact problem. Other applications on problems or higher dimensions have recently been studied (see, the masonry wall of Ferris and Tin Loi (2001), Nappi and Tin Loi (2001), or the related work of Baggio (2000)).

Semicoercive variational and hemivariational inequalities Existence results for classical variational inequality problems relevant to contact analysis have been developed in Fichera (1972) and further discussed, among others, in Antes and Panagiotopoulos (1992), pp. 38-42. They can be used for the solution of form and force grasping problems in robotics (see, Al-Fahed and Stavroulakis (1991)) or yield and stability questions in masonry structures, as it has been discussed in other parts of this text.

For hemivariational inequalities analogous results have been obtained by P.D. Panagiotopoulos and his co-workers (see also Goeleven (1996)). One such result for an abstract hemivariational inequality is given here, without proof. Here $a(\cdot,\cdot)$ is assumed to be semicoercive, i.e., $a(\cdot,\cdot)$ is continuous and symmetric but it has a nonzero kernel, i.e. $\ker a(\cdot,\cdot)=\{q : a(q,q)=0\}\neq\{0\}$. Moreover let $\ker a$ be finite dimensional. Let us We assume that the norm $||v||$ on V is equivalent to $|||v|||=p(\tilde{v})+|q|_2$, where $v=\tilde{v}+q$, $q\in\ker a$, $\tilde{v}\in\ker a^{\perp}$ (i.e. $(v,q)=0\ \forall q\in\ker a$) and $p(\tilde{v})$ is a seminorm on V such that $p(v)=p(v+q)\ \forall v\in V$, $q\in\ker a$ and let $a(v,v)\ \geq\ c\big(p(v)\big)^2$, $\forall v\in V$, $c=const>0$. This semicoercivity inequality replaces (51 of the coercive case. Moreover let q_+ and q_- be the positive and the negative parts of $\hat{q}$, where $\hat{q}=Lq$, i.e. $q_+=\max\{0,\hat{q}\}$, $q_-=\max\{0,-\hat{q}\}$. The following quantities will also be used: $\beta(-\infty)=\limsup_{\xi\to-\infty}\beta(\xi)$ and $\beta(\infty)=\liminf_{\xi\to\infty}\beta(\xi)$. On the assumption that

$$\beta(-\infty)\ \leq\ \beta(\xi)\ \leq\ \beta(\infty),\qquad \forall\xi\in R, \tag{128}$$

a necessary condition for the existence of a solution $u \in V$ of a semicoercive hemivariational inequality problem is the following inequality:

$$\int_{\Omega} [\beta(-\infty)q_+ - \beta(\infty)q_-]d\Omega \leq (l, q) \qquad \leq \int_{\Omega} [\beta(\infty)q_+ - \beta(-\infty)q_-]d\Omega \;\; \forall q \in \ker a. \quad (129)$$

If (128) holds strictly (with $<$ instead of $\leq$) then (129) also holds strictly. For variational inequality problems of the first or second kind, relation (129) is simplified analogously. The details can be found for the first kind problems in the original publications of Fichera (1972), in Panagiotopoulos (1985) or in Antes and Panagiotopoulos (1992), pages 38-40.

Besides the existence question, one may also find problems which admit multiple solutions. This point is certainly related to the use of 'nonmonotone' interaction relations (cf. the delamination or damaging interface outlined previously). A clear justification of multiplicity is provided by using the potential energy formulation of the mechanical problem: a nonconvex energy function may have several (local) minima and much more critical points which are all solutions of the mechanical problem. The next question is, of course, which one of these solutions are stable, or can be realized (attained) with the existing loading / external action. These links between nonconvex optimization and mechanics can be found, for example, in Mistakidis and Stavroulakis (1998). With boundary element formulations, which do not have an accompanying energy optimization principle, the situation is less favorable. The complementarity problem approach is quite general and can be extended to cover certain nonmonotone nonlinear interactions as well (thanks to Prof. Maier and his group, see, e.g., Bolzon et al. (1997b), Bolzon et al. (1997a), Bolzon and Maier (1998) or Tin-Loi and Li (2000)). Nevertheless, one has no information about stability of the calculated solution (besides the fact that finding multiple solutions may require, in the worst case, the use of some enumerative techniques).

In another context, consideration of more complicated friction laws (e.g. dynamical laws) may also lead to nonexistence or multiplicity problems. For details the reader is refered to the works given in the next paragraph.

6.2 More General Nonlinear Interaction and Nonsmooth Mechanics

The nonliner interaction mechanisms of contact and friction belong to the class of more general nonsmooth laws which can be described by means of convex and nonconvex analysis. They are one-dimensional boundary laws, but analogous techniques can be used for multi-dimensional boundary (e.g. anisotropic friction between three-dimensional bodies) or material laws. Some elements of convex and nonconvex analysis which demonstrate these links to nonsmooth mechanics and to the general theories of variational or hemivariational inequalities have been provided in section four. The interested reader may consult the spesialized literature, which includes the monographs of Panagiotopoulos (1985), Sewell (1987), Panagiotopoulos (1993), Demyanov et al. (1996), Mistakidis and Stavroulakis (1998) and Gao (2000). The term nonsmooth mechanics, firstly introduced by the late Professor P.D. Panagiotopoulos by following the example of the term nonsmooth analysis of Clarke (1983), appears in a number of recently published monographs with theoretical results and practical applications, such as those of Brogliato (1999), Christensen (2000), Gao et al. (2001), Fremond (2002) and Glocker (2001). This scientific area certainly provides a wealth of open problems for future research.

6.3 Stability of Solutions and path-following

Problems in mechanics whose governing relations can be obtained from a generally nondifferentiable and nonconvex, but quasidifferentiable (in the sense of V.F. Dem'yanov andA.M. Rubinov) potential function are considered. They consider a fairly general form for the modeling and the study of nonsmooth problems in mechanics Demyanov et al. (1996) and they cover certain classes of variational and hemivariational inequality problems of mechanics Panagiotopoulos (1985), Panagiotopoulos (1993). The notion of hemivariational inequalities has been introduced and thoroughly studied in mechanics by P.D. Panagiotopoulos . Moreover, there exists extensive theoretical support for the use of quasidifferentiable calculus and optimization techniques, see, e.g., Demyanov and Rubinov (1985), Demyanov et al. (1996). For methods and heuristic algorithms of nonconvex optimization in computational mechanics, see Mistakidis and Stavroulakis (1998). In this short note some techniques for treating stability problems for nonsmooth structures are outlined. This way results for classical, smooth structures (e.g., Bazant and Cedolin (1991), Li (1991), Li (1994)) can be extended to cover nonsmooth ones (cf., also Kurutz (1993), Kurutz (1994)). This work and the preliminary results outlined here are based on Stavroulakis and Rohde (1999). Further details concerning frictionless contact problems can be found in Klarbring (2002)

All previously mentioned potentials are piecewise–differentiable and may be described, in general, as continuous selections of differentiable functions. In turn, the structural analysis problem results from minimality or in general critical point conditions of the potential (see examples in Curnier et al. (1995), Jongen and Pallaschke (1988), Kuntz and Scholtes (1995), Panagiotopoulos (1985), Panagiotopoulos (1993)).

Results from stability analysis of parametric optimization problems for nondifferentiable functions are used for the study of a stepwise holonomic, incremental structural analysis problem. In particular the systematic first and second order linearizations proposed in respectively, and the arising normal forms are adopted for the potential energy function.

The techniques outlined here may be useful both for the analysis of the stability of structures which involve nonmonotone and possibly multivalued nonlinearities (in a holonomic or a stepwise holonomic setting) and for the design of incremental–iterative algorithms for structural analysis purposes.

Further work in this direction will generalize the computational mechanics techniques for the tracing of post-buckling equilibria in nonsmooth mechanics' applications. Theoretical support will be provided by the theory of parametric optimization (cf., e.g., Guddat et al. (1990) and the applications in contact mechanics Rohde and Stavroulakis (1995), Rohde and Stavroulakis (1997)). A recent numerical investigation of buckling problems for a von Karman plate problem with unilateral contact effects has been presented in Muradova (2004).

6.4 Multiphysics Problems

Certain semipermeability or temperature control problems in thermoelasticity, which may be combined with analogous mechanical unilateral contact effects, can be formulated and studied in a unified framework by nonsmooth modeling techniques Panagiotopoulos (1985). For monotone relations the notions of convex analysis (namely, the subdifferential) can be used for the modeling of unilateral relations of the thermal or the elastic part, or even for mixed relations. The arising variational inequality problems can be formulated and studied (see, for instance, our recent publication Foutsitzi and Stavroulakis (2004). Analogously, nonmonotone relations can be considered and lead to hemivariational inequality problems Panagiotopoulos (1985), Haslinger et al. (1993). Here a model problem with coupled thermal and kinematical nonconvex unilateral effects will be formulated by using the quasidifferentiable optimization approach Demyanov et al. (1996).

6.5 Dynamical Problems

It is known that frictional effects are usually formulated in terms of relative tangential velocities (instead of absolute values, as in relation (61)). After time discretization appropriate quasi-static problems, like the ones proposed in previous part of this work, are actually considered and solved. Existence or multiplicity results may or may not change during this discretization. Relevant information is given in the following references: Klarbring (1990a), Klarbring (1990b), Martins et al. (1994), Doudoumis et al. (1995), Pang and Trinkle (1996), Trinkle et al. (1997) and Klarbring (1999), Klarbring (2002).

Roughly speaking, unilateral contact phenomena, even with Coulombian friction with some sufficiently small friction coefficient and involving elastic materials with common material constants, and for suitable supported structures (or parts of them) so that no rigid body motions are permitted (coercive problems) should not lead in practice to nonexistence or multiplicity results. Every more complicated interaction law (i.e., slip-dependent friction coefficients or Stribeck friction laws, delamination or damaging interfaces etc.) may lead to serious existence and uniqueness problems. The critical question is how one finds that failure of a numerical scheme is due to these effects and is not caused from some other numerical instabilities and, then, how one interprets the numerical solution(s).

Further information on stability of unilateral frictional contact problems can be found in Martins et al. (2002), Raous et al. (2002).

7 Indicative Recent Applications

Large-scale unilateral contact applications on cracked masonry bridges have been presented in Stavroulaki et al (2002), Leftheris (2004) and Drosopoulos et al. (2006). A detailed unilateral contact analysis of a masonry bridge (Prestwood bridge) using a path-following method, from initial loading up to collapse has been presented in Drosopoulos et al. (2006). Lack of a finite solution indicates collapse (see Figure 1). The numbers in the figure indicate the activation of hinges up to the formation of a mechanism and collapse. The results are in agreement with published experimental data and with the results of more classical approaches based on Heyman's mechanism method. Further information can be found in Drosopoulos (2006).

An attempt to use QD and DC optimization for the pseudoelastic modelling of shape memory alloys has been described in Polyakova and Stavroulakis (2000). Delamination modelling has been studied with the theory of hemivariational inequalities, see Mistakidis and Stavroulakis (1998).

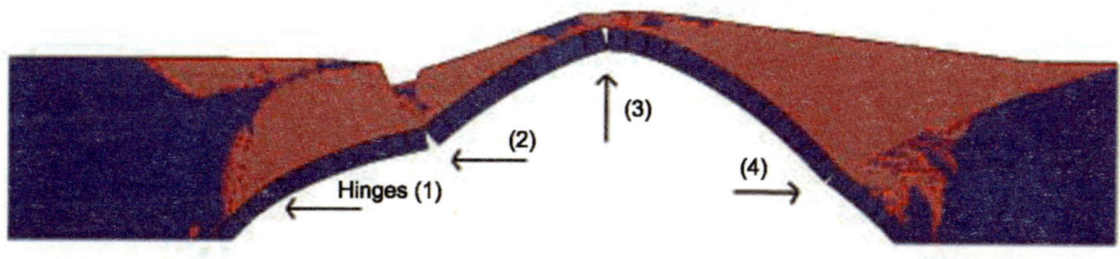

Figure 1. Collapse mode of a masonry bridge predicted by a unilateral contact model.

Acknowledgements

The support of the Greek Secretariat of Research and Technology, within the framework of bilateral cooperation projects with Italy and Hungary, and of the Greek Fellowships Foundation (IKY), within a cooperation project with Germany, are greatfully acknowledged. Thanks are also due to Dr G.A. Drosopoulos, for providing the numerical results and the figure of the last section.

References

Adly, S. and Goeleven, D. (2000): A discretization theory for a class of semi-coercive unilateral problems. *Numerische Mathematik*, 87:1-34.

Al- Fahed, A.M., Stavroulakis, G.E., and Panagiotopoulos, P.D. (1991): Hard and soft fingered robot grippers. *ZAMM, Z. Angew. Math. Mech.*, 71:257-266.

Al-Fahed, A.M., Stavroulakis, G.E., and Panagiotopoulos, P.D. (1992): A linear complementarity approach to the frictionless gripper problem. *International Journal of Robotics Research,* 11:112-122.

Al- Fahed, A.M., and Stavroulakis, G.E. (2000): A complementarity problem formulation of the frictional grasping problem. *Computer Methods in Applied Mechanics and Engineering*, 190:941-952.

Antes, H. (1998): *Anwendungen der Methode der Randelemente in der Elastodynamik und der Fluiddynamik.* B.G. Teubner Verlag, Stuttgart.

Antes, H., and Panagiotopoulos, P.D. (1992): *The Boundary Integral Approach to Static and Dynamic Contact Problems. Equality and Inequality Methods.* Birkhäuser, BaselBoston- Berlin.

Antes, H., and Steinfeld, B. (1991): Unilateral contact with friction by time domain BEM. In P. Wriggers and W. Wagner, editors, *Nonlinear Computational Mechanics–State of the Art*, pages 193–211, Berlin Heidelberg, Springer Verlag.

Antes, H., and Steinfeld, B. (1992): A boundary formulation study of massive structures static and dynamic behaviour. In *14th Boundary Element International Conference*. Sevilla, Spain.

Antes, H., Steinfeld, B., and Tröndle, G. (1991): Recent developments in dynamic stress analyses by time domain BEM. *Engineering Analysis with Boundary Elements*, 8:176–184.

Baggio, C. (2000): Collapse behaviour of three-dimensional brick-block systems using non-linear programming. *Structural Engineering and Mechanics* 10:181-195.

Baiocchi, C., and Capelo, A. (1984): *Variational and Quasivariational Inequalities. Applications to Free Boundary Problems*. J. Wiley and Sons, Chichester.

Barber, J.R., and Ciavarella, M. (2000): Contact mechanics. *International Journal of Solids and Structures*, 37:29-43

Bathe, K. J. (1996): *Finite Element Procedures*. Prentice– Hall, New Jersey.

Bathe, K.J. , and Bouzinov, P.A. (1997): On the constraint function method for contact problems. *Computers and Structures*, 64:1069–1086.

Bazant, Z.P., and Cedolin, L. (1991): *Stability of structures. Elastic, inelastic, fracture and damage theories*. Oxford University Press, New York - Oxford.

Billups, S.C., and Murty, K.G. (2000): Complementarity problems. *Journal of Computational and Applied Mathematics*, 124:303–318, 2000.

Bisegna, P., Lebon, F., and Maceri, F. (2001): D-PANA: a convergent block-relaxation solution method for the discretized dual formulation of the Signorini-Coulomb contact problem. *Comptes R. Acad. Sci. Serie II.*, 333:1053-1058, 2001.

Blanze, C., Champaney, L., Cognard, J.-Y., and Ladeveze, P. (1996): A modular approach to structure assembly computations. Application to contact problems. *Engineering Computations*, 1:15-32.

Bolzon, G., Ghilotti, D., and Maier, G. (1997): Parameter identification of the cohesive crack model. In H. Sol and C.W.J. Oomens, editors, *Material identification using mixed numerical experimental methods*, pages 213–222, Dordrecht, Kluwer Academic Publishers.

Bolzon. G., and Maier, G. (1998): Identification of cohesive crack models for concrete on the basis of three-point- bending tests. In R. deBorst, N. Bicanic, H. Mang, and G. Meschke, editors, *Computational Modelling of Concrete Structures*, pages 301–310, Dordrecht, Kluwer Academic Publishers.

Bolzon, G., Maier, G. and Tin-Loi, F. (1997): On multiplicity of solutions in quasibrittle fracture computations. *Computational Mechanics*, 19:511- -516.

Brézis, H. (1972): Problèmes unilatéraux, *J. Math. pures et appl.*, 51:1-168.

Brogliato, B. (1999): *Nonsmooth Mechanics. Models, Dynamics and Control.* Springer Verlag, Berlin, second edition.

Chabrand, P., Dubois, F., and Raous, M. (1998): Various numerical methods for solving unilateral contact problems with friction. *Mathematical and Computer Modelling*, 28:109-120.

Champaney, L. , Cognard, J.Y., Rureisseix, D., and Ladevèze, P. (1997): Large scale applications on parallel computers of a mixed domain decomposition method. *Computational Mechanics*, 19:253-263.

Christensen, P.W., and Klarbring, A. (1999): Newton's method for frictional contact problems. In P. Wunderlich, editor, *ECCM'99 European Conference on Computational Mechanics August 31 September 3*, Munich Germany, Technical University of Munich.

Christensen, P.W., Klarbring, A., Pang, J.S., and Strömberg, N. (1998): Formulation and comparison of algorithms for frictional contact problems. *International Journal for Numerical Methods in Engineering*, 42:145–173.

Christensen, P.W. (2000): *Computational Nonsmooth Mechanics : Contact, Friction and Plasticity.* Linköping University, Sweden.

Clarke, F.H. (1983): *Optimization and Nonsmooth Analysis*. J. Wiley, New York.

Coda, H.B., Venturini, W.S., and Aliabadi, M.H. (1999): A general 3D BEM/FEM coupling applied to elastodynamic continua/frame structures interaction analysis. *International Journal for Numerical Methods in Engineering*, 46:695–712.

Curnier, A., He, Q.-C., and Zysset, Ph. (1995): Conewise linear elastic materials, *Journal of Elasticity*, 37:1-38.

Curnier, A., (1999): Unilateral Contact. Mechanical Modelling. In P. Wriggers and P. Panagiotopoulos, editors, *New Developments in Contact Problems*, CISM Courses and Lectures No. 384, pages 1–54. Springer, Wien.

Czekanski, A., and Meguid, S.A. (2001): Solution of dynamic frictional contact problems using nondifferentiable optimization. *Intern. Journal of Mechanical Sciences*, 43:1369-1386.

Dafermos, S. (1980): Traffic equilibrium and variational inequalities. *Transportation Science* 14(1):42-54.

Dem'yanov, V. F., Stavroulakis, G.E, Polyakova, L.N., and Panagiotopoulos, P.D. (1996): *Quasidifferentiability and Nonsmooth Modelling in Mechanics, Engineering and Economics.* Kluwer Academic, Dordrecht.

Demyanov, V.F., and Rubinov, A.M. (1985): *Quasidifferentiable calculus*, Optimization Software, New York.

Demyanov, V.F., and Rubinov, A.M. (1995): *Introduction to constructive nonsmooth analysis*, Peter Lang Verlag, Frankfurt-Bern-New York.

Demyanov, V., and Rubinov, A. (2000): (Eds.), *Quasidifferentibility and Related Topics*, Kluwer Academic Publishers, Dordrecht.

Dominguez, J. (1993): *Boundary Elements in Dynamics.* Computational Mechanics Publications and Elsevier Applied Science, Shouthampton and New York.

Doudoumis, I., Mitsopoulou, E., and Charalambakis, N. (1995): The influence of the friction coefficients on the uniqueness of the solution of the unilateral contact problem. In M. Raous, M. Jean, and J.J. Moreau, editors, *Contact Mechanics*, pages 79–86, New York and London, Plenum Press.

Drosopoulos, G.A., Stavroulakis, G.E., and Massalas, Ch. (2006): Limit analysis of a single-span masonry bridge with unilateral frictional contact interfaces. *Engineering Structures* (in press).

Drosopoulos, G.A. (2006): *Unilateral contact analysis of masonry bridges*, Doctoral Dissertation (in Greek), University of Ioannina, Department of Material Science and Technology.

Duvaut, G., and Lions, J.L., (1972): *Les inéquations en méchanique et en physique*, Dunod, Paris.

Ekeland, I., and Temam, E. (1976): *Convex Analysis and Variational Problems.* NorthHolland, Amsterdam.

Facchinei, F., Fischer, A., and Kanzow, C. (1996): Inexact Newton methods for semismooth equations with applications to variational inequality problems. In G. DiPillo and F. Giannessi, editors, *Nonlinear Optimization and Applications*, pages 125- -149, New York, Plenum Press.

Ferris, M.C., and Tin-Loi, F. (2001): Limit analysis of frictional block assemblies as a mathematical program with complementarity constraints. *International Journal of Mechanical Sciences*, 43:209–224.

Fichera, G. (1972): Boundary value problems in elasticity with unilateral constraints. *Enzyclopedia of Physics*, Ed. S. Flügge, Vol. VI a/2, Springer Verlag, Berlin.

Fischer-Cripps, A.C. (2000): *Introduction to Contact Mechanics.* Springer Verlag, Berlin.

Fischer, A. (1992): A special Newton- type optimization method. *Optimization*, 24:269–284, 1991.

Foutsitzi, G., and Stavroulakis, G.E. (2004): A variational inequality approach to thermoviscoelasticity with monotone unilateral boundaries of kinematical and thermal type. *Nonlinear Analytis. Theory, Methods and Applications*, 57(5-6):743–771.

Frémond, M. (2002): *Nonsmooth Thermomechanics.* Springer Verlag, Berlin.

Fukushima, M. (1992): Equivalent differentiable optimization problems and descent methods for asymmetric variational inequality problems. *Mathematical Programming*, 53:99–110.

Gao, D.Y. (2000): *Duality Principles in Nonconvex Nonsmooth Systems: Problems, Methods and Solutions.* Kluwer Academic Publishers, Dordrecht.

Gao, D.Y., Ogden, R., and Stavroulakis, G.E. (Eds.) (2001): *Nonsmooth & Nonconvex Mechanics: Modeling, Analysis and Numerical Methods.* Kluwer Academic Publishers, Dordrecht, Boston, London.

Glocker, Ch. (2001): *Set-valued Force Laws: Dynamics of Non-smooth Systems.* Springer Verlag, Berlin.

Goeleven, D., (1996): *Noncoercive variational problems and related results*, Addison Wesley Longman.

Goeleven, D. Stavroulakis, G.E., Salmon, G., and Panagiotopoulos, P.D. (1997): Solvability theory and projection methods for a class of singular variational inequalities. Elastostatic unilateral contact applications. *Journal of Optimization Theory and Applications*, 95(2):263-294.

Gonzalez, J.A., and Abascal, R. (2000): Solvind 2D rolling problems using NORM-TANG iteration and mathematical programming. *Computers and Structures*, 78:149-160.

Guddat, J., Guerra Vasquez, F., and Jongen, H.Th., (1990): *Parametric Optimization: Singularities, path following and jumps*, Teubner and J. Wiley, Stuttgart.

Harker, P.T., and Pang, J.S. (1990): Finite dimensional variational inequality and nonlinear complementarity problems: a survey of theory, algorithms and applications. *Mathematical Programming*, 48:161–220.

Haslinger, J., Baniotopoulos, C.C., and Panagiotopoulos, P.D. (1993): A Boundary Multivalued Integral "Equation" Approach to the Semipermeability Problem, *Applications of Mathematics*, 38:39–60.

Haslinger, J., Miettinen, M., and Panagiotopoulos, P.D. (1999): *Finite Element Method for Hemivariational Inequalities. Theory, Methods and Applications*. Kluwer Academic Publishers, Dordrecht.

Hilding, D., Klarbring, A., and Petersson, J. (1999): Optimization of structures in unilateral contact. *ASME Applied Mechanics Review*, 52(4):139–160.

Jean, M. (1999): The non-smooth contact dynamics method. *Computer Methods in Applied Mechanics and Engineering*, 177:235-257.

Johnson, K.L. (1987): *Contact Mechanics*. Cambridge University Press.

Jongen, H.T., and Pallaschke, D. (1988): On linearization and continuous selection of functions,, *Optimization*, 19(3):343-353.

Kanzow, C. (1994): Some equation-based methods for the nonlinear complementarity problem. *Optimization Methods and Software*, 3:327–340.

Kanzow, C. (1996): Nonlinear complementarity as unconstrained optimization. *Journal of Optimization Theory and Applications*, 88:139-155.

Kikuchi, N. (1988): *Contact Problems in Elasticity: A Study of Variational Inequalities and Finite Element Methods*. SIAM, Philadelphia.

Kim, J.O., and Kwak, B.M. (1996): Dynamic analysis of two-dimensional frictional cotnact by linear complementarity problem formulation. *International Journal of Solids and Structures*, 33:4605-4624.

Kim, J.O., and Kwak, B.M. (1996): Numerical implementation of dynamic contact analysis by a complementarity formulation and applications to a valve-cotter system in motorcycle engines. *Mechanics of Structures and Machines*, 28:281-301.

Klarbring, A. (1986): A mathematical programming approach to three– dimensional contact problems with friction. *Comput. Meth. Appl. Mech. Engng.*, 58(2):175– 200.

Klarbring, A. (1990a): Derivation and analysis of rate boundary- value problems. *European Journal of Mechanics A/Solids*, 9(1):53– 86.

Klarbring, A. (1990b) Examples of non-uniqueness and non- existence of solutions to quasistatic contact problems with friction. *Ingenieur Archiv*, 60:529–541.

Klarbring, A., (1999): Contact, Friction, Discrete Mechanical Structures and Mathematical Programming. In P. Wriggers and P. Panagiotopoulos, editors, *New Developments in Contact Problems*, CISM Courses and Lectures No. 384, pages 55–100. Springer, Wien.

Klarbring, A. (2002): Stability and critical points in large displacement frictionless contact problems. In J.A.C. Martins and M. Raous, editors, *Friction and Instabilities*, CISM Courses and Lectures No. 457, pages 39–64. Springer, Wien.

Klarbring, A., and Björkman, G. (1998): A mathematical programming approach to contact problem with friction and varying contact surface. *Computers and Structures*, 30:1185– 1198.

Klisch, T. (1999): Contact mechanics in multibody systems. *Mechanism and Machine Theory*, 34:665-675.

Kortesis, S., and Panagiotopoulos, P.D. (1993): Neural networks for computing in structural analysis. Methods and prospects of applications. *International Journal for Numerical Methods in Engineering,* 36:2305-2318.

Kosior, F., Guyot, N., and Maurice, G. (1999): Analysis of frictional contact problem using boundary element method and domain decomposition. *International Journal for Numerical Methods in Engineering*, 46:65–82.

Kuntz, L., and Scholtes, S. (1995): Qualitative aspects of the local approximation of a piecewise differentiable function, *Nonlinear Analysis, Theory Methods and Applications*, 25(2):197-215.

Kurutz, M., (1993): Stability of structures with nonsmooth nonconvex energy functionals. The one dimensional case, *European Journal of Mechanics, A Solids*, 12(3):347-385.

Kurutz, M. (1994): Equilibrium paths of polygonally elastic structures, *Mechanics of Structures and Machines*, 22(2):181-210.

Ladèveze, P. (1995): *Mecanique des Structures Nonlineaires*. Hérmes, Paris.

Ladeveze, P., Lemoussu, H., and Boucard, P.A. (2000): A modular approach to 3-D impact computation with frictional contact. *Computers and Structures*, 78:45-51.

Laursen, T.A. (2002): *Computational Contact and Impact Mechanics*. Springer Verlag, Berlin.

Leblond, J.-B. (2000): Basic results for elastic fracture mechanics with frictionless contact between the crack lips. *European Journal of Mechanics, A/Solids*, 19:633-647.

Leftheris, B.P., Stavroulaki, M.E., Sapounaki, A.K., and Stavroulakis, G.E. (2006): *Computational mechanics for heritage structures* WIT Computational Mechanics Publications, Southampton.

Leung, A.Y.T. , Guoqing, Ch., and Wanji, Ch. (1998): Smoothing Newton method for solving twoand three-dimensional frictional contact problems. *International Journal for Numerical Methods in Engineering*, 41:1001–1027.

Li, L.-Y., (1991): The criteria of identifying the type of critical points, *Archive of Applied Mechanics*, 61:231-235.

Li, L.-Y., (1994): Determination of stability in nonlinear analysis of structures, *Archive of Applied Mechanics*, 64:119-126.

Luo, Z. Q., Pang, J. S., and Ralph, D. (1996): *Mathematical Programs with Equilibrium Constraints*. Cambridge University Press, Cambridge.

Maier, G., and Novati, G. (1990): Extremum theorems for finite-step backward-difference analysis of elastic-plastic nonlinearly hardening solids, *International Journal of Plasticity*, 6:1-10.

Martins, J.A.C., Marques, M.D., and Gastaldi, F. (1994): On an example of non-existence of solution to a quasistatic frictional contact problem. *European Journal of Mechanics A/Solids*, 13:113– 133.

Martins, J.A.C., Pinto de Costa, A. and Simoes, F.M.F., (2002): Some notes on friction and instabilities. In J.A.C. Martins and M. Raous editors, *Friction and Instabilities*, CISM Courses and Lectures No. 457, pages 1–136. Springer, Wien.

Mijar, A.R., and Arora, J.S. (2000a): Review of formulations for elastostatic frictional contact problems. *Structural and Multidisciplinary Optimization*, 20:167–189.

Mijar, A.R., and Arora, J.S. (2000b): Study of variational inequality and equality formulations for elastostatic frictional contact problems. *Archives of Computational Methods in Engineering*, 7(4):387-449.

Mistakidis, E.S., and Panagiotopoulos, P.D. (1998): A multivalued boundary integral equation for adhesive contact problems. *Engineering Analysis with Boundary Elements*, 21(4):317-327.

Mistakidis, E.S., and Stavroulakis, G.E. (1998): *Nonconvex Optimization in Mechanics. Algorithms, Heuristics and Engineering Applications by the F.E.M.* Kluwer Academic Publishers, Dordrecht.

Moreau, J.J. (1999): Numerical aspects of the sweeping process. *Computer Methods in Applied Mechanics and Engineering*, 177:329-349.

Motreanu, D. and Radulescu, V., (2003): *Variational and non-variational methods in nonlinear analysis and boundary value problems*, Kluwer Academic Purlishers, Dordrecht.

Muradova-Contadaki, A., and Stavroulakis, G.E. (2004): Computation of unilaterally supported elastic plates. *European Congress on Computational Methods in Applied Sciences and Engineering ECCOMAS 2004, CD-ROM Proceedings* Neitaanmaaki, P. et al. (eds), Jynäskylä, Finnland, June 24-28, 2004.

Murty, K. G. (1988): *Linear Complementarity, Linear and Nonlinear Programming*. Heldermann, Berlin.

Nagurney, A. (1994): Variational inequalities in the analysis and computation of multisector, multi-instrument financial equlibria. *Journal of Economic Dynamics and Control*, 18(1):161-184.

Nagurney, A., and Siokos, S. (1999): Dynamic multi-sector, multi-instrument financial networks with futures: Modeling and computation. *Networks* 33(2):93-108.

Naniewicz, Z., and Panagiotopoulos, P.D., (1995): *Mathematical theory of hemivariational inequalities and applications*, Marcel Dekker.

Nappi, A., and Tin-Loi, F. (2001): Numerical model for masonry implemented in the framework of a discrete formulation. *Structural Engineering and Mechanics*, 11(2):171– 184.

Oden, J.T., and Kikuchi, N. (1980): Theory of variational inequalities with applications to problems of flow through porous media. *International Journal of Engineering Science,* 18(10):1173-1284.

Outrata, J., Kočvara, M., and Zowe, J. (1998): *Nonsmooth Approach to Optimization Problems with Equilibrium Constraints: Theory, Applications, and Numerical Results.* Kluwer Academic Publishers, Dordrecht.

Pagano, S., and Alart, P. (1999): Solid-solid phase transition modelling: relaxation procedures, configurational energies and thermomechanical behaviours. *Journal of Engineering Science* 37:1821-1840.

Pallaschke, D., and Rolewicz, S., (1997): *Foundations of mathematical optimization. Convex analysis without linearity,* Kluwer Academic Publishers, Dordrecht-Boston-London.

Panagiotopoulos, P.D. (1985): *Inequality Problems in Mechanics and Applications. Convex and Nonconvex Energy Functions.* Birkhäuser, Basel Boston - Stuttgart.

Panagiotopoulos, P.D., and Stavroulakis, G.E. (1992): New type of variational principles based on the notion of quasidifferentiability. *Acta Mechanica,* 94:171-194.

Panagiotopoulos, P.D. (1993): *Hemivariational Inequalities. Applications in Mechanics and Engineering.* Springer, Berlin - Heidelberg - New York.

Panagiotopoulos, P.D., and Glocker, Ch. (1998): Analytical mechanics. Addendum I: Inequality constraints with elastic impact. The convex case. *ZAMM* 78(4):219-230.

Pang, J. S., and Trinkle, J.C. (1996): Complementarity formulations and existence of solutions of dynamic multi-rigid-body contact problems with Coulomb friction. *Mathematical Programming,* 73:199–226.

Pang, J.-S., and Stewart, D.E. (1999): A unified approach to discrete frictional contact problem. *International Journal of Engineering Science* 37:1747-1768.

Park, J. K., and Kwack, B.M. (1994): contact analysis using the homotopy method. *ASME Journal of Applied Mechanics,* 61:703-709.

Pfeiffer, F., and Hajek, M. (1992): Stick-slip motion of turbine blade dampers. *Phil. Trans. of the Royal Society of London, Ser. A,* 338:503-517.

Pfeiffer, F., and Glocker, Ch. (1996): *Multibody Dynamics with Unilateral Contacts.* John Wiley, New York.

Pfeiffer, F. (1998): Robots with unilateral constraints. *Annual Reviews in Control,* 22:121-132.

Pfeiffer, F., and Glocker, Ch. (2000): Contacts in multibody systems. *Journal of Applied Mathematics and Mechanics,* 64:773-782.

Polyakova, L.N., and Stavroulakis, G.E. (2000): QD and DC optimization for pseudoelastic modeling of Shape Memory Alloys. In: *Quasidifferentibility and Related Topics,* Chapter 9, pp. 215-233, Eds. V. Demyanov, A. Rubinov, Kluwer Academic Publishers.

Raous, M., Barbarin, S., and Vola, D. (2002): Numerical characterization and computation of dynamic instabilities for frictional contact problems. In J.A.C. Martins and M. Raous editors, *Friction and Instabilities,* CISM Courses and Lectures No. 457, pages 233–291. Springer, Wien.

Refaat, M.H., and Meguid, S.A. (1994): On the elastic solution of frictional contact problems using variational inequalities. *International Journal of Mechanical Science,* 36:329-342.

Rockafellar, R.T., (1981): *The theory of subgradients and its applications to problems of optimization: Convex and nonconvex functions,* Helderman Verlag, Berlin.

Rohde, A., and Stavroulakis, G.E., (1995): Path–following energy optimization in unilateral contact problems, *Journal of Global Optimization,* 6:347-365.

Rohde, A., and Stavroulakis, G.E. (1997): Genericity analysis for path-following methods. Application in unilateral contact elastostatics. *ZAMM,* 77(10):777–790.

DeSaxcé, G., and Feng, Z.Q. (1998): The bipotential method: A constructive approach to design the complete contact law with friction and improved numerical algorithms. *Mathematical and Computer Modelling,* 28:225-245.

Sewell, N.J. (1987): *Maximum and Minimum Principles. A Unified Approach with Applications.* Cambridge University Press, Cambridge.

Simunovic, S., and Saigal, S. (1992): Frictionless contact with BEM using quadratic programming. *ASCE Journal of Engineering Mechanics* 118:1876-1891.

Song, P., Kraus, P., Kumar, V., and Dupont, P. (2001): Analysis of rigid-body dynamic models for simulation of systems with frictional contacts. *ASME Journal of Applied Mechanics*, 68:118-128.

Stamos, A.A., and Beskos, D.E. (1995): Dynamic analysis of large 3D underground structures by the BEM. *Earthquake Engineering and Structural Dynamics*, 24:917–934.

Stavroulakis, G.E. (1993): Convex decomposition for nonconvex energy problems in elastostatics and applications. *European Journal of Mechanics A / Solids*, 12:1–20.

Stavroulakis, G.E. (1995): Optimal prestress of cracked unilateral structures: finite element analysis of an optimal control problem for variational inequalities. *Computer Methods in Applied Mechanics and Engineering*, 123:231-246.

Stavroulakis, G.E. (1999): Impact-echo from a unilateral interlayer crack. LCP- BEM modelling and neural identification. *Engineering Fracture Mechanics*, 62:165–184.

Stavroulakis, G.E., and Antes, H. (1999): Nonlinear boundary equation approach for inequality 2- D elastodynamics. *Engineering Analysis with Boundary Elements*, 23:487–501.

Stavroulakis, G.E., and Antes, H. (2000): Nonlinear equation approach for inequality elastostatics. A 2-D BEM implementation. *Computers and Structures*, 75(6):631-646.

Stavroulakis, G.E., Panagiotopoulos, P.D., and Al-Fahed, A.M. (1991): On the rigid body displacements and rotations in unilateral contact problems and applications. *Computers and Structures*, 40:599–614.

Stavroulakis, G.E. (2000): *Inverse and Crack Identification Problems in Engineering Mechanics.* Kluwer Academic Publishers, Dordrecht. Habilitation Thesis, Technical University of Braunschweig, Germany, 2000.

Stavroulakis, G.E., and Antes, H. (1997): Nondestructive elastostatic identification of unilateral cracks through BEM and neural networks. *Computational Mechanics*, 20:439- 451.

Stavroulakis, G.E., and Antes, H. (2000): Unilateral crack identification. a filter-driven, iterative, boundary element approach. *Journal of Global Optimization*, 17:339–352.

Stavroulakis, G.E., Antes, H., and Panagiotopoulos, P.D. (1999): Transient elastodynamics around cracks including contact and friction. *Computer Methods in Applied Mechanics and Engineering*, 177:427–440.

Stavroulakis, G.E., and Rohde, A., (1999): Normal forms and stability in nonsmooth potential elastostatics, *Mechanics Research Communications*, 26(2):185–190.

Stavroulakis, G.E., and Antes, H. (2003): Classical and unilateral contact analysis in statics and dynamics, In: D. Beskos and G. Maier (eds), *Boundary Element Advances in Solid Mechanics*, CISM Lecture Notes Vol. 440, pp. 205–244.

Stavroulaki, M.E., Sapounaki, A.K., Leftheris, B.P., and Stavroulakis, G.E. (2002): Nonlinear finite elements for the damage analysis of the Plaka stone bridge in Epirus, *CD-ROM Proceedings of the 4th GRACM Congress on Computational Mechanics, Patras, Greece, 27-29 June 2002.*

Telega, J.J. (1988): Topics on unilateral contact problems in elasticity and inelasticity. In: *Nonsmooth Mechanics and Applications*, Eds. J.J. Moreau and P.D. Panagiotopoulos, CISM Lecture Notes No. 302, pp. 341-462, Springer Verlag, Wien.

Tin-Loi, F., and Li, H. (2000): Numerical simulations of quasibrittle fracture processes using the discrete cohesive crack model. *International Journal of Mechanical Sciences*, 42:367–379.

Tin-Loi, F., and Xia, S.H. (2001): Nonholonomic elastoplastic analysis involving unilateral frictionless contact as a mixed complementarity problem. *Computer Methods in Applied Mechanics and Engineering*, 190:4551-4568.

Trinkle, J. C., Pang, J. S., Sudarsky, S., and Lo, G. (1997): On dynamic multi-rigid-body contact problems with Coulomb friction. *Zeitschrift für Angewandte Mathematik und Mechanik (ZAMM)*, 77:267-280.

Tzaferopoulos, M.Ap., and Panagiotopoulos, P.D. (1993): Delamination of composites as a substationarity problem: Numerical approximation and algorithms. *Computer Methods in Applied Mechanics and Engineering* 110:63-86.

Vola, D., Pratt, E., Jean, M., and Raous, M. (1998): Consistent time discretization for a dynamical frictional contact problem and complementarity techniques. *Revue Européenne des éléments Finis*, 7:149-162.

Wolfsteiner, P., and Pfeiffer, F. (2000): Modeling, simulation and verification of the transportation process in vibratory feeders. *ZAMM*, 80:35-48.

Wriggers, P. (1995): Finite element algorithms for contact problems. *Archives of Computational Methods in Engineering,* 2:1-49.

Wriggers, P. (2002): *Computational Contact Mechanics*. J. Wiley and Sons.

Zavarise, G., Wriggers, P., and Schrefler, B.A. (1998): A method for solving contact problems. *International Journal for Numerical Methods in Engineering*, 42:473–498.

Zhong, Z.-H. (1993): *Finite Element Procedures for Contact-Impact Problems*. Oxford University Press, Oxford.

Zozulya, V.V., and Gonzalez-Chi, P.I. (2000): Dynamic fracture mechanics with contact interaction at the crack edges. *Engineering Analysis with Boundary Elements*, 24:643-659.